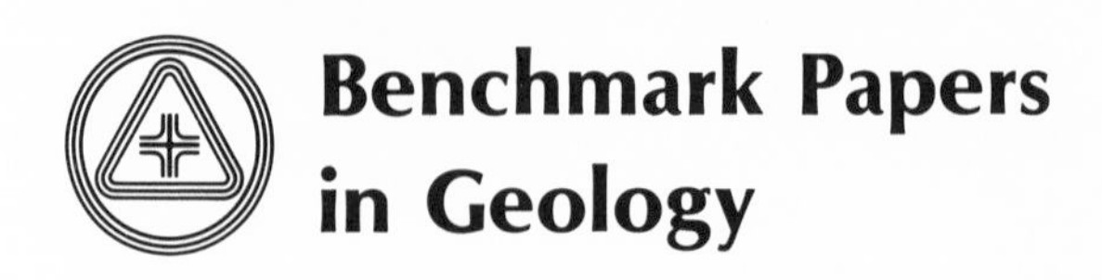

Benchmark Papers in Geology

Series Editor: Rhodes W. Fairbridge
Columbia University

Volume

1 ENVIRONMENTAL GEOMORPHOLOGY AND LANDSCAPE CONSERVATION, Volume I: Prior to 1900 / *Donald R. Coates*
2 RIVER MORPHOLOGY / *Stanley A. Schumm*
3 SPITS AND BARS / *Maurice L. Schwartz*
4 TEKTITES / *Virgil E. Barnes and Mildred A. Barnes*
5 GEOCHRONOLOGY: Radiometric Dating of Rocks and Minerals / *C. T. Harper*
6 SLOPE MORPHOLOGY / *Stanley A. Schumm and M. Paul Mosley*
7 MARINE EVAPORITES: Origin, Diagenesis, and Geochemistry / *Douglas W. Kirkland and Robert Evans*
8 ENVIRONMENTAL GEOMORPHOLOGY AND LANDSCAPE CONSERVATION, Volume III: Non-Urban / *Donald R. Coates*
9 BARRIER ISLANDS / *Maurice L. Schwartz*
10 GLACIAL ISOSTASY / *John T. Andrews*
11 GEOCHEMISTRY OF GERMANIUM / *Jon N. Weber*
12 ENVIRONMENTAL GEOMORPHOLOGY AND LANDSCAPE CONSERVATION, Volume II: Urban Areas / *Donald R. Coates*
13 PHILOSOPHY OF GEOHISTORY: 1785–1970 / *Claude C. Albritton, Jr.*
14 GEOCHEMISTRY AND THE ORIGIN OF LIFE / *Keith A. Kvenvolden*
15 SEDIMENTARY ROCKS: Concepts and History / *Albert V. Carozzi*
16 GEOCHEMISTRY OF WATER / *Yasushi Kitano*
17 METAMORPHISM AND PLATE TECTONIC REGIMES / *W. G. Ernst*
18 GEOCHEMISTRY OF IRON / *Henry Lepp*
19 SUBDUCTION ZONE METAMORPHISM / *W. G. Ernst*
20 PLAYAS AND DRIED LAKES: Occurrence and Development / *James T. Neal*
21 GLACIAL DEPOSITS / *Richard P. Goldthwait*
22 PLANATION SURFACES: Peneplains, Pediplains, and Etchplains / *George F. Adams*
23 GEOCHEMISTRY OF BORON / *C. T. Walker*
24 SUBMARINE CANYONS AND DEEP-SEA FANS: Modern and Ancient / *J. H. McD. Whitaker*
25 ENVIRONMENTAL GEOLOGY / *Frederick Betz, Jr.*
26 LOESS: Lithology and Genesis / *Ian J. Smalley*

27 PERIGLACIAL PROCESSES / *Cuchlaine A. M. King*
28 LANDFORMS AND GEOMORPHOLOGY: Concepts and History / *Cuchlaine A. M. King*
29 METALLOGENY AND GLOBAL TECTONICS / *Wilfred Walker*
30 HOLOCENE TIDAL SEDIMENTATION / *George deVries Klein*
31 PALEOBIOGEOGRAPHY / *Charles A. Ross*
32 MECHANICS OF THRUST FAULTS AND DÉCOLLEMENT / *Barry Voight*
33 WEST INDIES ISLAND ARCS / *Peter H. Mattson*
34 CRYSTAL FORM AND STRUCTURE / *Cecil J. Schneer*
35 OCEANOGRAPHY: Concepts and History / *Margaret B. Deacon*
36 METEORITE CRATERS / *G. J. H. McCall*
37 STATISTICAL ANALYSIS IN GEOLOGY / *John M. Cubitt and Stephen Henley*
38 AIR PHOTOGRAPHY AND COASTAL PROBLEMS / *Mohamed T. El-Ashry*
39 BEACH PROCESSES AND COASTAL HYDRODYNAMICS / *John S. Fisher and Robert Dolan*
40 DIAGENESIS OF DEEP-SEA BIOGENIC SEDIMENTS / *Gerrit J. van der Lingen*
41 DRAINAGE BASIN MORPHOLOGY / *Stanley A. Schumm*
42 COASTAL SEDIMENTATION / *Donald J. P. Swift and Harold D. Palmer*
43 ANCIENT CONTINENTAL DEPOSITS / *Franklyn B. Van Houten*
44 MINERAL DEPOSITS, CONTINENTAL DRIFT AND PLATE TECTONICS / *J. B. Wright*
45 SEA WATER: Cycles of the Major Elements / *James I. Drever*
46 PALYNOLOGY, PART I: Spores and Pollen / *Marjorie D. Muir and William A. S. Sarjeant*
47 PALYNOLOGY, PART II: Dinoflagellates, Acritarchs, and Other Microfossils / *Marjorie D. Muir and William A. S. Sarjeant*
48 GEOLOGY OF THE PLANET MARS / *Vivien Gornitz*
49 GEOCHEMISTRY OF BISMUTH / *Ernest E. Angino and David T. Long*
50 ASTROBLEMES—CRYPTOEXPLOSION STRUCTURES / *G. J. H. McCall*
51 NORTH AMERICAN GEOLOGY: Early Writings / *Robert Hazen*
52 GEOCHEMISTRY OF ORGANIC MOLECULES / *Keith A. Kvenvolden*
53 TETHYS: The Ancestral Mediterranean / *Peter Sonnenfeld*
54 MAGNETIC STRATIGRAPHY OF SEDIMENTS / *James P. Kennett*
55 CATASTROPHIC FLOODING: The Origin of the Channeled Scabland / *Victor R. Baker*
56 SEAFLOOR SPREADING CENTERS: Hydrothermal Systems / *Peter A. Rona and Robert P. Lowell*
57 MEGACYCLES: Long-Term Episodicity in Earth and Planetary History / *G. E. Williams*
58 OVERWASH PROCESSES / *Stephen P. Leatherman*
59 KARST GEOMORPHOLOGY / *M. M. Sweeting*
60 RIFT VALLEYS: Afro-Arabian / *A. M. Quennell*
61 MODERN CONCEPTS OF OCEANOGRAPHY / *G. E. R. Deacon and Margaret B. Deacon*

A BENCHMARK® Books Series

TETHYS
The Ancestral Mediterranean

Edited by
PETER SONNENFELD
University of Windsor

Hutchinson Ross Publishing Company
Stroudsburg, Pennsylvania

Benchmark Papers in Geology, Volume 53
Library of Congress Catalog Card Number 80-13974
ISBN: 0-87933-355-3

83 82 81 1 2 3 4 5
Manufactured in the United States of America.

LIBRARY OF CONGRESS CATALOGING IN PUBLICATION DATA
Main entry under title:
Tethys, the ancestral Mediterranean.
(Benchmark papers in geology; v. 53)
Includes indexes.
1. Tethys (Paleography)—Addresses, essays, lectures. I. Sonnenfeld, Peter.
QE65.T47 551.7'09182'2 80-13974
ISBN 0-87933-355-3

Distributed world wide by Academic Press,
a subsidiary of Harcourt Brace Jovanovich,
Publishers.

CONTENTS

SERIES EDITOR'S FOREWORD

The philosophy behind the "Benchmark Papers in Geology" is one of collection, sifting and rediffusion. Scientific literature today is so vast, so dispersed, and, in the case of old papers, so inaccessible for readers not in the immediate neighborhood of major libraries that much valuable information has been ignored by default. It has become just so difficult, or so time consuming, to search out the key papers in any basic area of research that one can hardly blame a busy person for skimping on some of his or her "homework."

This series of volumes has been devised, therefore, to make a practical contribution to this critical problem. The geologist, perhaps even more than any other scientist, often suffers from twin difficulties—isolation from central library resources and immensely diffused sources of material. New colleges and industrial libraries simply cannot afford to purchase complete runs of all the world's earth science literature. Specialists simply cannot locate reprints or copies of all their principal reference materials. So it is that we are now making a concerted effort to gather into single volumes the critical materials needed to reconstruct the background of any and every major topic of our discipline.

We are interpreting "geology" in its broadest sense: the fundamental science of the planet Earth, its materials, its history, and its dynamics. Because of training in "earthy" materials, we also take in astrogeology, the corresponding aspect of the planetary sciences. Besides the classical core disciplines such as mineralogy, petrology, structure, geomorphology, paleontology, and stratigraphy, we embrace the newer fields of geophysics and geochemistry, applied also to oceanography, geochronology, and paleoecology. We recognize the work of the mining geologists, the petroleum geologists, the hydrologists, and the engineering and environmental geologists. Each specialist needs a working library. We are endeavoring to make the task of compiling such a library a little easier.

Each volume in the series contains an introduction prepared by a specialist (the volume editor)—a "state of the art" opening or a summary of the object and content of the volume. The articles, usually some twenty to fifty reproduced either in their entirety or in significant extracts, are selected in an attempt to cover the field, from the key papers of the last century to fairly recent work. Where the original works are in foreign languages, we have endeavored to locate or commission translations. Geologists, because of their global subject, are often acutely aware of the oneness of our world. The selections cannot, therefore, be restricted to any one country, and whenever possible an attempt is made to scan the world literature.

To each article, or group of kindred articles, some sort of "highlight commentary" is usually supplied by the volume editor. This commentary should serve to bring that article into historical perspective and to emphasize its particular role in the growth of the field. References, or citations, wherever possible, will be reproduced in their entirety—for by this means the observant reader can assess the background material available to that particular author, or, if desired, he or she, too, can double check the earlier sources.

A "benchmark," in surveyor's terminology, is an established point on the ground, recorded on our maps. It is usually anything that is a vantage point, from a modest hill to a mountain peak. From the historical viewpoint, these benchmarks are the bricks of our scientific edifice.

RHODES W. FAIRBRIDGE

PREFACE

When I was asked in 1974 to compile a volume on the Tethys Sea, I refused the assignment at first. There are many experts available who have devoted a lifetime of research to Tethyan problems whereas I could not make such a claim. However, I had had the opportunity to gain some first-hand knowledge of field relationships at several localities between Central Europe, Turkey, Iran, Afghanistan, Pakistan, the Deccan, northern Nepal, Malaysia, and Indonesia and had delved into Miocene salt formation in the Mediterranean and adjacent seas.

"Rocks do not lie" is an old adage. All rock outcrops described by generations of field geologists must fit into any omnibus framework of the history of the area. Otherwise, the interpretation of these rocks and their relationships would seem to be ephemeral. It is nonetheless surprising how many of our present concepts were first formulated 60-100 years ago.

A tremendous number of detailed reports is available on local areas. They are written in a host of languages. In contrast, comprehensive summaries for large portions of the Tethys Sea are only few in any language. They cover neither all the Tethys nor all periods of earth history. For this volume, a cross section of regional reviews was selected for each era of Phanerozoic history. A benchmark volume must concentrate on a bird's eye view of the subject.

An anthology can illuminate a subject from many angles. The papers selected for inclusion were chosen mainly to present an account of the paleogeography of the Tethys Sea as recorded in the marine sediments, in the material transported from adjacent lands, and in the rocks on the shores of this ancient body of water. The selection was not based on any particular school of thought except to weigh in favor of geological field relationships rather than geophysical interpretations.

The selection of papers represents only a portion of our total accumulated knowledge on the subject. An attempt has been made to select articles that will excite the reader to do additional work, promote new answers in other regions, or stimulate further discussion. If that happens even in only one instance, the purpose of the book has been achieved.

I would like to thank here the two correspondents, Dr. P. P. Hudec,

University of Windsor, Ontario, and Mr. G. J. Genik, Esso Exploration Inc., Walton-on-Thames, United Kingdom, for the many hours of labor put into helping with the selection and the arrangement of the anthology.

PETER SONNENFELD

CONTENTS BY AUTHOR

TETHYS

Part I

ORIGINS OF THE TETHYS CONCEPT

Editor's Comments on Papers 1 and 2

1 **SUESS**
Are Great Ocean Depths Permanent?

2 **STEVANOVIĆ**
50 Jahre seit der Proklamation der Paratethys

The Tethys Sea is that ancestral sea which covered the areas of southern Europe and Asia where Mesozoic marine sediments can be found. In the same areas, Paleozoic marine sediments also occur; the sea in which these were deposited is either called Paleotethys, Paleozoic Tethys, or Prototethys. The southern shores of this sea were found in central Europe, southern Russia, and Central Asia.

The first geologists mapping Tethyan marine sequences were mainly interested in the paleontology of the ubiquitous Mesozoic marine fauna. Structural setting and stratigraphic position of these marine sediments soon elicited comparisons with the modern Mediterranean Sea. To call the marine basin in which these sediments had accumulated the "Ancestral Mediterranean," as Melchior Neumayr had done, proved to be unwieldy. And so, the name "Tethys Sea" was coined within the Austro-Hungarian Geological Survey.

Deep- and shallow-water sediments were found in this sea. In precise analogy to the Mediterranean Ridge in the modern eastern Mediterranean Sea, a geanticline, or linear set of shoals, was found to extend from the Alps and Dinarides to northern Turkey, Afghanistan, and the Pamirs. North of these shoals, the sea had the character of a shelf sea or a miogeosyncline. Eventually, this northern portion was given the name "Paratethys." The term is usually applied to Neogene marine sediments because these are the sediments studied most intensely. A marginal sea of similar proportions existed in the territory of the Paratethys throughout the Phanerozoic eon. Its history was similar to, yet distinct from, that of the Tethys.

Editor's Comments on Papers 1 and 2

No better preamble of this collection of papers on the Tethys could be provided than the reproduction of Eduard Suess' address to the Geological Society of London in 1893 in which he placed the term "Tethys" for the first time on record before the geological fraternity outside Austria-Hungary (Paper 1).

This is followed by Peter Stevanovic's recount of the origins of the term "Paratethys," initially conceived to designate only a Neogene body of water although much of the same area was covered by a sea throughout nearly all of Phanerozoic time (Paper 2).

1

Reprinted from *Natural Sci.* **1**:180–187 (1893)

ARE GREAT OCEAN DEPTHS PERMANENT?

E. Suess

IT has been suggested by the editor that English readers would like to hear my views on that much-debated question, the Permanence of Ocean Basins. I am rather at a loss how to deal with the subject, because this question involves so many difficult chapters in the history of our planet, and because I regret to see that discrepancy of views exists on fundamental principles.

Mr. Wallace begins by arguing from the principle that "on any large scale, elevation and subsidence must nearly balance each other, and thus, in order that any area of continental magnitude should rise from the ocean floor, ... some corresponding area must sink to a like amount." I venture with deference to reply that I cannot agree to this. It seems, on the contrary, as if two different types of movement had been going on since the first formation of the terrestrial crust. In the first place, there is folding, recently explained in a masterly way by Professor Lapworth, in his Address to the British Association. Secondly, there is the sagging-down or "effondrement" of smaller or greater parts of the crust, caused by the progressive diminution of the planet's radius. *This descent of parts of the earth's crust seems to be the true origin of the great oceanic basins.*

Sometimes the contour of the sunken area follows the trend of a folded mountain chain; at another time it may cut right across it. In smaller examples the outline very often takes a more or less irregularly circular or elliptical form. The descent of a considerable area, forming a large new depression, demands a certain part of the existing volume of oceanic waters for the filling of the new depth. The consequence is the sinking of the oceanic surface all over the planet, and the *apparent step-like rising of coast lines.* Thus is explained the apparently episodic elevation of whole continents, without any disturbance of horizontality, or the least alteration of the net of watercourses spread over the land. It is in this sense alone that a certain balance of "elevation" and "subsidence" might be conceded.

In the entire Pacific region the limits of the oceanic basin are traced out by the trend of long mountain folds. So it is from New Zealand and New Caledonia to the borders of Eastern Asia, to the Aleutians, and all along the western coast of both Americas. This is not the case in the Atlantic, nor in the Indian Ocean; here the coasts are not outlined by folded structure, except in the arch of the Lesser Antilles, and in the corresponding short arch passing through Gibraltar, which serves to connect the mountains of north-western Africa with those of the south of Spain. These are what we may call the Pacific and the Atlantic types of oceanic regions.

Indian geologists have shown how the immense Asiatic mountain waves, moving southwards against the Peninsula, have been dammed back by the resistent Peninsular mass, the Korána Hills, and the wedge-shaped mass of Assam; so that they actually form distinct arches, separated by deep angles receding to the neighbourhood of Tank, north of Dera Ismail Khan, to the Upper Jhelum and Brahmaputra Valley. In this case we call the Peninsula the "Vorland."

Now cast a look on the map of the North Pacific, and compare the receding angles which mark the western and the eastern ends of the Aleutian arch, where it abuts against Kamschatka and North-West America. You will remark that this part of the Pacific is a "vorland," and homologous to the Indian Peninsula, whilst the Yellow Sea, Behring Sea, and others lie within the folded region. You may also examine the Mediterranean, and observe that the western half lies wholly within a curved and folded mountain chain (Apennines, Sicily, North Atlas, Gibraltar, Andalusian Cordillera), and that in the eastern half all the part south of Crete and Cyprus and east of Sicily lies on the African "vorland," and the rest on the sunken parts of the Tauro-Dinaric arch.

In the Atlantic region the mountain folds, as a rule, break off against the ocean (*e.g.*, Brittany coast, Devon and Cornwall, south-west Ireland), or else have their folds facing away from the ocean, as in the case of the Alleghanies, and all other folded zones on the eastern side of North America as far as Newfoundland. The folds disappearing in south-west Ireland and in Brittany so very much resemble those rising from beneath the Atlantic on the coasts of Nova Scotia and New Brunswick, that M. Marcel Bertrand has ventured to publish a sketch-map, showing these chains trending right across the Atlantic.

There exists a curious tendency for a depression or a sort of valley to form in front of the great folds facing the "vorland." For instance, the depressions of the African desert in front of eastern Atlas, the river valleys in front of the high Indian chains, and the Persian Gulf in front of the Zagros chains. Quite recently the Austrian exploring ship "Pola" found a depth of 4,400 metres near the south-west coast of Greece, near the front of the Dinaric arch; and some of the greatest oceanic depths show exactly the same position in front of the arches of Japan, the Kuriles, and the Aleutians with Alaska. This is the homology, for example, between the Ganges valley and the Tuscarora depth, both marking the limit of the folded ranges and the "vorland."

The structure of the earth's crust does not, therefore, tell us that the great depths must be very old or permanent; so let us compare the character of the sediments near the existing coasts.

The Old Red Sandstone is an extra-marine formation; yet the Old Red Sandstone runs out into the sea in the Orkneys, reappears in the Shetlands, and the same palæontological and extra-marine character is known in Spitzbergen. Old Red exists on the north coast of Lapland, and on the White Sea. The plant-bearing beds of the Karoo run out to sea in British Kaffraria; they are repeated in India and Australia. The fresh-water beds of the Wealden pass over from England to the Continent; they not only reappear in Hanover, they run out into the Atlantic in the lower Charente, and on the coasts of Spain and Portugal. Why must the continent which formerly bounded all these vast fresh-water basins have been limited within the existing 1,000- or 1,500- or 2,000- fathom line? The breaking down of the bed of the Ægean Sea, described by Spratt and Neumayr, is of Post-pliocene date, for Pliocene fresh-water deposits form parts of the coast; and yet the depths go far beyond 1,000 fathoms. In 1892 the "Pola" measured 3,591 metres in lat. 35° 52 36″, long. 29° 1′ 24″ E., quite near the south-west coast of Asia Minor, and close to the mighty Ak Dagh (10,000 feet); and this although the separation of the neighbouring island of Rhodes is so recent, that not only do the Pliocene fresh-water beds pass over from the Continent, but according to Bukowski also considerable masses of Middle Pliocene fluviatile conglomerates, originating in Asia Minor, have been deposited by a great river on this island.

Now suppose the existing quantity of oceanic water to decrease, say by evaporation into the ether, as Zöllner once thought, or in any other way; we might by this gradual diminution of the entire quantity attain a beach-line 500 fathoms below the present shores. The continents would appear so much higher, and dry land would extend. Plains would successively appear, more or less similar to Holland, and our present rivers Shannon, Seine, Loire, &c., would flow through these plains. In most cases the rivers would be caused to cut back their valleys, new transverse and parallel lines of erosion would appear, and the plain would be diversified into hills and valleys. The hills south of the Shannon would probably show the rest of those anticlines and synclines which dip below the ocean in south-west Ireland, and we should be able to see a greater part of the northern Armorican arch. The Scilly Islands would appear as another granitic laccolite within the continued Armorican region of Cornwall. The gneiss of Eddystone would come up within this northern Armorican arch, exactly as the gneisses of the Alps stand up within and behind the folded arches. In a similar way, in the south, the anticlines and synclines of French Armorica, which disappear north and south of Brest, would begin to be visible; but in the north-west of Ireland we should see a plateau, ending in a steep cliff, the abrupt boundary of a deep channel, separating the great island of Rockall from the European continent. And all this varied new-born land would be green and full of life, and people would not be at all willing to believe that it ever was not so.

Then we would descend to the new shore, and one of our great masters would tell us that ocean depths are permanent, that is to say, from 1,500 fathoms downward, or from 2,000 fathoms. The general impression resulting from the study would be the same as now, but the assumed permanent level would be reduced by 500 fathoms.

We might invite our master to undertake an excursion with us; we would go to Scotland. An isthmus, some 1,200 feet high at the narrowest part—Sir Wyville Thomson's ridge—would lead us first to some isolated peaks, nearly 3,000 feet high, and over a rising country to the high peaks of the Faröes. We would observe land-born coal-beds between the great coulées of basalt. Proceeding further, we would travel to the north-west, over a broad tract of rocky land some 1,200 or 1,300 feet high; then first meet an isolated mountain of about 1,800 feet, and from there ascend to the volcanic mass of Iceland, where we should see those vast fields of lava, dotted with active volcanoes; and observe the long faults and open rents cutting through the masses of lava and trending across Iceland in a broad arch, first from south-west to north-east, then northward, beyond the volcano Askja to Husavik; and beyond this broken and breaking zone we might gain the great "effondrement" or "Kessel-bruch" of Faxafjord, beset all round by volcanoes and hot springs, from Snaefells Jokull to Reykjavik. In following the rents so well described by our indefatigable Thoroddsen, we might detect faulted plant-bearing beds, and recognise the equivalents of the Faröe coal-seams. If some younger and more impressible student be in our company he might well exclaim, in face of these plant-bearing beds, stretching on to Greenland and showing the existence of a vast dry land in late times, and in face of these rents and volcanoes: "Verily, Professor, the sagging-down of the North Atlantic is the most recent event; it is going on before our eyes; and as the highest mountain chains are the youngest, so also are the deepest parts of the planet the most recent." I fear I should not know how else to answer the student than, "Really, I do not know."

Now let us quit the coasts and examine the interior of a great continent.

Modern geology permits us to follow the first outlines of the history of a great ocean which once stretched across part of Eurasia. The folded and crumpled deposits of this ocean stand forth to heaven in Thibet, Himalaya, and the Alps. This ocean we designate by the name "Tethys," after the sister and consort of Oceanus. The latest successor of the Tethyan Sea is the present Mediterranean.

I asked Dr. Diener, recently returned from India, to give me his estimate of the thickness of the deposits in the Silakank region. Dr Diener answers: · From Dhauli-Ganga Valley, between Gweldung and Pethathali encamping ground above Silakank (19,265 English feet), to Sirkia River in Hundes; a complete section from (Cambrian?) Haimantas to the Gieumal Sandstone (Cretaceous), without a great discordance, gives, according to Dr. Griesbach:—

Haimantas	3,000—4,000 feet.
Lower Silurian	200 ,,
Upper Silurian	1,100 ,,
Devonian..	700 ,,
Carboniferous	1,200—1,400 ,,
Permian and Trias	3,600—3,900 ,,
Lias and Spiti Shales	1,400? ,,
Gieumal Sandstone	1,200—1,500 ,,
	12,400—14,200 feet.

The determination of the thickness of the Spiti Shales and Gieumal Sandstone is difficult, because these less-resistent beds are crumpled into local folds.

A parallel section across the Kurkutidhár range (Chor Hoti, about 18,000 feet), Shalshal encamping ground and Shalshal Pass (16,390 feet) to Hundes gives, without the Haimantas:—

Silurian	1,200—1,400 feet.
Devonian and Carboniferous	1,800—2,300 ,,
Permian and Lower Trias	200 ,,
Middle Trias	100 ,,
Upper Trias to Dachstein Limestone ..	1,400 ,,
Dachstein Limestone, Rhætic and Lias ..	2,200 ,,
Spiti Shales	1,000? ,,
Gieumal Sandstone	1,500? ,,
	9,400—10,100 feet.

These figures show that a great and deep ocean has been incorporated into the continent, and that the deposits of this ocean form part of the highest mountain ranges.

It may be remarked that within the eastern Alps Mesozoic limestones of different ages contain deep-water radiolarian chert. But the great and well-bedded masses of white Rhætic limestones of Austria betray distinct proofs of a continuous rising of the shore-line. It is also true that certain bright red enclosures within these white limestones seem clearly not to be red deep-sea clay, but true *terra rossa*, formed by atmospheric decomposition of the limestone; so that these beds must have formed reefs in the ocean. Therefore it is at present difficult to say whether in the Alps the Tethyan Ocean did at any time attain the total depth of, say, 2,000 fathoms; or whether deposits followed so rapidly and depression was so continuous that this was not the case.

The later Tethyan history, the recapitulation of the vicissitudes which led to the formation of the existing Mediterranean, forms certainly one of the most attractive chapters of historical geography. Marine deposits of Mediterranean type (Erste Mediterranstufe, Miocène inférieur) enter the Rhone Valley, surround the present site of the Alps, and continue far away to the East, to Persia, and have been met with by Griesbach near Balk, on the Oxus. Then these deposits were folded, in the Taurus, in Asia Minor, in Switzerland. Afterwards came the sagging-down of certain parts of these folds, near Vienna, in Hungary, &c., and all that varied series of consequent events. After the first Mediterranean came the formation of an immense horizon of salt deposits, stretching from Wieliczka to Persia; then a second Mediterranean reaching far into the newly-formed depressions; then the appearance of vast fresh-water lakes, lasting through a long period of time till the breaking down of the Ægean land and the re-conquest of the Black Sea.

Look at smaller examples of such partial subsidences; see Margerie's instructive paper on the Corbières, showing the sinking down of the Pyrenees, Miocene beds passing beyond Narbonne, while south of Cape Leucate two more recent Pliocene "effondrements" form the Rousillon, described by Depéret, and the Golfe de Rosas.

But this is only part of the Tethyan history. Michelin's and Duncan's palæontological studies in the West Indies have revealed the European character of certain deposits. It is the "Gosau type" of the Cretaceous which appears in Jamaica, and the Castel-Gomberto horizon of Upper Oligocene (warm type of Sables de Fontainebleau) is known in several other isles. In regions still further off, one of our first masters, the venerable Dr. Philippi, has shown that the present molluscan fauna of the Chilian coasts is of quite recent origin, and that until the beginning of Quaternary times the European Mediterranean molluscan types stretched far down the western coast of South America. At the same time the Mesozoic deposits of Chili, and those recently discovered at Taylorville in California,[1] show purely European characters, and the Neocomian of Bogotá is the exact equivalent of that of Barrême.

These facts teach us that an ocean-bed existed, but that some coast-line, maybe only an interrupted line, once stretched across the present Atlantic, and permitted the Gosavian and Oligocene corals, and the Miocene shells also, to cross. I do not overlook the fact that Dr. Philippi himself, struck by the analogies existing between the flora of Chili and that of Europe, recently refused to accept the hypothesis of a "bridge" to Europe, and preferred to suppose that identical climatic and other external causes produced analogous and even identical species of terrestrial plants. I refer to what has been excellently said by Mr. Blanford on this theory, a few years ago, in his address to the Geological Society of London. I believe that the parallel correspondence of the marine faunas up to the Quaternary period gives a more correct clue to the correspondence of the existing terrestrial floras in Chili and in Europe.

So I think that we must not only concede the extinction of a great

[1] See Hyatt and Dillen on the Jura and Trias at Taylorville, California. *Bull. Geol. Soc. America*, 1892, pp. 369—412.

Palæozoic, Mesozoic, and Tertiary ocean in south-western Eurasia, but admit also great recent changes in the middle or southern Atlantic. Geological evidence, therefore, does not prove, nor even point to, a permanence of the great depths, at least in the oceans of the Atlantic type.

Let me remark in a few words that, although I believe in the possibility of the formation of large new depressions, I do not hold with the old opinion, lately taken up again by M. Faye, that the continued sinking of the ocean beds may force chains of mountains to appear all round. This view could only be propounded for the Pacific basin; but the Pacific chains are folded in the direction *towards* the ocean, and not *from* the ocean. They are easily divided into arches, each of which presents the *convex* side to the ocean, so that the Pacific everywhere presents the character of a "vorland."

Let me, at the end of this long note, allude to a broad biological fact. In the higher beings we see *lungs* always preceded by gills; so it is even with the human child. The adaptation for breathing our atmosphere is of a later date; and we conclude that the whole terrestrial air-breathing fauna is a *derived* fauna, derived from amphibious forms quitting shallow water. This fact evidently points to a long existence of dry land, long enough to permit this accommodation to be effected; the accommodation clearly has been going on since Palæozoic times. Still there exists no proof that individual continents always remained the same, and we even know for certain that such was not the case with by far the greater part of these continents.

A similar lesson is also taught by the *eyes* in all the higher organised beings of the deep sea. The optical apparatus of abyssal species is profoundly modified by the exceptional environment, while the normal types of eyes are met with in the same genera within moderate depths. Therefore, we must also regard these deep-sea forms as *derived* forms. The blind and blinded Trilobites of Cambrian beds, the blind Trinuclei and the widely-expanded eyes in certain species of Aeglina in Lower Silurian strata teach the same lesson. At the same time, they show that deep-water must have existed over Bohemia, and over a good many other Palæozoic tracts, and that the depths were considerable enough to call forth these same abyssal metamorphoses of the eye.

We might, therefore, rather be induced to infer that in Pre-palæozoic times there may have existed a universal hydrosphere or panthalassa covering the whole of the planet. Only with the first appearance of dry land began the deposition of clastic sediments. The higher forms of life may have been developed in waters of moderate depth and may successively have spread to the sun-lit continents, and to the dark depths, while the slow elaboration of the existing inequalities of the terrestrial surface was going on.

But this elaboration is still in progress. I believe with Reyer,

Fisher, Jukes, and many others that the great depths are mostly covered with volcanic products, with lava and ashes forming immense plains and overlain eventually by the deposits of the abyss; but I see no reason why parts of the ocean or even of the dry land may not to-morrow sink to form new depths, or why we should believe that all the great ocean basins have been continuously covered by water since panthalassic times. So far as the Atlantic is concerned, there even exists some evidence to the contrary.

But all this is unripe fruit. Our scholars will some day know more than their masters do now; so let us patiently continue our work and remain friends.

2

Reprinted from *Geol. Zborník (Geol. Carpathica Slowak. Akad. Wiss.)* **26**: 170–171 (1975)

50 JAHRE SEIT DER PROKLAMATION DER PARATETHYS

P. Stevanović

Motto: Repetitio est mater studiorum

Dieses Jahr ist das 50. jährige Jubiläum von der Einführung des Namens Paratéthys für das Mittel — und Südosteuropäische neogene Meer, das von den Alpen bis zum Aralbecken sich erstreckte.

Vor 50 Jahren, am 10. April 1924, hielt Professor der Belgrader Universität Vladimir Laskarev in der Serbischen Geologischen Gesellschaft einen Vortrag unter dem Namen „Über die Tektonik der Umgebung von Belgrad", der in demselben Jahr in der erweiterten Form auf französisch unter dem Titel „Sur les équivalents du Sarmatien Supérieur en Serbie" veröffentlicht wurde (s. Récueil des travaux offért a M. J. Cvijić, 1924, Beograd). Dabei wurde zum ersten Mal der Name „Paratéthys" gebraucht.

Nachdem die erste Definition der Paratéthys in einer wenig bekannten und schwer zugänglichen Publikation gegeben wurde, drang der Name in den Gebrauch nicht leicht ein. Erst mit der Gründung des „Commitée für das méditerrane Neogen" (Aix en Provence. 1958) und nach seinem ersten Kongress in Wien (1959) fand Paratéthys eine allgemeine Anerkennung. Mit der Bildung der „Arbeitsgruppe Paratéthys" in Bologna (1967) wurde der Name überall bekannt und in allen Kreisen aufgenommen.

Paratéthys wurde von V. Laskarev auf französisch folgenderweise begründet: „Une des conséquences les plus importantes, qui suivirent en Europe l'élévation du système alpin fut la formation au commencement du miocène d'une vaste mer qui fut séparée de la Téthys par des procédés tectoniques alpins. Cette mer, en se tortillant parmi les diverses parties du Système alpin, s'étendait parallelement à Téthys, du bassin du Rhône jusqu'aux régions transcaspiennes, en passant par la Suisse, la Bavière, par les bassins de Vienne et Pannonien, par la Serbie, la Roumanie et la Russie méridionale. On peut proposer le nom de Paratéthys pour cette mer (V. Laskarev lit. c. 1924, p. 1)...

....L'histoire géologique de l'Europe méridionale (surtout du Sud-Est de l'Europe) était étroitement liée a cette mer, dont l'emplacement est occupé à present par le système du Danube,... et par les restes de la Paratéthys — les mers Noire, d'Azov et Caspienne".

...„Les données palaeogéographiques nous permettent de conclure qu'une communication entre Téthys et Paratéthys existait pendant le miocène inférieur (burdigalien — I. étage méditerranéen) et moyen (vindobonien — II. étage méditerranéen), ce qui prouve l'uniformité de leur faune. Cette communication très limitée s'accomplissait par le bassin Rhône — Suisse — Bavière et le sillon transégéen de M. E. Haug. Il parait très vraisemblable que la Téthys, pénetra au commencement du miocène en Styrie par l'Italie du Nord" ... „Vers la fin du Miocène le mouvement téctoniques ont provoqué la rupture de ces voies de communication: il s'instala un régime continental avec des lacs d'eau douce ou saumâtre. De sorte qu'au miocène supérieur (sarmatien) la communication fut interrompue, la Paratéthys se transforma en une mer, complètement isolée de l'océan et commenca une vie séparée".

...„Dans le rapport tectonique l'immence territoire, que la Paratéthys a couvert, était composé:

1° de trois effondrements principaux: a) bassin Pannonien..., b) bassin Pontique (avec des branches en Galicie, en Roumanie, en Bulgarie), c) bassin de la Mer caspienne et

2° de quatre voies de communication principales entre ces effondrements et entre eux et la Téthys"...

„Les voies de communication les plus importantes entre Téthys et Paratéthys (bassin Rhône — Suisse — Bavière et Sillon transégéen, en même temps les détroits de la Silésie et d'Oltu) cessèrent les premières de fonctionner à la fin du miocène moyen, ce qui amena l'isolation de la Paratéthys de l'océan et sa transformation en mer sarmatique".

Wie aus dem angeführten Originaltext ersichtlich ist, schied V. Laskarev drei Hauptbecken der Paratéthys aus: Panonisches, Pontisches oder Euxinisches und Kaspisches Becken alle drei mit Randbecken und Verzweigungen (Buchten, sekundären Meerengen usw.). Sowohl aus dem erwähnten als auch aus allen späteren, zwischen 1924—1954 Jahren veröffentlichten Arbeiten, sowie Mitteilungen und auf dem Lehrstuhl gehaltenen Vorträgen, vertritt V. Laskarev die Meinung, dass Paratéthys im Westen am Ostrand der Alpen bzw. bei Wien beginnt. Alles was sich westlich und nördlich davon in der perialpin — perikarpatischen Zone befand gehört den „voie de communications" zwischen den Hauptbecken der Paratéthys bzw. zwischen Téthys und Paratéthys.

Auf der palaeogeographischen Karte in der Beilage (V. Laskarev 1924) wurde „Reconstruction de la Téthys et de la Paratéthys du miocène moyen" gegeben. In der Legende wurden: „les régions effondrées" und „les communications entre la Téthys et Paratéthys" gegeben. Unter den „communications" befinden sich „les bassins du Rhône, de la Suisse et de la Bavière", die in den späteren Arbeiten (J. Seneš 1959) als „Westparatéthys" bennant wurden. Jugoslawische Geologen sind auch heute, wie V. Laskarev (1924), der Meinung, dass die echte Paratéthys bei Wien bzw. St. Pölten, nicht aber an der Rhône — Mündung beginnt. Westlich davon, durch Bayern, Schweizerisches Niederland und Rhône-Becken, ist ein breiter, von geosinklinalem Charakter „voie de communication", der mit Recht sowohl als Téthys als auch Paratéthys betrachtet werden kann, ja sogar faziell eher zu Téthys als zu Paratéthys gehört: In dieser Zone gibt es weder Torton (mit der ausnahme von Rhône-Becken) noch Sarmat, die für die Paratéthys s. str. so bezeichnend sind.

Es ist wohl bekannt, dass im Laufe der letzten 20—30 Jahre ein grosser Fortschritt bei den Studien der Stratigraphie, Tektonik, Palaeogeographie und Faunistik in dem Paratéthys-Gebiet erzielt wurde. Dieser Fortschritt war möglich sowohl, dank der erdölgeologischen Forschung als auch den organisierten Arbeiten im Rahmen des Commitée's für das Mediterrane Neogen. Arbeitsgruppe „Paratéthys", als Hauptoperative des Commitée's, steht am Kopfe des ganzen Unternehmens. Die Meinungsunterschiede in Rahmen der Arbeitsgruppe haben unserer Meinung nach zu diesen Fortschritt beigetragen, nicht aber ihn aufgehalten.

Nächste Forschungen werden hoffentlich zeigen wo Fehler gemacht wurden, so wie die bisherige Arbeit vieler Neogenforscher schon bewiesen hat, wie bedeutungsvoll und glücklich die Idee V. Laskarev's var, das Innereuropäische Neogenmeer mit dem Namen Paratéthys zu benennen.

Part II

AN OVERVIEW

Editor's Comments on Paper 3

3 SONNENFELD
The Phanerozoic Tethys Sea

The Tethys Sea has been the subject of many studies in the Alps and, of late, in Afghanistan and the Himalayas. Two schools of thought arose. One proclaimed that there had been vast expanses of ocean between Europe and Asia on one side and Afro–Arabia and India on the other of the order of 45° of arc. A second school maintained that crustal shortening in the Alps and in other related mountain chains did not account for more than a distance of 300–500 km of lost surface, that is, less than 5° of arc.

The former school accounted for the highlands of Tibet as an Asiatic landmass underthrust by the subcontinent of India. A powerful assist was given to this opinion when it was found that Indian rocks bear some relationship to those of East Africa and Australia. A second line of support was the discovery that paleomagnetic data from Africa and from Europe do not give pole positions 180° of arc apart. Thus, the Tethys Sea was a convenient place to call for a subduction zone that had swallowed up the supposedly missing vast expanses of surface area. Based on the occurrence of pillow lavas in ophiolite sequences scattered throughout the realm of the Tethys, an analogy could be drawn to modern oceanic basalts. That the tuffs, which are so frequently associated with ophiolitic pillow lavas, could not have originated under the hydrostatic pressure of deep oceans was overlooked.

Detailed fieldwork by oil company personnel and by state geological surveys straddled both margins of the Tethys and found surprising similarities between north and south shores. Any purportedly lost surface area was always placed well away from the territory studied. When gathering selections for the present anthology, the following paper grew out of an annotated bibliography in an attempt to catalogue the available literature on the Tethys and to sort out apparent contradictions (Paper 3).

The term "Tethys" is rarely entered into computerized literature files as a key word and thus is not helpful in a literature search. However, a very good coverage of literature published since 1972 is found in the monthly *Bulletin Signaletique,* Series E:224 Stratigraphy and Regional Geology and Series F:225 Tectonics, published by the Centre National de la Recherche Scientifique, Bureau de Recherches Géologiques et Minières, Paris, France.

3

This article was prepared expressly for this Benchmark volume

THE PHANEROZOIC TETHYS SEA

Peter Sonnenfeld
University of Windsor

INTRODUCTION

The notion of a vast body of water having occupied at one time the lands from the Alps to the Himalayas was first entertained by various members of the Austro-Hungarian Geological Survey. Melchior Neumayr referred to it as "Ancestral Mediterranean" and Eduard Suess, at first as Thetys. After some argument with his colleague, Alexander Bittner, Suess changed the spelling to Tethys. In 1893, Eduard Suess summarized his ideas in a lecture in London, England, reproduced elsewhere in this volume. If he surmised that "Asian mountain waves collided with the crustal slab of the Indian Peninsula and Assam," he anticipated by more than three decades Argand (1924) who is generally considered to be the father of the idea of India underriding Tibet. If he further surmised that "folds disappearing in southwest (SW) Ireland and in Brittany find their continuation on the coasts of Nova Scotia and New Brunswick" or that "Cretaceous beds in Jamaica and the western Americas are closely related to European ones," he anticipated Wegener's continental drift and modern ideas of seafloor spreading. His suggestion that the Mediterranean Ridge lies on parts of Africa is echoed today in many models of plate tectonics.

A generation after Suess, it was found that another body of water existed, separate and distinct from the Tethys and to the north of it, extending from west-central Europe to the Pamirs. This set of sedimentary basins or portions of them have variously been called the "Southern Russian Miogeosynclinal," the "northern Tethys," the "Donets-Mangyshlak Basin" (Shevchenko et al., 1974), the "L'vov-Dnepr-Donets Depression" (Eynor, 1970), the "pre-Black Sea" (Archangel'skiy and Shatskiy, 1933), the "Donets Basin Marginal Trough" (Klemme, 1958), the "Siret Ocean" (Herz and Savu, 1974; Bechstädt et al., 1976), the "Ancestral Black Sea," the "Lac-Mer," or the "Paratethys" (Laskarev, 1924). Although the last name was originally coined only for a Neogene body of alternating fresh and saline waters, the term should be expanded to cover also the older equivalents. Thick marine and brackish sediments ranging from Cambrian to Tertiary age and from the Pamir Plateau into Central Europe portray a depositional history distinct from the Tethys proper. However, the histories of the Tethys and Paratethys seas frequently show similarities caused by repeated opening of multiple connecting passages.

The Tethys and Paratethys were separated by a ridge of lesser subsidence, or at times minor uplift, traceable at least since Ordovician time from northwest (NW) Afghanistan through northern Iran (Wolfart, 1972)

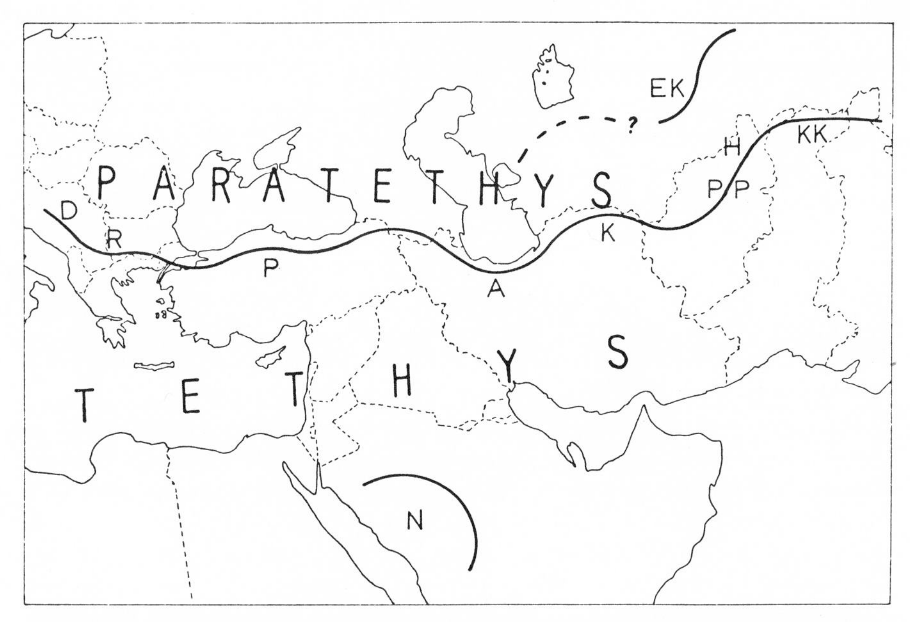

Figure 1. The Mediterranean macroisthmus (from Sonnenfeld 1978a; copyright © 1978 by E. Schweizerbart'sche Verlagsbuchhandlung [Nagele u. Obermiller] Stuttgart).
D, Dinarides; R, Rhodope; P, Pontides; A, Alborz; K, Kopet; PP, Paropamismus; H, Hindu Kush; KK, Karakoram; N, Nubian Shield; EK, East Kazakh macroisthmus

into southeast (SE) Turkey and thence possibly into northern Italy (Dercourt, 1972; Havlíček, 1974). This geanticline, dorsal (Seyfert, 1978), or spine (Klemme, 1958) was also called the "Mid-Tethyan Ridge" (Dercourt, 1972), the "zone médiane," or the "Mediterranean macroisthmus" (Nalivkin, 1973a). It also went under several local names, such as "central crystalline axis" (Hashimoto et al., 1973), "Himalayan Ridge" (Fuchs, 1967, 1977), "Vindelician swell" (Megard-Galli and Baud, 1977), and many more (Fig. 1).

A significant faunal and sedimentological boundary also formed from time to time at right angles to this Mediterranean macroisthmus (so termed in order not to introduce another name) through southern Turkey, Iraq, Syria, and Jordan into the Al Jawf area of Saudi Arabia. It may have represented another partially submerged ridge capped by several small islands. Facies differences in rocks of the same age on either flank of this arch suggest it to have been a boundary between an ancestral Mediterranean Sea and the remainder of the Tethys (Sander, 1968). From Ordovician to Upper Cretaceous time, the area along the Dead Sea axis was dry land (Bender, 1974) and continued northward as a chain of islands.

A second such transversal barrier was located in about northern Pakistan and western Kashmir and thence into Turkmenistan. It persisted from early Paleozoic into Jurassic times (Voskresenskiy et al., 1971); Permian, Triassic, and Jurassic deposits thin against it (Petrushevskiy, 1970b).

THE CAMBRIAN TETHYS

A basal Cambrian transgression allowed East Asiatic faunas to penetrate the Dead Sea area, Morocco, and Spain (Brinkmann, 1969). A tidal or deltaic sandstone overlain by fine-grained silty and micaceous sandstones marks this transgression in Libya and Tunisia (Burollet et al., 1978) and in Egypt (El Shazly, 1977). A conspicuous lithologic unit, the purple-to-red Lalun Sandstone, is found from Transcaucasia (the Paratethys region) to eastern and central Iran and into the Zagros Mountains, over a distance of a thousand kilometers (Flügel, 1972). It may extend as far as Pakistan, Turkey, Jordan, and Saudi Arabia (Stöcklin, 1968; Demirtasli, 1975) and onto the north rim of Africa. The sandstone represents the nearshore facies of an essentially very shallow set of basins.

The clastics interfinger basinward with shallow marine carbonates that occur from the Black Sea area to northeast (NE) Iran and the Pamir Plateau, apparently fringing a series of islands representing the only partially submerged Mediterranean macroisthmus. In turn, the carbonate facies grades into thick halite deposits in the Salt Ranges of Pakistan, Rajasthan (India), and central Afghanistan and in a belt from central Iran to the Persian Gulf. From eastern Iran, the evaporitic facies also continues to Oman and turns to gypsum in Hadramawt (South Yemen) in the Arabian Peninsula, a distance of 1,500 km (Gee, 1965; Powers et al., 1966; Wolfart and Kürsten, 1974). This evaporite facies is correlative to the salt-bearing formation mapped in the Tarim Basin of Central Asia (Bing et al., 1965) and extends thus from the Tethys beyond the Mediterranean macroisthmus (Fig. 2).

The source of the salt appears to have been an influx of seawater from the Arctic Ocean through the narrow Uralian geosyncline to the shores along the southern margins of the Pakistani, Iranian, and Arabian Tethys. Clastic sediments of continental provenance suggest here the presence of some landmass farther to the south. The salts indicate that the Cambrian-Ordovician arid belt was about 50–60° of arc from a pole in West Africa, which is in accord with paleomagnetic data. These evaporites fall within the Horse Latitudes in a worldwide reconstruction of Ordovician climatic belts (Ronov et al., 1974, 1976).

The nearly continuous presence of time-equivalent evaporite sequences from SW Arabia (then part of Gondwana) to NW China (then part of Laurasia) contradicts any suggestion that vast distances of ocean have existed between these two supercontinents, distances that are now subducted or otherwise swallowed without a trace. A frequently repeated suggestion (Holmes, 1965; Jain and Kanwar, 1970; Drake and Nafe, 1972; Mascle, 1977) that the Tethys had been a broad and deep ocean, as much as 5–7,000 km wide, cannot be maintained. That distance represents 45–60° of arc and no evaporite belt or any other climatic zone is that wide across. On the other hand, Falcon (1967) concluded that no angular crustal displacement could have occurred in the Tethys since Precambrian times. Maxwell (1970) observed that a mismatch in rock types and a dis-

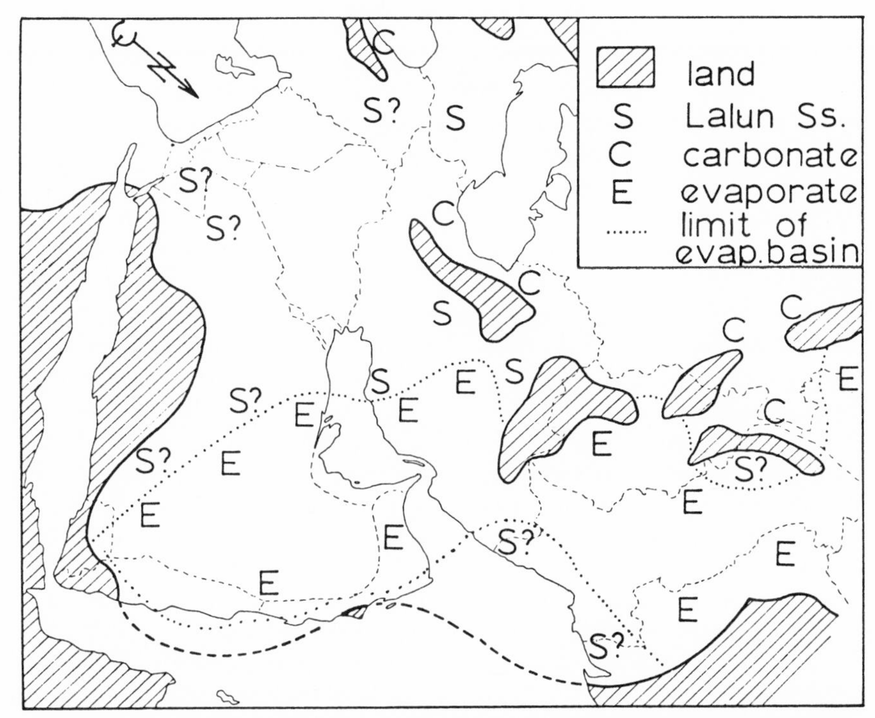

Figure 2. Cambrian facies map (from Sonnenfeld 1978a; copyright © 1978 by E. Schweizerbart'sche Verlagsbuchhandlung [Nagele u. Obermiller] Stuttgart).

similarity in geologic history should be observable in continents that had travelled toward each other for several thousand kilometers. This is not the case in the Paleozoic Tethys as shown by the match of facies against facies across that sea on paleogeographic maps constructed for several time intervals in each of the Paleozoic and Mesozoic periods (Ronov et al., 1954–1977; Grossgeym and Khain, 1975).

An early Middle Cambrian sea submerged much of Iran, Iraq, and the Levant where Atlantic and West-Pacific faunas met. A Middle Cambrian carbonate sequence occurs in Morocco; it has also been mapped from Egypt (Sadek, 1977) to Turkey, Jordan, Israel, eastern and nothern Iran (Flugel, 1972), and into the Himalayas (Bordet et al., 1971). Atlantic faunas continued to flourish in the Paratethys as far as the Himalayas, but were shut out of the central and southern Tethys (Wolfart, 1972). A landmass must have emerged west of Thailand because Upper Cambrian marine sandstones, which resemble the slightly older Lalun Sandstone, indicate a western source (Suensilpong et al., 1978).

THE ORDOVICIAN TETHYS

Ordovician marine beds were deposited concordantly on Cambrian sediments along the whole southern flank of Eurasia while the sea retreated from Central Europe and from North Africa (Sdzuy, 1972). A highland arose along the Mediterranean macroisthmus in early Caledonian times; only gentle uplifts without folding can be documented (Wolfart, 1967). Erosion of the more elevated portions gave rise to substantial thicknesses of sediments in the Balkan Peninsula (Klemme, 1958), produced an unconformable contact between Ordovician and Devonian beds near the Bosporus, and uncovered Ordovician granitic gneisses in the Caucasus. This unconformity has also been observed in the Alps (Holl et al., 1978).

Within the Tethys, Selley (1970) recognized an Ordovician littoral or deltaic facies in Jordan. Southern Anatolia and northern Iraq were covered by sublittoral or bathyal sediments of a mobile shelf (Seilacher, 1963). The southern rim ended up in a coldwater regime as evidenced by Late Ordovician tillites in Saudi Arabia (McClure, 1978).

The eastern Tethys in Afghanistan and Iran maintained a tropical fauna (Wolfart, 1972) about 50–70° of arc away from the Ordovician pole located in the western Sahara. The tropical conditions extended deep into Central Asia (Ronov et al., 1976) and as far as Thailand (Kobayashi, 1964). At the same time, the European Tethys was dominated by a coldwater fauna occasionnally intruded by warmwater Baltic species (Havlíček, 1974).

Glacial features in central Germany and Bohemia, presently about 30° of arc from the Ordovician pole position in the Sahara, also place a limit on the possible overall movement (compression less distension) between Africa and Europe since the Ordovician.

THE SILURIAN TETHYS

The Tethys Sea shrank in Silurian time and became gradually shallower and narrower in the Himalayas (Gupta, 1977) between the highlands of northern Iran and those of western and southern Arabia (Wolfart, 1967) where some uplift seems to have occurred (Sander, 1968). In Libya and Tunisia, Silurian rocks grade from well-laminated black shales to fine-grained sandstones; they are northward time transgressive and are separated by an unconformity from Devonian rocks (Burollet et al., 1978), confirming the general regression of the Silurian Tethys Sea.

In the Carnic Alps of Austria, Silurian limestones show considerable redissolution, burrowing and boring, hematite staining, and phosphate and manganese nodule development (Schonlaub, 1970). No evidence of deep water exists, however, and the lithology may indicate nothing more than a large influx of humic acids and groundwater solutes from nearby humid lands.

In contrast to the regressive phase in the Tethys, the Paratethys subsided as evidenced by extensive sedimentation that continued right into early Devonian time. All along the south rim from northern Turkey to the Pamir Plateau, including the East Afghan Trough, there are thick wedges of Silurian sediments. Graptolite shales and occasionally intercalated limestone banks occur from Bulgaria (Pol'ster et al., 1977) to the Himalayan Tethys (Bordet et al., 1971).

THE DEVONIAN TETHYS

The Lower Devonian seas exhibited a northern shoreline facies from the Caucasus through the Balkans to Bohemia, Belgium, and southern England and a southern shoreline facies from southern Turkey to northern Tunisia and Algeria (Erben, 1962, 1964) with pelagic sediments and island chains in between. In the west, the sea washed against the Transcontinental Arch of North America.

Lower Devonian clastics of an arid environment are covered by gypsum in eastern Iran and anhydrite in Bulgaria in embayments of the Mediterranean macroisthmus. An equivalent series occurs in the Al Jawf region of northern Saudi Arabia (Wolfart, 1967) whereas clastics in Egypt carry authigenic anhydrite (El Shazy, 1977). Both areas represented the southern rim of the Tethys Sea. Again, the arid belt remained 45–50° of arc from a Devonian pole position, which is consistent with paleomagnetic data. However, it again offers little room for separating western Saudi Arabia from northern Iran by vast expanses of sea.

Lower and Middle Devonian seas at first advanced onto the African craton in both Libya and Tunisia, but a regression occurred before the end of the Devonian (Burollet et al., 1978). A Middle-to-Upper Devonian transgression also inundated most of the Mediterranean macroisthmus (Wolfart, 1972; Brice et al., 1978). This coincided with a regression on the margins, on the north rim of the Nubian–Arabian Shield, and in Spain, and also in the Caucasus (Belov, 1972) on the other side of the macroisthmus in the then Paratethys area.

THE CARBONIFEROUS TETHYS

The Paratethys was submerged in early Carboniferous time (Eynor, 1970), reaching into the lower Danube River valley and into NE Romania (Sander, 1968). On the north rim of the Hindu Kush, the clastics of the transgressive Carboniferous sea are followed by carbonate banks with increasing intercalations of radiolarian cherts followed by nearshore sands and microconglomerates, a second development of limestones, and a final series of regressive Lower Permian conglomerates without visible interruption in sedimentation (Boulin et al., 1975).

The Mediterranean macroisthmus underwent some early Hercynian vertical movements in Greece, western Turkey, and the SE Pamirs. End-Devonian nodular limestones are followed by Lower Carboniferous radiolarian cherts east of the Bosporus (Abdusselamoglu, 1963). The macroisthmus was later rejuvenated along its whole length (Wolfart, 1972). Extensive erosion, braded rivers, and paludal and lacustrine deposits mark the emergence in Bulgaria (Pol'ster et al., 1977).

The connection to a Proto-Arctic Ocean through NE Europe closed in early Carboniferous time (Ramsbottom, 1971; Grunt, 1970) and a Siberian fauna invaded the southern shores of the Tethys from the Sahara to the Himalyas and Australia (Ali Meissami et al., 1978). The NE European seaway was replaced by one through the Russian Platform west of the

Urals in Permian time and this, by a seaway through the West Siberian Platform east of the Urals in Triassic time. The seaway passage thus shifted eastward in time.

The Tethys Sea extended westward through southern Turkey and Syria into coastal Egypt and SE Tunisia. A gradual subsidence of the African craton in the early Carboniferous allowed the seas to advance, but by the middle of that period, the rim of North Africa turned into lagoons on the margins of dry land. Mid-Carboniferous faults produced troughs on the African craton that were filled with bioclastic limestones (Burollet et al., 1978). Narrow inlets extended through the Adriatic Sea into the Po River valley and through the Straits of Sicily to the Balearic Islands.

Both north and south shores of the western and central Tethys were inhabited by tropical faunas, but coldwater faunas are reported from sites in Oman, Afghanistan, and Pakistan (Termier and Termier, 1974). Boulder beds of in part ice-rafted erratics in Saudi Arabia, Oman, and Pakistan also point toward a probable cold temperature regime (Wolfart, 1972; Roland, 1978) possibly derived from mountain glaciation in northern Ethiopia and adjacent Yemen (Beauchamp, 1978). Upper Carboniferous tillites in the eastern Tyan Shan Mountains (Norin, 1937) or in the Ladakh area of Kashmir (Gansser, 1974) and in the eastern Himalayas (Colchen, 1978) suggest plateau or mountain glaciations in Central Asia and thus high altitudes similar to present ones because the paleolatitudes were nearly equatorial. They are not arctic sites because the tillites in the Himalays occur both in the west and in the east with the warm-climate *Gangamopteris* vegetation rather than the circum-polar *Glossopteris* flora (Colchen, 1978). One source of ice was in the SW Deccan; the direction of striae and of glacial gravels is away from there toward the Pamirs and Assam (Pomerol, 1975).

THE PERMIAN TETHYS

A Permian transgression restored the early Paleozoic extent of the Tethys Sea. This transgression has not only been documented in Yugoslavia (Pomerol and Babin, 1977) and in the Near East (Wolfart, 1967) but also in the Alburz Mountains (Jenny and Stampfli, 1978) and in the Himalayas (Fuchs, 1977; Colchen, 1978). This transgression brought an Uralian fauna into Sicily (Pomerol and Babin, 1977).

The Mediterranean macroisthmus continued to separate a northern Paratethys (in Central Europe, northern Turkey, the Black Sea, and the Caucasus) from a Tethys proper to the south (Mokrinskiy, 1939) that in turn faced an epicontinental platform in western Arabia and in South Yemen (Saint-Marc, 1978). This separation can be followed into Afghanistan (Montenat et al., 1977).

Lower Permian nearshore sediments in southern Austria are overridden from the south by Middle and Late Permian marine carbonates extending from the Alps (Cadet, 1978) to Oman, Iran, Afghanistan, and Nepal (Bruggey, 1973). Local embayments in the Alps, Dinarides, and

Balkans became the sites of deposition for Upper Permian evaporites and bituminous dolomites behind more open marine carbonates (Buggisch et al., 1976; Erkan, 1977; Pol'ster et al., 1977; Cadet, 1978).

Continental conditions prevailed only in Bulgaria, northern Greece, northernmost Turkey, and the northern Caucasus, that is, the shores of the Paratethys. Waters eventually also entered here from the north, but by Late Permian time the Paratethys was again reduced to a narrow channel extending north of and parallel to the Mediterranean macroisthmus (Flügel, 1972; Grunt, 1970). Terminal Permian erosion of that macroisthmus is evidenced by a widespread angular marine sandstone in the Salt Range of Pakistan (Kummel and Teichert, 1966) possibly related to a disconformity over Permian fringing reefs (Ganss, 1964) in the Hindu Kush (Boulin et al., 1975).

The western margin of the Tethys appears to have been located near present shorelines of the Balearic Sea (Kamen-Kaye, 1976); the west coast lay somewhere north of the city of Algiers between the marine carbonates outcropping in Tunisia and the nonmarine sands found in the Anti-Atlas and on the flanks of the Spanish Meseta (Klemme, 1958). To the east, the Permian Tethys initially extended from the Mediterranean Sea between northern Turkey and central Egypt through Central Asia north of the Tarim Platform to the North China coast and to the south of that platform through Thailand to the Indochina coast (Kobayashi, 1964; Grunt, 1970; Petrushevskiy, 1975; Termier and Termier, 1976). Thus emerges a picture of a Permian Tethys narrowed in the east to two progressively closing channels, that is, an epicratonic inland sea separated both from the proto-Atlantic and proto-Pacific oceans.

The northeastern, Sino-Korean connecting channel to an ancestral Pacific Ocean progressively narrowed and was gradually cut off during Permian time. Similarly, the connection to a sea covering the West Siberian Lowlands was severed, and the latter turned into an arm of the Arctic Ocean of progressively smaller dimensions. This represents a Late Permian regression that is also documented in Thailand (Kobayashi, 1964), in the vicinity of the Mediterranean macroisthmus in northern Iran (Jenny and Stampfli, 1978), and in the Zagros Mountains (Setuhdenia, 1978). The latter were uplifted and remained thereafter a positive shoal area throughout the Mesozoic era. Carbonates (often reefal) accumulated on top while Triassic or Jurassic evaporites precipitated on either side of the Zagros shoal.

Middle Permian *Glossopteris* elements in southern Turkey (Schmidt, 1964) suggest a Gondwana affinity on Laurasian territory, again showing that faunal differences across the Tethys remained at a minimum.

SUMMARY OF THE PALEOZOIC TETHYS

The Paleozoic Tethys was a relatively narrow epicontinental sea as inferred from all lithologic evidence. Nowhere did it develop any oceanic deeps or even a continental slope. Faunal variability remained much greater in an east–west direction than from south to north, that is, from Gondwana to Laurasian territory. Climatic belts again and again extended across the Tethys Sea, but changed along the length of it, which suggests a rather limited width. To the north of the Tethys, separated by a string of islands, was the Paratethys Sea. The Paratethys extended from Central Europe to the margins of Central Asia. At times, it had connections to the European Sea and to the Ural Sea.

The suggestion that the Tethys Sea was an early Jurassic creation (Hsu and Bernoulli, 1978) cannot be maintained unless one restricts the term to that period of earth history when block faulting, ultramafic intrusions, basaltic extrusions, and attendant specialized sedimentation, that is, the formation of ophiolitic rock suites, had commenced. There are no ophiolites known in the Paleozoic Tethys, yet radiolarian cherts, siliceous limestones, and black noncalcareous shales were at times quite widespread. Fault and graben or horst structures are known today from the northern Aegean Sea and from Lake Superior without creating either an ocean or faunal provinciality.

On the other hand, there are Paleozoic ophiolites known throughout Laurasia generally younging away from ancient platforms and along strike. One such trend runs from the East Indies along the southern margin of the Tarim Platform through Tibet, meets the one ringing the Russian Platform along the later Ural Mountains, and continues through the Ukraine, Central Europe, the Alps, and southern England to NW Spain, progressively changing from early to late Paleozoic in age (Sonnenfeld, 1978a) (Fig. 3).

The Upper Carboniferous ophiolite trend splits in Cornwall: one branch veers off into Scotland, the other makes a U-turn into Spain as if the attendant fractures had been unable to continue from Cornwall and cross the annealed Taconic–Acadian belt of Appalachia. The bridge to Alpine ophiolites is made by Permian and Triassic ophites in Spain where ultrabasic components increase away from the craton (Gölz, 1978).

With the onset of fracturing of the Tethys Sea in Triassic time, the first Tethyan ophiolites formed in the western Alps. They are progressively younger in an easterly direction, exactly opposite to the direction taken by the Paleozoic ophiolites at the south rim of the Russian Platform. At that, they are using the older ophiolite belts as guard rails as Schwinner (1919) observed.

The Tethys–Paratethys seas thus never became a geosyncline in Paleozoic time despite the aggregate accumulation of very thick sedimentary sequences, but they probably acted as a miogeosyncline to a system of Paleozoic geosynclines developing immediately to the north. These were gradually exposed, from east to west, to Hercynian tectogenesis. Only minor foreland diastrophism, such as some folding in Bulgaria and the Bosporus area, spilled over into the Tethyan realm (Fourquin, 1975; Biju-Duval et al., 1978). In the Himalayas, no Caledonian or Hercynian orogenies can be seen; all Paleozoic strata are conformable (Bordet et al., 1971).

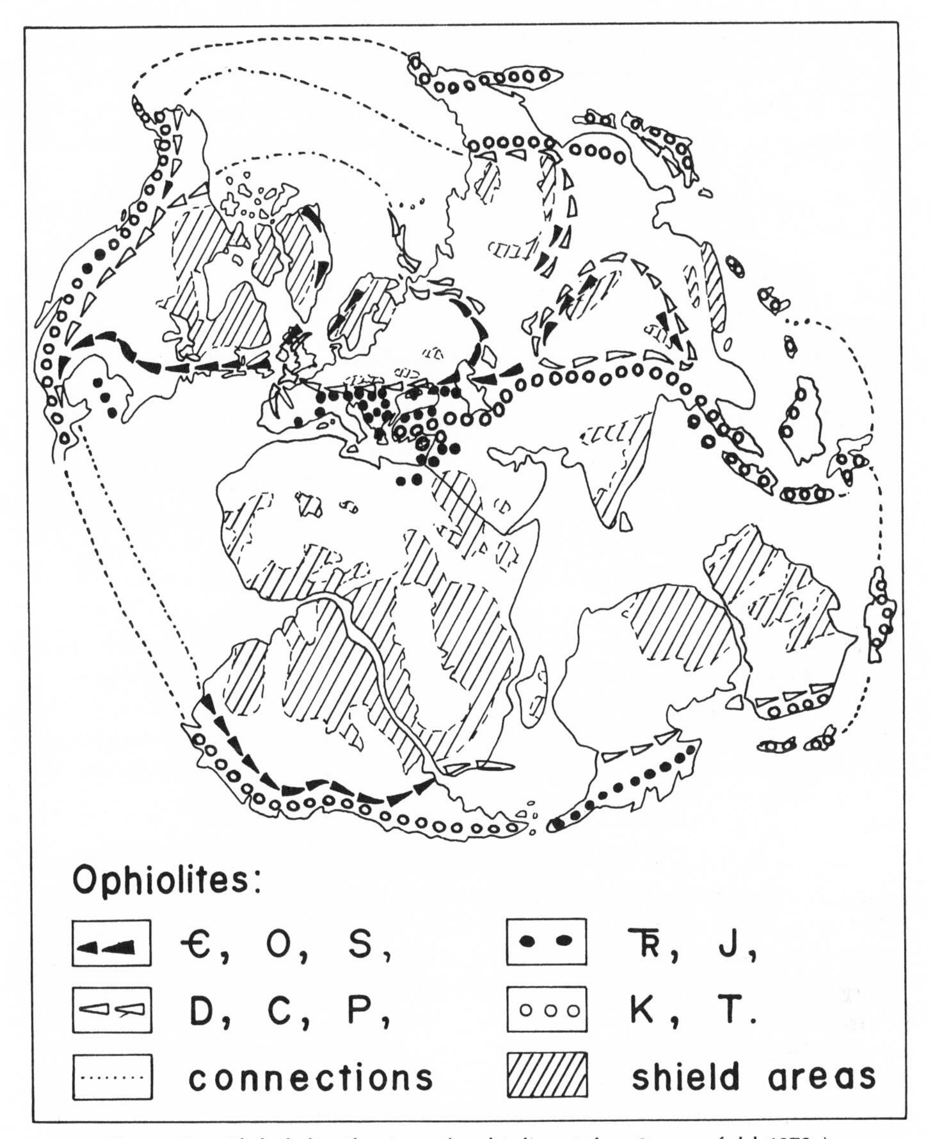

Figure 3. Global distribution of ophiolites (after Sonnenfeld 1978a).

THE TRIASSIC TETHYS

Overall, the Triassic landscape was marked by a very low relief in Europe and Africa (Trümpy, 1971). The uplands of Central Asia remained a provenance area of sediments (Rukhin, 1960) mainly because of early Triassic orogenic events that cemented the Siberian, Tarim, and Sino-Korean platforms (Huang, 1978). They caused a disconformity in the Persian Gulf area (Setuhdenia, 1978), in Baluchistan (Ganss, 1966), in East Afghanistan (Bruggey, 1973), in the Hindu Kush (Boulin et al., 1975), and in Thailand (Kobayashi, 1964) and a paraconformable hiatus in the Salt

Ranges of Pakistan (Kummel and Teichert, 1966). In the Himalayas of Nepal, however, the Triassic sediments lie concordantly on Paleozoic deposits (Bordet et al., 1971). The north shore of the eastern Tethys was in the vicinity of the Tsinling-Kunlun mountain chain (Chinese Academy of Geological Sciences [C.A.G.S.], 1977).

The Tethys gained ground westward, assuming the shape of a gulf (Aubouin et al., 1977). The shoreline advanced over the eastern half of Spain (Virgili, 1977; Seyfert, 1978), and a semideltaic clastic sequence developed in Morocco (Klemme, 1958). Interbedded sands and shales were deposited in Tunisia and Libya (Burollet et al., 1978). Shallow-water marine sedimentation prevailed in most of the western Tethys (Trümpy, 1971; Biju-Duval et al., 1978).

Early Triassic clastic sedimentation in the western Po River valley indicates a marine ingression from the east (Kälin and Trümpy, 1977). A similar westward transgression occurred into the Paris Basin (Pomerol, 1975). The same also happened in the Paratethys, which advanced westward and northward from Iran into the northern Caspian region.

Carbonate banks became more widespread in Middle Triassic time when the sea invaded the greater part of the European platform. These limestone deposits covered the Mediterranean macroisthmus that was flanked by reefal carbonates (Klemme, 1958; Fourquin, 1975), such as those from eastern Algeria and from the Dinaric Alps (Lucas, 1952; Sander, 1970). Extremely thick carbonates also accumulated in Afghanistan (Lapparent, 1972). No faunal provinciality developed within the Tethys; the exchange of species was uninhibited at this time (Kummel, 1966; Bruggey, 1973).

The Upper Triassic interval is marked by evaporite sequences in the Algerian and Moroccan Sahara, in Tunisia, in the Lusitanian basin of Portugal, in Spain, Italy, Austria, western Greece, SE Anatolia, in the Gulf of Suez area, in the NE of the Caucasus, in Iran, Iraq and Arabia, and in southern China (Klemme, 1958; Sikosek and Medwenitsch, 1965; James and Wynd, 1965; Kozary et al., 1968; Querol, 1969; Sander, 1970; Belov, 1972; Busson, 1975; El-Shazy, 1977; C.A.G.S., 1977; Setuhdenia, 1978; Burollet et al., 1978). Locally, gypsum occurrs in mid-Triassic sediments (Bechstadt et al., 1976). Again, the whole length of the Tethys can be shown to lie in one climatic belt that affected north and south shores alike as it had during the Paleozoic era.

Synsedimentary dilatational fracturing of the underlying crust (Trümpy, 1971) produced widely variable water depths over very short distances and with them, local pelagic sedimentation in the western Tethys and similar horst-and-graben structures even in East Africa (Beauchamp, 1978). Some of the blocks became source areas of plant-bearing coarse clastics derived from an exposure of the crystalline basement stripped of sedimentary cover (Bechstädt et al., 1976).

A series of worldwide fractures developed toward the end of the Triassic period and into the Early Jurassic. One trend gave rise to basic volcanism from western North America along the Taymyr Peninsula and

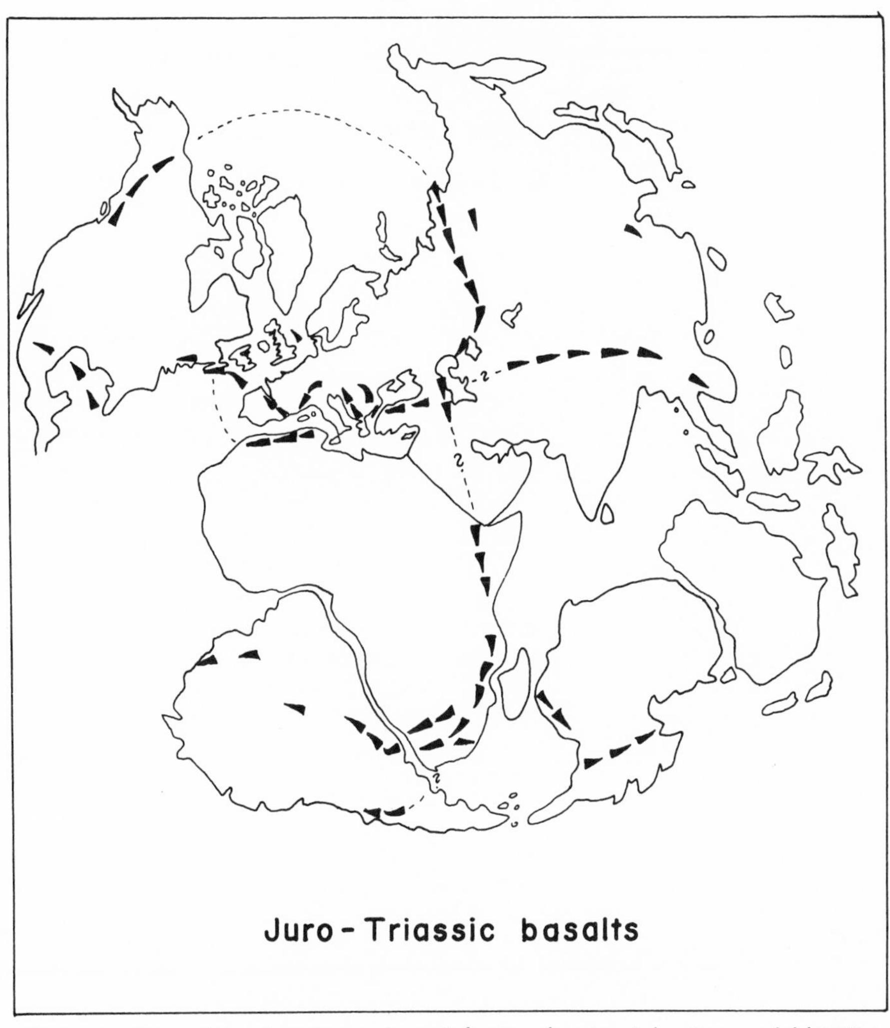

Figure 4. Upper Triassic to Lower Jurassic basic volcanism (after Sonnenfeld 1978a).

the West Siberian Lowlands to the Talish Mountains south of the Caucasus. Centers of subsidence thereby migrated from the NE to the western and southern edges (Rudkevich, 1976). The trend continues through Ethiopia to the Drakensberg Mountains of South Africa (Ronov et al., 1961) and farther into southern and central Brazil where the extrusions are Jurassic in age (Fig. 4). A second line runs from Triassic terrestrial basalts of New England (located at the younger end of the Appalachian orogeny) through Nova Scotia to the north rim of the Atlas Mountains, the Dinarides and Pontides, the Caucasus, the Karakoram Range of Kashmir, the Trans-Himalayas and southeastern Tibet and thence into Thailand, or possibly to the mouth of the Yellow River (Sonnenfeld, 1978a). This line contains Jurassic sites at both ends, in eastern Asia and in the Caribbean Sea.

Van Bemmelen (1972) suggested that the Antarctic, African, and South American Jurassic–Triassic basalts represented a megatumour; if that is so, a second such megatumour must then have existed in or near the Caucasus where a similar crossing of volcanic fractures occurred.

THE JURASSIC TETHYS

The low relief of adjacent lands in Europe and in Africa in Triassic times continued to prevail in the Jurassic period, and the influx of terrigenous clastics decreased to a minimum. Terrigenous facies developed only locally, notably in southern Spain and North Africa (Biju–Duval et al., 1978).

There was a general regression and reduction of water depth in the eastern Alps and a concurrent decrease in clastics (Neumayr, 1886–1887). In the western Alps, one finds bauxites and coals. The regression continued as far as Arabia (Saint–Marc, 1978) and Afghanistan (Ganss, 1966; Lapparent, 1972). Evaporites accumulated in some marginal bays of the Tethys and Paratethys (Kozary et al., 1968), for example, in the Persian Gulf (James and Wynd, 1965), in central Arabia (Saint–Marc, 1978), and in SE Anatolia. Thick carbonate sequences accumulated in the open waters. They are reported from the Alps, from central Arabia, Afghanistan, and Tibet. Limestones with thick–shelled, wave–resistant mussels are associated with coral reefs from France and the Austrian Alps through Iran and Oman to the Himalayas and Timor (Geyer, 1977). The widely variable water depths produced a local gradation of Upper Jurassic carbonates to radiolarian cherts in the eastern Alps, the Dinarides, and Turkey.

Alternating bioclastic limestones and black shales in Nepal suggest mobility above and below wave base (Bordet et al., 1971). Because carbonate deposition could not keep up with increasing rates of subsidence in the Dinarides (Cadet, 1978), sedimentation became more pelagic, but limestone boulders and calcareous turbidites prove the continuation of shallow–water sedimentation on adjacent crustal blocks.

However, the occurrence of Jurassic stromatolites in Sicily, northern Italy, Poland, Hungary, and Yugoslavia (Jenkyns, 1971) on shoals or "intrageanticlinal ridges" within the geosyncline is invariably associated with reworked faunas, stratigraphic lacunae, and sedimentary iron enrichment. Abutting such current–scoured submarine shoals, one finds condensed sections of dissolution horizons in the form of nodular limestone (Hollmann, 1962, 1964; Jenkyns, 1971) at times associated with radiolarites (Grunau, 1965).

Neumayr (1886–1887) found that the Tethyan Jurassic fauna showed a more endemic aspect than the fauna of the Paratethys, which maintained its worldwide affinities. He contrasted an "Alpine" fauna, extending from SW Spain to Ethiopia and India, with a "Central European" fauna to the north of the Mediterranean macroisthmus, from northern Spain through France to the north rim of the Caucasus. Only the latter fauna is also found on other continents, possibly migrating via straits leading into boreal seas in European Russia and in NW Europe. This contrast in faunas has been reaffirmed more recently in Anatolia (Fourquin, 1975). Paralic sediments in Thailand (Suelsilpong et al., 1978) mark the eastern extent of the Tethys at this time.

In Late Jurassic times, the rates of subsidence increased (Bourbon et al., 1977), producing a sea of diverse depths in the central Alps and Dinarides (Trümpy, 1960, 1965, 1971). Radiolarites show abrupt lateral differences in thickness, indicating a very uneven seafloor (Folk and McBride, 1978). This unevenness was due to block faulting (Cadet, 1978; Elmi, 1978), some of which had already been initiated in Late Triassic times (Kälin and Trümpy, 1977). This block faulting extended as far as the Caucasus on the one side of the waters (Beloussov, 1962), into the northern Arabian Peninsula, and into Ethiopia and Somalia on the other side (Saint-Marc, 1978) where it was preceded by a land rise (doming?) indicated by regression of the Tethys. The resulting horst-and-graben topography of the Tethys seafloor (Bernoulli and Jenkyns, 1974) was a mark of tension, dilation, and local lack of support of the crust by the underlying mantle.

Peters (1969) saw the early expansion phase in the Austrian Alps as a diapiric rise of upper mantle material whereby the crust thinned and distended. In analogy, Krestnikov and Nersesov (1964) observed on Neogene vertical movements in the Tyan Shan Mountains that thinned crustal slabs descend, thicker ones rise.

Jurassic ophiolites in part follow Triassic basalts; they occur in the central and eastern Alps, the Apennines, Dinarides, Carpathians, the Crimea, northern Turkey, and the Caucasus. No Jurassic volcanism (basic, intermediate or acidic) has so far been noted between Iran and Indonesia.

The fractures also extended westward, and thus, one finds Jurassic intrusions and basic volcanism in Newfoundland (Helwig et al., 1974), in the Greater Antilles, and in Mexico. An initial transgression into Mexico and Venezuela from the east was caused by the rapid subsidence of graben depressions filled with black (plant-bearing) nearshore shales alternating with turbidites. The fauna was impoverished and partly endemic under the euxinic conditions of restricted circulation (Schmidt-Effing, 1977). However, a passage to the ancestral Pacific Ocean may have been breached at times (Hallam, 1977).

The Caribbean thus became the westernmost gulf of an advancing Tethys with widespread evaporites in southern and eastern Mexico and in the southern United States (Aubouin et al., 1977). Later opening of Atlantic fractures destroyed this Tethyan connection of Central America. The westward advance of the Tethys coincided with a transgression into Arabia and onto the African craton.

Compression followed immediately after the period of distension and attendant magmatic activity followed compression, so in the central Dinarides, for example, ophiolites are immediately followed by flysch deposition (Aubouin, 1977), the product of denudation of rising mountains.

THE CRETACEOUS TETHYS

Carbonate deposition was less widespread in the Cretaceous Tethys than it was before. Massive erosion of surrounding lands supplied great quantities of shales and siltstones to advancing seas. Flysch deposits became common in the eastern Mediterranean region. Local emergence in the Dinarides led to bauxite formation; end-Cretaceous uplift produced vigorous erosion here (Cadet, 1978).

The Cretaceous seas not only inundated vast areas of the African craton, they also penetrated deep into Europe and the Near and Middle East as far as Pakistan. Because of the irregular seafloor, both shoal facies and condensed series are common along the North African craton (Burollet et al., 1978).

In the Caspian region and in various marginal seas of the Mediterranean area, evaporites precipitated in Lower and Middle Cretaceous time. Maastrichtian-to-Paleocene gypsum is known from an extensive interior basin of southern Iran (James and Wynd, 1965). Campanian-to-Paleocene halites are missing worldwide (Sonnenfeld, 1978b), and thus, their absence has climatic rather than geologic significance.

The lower Cretaceous sandstones are continental in the Himalayas (Bordet et al., 1971), indicating a brief regression phase (doming?) prior to Upper Cretaceous fracturing of the local Tethyan seafloor. This fracturing and tilting of blocks can be followed into Afghanistan (Bruggey, 1973).

The southern shores of the Tethys were in Arabia where there is a gradual facies change from Iranian carbonates southwestward into Arabian sandstones (Setuhdenia, 1978). Progressive inundation is found in Anatolia (Fourquin, 1975) and in Afghanistan where incursions came both from an "ancestral Arabian Sea" to the SW, that is, the Tethys, and from the north, that is, the Paratethys (Lapparent, 1972). The Hindu Kush zone marks the dividing line between the two resulting sedimentary provinces. Again, the Paratethys maintained connections to the world oceans at first through European Russia and later through the West Siberian Lowlands. Continental and paralic sediments in Thailand (Suensilpong et al., 1978) mark the eastern extent of the Tethys.

RADIOLARIAN CHERTS

Whereas early Mesozoic radiolarian cherts are found in the western Tethys, Cretaceous ones occur only in the central Tethys in Turkey and Iran, and Tertiary ones occur in the eastern Tethys. Concurrently, Jurassic manganese nodules give way to Cretaceous phosphate nodules. Both phenomena correspond to the gradual turning away of Eurasia from the South Pole, as documented also by African paleomagnetic data. In other words, they are latitude or climate dependent.

Against prevailing opinion that ophiolites and associated radiolarian cherts signify great oceanic depths, Sonnenfeld (1978a) argued the opposite on the following grounds:

1. Lateritic soil profiles occur on ophiolites of the eastern Alps (Oberhauser, 1968).
2. Extensive flat surfaces are common on ophiolites, frequently beneath lateritic soils. Coleman (1977) queried whether these have an erosional origin. Unconformable contacts are also found between radiolarites and underlying ophicalcites (Folk and McBride, 1978) or serpentinized gabbros (Passerini, 1965).
3. A caliche deposit occurs on Jurassic ophiolites in Italy (Folk and McBride, 1976), and bioclastic limestone is subaerially exposed with the associated radiolarite (Schlager, 1974).
4. High-energy environment deposits, such as oolitic and reefal limestones, are frequently interbedded or in juxtaposition with radiolarian cherts in Sardinia (Abbate et al., 1973), in Oman and SE Turkey (Grunau, 1965), in the Dolomitic Alps (Tromp, 1948), in SW Turkey (Graciansky, 1973), and in the southern Ural Mountains (Zavaritskiy, 1960).
5. Radiolarian cherts frequently rest on tuff beds, but do not often overlie lava flows. Pyroclastic rocks as products of explosive eruptions are unlikely at depths greater than 500 m (for references see Kanmera, 1974) because of the rapid increase in confining water pressure.
6. The absence of calcite within the chert is not necessarily to be equated with great water depths but can also be caused by an acidic environment. the carbonate compensation depth migrated into the photic zone in Late Cretaceous time, moved downwards again in the Cenozoic era, and has accelerated its descent since the Oligocene. Even today, it varies in the oceans from 300–4,000 m.

   The close proximity of siliceous limestones and of serpentinized carbonates reinforces the notion that carbonate-free cherts are a local phenomenon. At times, one can recognize originally aragonitic skeletal fragments of ostracods, foraminifera, crinoids, belemnites, and ammonites (Seyfert, 1978). This indicates post-depositional silicification. Such shell fragments would not have become available for silicification beneath an aragonite compensation depth. Bore holes filled with bioclastic material suggest depths within the photic zone; boring animals are rare beneath that zone.
7. The common red color (Red Beds of Klemme, 1958; Scaglia Rossa of Cadisch, 1953; Couches Rouges of Oberhauser, 1968) is due to a terrigenous admixture of ferric iron liberated on nearby shores by humic acids (Grunau, 1965; Seyfert, 1978). Only radiolarites and limestones deposited on submarine ridges are red; those deposited in troughs are gray (Diersche, 1978; Jurgan, 1969; Fabricius, 1966; Jacobshagen, 1964; Schütz, 1979).

8. Interbedded plant remains occur in Malaya (Tromp, 1948).
9. Ophiolitic radiolarite zones in Turkey frequently have a blind ending analogous to the bayhead of the modern Gulf of California (Brinkmann, 1972). A post-Eocene delta of the Irrawaddy River prograded southward into a narrow gulf in Burma and progressively covered the submarine basalt flows (Rodolfo, 1969).

CRETACEOUS OPHIOLITES

Cretaceous ophiolites are not found in Europe; only some of the Jurassic ophiolite emplacement may have continued into Cretaceous time. Middle and Upper Cretaceous ophiolites occur in Turkey and Iran, and these appear to document a bilateral symmetry of age of formation. Between Jurassic ophiolites of northern Turkey and the Caucasus and ultramafic intrusions of southern Israel and Jordan, Middle Cretaceous ophiolites are placed in central Turkey, and Upper Cretaceous ones, in southern Turkey, on the island of Cyprus, and in NW Syria. The Jurassic occurrences in Israel and Jordan, in turn, are north of the Triassic and Permian occurrences in Egypt (El Shazly, 1977).

The intrusive nature of Turkish ophiolites is demonstrated by the match of sialic country rocks from one side of the suites to the other (Brinkmann, 1972). A continental crust was adduced as basement for the Paratethys by Vylov (1967); for the western Alps by Nicolas (1966), Trümpy (1971), and Bourbon et al. (1977); for the Balkan Mountains by Boncev (1976); and for the Tethys Sea by De Booy (1966) and Belov (1967). It had been noted in Anatolia by Kaya (1972) and in Iran by Pilger (1971) and Gansser (1974). This continental crust was not situated at the continental rise from a floor of sima, but on top of a continent (De Jong, 1973). Ophiolites are found in narrow belts of vertical instability between more rigid platforms. They are emplaced during complex high-angle wrench faulting through a submerged sialic crust and do not form in association with subduction zones (Brookfield, 1977).

The younging away from the Russian and Arabian platforms has its corollary in a Paleozoic younging of ophiolites in the Mongolian geosyncline. Here, Cambrian ophiolites abut against the Angaran and Tarim platforms and are flanked by progressively younger ophiolites until Permian ones occupy the axis of the geosyncline (Zonenschein, 1973). In both examples, the ratio of ultramafics to mafics increases away from the shores, suggesting progressively deeper sources.

End-Cretaceous-to-Tertiary ophiolites are restricted to Afghanistan, the Himalayas, Burma, and Indonesia and eventually become Neogene ophiolites towards Timor and Taiwan. In the Himalayas, they again parallel the Triassic volcanic fracture zone they had followed through Yugoslavia and Turkey in Jurassic times. They also appear at the south end of a Paleozoic sequence of ophiolites younging away from the Tarim Platform through Tibet.

Tethyan ophiolites thus occur always along the southern margin of Paleozoic ophiolites, but young eastward in contrast to their Paleozoic predecessors, which young in the opposite direction (Fig. 3).

In the same manner, the onset of Lower Cretaceous tectogenesis in parts of Spain, in the Alps, and in Yugoslavia corresponds to an end-Cretaceous phase in Turkey and Egypt (Biju-Duval et al., 1978) and to still younger phases to the east.

Ancient massifs in the path of high-angle faults and of attendant ophiolite emplacement were rotated counterclockwise in the Alps and in Iran, at times deforming the related fault planes (Kraus, 1951; Wellmann, 1966; Dietrich, 1976; Fürst, 1976; Sborschikov, 1976; Heller, 1977).

SUMMARY OF THE MESOZOIC TETHYS

The Mesozoic Tethys seafloor fractured into small blocks of different buoyancy that gave rise in all segments to a submarine horst-and-graben topography (Trümpy, 1960, 1965;, Falcon, 1967;, Friedrich, 1968; Stocklin, 1968; Svoboda et al., 1966–1968; Kronber, 1969; Schlager, 1969; Tollmann, 1969; Pilger, 1971; Brinkmann, 1972; Bernoulli and Jenkyns, 1974; Plotnikov, 1964; Sugisaki et al., 1972) and locally variable water depths. This indicates lack of support of the crust by the underlying mantle. Tollmann (1970) related the Triassic longitudinal cracks in the Alpine geosyncline to a developing magma deficit, that is, to the wandering away of subcrustal materials.

The fracturing progressed from west to east with time and followed a Late Triassic basalt-filled fracture system that developed on the southern margin of a Hercynian ophiolite and orogenic belt. The fractures allowed mafic and ultramafic mantle material to rise along fracture planes; frequently associated chaotic breccias or melanges betray the streaming of explosive gas that preceded ophiolite emplacement (Sonnenfeld, 1978a).

Later fractures occurred inside the older ones as can be expected in a crust stretched under tension. This gave rise to a younging of fractures along strike and to a bilateral symmetry of ages away from ancient platforms. Rigid blocks in the path of the propagating fractures were rotated counterclockwise, suggesting Coriolis effects in the mantle.

The depressions filled rapidly with vast amounts of clastics, indicating that lands adjacent to the fractures had domed up and were eroded. The marine flysch often grades into deltaic deposits.

Toward the end of the Cretaceous period, the supracrustal infill was uplifted and deformed; some of these deposits were thrust out over adjacent blocks. The thrust faults have a shallow ultimate angle and do not cut through the crust, unlike the earlier high-angle block faults. The apparent crustal shortening produced by the deformation and the thrusting of supracrustal infill is variously calculated as a maximum of 1° of arc in the Zagros Mountains (Falcon, 1967), 4½° (Cadisch, 1953) or

6° of arc (Tollmann, 1963; Trümpy, 1971) in the Alps, and 3° in the Himalayas (Gansser, 1977).

Diastrophism also progressed from west to east in distinct phases as the mantle returned to its previous state. By the end of the Mesozoic era, high-angle fracturing had reached the eastern Tethys; diastrophism was still restricted to the westernmost parts.

Throughout the Mesozoic era, Arabia and North Africa were shelf areas in the Tethys Sea. The other shelf area existed in Spain, southern France, the Balearic Islands, and Sardinia (Fourcade et al., 1977) and continued into the northern Black Sea region, the domain of the Paratethys (Chekunov, 1974). An onlap of pelagic sediments from the south was known for the Austrian Alps over a century ago (Suess, 1875). Stratigraphic cross sections across the Tethys Sea (Lees, 1950; Dunnington, 1958; Greig, 1958) show a continuity of facies from one margin to the other.

Because of its frequently endemic fauna, the Tethys Sea really does not deserve to be called an "ocean" in Mesozoic time; the latter implies worldwide affinities and distinct faunal provinces. In contrast, the Paratethys maintained its connections to the world oceans either through northeastern Europe or around the north rim of Central Asia and fed this fauna at times into the Tethys. Only in Cretaceous times did oceanic faunas enter directly into the Tethys, at first through central America and later through the opening Atlantic Ocean.

THE PALEOGENE TETHYS AND THE ANCESTRAL MEDITERRANEAN SEA

Most of the western Tethys became land during Paleogene times. A peak in diastrophism occurred in the Eocene in Spain, in the Atlas, in Calabria, in southern Turkey (Biju-Duval et al., 1978), and in the Zagros Mountains (James and Wynd, 1965). As a result of this crustal shortening, only a relatively narrow sea (the ancestral Mediterranean Sea) remained along the south shore as shelf area of the African craton. At the beginning of the Eocene, the waters briefly transgressed onto the African craton from Egypt to Morocco, only to retreat again in middle Eocene time (Furon, 1968). The same transgression is also noted in the Dinarides (Cadet, 1978).

The Mediterranean Sea thus started out as an epicontinental sea, a smaller replica of the Paleozoic Tethys Sea.

The Baluchistan-Indus geosyncline broke up into a stable carbonate shelf in the south and a fractured trough in north composed of tilted blocks covered by thick flysch (Ganss, 1966). Here, too, there is a craton to the south. The Indochina and Malayan peninsulas were peneplained at this time prior to Neogene crustal movements (Kobayashi, 1964).

The Paratethys, as northern miogeosynclinal shelf area to the Tethys, had largely been spared the fracturing, ophiolite emplacement, and subsequent diastrophism of the Tethys. The exceptions form the chains comprising the Carpathians, the Balkans, the Crimea, and the Caucasus. The rotation of Transylvania (Crawford, 1977) and the southward displacement of the Balkans produced periodic strictures on connections between the Paratethys and the Mediterranean Sea. Commencing diastrophism in the central Tethys of Iran and Afghanistan servered, step by step, the connections of the Paratethys with the eastern Tethys.

The counterclockwise rotation and resulting twist of the Carpathians-to-Balkans arc (Crawford, 1977) seems to have the same deep-seated causes as the apparent counterclockwise contortion of the western Alps-to-Apennines arc (Van Bemmelen, 1969; Wunderlich, 1973). Thrusting of nappes in the southern Alps of Switzerland is to the north, in the western Alps of France to the west and south, and in the Apennines to the NE (Scholle, 1970). Other examples include Malaya with the famous Banda arc in Indonesia (Holcombe, 1977) and a contortion of similar sense in the curvature of all mountain ranges south of the Pamirs, or north of Assam. A Paleozoic equivalent is the counterclockwise rotation of the ophiolite trends of NW Spain; they were interpreted as diapiric masses derived from a mantle plume (Van Calsteren and Den Tex, 1979) and thus would also have been subject to Coriolis effects.

In Eocene time, the Mediterranean Sea resumed, with parts of the remaining Paratethys, the deposition of evaporites in marginal bays of the north and south shores and carbonates on open shelves. Evaporites in some bays and open marine carbonates and marls are also noted from Oligocene times. Oligocene evaporites in SE Anatolia, however, pass into freshwater deposits (Lahn, 1950) akin to modern evaporites fed by brackish Caspian waters.

The Mediterranean Sea gained access to Atlantic waters at times through a strait in the Rif Atlas in Morocco and another passage in the Betic Cordillera of Spain. The Straits of Gibralter opened only in Pliocene time (Sonnenfeld, 1974, 1975).

THE NEOGENE TETHYS AND THE MODERN MEDITERRANEAN SEA

The remaining portions of the Tethys in Asia filled with flysch sediments and underwent folding and thrusting in late Tertiary and Quaternary times. New orogenic pulses affected the folded mountain chains in the western and central regions of the former Tethys. One peak of nappe formation occurred in early-to-middle Miocene time, another one in Pliocene Quaternary time. The latter produced increasing erosion and a change of sedimentation. Predominant marls of mid-Pliocene age give way in Tunisia to coarse sands and gravels in the late Pliocene (Fekin, 1975), and the same change has been observed in other circum-Mediterranean regions.

A new set of fractures crossed the old Tethyan domain in Neogene time, and for the first time in Phanerozoic history, such fractures and

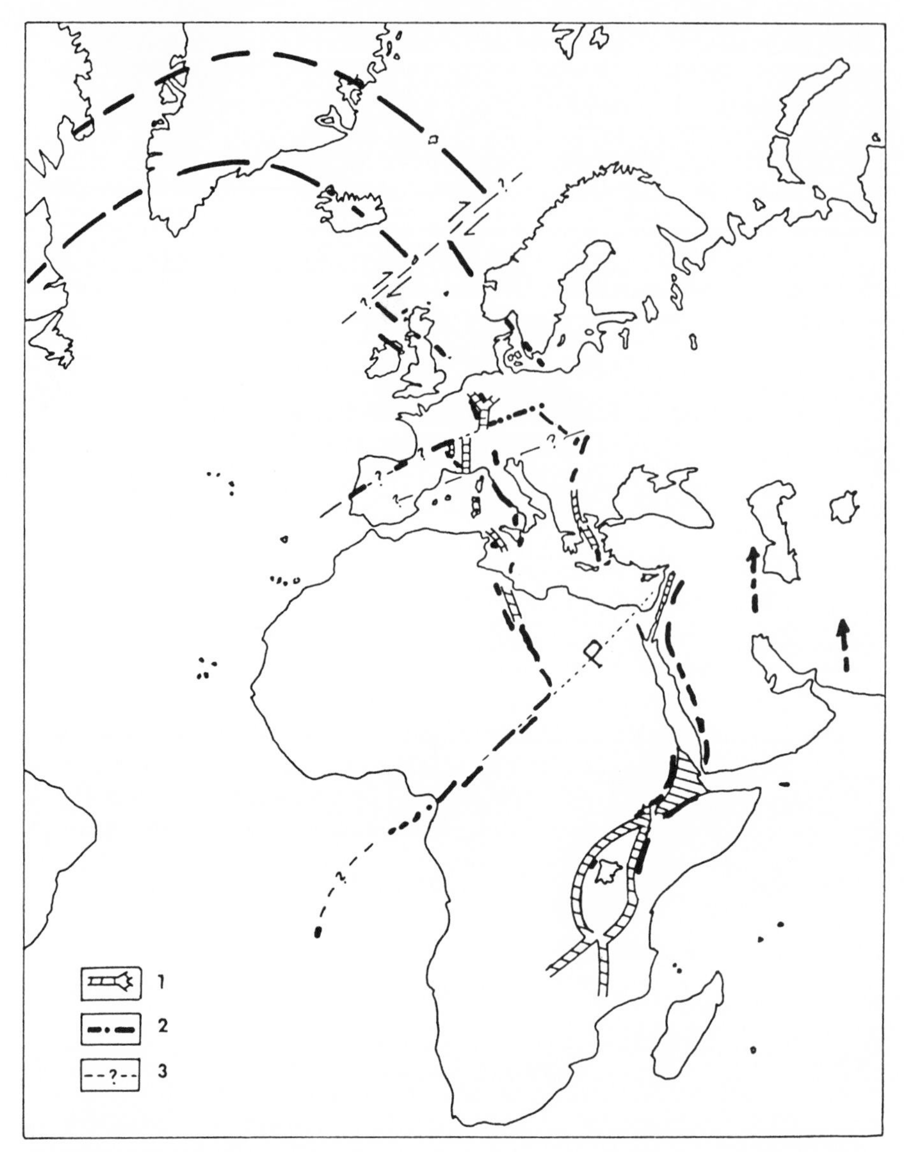

Figure 5. The double track of basic volcanism and doming (from Sonnenfeld 1978a; copyright © 1978 by E. Schweizerbart'sche Verlagsbuchhandlung [Nagele u. Obermiller] Stuttgart).

attendant basic volcanism crossed older ophiolite belts. These fractures resulted in a double track of basaltic effusions, mafic dikes, and maar eruptions, a belt that extends from Paleocene and older occurrences in Greenland and Canada to mid-Tertiary events in Central Europe and hence into Africa (Sonnenfeld, 1978a, 1980) (Fig. 5).

The double track is offset at intervals by several wrench faults that are marked by a linear arrangement of basaltic volcanism and maar eruptions. Most of these wrench faults strike NE or ENE, suggesting a principal west-to-east stress direction evidently related to the eastward push of Eurasia by the opening Atlantic Ocean. Only one such line of effusions in Turkey seems to be contorted by an apparent southward displacement of the Balkan Peninsula in respect to the Crimea and Turkey and consequent "crumpling" of the Turkish peninsula.

Along the track of eruptions, some ophiolitic intrusions have also been found in Arabia (Coleman, 1975) and in the Tyrrhenian Sea (Chumakov, pers. com.). Evidence of gas streaming has been observed in many places (King and McMillan, 1975; Baranyi, 1977; Mäussnest, 1974) akin to the ubiquitous development of explosive breccias or melanges around ancient ophiolites. The tracks are also accompanied by doming, graben formation, and counterclockwise rotation of crustal blocks. The doming has been described in Greenland (Brooks, 1973), in the Rhine River valley (Illies, 1975b), in the Carpathians (Crawford, 1974; Lexa and Konecny, 1974; Trunko, 1977), in southern Greece (Makris, 1976, 1977), in the southern Red Sea (Lowell and Genik, 1972), in East Africa (Belousov, 1962; Picard, 1970; Baker et al., 1972), and in the Cameroons (Burke et al., 1973).

The double track appears to be the product of an advancing set of rotating mantle plumes. The plumes emanating from the lower mantle (Sonnenfeld, 1980) rise on their margins, but descend in their center, as they degassified, cooled and became dense. Above the downdraft the crust was unsupported and subsided, producing distended, block-faulted basins. Their sedimentary infill is derived from adjacent exposed domes (Fig. 6).

Coincident with the initiation of the volcanism was the sudden collapse of the North Sea, of several grabens in continental Europe, and of a series of round or oval depressions in the Mediterranean Sea. The subsidence of individual basins of the Mediterranean Sea actually started in Oligocene time; after an upper Miocene respite, it continued at an accelerated rate. Topographic reversals of major proportions occurred in Tunisia (Burollet et al., 1978), but the shoreline remained north of the Sahara (Selley, 1969). Reversals of similar magnitude occurred also around the Gulf of Adana in SE Turkey and elsewhere (Sonnenfeld, 1975).

Lower-to-middle Miocene carbonates, marls, and local evaporites gave way to end-Miocene evaporites throughout the Mediterranean Sea and in the opening Red Sea. They occur today underneath the Mediterranean and Red seafloors, and also 1–3,000 m above sea level in Tunisia, central Italy, and southern Turkey (Sonnenfeld, 1974, 1975, 1977). They are found juxtaposed to remnants of Triassic evaporites in Spain, Tuscany, Tunisia, Greece, and Syria, making it probable that some of the precipitated salts were recycled.

Along the north rim of the upper Miocene evaporite basins, brackish water faunas (ostracods and pelecypods) of a Paratethys affinity

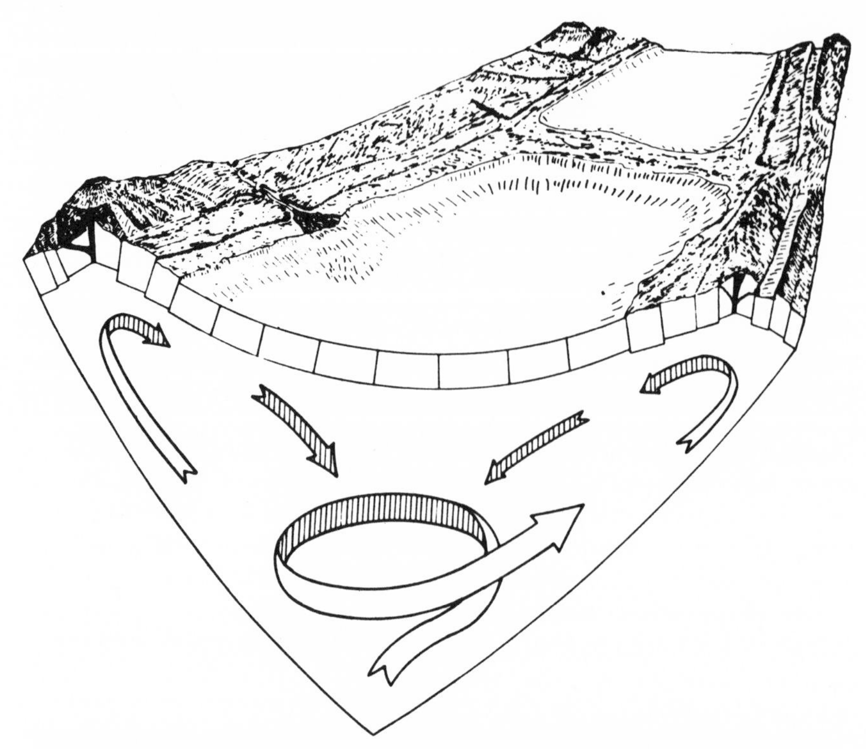

Figure 6. Artist's conception of rising, tandem, mantle plumes including crusted foundering inside double track, doming, and volcanism.

occur in shales associated with sandstones and conglomerates of westward-decreasing grain size. This indicates that the main water supply to this evaporite basin originated in the Paratethys Sea, at that time a brackish water body. Evaporite facies are known locally from the Aegean Sea (Lalechos and Savoyat, 1979), which throughout the Miocene was a narrow drainage channel of the Paratethys (Guernet, 1978) and not a Mediterranean embayment as today. In analogy, lower Miocene evaporites of Iran, deposited in a bay open to the NW (Stocklin, 1968) and thus fed by Paratethys waters, are associated with a brackish water fauna (James and Wynd, 1965). Paleogene evaporites in Anatolia also had a brackish water source. Gypsum deposits of Miocene age were strung along a current of brackish water from Soviet Armenia through Turkey to the Mediterranean coast (Sonnenfeld, 1974). The eastern limit of such nonlacustrine evaporites runs as a nearly straight line from the southern Ural Mountains along the western shore of the Aral Sea into eastern Iran (Blagovidov, 1978). Isolated lacustrine salt deposits formed at the same time throughout Central Asia.

It is noteworthy that the Balearic, Tyrrhenian, and Ionian basins; the Hellenic Trough; and the Nile Abyssal Plain have accelerated their subsidence in Quaternary time (Neumayr, 1886–1887; Cvijic, 1908; Selli and Fabbri, 1971; Sonnenfeld, 1974, 1975, 1977; Fabricius and Hieke, 1977) along block faults. The appearance of an Eastern Mediterranean Ridge in mid-Pleistocene time between the Nile Abyssal Plain and the Hellenic Trough may be a replica of the Mediterranean macroisthmus of earlier times; the African swell between Tunisia and Sicily resembles, then, the Tethyan meridional swell in the Levant.

The mainly Pliocene-Quaternary dilation of the Mediterranean Sea occurred in the face of increased compression in the Maghrebian Atlas and all along the northern margin of this sea from the Betic Cordillera to the Taurides. Compression reached a maximum in Quaternary times in Tangiers, Tunisia, and Cyprus, but even the Alps and Carpathians that are presently rising rapidly show that all parts of the orogenic system are affected by this compression.

Subduction zones have been postulated for the Pontides and Alburz Mountains (Smith, 1971) or alternatively for the Taurides and Zagros Mountains (Dewey et al., 1973). No specific line or belt of subduction can so far explain this tension juxtaposed next to compressional belts, or inside them (for example, the Alboran, Po River, Vienna, Styrian, and Pannonian basins), with rotated fault blocks of circular or lens-shaped outline. This feature must be related to the behavior of the mantle, that is, to the withdrawal of support of the crust in one place, and the reinstatement of support in another. If one replaces the concept of collision of two large plates and consequent crushing of intervening crust by interaction of many microplates or platelets, one only substitutes new terminologies for the older concept of tilting fault blocks interfering with each other.

On the other hand, Nalivkin (1973b) drew attention to the similarity between the shape of the Paleozoic Uralian geosyncline and the modern Red Sea. The Triassic and Jurassic Tethys advancing into Spain and the Caribbean area may also have had the shape of a long, narrow, finger-shaped gulf.

During the time of rapid subsidence of individual blocks of the Mediterranean Sea, the Paratethys Sea shrank even further. Alternating between a normal marine and a freshwater body, it produced massive evaporites as late as middle Miocene in several embayments; minor evaporite occurrences are even known from younger strata (Sonnenfeld, 1977). Eventually, the western Paratethys dried up leaving behind the muddy Lake Balaton in Hungary. The Caspian and Aral seas were cut-off remnants in the east; the Black Sea, as remnant of the central Paratethys, retained a passage to the Mediterranean Sea, albeit a connection squeezed to one third its Miocene width (Cvijić, 1908).

SUMMARY OF THE CENOZOIC TETHYS

Fracturing and ophiolite emplacement, typical of the western Tethys in Mesozoic time, became confined to the eastern Tethys in early Cenozoic time. Diastrophism followed along the whole length of the fracture zone in several phases, the last of which is still going on.

Due to the diastrophism, the Tethys, as a body of water, ceased to exist. In the west, the southern shelf area became the Mediterranean Sea, initially a shallow epicontinental body of water. Farther east, the Persian Gulf opened to the Indian Ocean. The Red Sea represents a new fracture zone at an angle to the older system.

Huang (1960) drew attention to the absence of thick geosynclinal sequences in the realm of the eastern Tethys, which is composed of relatively thin shelf sediments on a metamorphosed Paleozoic basement. Until Neogene time, the Himalayas belonged in the shelf area of the Indian Platform (Petrushevskiy, 1970a) and were facing Mesozoic folding around the Tarim Platform (Petrushevskiy, 1970b).

North-to-south shore correlations of lithologies and their facies are possible throughout the Phanerozoic history of the Tethys, as Maxwell (1970) observed. No correlation of rock types is possible assuming that giant wrench faults let Colombia and Panama, with Africa in tow, wander away from Burma in the Triassic and pass the Caspian Sea in Jurassic time and Turkey in the Cretaceous period (Van Hilten, 1964). This is referred to by Van Hilten (1964) as the "Tethys Twist." Similarly, it is not possible to assign to the Tethys a width of more than a few degrees of arc.

The Paratethys, cut off from its Tethyan water sources, became landlocked, eventually freshened, and shrank into the Aral, Caspian, and Black seas, of which only the latter has retained a connection to the world ocean through the Mediterranean Sea with nearly insignificant faunal and water exchange.

The Mediterranean Sea has broken up since Oligocene time into a series of differentially subsiding blocks with accompanying basic volcanism. In that respect, the Mediterranean Sea repeats in Cenozoic time the Mesozoic history of the western Tethys on a smaller scale.

CONCLUSIONS

Throughout the Paleozoic era, there existed an epicontinental sea between the Russian and Afro-Arabian platforms between Central Asia and India. An episodically emergent ridge capped by islands subdivided this body of water into a Tethys Sea between central Spain and Thailand and a Paratethys Sea of analogous but not identical history between the Pyrenees and Central Asia. Facies correlations are possible across this sea and show that both shores were in identical climatic belts. Climates varied, however, along the length of this sea. Ophiolites younging westward from Central Asia to Britain and Spain were embedded beyond the Paratethys, and that trend was then subjected to diastrophism.

At the beginning of the Mesozoic era, the Tethys began to break up into a horst-and-graben topography with resulting diverse water depths. The fractures progressively extended both westward into the Caribbean area and eastward to Indonesia along the guard rail of the dividing ridge. While the Paratethys maintained its connections to the world oceans, the Tethys repeatedly became the site of an endemic fauna. The propagating fractures were followed by ophiolite emplacements that young eastward. The ophiolites and their attendant sediments were capped by flysch infill eroded from the flanks of the Tethys and were, thereafter, followed by diastrophism.

A bilateral symmetry of ages younging away from platforms, an ultrabasicity increasing in the same direction, and the rotation of ancient massifs in the path of ophiolite trends can be demonstrated. A new trend of fracturing, mantle effusions, doming, and subsidence propagated in the Cenozoic era and, for the first time in Phanerozoic history, crossed older annealed ophiolite trends. The activity is responsible for Quaternary volcanism in Europe and Africa, for the subsidence of lenticular and round blocks of distended crust, and for rotation of thicker crustal massifs. One can adduce the passing of a set of rotating mantle plumes that leave the crust unsupported in the centre. They are responsible for the fracturing, doming on either flank, volcanism, gas streaming, and heat radiation. They have converted the Mediterranean Sea from a residual shelf sea on the African craton into a set of differentially subsiding basins and ridges analogous to the irregular Mesozoic Tethys seafloor.

REFERENCES

Abbate, E., V. Bartolotti, and P. Passerini. 1973. Paleogeographic and tectonic considerations on the ultramafic belts in the Mediterranean area. *Soc. Geol. Italiana Boll.* **91**(20):239–282, **92**(1):33–35.

Abdusselamoglu, S. 1963. Nouvelles observations stratigraphiques et paléontologiques sur les terrains paléozoiques affleurant à l'est du Bosphore. *Miner. Res. Explor. Inst. Ankara Bull.* **60**:1–6.

Ali Meissami, H. Termier, and G. Termier. 1978. La phase transgressive morabakienne (Tournaisien-Viséen) sur la bordure meridionale de la Tethys. *Acad. Sci. Comptes Rendus,* sér. D, **285**(12):1163–1165.

Archangel'skiy, A. D., and N. S. Shatskiy. 1933. Skhema tektoniki SSSR. *Moskov. Obshch. Ispytoteley Prirody Byull., Otdel Geol.* **12**(4):1–88.

Argand, E. 1924. La tectonique de l'Asie. *Internat. Geol. Congr., 13th, Liege, 1922, Proc.,* pp. 171–372.

Aubouin, J. 1977. Méditerranée orientale et Méditerranée occidentale: esquisse d'une comparison du cadre alpin. *Soc. Géol. France Bull.* **29**(3):421–435.

Aubouin, J., R. Blanchet, J. P. Cadet, P. Celet, J. Charvet, J. Chorowicz, M. Cousin, and J. P. Rampnoux. 1970. Essai sur la géologie des Dinarides. *Soc. Geol. France Bull.* **12**(6):1060–1095.

Aubouin, J. R. Blanchet, J. F. Stephan, and M. Tardy. 1977. Téthys (Mésogée) et Atlantique: données de la géologie. *Acad. Sci. Comptes Rendus,* sér. D., **285**(10):1025–1028.

Baker, B. H., P. A. Mohr, and L. A. J. Williams. 1972. *Geology of the Eastern Rift System of Africa.* Geol. Soc. America Spec. Paper 136, pp. 1–67.

Baranyi, I. 1977. Petrographie und Geochemie der subvulkanen Brekzien des Kaiserstuhls. *Neues Jahrb. Mineralogic Abh.* 128(3):254–284.

Beauchamp, J. 1978. L'évolution sédimentaire de l'Ethopie, du Carbonifère au Crétacée. *Soc. Géol. Nord Annales* **97**(4):329–335.

Bechstädt, T., R. Brandner, H. Mostler. 1976. Das Frühstadium der alpinen Geosynklinalentwicklung im westlichen Drauzug. *Geol. Rundschau* **65**(2): 616–648.

Belousov, V. V. 1962. *Problems in Basic Tectonics*. New York: McGraw-Hill, 816 pp.

Belov, A. A. 1967. Tectonic development of the alpine folded belt during the Paleozoic (Balkan Peninsula—Iranian Plateau—Pamirs). *Geotectonics* **1**(2): 145–152.

Belov, A. A. 1972. The Alps, the Balkans and the Greater Caucasus in the Paleozoic (a comparative analysis of their tectonic development). *Internat. Geology Rev.* **14**(9):1321–1337.

Bender, F. 1974. Explanatory notes on the geologic map of the Wadi Araba, Jordan. *Geol. Jahrb.* **B10**:3–62.

Bernoulli, D., and H. C. Jenkyns. 1974. Alpine, Mediterranean and Central Atlantic Mesozoic facies in relation to the early evolution of the Tethys. In Dott, R. H., Jr., and R. H. Shaver, eds., *Modern and Ancient Geosynclinal Sedimentation*. Soc. Econ. Paleontologists and Mineralogists Spec. Pub. 19, pp. 129–160.

Biju-Duval, B., J. Letouzey, and L. Montadert. 1978. Structure and evolution of the Mediterranean basins. In Hsu, K. J., L. Montadert et al., eds., *Initial Reports of the Deep Sea Drilling Project*, vol. 42, pp. 951–984.

Bing, H., W. Jing-Bin, G. Zhen-Jia, and F. Xiao-Di. 1965. Problems of the Paleozoic of Tarim Platform. *Internat. Geology Rev.* **11**(6):650–665.

Blagovidov, V. V. 1978. Neogenovye solenosnye formatsii sredney tsentral'noy Azii (Neogene saliferous formations of the middle of Central Asia). *Moscov. Geologorazvedoch. Inst. Trudy* **394**, 149pp.

Boncev, E. 1976. Lineament-geosynclinal zones—zones of impermanent riftogenesis. *Geologica Balcanica* **6**(1):83–101.

Bordet, P., M. Colchen, D. Krummenacher, P. Lefort, R. Mouterde, and M. Remy. 1971. *Recherches géologiques dans l'Himalaya de Népal, région de la Thakkala*. Paris: C.N.R.S., 280pp.

Boulin, J., E. Bouyx, A. F. de Lapparent, M. Lys, and P. Semenoff. 1975. La transgression du Paléozoique supérieur dans la versant nord de l'Hindou-Kouch occidental en Afghanistan. *Acad. Sci. Compte. Rendus* sér. D, **281**:495–502.

Bourbon, M., J. M. Caron, P. C. de Graciansky, M. Lemoine, J. Megard-Galli, and D. Mercier. 1977. Mesozoic evolution of the western Alps: birth and development of part of the spreading oceanic Tethys and of its European continental margin. In Biju-Duval, B., and L. Montadert, eds., *Structural History of the Mediterranean Basins*. Paris: Editions Technip, pp. 19–34.

Brice, D., J. Jenny, G. Stampfli, and F. Bigey, 1978. Le Dévonien de l'Elbourz oriental: stratigraphie, paléontologie (brachiopodes et bryozoaires), paléogéographie. *Riv. Italiana Paleontologia* e *Stratigrafig* **84**(1):1–55.

Brinkmann, R. 1969. *Geologic Evolution of Europe*, 2nd ed. Stuttgart: Ferd. Enke, 161pp.

Brinkmann, R. 1972. Mesozoic troughs and crustal structure of Anatolia. *Geol. Soc. America Bull.* **83**(3):819–826.

Brookfield, M. E. 1977. The emplacement of giant ophiolite nappes. I. Mesozoic–Cenozoic examples. *Tectonophysics* **37**(4):247–303.

Brooks, C. K. 1973. Rifting and doming in southeastern Greenland. *Nature (Phys. Sci.)* **244**(131):23–25.

Bruggey, J. 1973. Mesozoikum und Alttertiär in Nord-Paktia (SE-Afghanistan). *Geol. Jahrb.* **B3**:3–61.

Buggisch, W., E. Flugel, F. Leitz, G. F. Tietz. 1976. Die fazielle und paläogeographische Entwicklung im Perm der karnischen Alpen und in den Randgebieten. *Geol. Rundschau* **65**(2):649–690.

Burke, K., and J. A. Whiteman, 1973. Uplift, rifting and the break-up of Africa. In Tarling, D. H., and S. K. Runcorn, eds., *Implications of Continental Drift to the Earth Sciences*, vol. 2. New York: Academic Press, pp. 735–756.
Burollet, P. F., J. M. Mugniot, and P. Sweeney. 1978. The geology of the Pelagian block: the margins and basins off southern Tunisia and Tripolitania. In Nairn, A. E. M., W. H. Kanes, and F. G. Stehli, eds., *The Ocean Basins and Margins*, vol. 4B. New York: Plenum Press, pp. 331–360.
Busson, G. 1975. Le trias evaporitique d'Afrique du Nord et d'Europe occidentale: données sur la paléogéographie et les conditions de dépôt. *Soc. Géol. France Bull.*, **16**(6):653–665.
C. A. G. S. (Chinese Academy of Geological Sciences). 1977. An outline of the geology of China. *Geol. Jahrb.* **B27**:7–30.
Cadet, J. P. 1978. Essai sur l'évolution alpine d'une paléomarge continentale: Les confines de la Bosnie—Hercégovine et du Monténégro (Yougoslavie). *Soc. Géol. France Mém.* **57**(133):1–83.
Cadisch, J. 1953. *Geologie der Schweizer Alpen.* Basel: Wepf & Co., 480pp.
Chekunov, A. V. 1974. Main stages in geotectonic development of northern Black Sea region. *Internat. Geology Rev.* **16**(6):495–508.
Colchen, M. 1978. Les charactéres gondwaniens, et téthysiens des séries himalayennes. Implications paléogéographiques. *Soc. Géol. Nord, Annales* **97**:279–286.
Coleman, R. G. 1975. Miocene ophiolite on the Red Sea coastal plain. *EOS (Am. Geophys. Union Trans.)* **56**(12):1080.
Coleman, R. G. 1977. *Ophiolites: Ancient Oceanic Lithosphere.* New York: Springer-Verlag, 229pp.
Crawford, A. R. 1974. The Indus Suture Line, the Himalaya, Tibet and Gondwanaland. *Geol. Mag.* **111**(5):369–383.
Crawford, A. R. 1977. Danubian deviations and mantle diapirism: a possible origin of the Carpathian arc. *Geol. Mag.* **114**(2):115–125.
Cvijić, J. 1908. Grundlinien zur Geographie und Geologie von Mazedonien und Altserbien. *Petermanns Mitt. Erg. H.* **162**:1–392.
De Booy, T. 1966. Neue Daten fur die Annahme einer sialischen Kruste unter den frühgeosynklinalen Sedimenten der Tethys. *Geol. Rundschau* **56**(1): 94–102.
De Jong, K. A. 1973. Mountain building in the Mediterranean region. In De Jong, K. A., and R. Scholten, eds., *Gravity and Tectonics.* New York: John Wiley & Sons, pp. 125–139.
Dercourt, J. 1972. The Canadian Cordillera, the Hellenides, and the sea-floor spreading theory. *Canadian Jour. Earth Sci.* **9**(6):709–743.
Demirtasli, E. 1975. Stratigraphic correlation of the lower Paleozoic rocks of Iran, Pakistan and Turkey. In Doruyan, S., ed., Congr. *Earth Sci. (on the occasion of the 50th anniversary of the Turkish Republic), Dec. 17–19, 1973, Papers.* Ankara, Turkey: Miner. Res. and Explor. Inst., pp. 210–227.
Dewey, J. F., W. C. Pitman III, W. B. F. Ryan, and J. Bonnin. 1973. Plate tectonics and the evolution of the Alpine system. *Geol. Soc. America Bull.* **84**(10): 3137–3180.
Diersche, V. 1978. Upper Jurassic radiolarites in the northern Calcareous Alps (upper Austroalpine unit). In Closs, H., D. Roeder, and K. Schmidt, eds., *Alps, Apennines, Hellenides.* Stuttgart: E. Schweizerbart, pp. 113–117.
Dietrich, V. N. 1976. Plattentektonik in den Ostalpen, eine Arbeitshypothese. *Geotektonische Forschungen* **50**:1–84.
Drake, C. L., and J. E. Nafe. 1972. Geophysics of the North Atlantic Region. *EOS, Amer. Geophys. Union Trans.* **53**(2):175–176.
Dunnington, H. V. 1958. Generation, migration, accumulation and dissipation of oil in northern Iraq. In Weeks, L. G., ed., *Habitat of Oil.* Tulsa: Am. Assoc. Petroleum Geologists, pp. 1194–1251.

Elmi, S. 1978. Polarité tectono-sédimentaire pendant l'effritement des marges septentrionales du bâti africain au cours du Mésozoique (Maghreb). *Soc. Géol. Nord, Annales* **97**(4):315–323.

El Shazly, E. M. 1977. The geology of the Egyptian region. In Nairn, A. E. M., W. H. Kanes, and F. G. Stehli, eds., *The Ocean Basins and Margins*, vol. 4A. New York: Plenum Press, pp. 379–444.

Erben, H. K. 1962. Zur Analyse und Interpretation der rheinischen und hercynischen Magnafacies des Devons. In Erben, H. E., eds., *Internat. Arbeitstagung Silur/Devon Grenze, II, Bonn-Brussels, 1960, Proc.*, pp. 42–61.

Erben, H. K. 1964. Facies development in the marine Devonian of the Old World. *USSHER Soc. Proc.* **1**(3):92–118.

Erkan, E. 1977. Uran- und gipsfuhrendes Permoskyth der östlichen Ostalpen. *Geol. Bundesanst. Jahrb.*, **120**(2):343–400.

Eynor, O. L. 1970. Carboniferous geology and paleogeography of the USSR. *Internat. Geology Rev.* **12**(2):105–113.

Fabricius, F. 1966. *Beckensedimentation und Riffbildung an der Trias/Jura-Wende in den Bayerisch-Tiroler Kalkalpen.* Leiden: Brill, 143pp.

Fabricius, F. H., and W. Hieke. 1977. Neogene to Quaternary development of the Ionian Basin (Mediterranean): Considerations based on a "dynamic shallow basin model" of the Messinian salinity event. In Biju-Duval, B., and L. Montadert, eds., *Structural History of the Mediterranean Basins.* Paris: Editions Technip, pp. 391–400.

Falcon, N. L. 1967. The geology of the northeast margin of the Arabian basement shield. *Adv. Sci.* **24**:1–12.

Fekin, M. 1975. Paleoecologie du Pliocène marin au nord de la Tunisie. *Annales Mines Géol. Tunisie 27;* 194pp.

Flügel, H. W. 1972. Zur Entwicklung der 'Prototethys' im Palaozoikum Vorderasiens. *Neues Jahrb. Geologie u. Palaontologie. Monatsh.* **10**:602–610.

Folk, R. L., and E. F. McBride. 1976. Possible pedogenic origin of Ligurian ophicalcite: A Mesozoic calichified serpentinite. *Geology* **4**(6):327–332.

Folk, R. L., and E. F. McBride. 1978. Radiolarites and their relation to subjacent "oceanic crust" in Liguria, Italy. *Jour. Sed. Petrology* **48**(4):1069–1102.

Fourcade, E., J. Azéma, G. Chabrier, P. Chauve. 1977. Liaisons paléogéographiques au Mésozoique entre les zones externes, bétiques, corso-sardes et alpines. *Rev. Géographie Phys. et Géologie Dynam.* **19**(4):377–388.

Fourquin, C. 1975. L'Anatolie du Nord-Quest, marge meridionale du continent europeen, histoire paleogeographique, tectonique et magmatique durant le Secondaire et le Tertiaire. *Soc. Geol. France Bull.* **17**(6):1058–1170.

Friedrich, O. M. 1968. Die Vererzung der Ostalpen, gesehen als Glied des Gebirgsbaues. *Archiv f. Lagerstättenforsch. Ostalpen, Leoben,* **8**:1–136.

Fuchs, G. 1967. *Zum Bau des Himalaya.* Österreichische Akad. Wiss., Math.-Naturw. Kl. Denkschr. 113, 211pp.

Fuchs, G. 1977. The geology of the Karnali and Dolpo regions, western Nepal. *Geol. Bundesanst. Jahrb.*, **120**(2):165–217.

Furon, R. 1968. *Géologie de l'Afrique.* Paris: Payot, 374pp.

Furst, M. 1976. Tektonik und Diapirismus der östlichen Zagrosketten. *Deutsch. Geol. Gesell. Zeitschr.* **127**(1):183–225.

Ganss, O. 1964. Geosynklinalbecken, Tektonik, Granite und junger Vulkanismus in Afghanistan. *Geol. Rundschau* **54**(2):668–698.

Ganss, O. 1966. Zur geologischen Geschichte der Belutschistan-Indus Geosynklinale. *Geol. Jahrb.* **82**:203–242.

Gansser, A. 1974. The ophiolite mélange, a world-wide problem on Tethyan examples. *Eclogae Geol. Helvetiae* **67**(3):429–507.

Gansser, A. 1977. The great suture zone between Himalaya and Tibet, a preliminary account. In *Himalaya.* Conseil Natl. Recherche Sci. Colloq. Internat., vol. 268, Actes. pp. 181–191.

Gee, E. R. 1965. Written contribution. In *Salt Basins around Africa*. London: Inst. of Petroleum, pp. 118–119.
Geyer, O. F. 1977. The "Lithiotis limestone" in the Lower Jurassic Tethys realm. *Neues Jahrb. Geologie Paläontologie Abh.* **153**(3):304–340.
Gölz, E. 1978. *Basische eruptiva ("ophite") im Mesozoikum von Murcia (SE-Spanien).* Stuttgart Tech. Hochschule Geol.-Paläont. Inst. Arb. 72, 102pp.
Graciansky, C. de. 1973. Le problème des 'coloured mélanges' à propos de formations chaotiques associées aux ophiolites, Lycie occidentale (Turquie). *Rev. Géographie Phys. et Géologie Dynam.* **15**(5):555–566.
Greig, D. A. 1958. Oil horizons in the Middle East. In Weeks, L. G., ed., *Habitat of Oil.* Tulsa: Am. Assoc. Petroleum Geologists, pp. 1182–1193.
Grossgeym, V. A., and V. Ye. Khain. 1975. *Paleogeografiya SSSR.* Moscow: Nedra, 4 vols.
Grunau, H. R. 1965. Radiolarian cherts and associated rocks in space and time. *Eclogae Geol. Helvetiae* **58**:157–208.
Grunt, T. A. 1970. K biogeografii brakhiopod permskogo tetisa. (Brachiopod biogeography of the Permian Tethys). *Moskov. Obshch. Ispytateley Prirody Byull., Otedl. Geol.* **45**(6):90–101.
Guernet, C. 1978. L'èvolution paléogéographique et tectonique de la Grèce au Miocène: un essai de synthése. *Rev. Geographic Phys. et Géologie Dynam.* **20**(1):95–107.
Gupta, V. J. 1973. *Indian Paleozoic Stratigraphy.* Delhi: Hindustan Publ. Co., 207pp.
Gupta, V. J. 1975. *Indian Mesozoic Stratigraphy.* Delhi: Hindustan Publ. Co., 267pp.
Gupta, V. J. 1976. *Indian Cenozoic Stratigraphy.* Delhi: Hindustan Publ. Co., 344pp.
Gupta, V. J. 1977. Paleozoic biostratigraphy and paleogegography of the Himalaya. In *Himalaya.* Conseil Natl. Recherche Sci. Colloq. Internat., vol. 268, Actes, pp. 197–202.
Hallam, G. 1977. Biogeographic evidence bearing on the creation of Atlantic seaways in the Jurassic (with discussion). In West, R. M., ed., *Paleontology and Plate Tectonics.* Milwaukee Public Mus. Pubs. Geology 2, pp. 23–39.
Hashimoto, S., et al. 1973. *Geology of the Nepal Himalayas.* Sapporo: Saikong Publ. Co., 286pp.
Havlícek, V. 1974. Some problems of the Ordovician in the Mediterranean region. *Geotechnica* **49**(6):343–348.
Heller, F. 1977. Paleomagnetic data from the western Lepontine area (central Alps). *Schweiz. Mineralog. u. Petrog. Mitt.* **57**(1):135–143.
Helwig, J., J. Aronson, and D. S. Day. 1974. A Late Jurassic mafic pluton in Newfoundland. *Canadian Jour. Earth Sci.* **11**(9):1314–1318.
Herz, N., and H. Savu. 1974. Plate tectonics history of Romania. *Geol. Soc. America Bull.* **85**(9):1429–1440.
Holcombe, C. J. 1977. How rigid are the lithospheric plates? Fault and shear rotation in southeast Asia. *Geol. Soc. London Quart. Jour.* **134**(3):325–342.
Höll, R., J. Loeschke, A. Maucher, and K. Schmidt. 1978. Early Paleozoic geodynamics in the eastern and southern Alps. In Closs, H., D. Roeder, and K. Schmidt, eds., *Alps, Apennines, Hellenides.* Stuttgart: E. Schweizerbart, pp. 124–127.
Hollmann, R. 1962. Über Subsolution und die "Knollenkalke" des Calcare Ammonitico Rosso Superiore in Monte Baldo (Malm, Norditalien). *Neues Jahrb. Geologie u. Paläontologie Monatsh.* **4**:163–179.
Hollmann, R. 1964. Subsolutions-Fragmente (Zur Biostratonomie der Ammoiden im Malm des Monte Baldo, Norditalien). *Neues Jahrb. Geologie u. Palaontologie Abh.* **119**:22–82.
Holmes, A. 1965. *Principles of Physical Geology,* 2nd ed. New York: Ronald Press, 1288pp.

Hsu, K. J., and D. Bernoulli. 1978. Genesis of the Tethys and the Mediterranean. In Hsu, K. J., L. Montadert et al., eds., *Initial Reports of the Deep Sea Drilling Project,* vol. 42; pp. 951–984.
Huang, C. (T. K.). 1960. Die geotektonischen Elemente im Aufbau Chinas. *Geologie* **9**(7):715–733, **9**(8):841–866.
Huang, C. (T., K.) 1978. An outline of the tectonic characteristics of China. *Eclogae Geol. Helvetiae* **71**(3):611–635.
Illies, J. H. 1975a. Intraplate tectonics in stable Europe as related to plate tectonics in the Alpine system. *Geol. Rundschau* **64**(3):677–699,
Illies, J. H. 1975b. Recent and paleo-intraplate tectonics in stable Europe and the Rhine Graben rift system. *Tectonophysics* **29**(1–4):251–264.
Jacobshagen, V. 1964. Lias und Dogger in Westabschnitt der nördlichen Kalkalpen. *Geol. Romana* **3**:303–318.
Jain, S. P., and R. C. Kanwar. 1970. Himalayan Ridge in the light of the theory of continental drift. *Nature* **227**(5260):829.
James, G. A., and J. G. Wynd. 1965. Stratigraphic nomenclature of Iranian Oil Consortium agreement area. *Am. Assoc. Petroleum Geologists Bull.* **49**(12): 2182–2245.
Jenny, J., and G. Stampfli. 1978. Lithostratigraphie du Permien de l'Elbourz oriental en Iran. *Eclogae Geol. Helvetiae* **71**(3):551–580.
Jenkyns, H. C. 1971. The genesis of condensed sequences in the Tethyan Jurassic. *Lethaia* **4**(3):327–352.
Jurgan, H. 1969. Sedimentologie des Lias der Berchtesgadener Kalkalpen. *Geol. Rundschau* **58**(2):464–501.
Kälin, O., and D. M. Trümpy. 1977. Sedimentation und Paläotektonik in den westlichen Südalpen: Zur triassisch-jurassischen Geschichte des Monte Nudo—Beckens. *Eclogae Geol. Helvetiae* **70**(2):295–350.
Kamen-Kaye, M. 1976. Mediterranean Permian Tethys. *Am. Assoc. Petroleum Geologists Bull.* **60**(4):623–626.
Kanmera, K. 1974. Paleozoic and Mesozoic geosynclinal volcanism in the Japanese islands and associated chert sedimentation. In Dott, R. H., Jr., and R. H. Shaver, eds., *Modern and Ancient Geosynclinal Sedimentation.* Soc. Econ. Paleontologists and Mineralogists Spec. Pub. 19, pp. 161–173.
Kaya, O. 1972. Aufbau und Geschichte einer anatolischen Ophiolit-Zone. *Deutsch. Geol. Gesell. Zeitschr.* **123**(3):491–501.
Khain, V. Ye., A. B. Ronov, and A. N. Balukhovskiy. 1975. Cretaceous lithologic associations of the world. *Sovetskaya Geologiya* **11**:10–39; transl. *Internat. Geol. Rev.* **18**(11):1269–1295 (1976).
Khain, V. Ye., A. B. Ronov, and K. B. Seslavinskiy. 1977. Silurian lithologic associations of the world. *Sovetskaya Geologiya* **5**:21–43; transl. *Internat. Geological Rev.* **20**(3):249–268 (1978).
King, A. F., and N. J. McMillan. 1975. A mid-Mesozoic breccia from the coast of Labrador. *Canadian Jour. Earth Sci.* **12**(1):44–51.
Klemme, H. D. 1958. Regional geology of circum-Mediterranean region. *Am. Assoc. Petroleum. Geologists Bull.* **42**(3):447–512.
Kobayashi, T. 1964. Geology of Thailand. In Kobayashi, T., ed., *Geology and Paleontology of Southeast Asia,* vol. 1. Tokyo: Univ. Tokyo Press, pp. 1–16.
Kozary, M. T, J. C. Dunlap, and W. E. Humphreys. 1968. Incidents of saline deposits in geologic time. In Mattox, R. B., W. T. Holser, H. Odé, W. L. McIntire, N. M. Short, R. E. Taylor, and D. C. van Siclen, eds., *Saline Deposits.* Geol. Soc. America Spec. Pub. 88, pp. 43–57.
Kraus, E. 1951. *Die Baugeschichte der Alpen,* 2 vols. Berlin:Akademie-Verlag.
Krestnikov, V. N., and I. L. Nersevov. 1964. Relationship of the deep structure of the Pamirs and Tien Shan to their tectonics. *Tectonophysics* **1**(1):183–195.
Kronberg, P. 1969. Bruchtektonik im Ostpontischen Gebirge (NE Türkei). *Geol. Rundschau* **59**(1):257–264.

Kummel, B. 1966. The lower Triassic formations of the Salt Range and the trans-Indus ranges, West Pakistan. *Harvard Univ. Mus. Comp. Zoology Bull.* **134** (10):361–419.
Kummel, B., and C. Teichert. 1966. Relation between the Permian and Triassic formations in the Salt Range and Trans-Indus Ranges, West Pakistan. *Neues Jahrb. Geologie u. Paläontologie Abh.* **125**:297–333.
Lahn, E. R. 1950. La formation gypsifère en Anatolie (Asie Mineure). *Soc. Géol. France Bull.*, sér. 5, **20**(7–9):451–457.
Lalechos, N., and E. Savoyat. 1979. La sédimentation Nèogéne dans le Fosse Nord Egéen. *Colloq. Geology of the Aegean Region, 6th, Athens, 1977, Proc.*, **2**: 591–603.
Lapparent, A. F. de. 1972. Esquisse géologique de Afghanistan. *Rev. Géographic Phys. et Geologie Dynam.* **14**(4):327–344.
Laskarev, V. D. 1924. Sur les équivalents du Sarmatien Superieur en Servie. *Recueil des travaux offert à M. J. Cvijic*, pp. 73–85.
Lees, G. M. 1950. Some structural and stratigraphic aspects of the oilfields of the Middle East. *Congr. Internat. Géol. 18th, London, 1948, Proc.*, pt. 6, sec. 8, pp. 26–33.
Lemoine, M. 1978. *Geological Atlas of Alpine Europe and Adjoining Areas*. New York: Elsevier, 584pp.
Lexa, J., and V. Konečny. 1974. The Carpathian volcanic arc: a discussion. *Acta Geol. Hungarica* **18**:279–283.
Lowell, J. D., and G. J. Genik. 1972. Sea-floor spreading and structural evolution of southern Red Sea. *Am. Assoc. Petroleum Geologists Bull.* **56**(2):247–259.
Lucas, G. 1952. Bordure nord des Hautes Plaines dans l'Algérie occidentale. *Congr. Géol. Internat., 19th, Algiers, 1952, Mon. Rég.*, sér. Algérie, **21**:1–139.
Makris, J. 1976. A dynamic model of the Hellenic arc deduced from geophysical data. *Tectonophysics* **36**:339–346.
Makris, J. 1977. "Plate tectonics" versus "plumes" in the Aegean tectonics. *Colloq. Geology of the Aegean Region, 6th, Athens, 1977, Abstr.*, p. 9.
Mascle, G. 1977. Quelques réflexions sur la question de l'Océan téthysien. In *Himalaya*. Conseil Natl. Recherche Sci. Colloq. Internat., vol. 268, Actes, pp. 251–260.
Mäussnest, O. 1974. Die Eruptionspunkte des schwäbischen Vulkans. *Deutsch. Geol. Gesell. Zeitschr.* **125**:23–54, 277–353.
Maxwell, J. C. 1970. The Mediterranean ophiolites and continental drift. In Johnson, H., and B. L. Smith, eds., *The Megatectonics of Continents and Oceans*. New Brunswick, N.J.: The Rutgers Univ. Press, pp. 167–193.
McClure, H. A. 1978. Early Paleozoic glaciation in Arabia. *Paleogeography, Paleoclimatology, Paleoecology* **25**(4):315–326.
Megard-Galli, J., and A. Baud. 1977. Le Trias moyen et supérieur des Alps nord-occidentales et occidentales: données nouvelles et corrélations stratigraphiques. *Bur. Recherches Géol. et Miniere Bull. (France), ser. 4, no. 3*, pp. 233–250.
Mokrinskiy, V. V. 1939. La distribution stratigrapho-géographique de terrains à charbon Mésozoique dans la province Crimeo-Caucaso-Caspienne. *Congr. Géol. Internat, 17th, Moscow, 1937, Comptes Rendu* **1**:503–518.
Montenat, C., D. Vachard, and G. Termier. 1977. L'Afghanistan et le domaine gondwan. Différentiation paléogéographique au Permo-Carbonifére. *Soc. Géol. Nord Annales* **97**(4):287–296.
Nalivkin, D. V. 1973a. *Geology of the U.S.S.R.* Edinburgh: Oliver & Boyd, 855pp.
Nalivkin, D. V. 1973b. Paleogeography of the Urals Geosyncline in the Paleozoic. *Internat. Geology Rev.* **15**(5):585–590.
Neumayr, M. 1886–1887. Erdgeschichte, 2 vols. Leipzig: Bibliogr. Inst.
Nicolas, A. 1966. Interprétation des ophiolites piémontaises entre le Grand Paradis et al Dora Maïra. *Schweizer. Mineralog. u. Petrog. Mitt.* **46**:25–41.

Norin, E. 1937. Geology of western Quruq Tagh, eastern T'ien Shan. In *Reports*

Norin, E. 1937. Geology of western Quruq Tagh, eastern T'ien Shan. In *Reports, Sino-Swedish Scientific Expedition in the Northwestern Provinces of China under Sven Hedin, Part III: Geology*, vol. 1, Stockholm: AB Thule, pp. 1–195.

Oberhauser, R. 1968. Beiträge zur Kenntnis der Tektonik und der Paläogeographie während der Oberkreide und dem Palaogen im Ostalpenraum. *Geol. Bundesanst. Jahrb.* **111**:115–145.

Passerini, P. 1965. Rapporti fra le ofioliti e le formazioni sedimentarie fra Piacenze edil Mare Tirreno. *Soc. Geol. Italiana Boll.* **84**:93–176.

Peters, T. J. 1969. Rocks of the Alpine ophiolite suite. *Tectonophysics* **7**(5–6): 507–509.

Petrushevskiy, B. A. 1970a. O tektonicheskoy prirode Gimalayev (On the tectonic nature of the Himalayas). *Moskov. Obshch. Ispytateley Prirody Byull. Otedl. Geol.* **45**(1):5–30.

Petrushevskiy, B. A. 1970b. O geotektonicheskom sootnoshenii sredizemnomorskogo i tikhookeanskogo skladchatykh poyasov (The geotectonic relationship between Mediterranean and Pacific folded mountain belts). *Moskov. Obshch. Ispytateley Prirody Byull. Otedl. Geol.* **45**(2):31–80.

Petrushevskiy, B. A. 1975. Problems in geology of the Himalayas. *Internat. Geology Rev.* **17**(6):712–724.

Picard, L. 1970. On Afro-Arabian graben tectonics. *Geol. Rundschau* **58**(2):337–381.

Pilger, A. 1971. Die zeitlich-tektonische Entwicklung der iranischen Gebirge. *Clausthaler Geol. Abh.* **8**:1–27.

Plotnikov, L. M. 1964. Tectonic formation conditions of the trap intrusions in the Siberian Platform. *Internat. Geology Rev.* **6**(10):1805–1810.

Pol'ster, L. A., S. Yanev, L. G. Shustova, N. Gnoyevaya, B. M. Ulizlo. 1977. History of geological development of northern Bulgaria during the Paleozoic. *Internat. Geology Rev.* **19**(6):633–646.

Pomerol, C. 1975. *Era Mésozoique*. Paris: Doin, 384pp.

Pomerol, C., and C. Babin. 1977. *Précambrien, Ere Paléozoique*. Paris: Doin, 430pp.

Powers, R. W., L. F. Ramirez, C. D. Redmond, and E. I. Elberg, Jr. 1966. *Geology of the Arabian Peninsula. Sedimentary Geology of Saudi Arabia*. U.S. Geol. Survey Prof. Paper 560-D, 147pp.

Querol, R. 1969. Petroleum exploration in Spain. In Hepple, P., ed., *The Exploration for Petroleum in Europe and North Africa*. London: Inst. of Petroleum, pp. 49–72.

Ramsbottom, W. H. C. 1971. Paleogeography and goniatite distribution in the Namurian and early Westphalian. *Congr. Stratigr. Geol. Carbonif., 6th, Sheffield, 1967, Proc.*, pp. 1395–1400.

Rodolfo, K. S. 1969. Bathymetry and marine geology of the Andaman Basin, and tectonic implications for southeast Asia. *Geol. Soc. America Bull.* **80**(7): 1203–1230.

Roland, N. W. 1978. Jungpaläozoische Glazialspuren auf dem arabischen Schild. *Eiszeitalter u. Gegenwart* **28**:133–138.

Ronov, A. B., and V. Ye. Khain. 1954. Devonian lithologic associations of the world. *Sovetskaya Geologiya (Sbornik)* **41**:46–76.

Ronov, A. B., and V. Ye. Khain. 1954. Permian lithologic associations of the world. *Sovetskaya Geologiya (Sbornik)* **54**:20–36.

Ronov, A. B., and V. Ye. Khain. 1955. Carboniferous lithologic associations of the world. *Sovetskaya Geologiya (Sbornik)* **48**:92–117.

Ronov, A. B., and V. Ye. Khain. 1961. Triassic lithologic associations of the world. *Sovetskaya Geologiya* **1**:27–48.

Ronov, A. B., and V. Ye. Khain. 1962. Jurassic lithologic associations of the world. *Sovetskaya Geologiya* **1**:3–34.

Ronov, A. B., K. B. Seslavinskiy, and V. Ye. Khain. 1974. Cambrian lithologic associations of the world. *Sovetskaya Geologiya* **12**:10–33; transl. *Internat. Geological Rev.* **19**(4):373–394 (1977)

Ronov, A. B., V. Ye. Khain, and K. B. Seslavinskiy. 1976. Ordovician lithologic associations of the world. *Sovetskaya Geologiya* **1**:7–27; transl. *Internat. Geol. Rev.* **18**(12):1395–1413 (1976).
Rudkevich, M. Ya. 1976. The history and the dynamics of the West Siberian Platform. *Tectonophysics* **36**(1–3):275–287.
Rukhin, L. B. 1960. Paleogeografiya Aziatskogo materika v mesozoe (Paleogeography of the Asian land mass in the Mesozoic). *Internat. Congr. Géol., 21st, Copenhagen, 1960, Doklady Soviet Geol.*, problem 12, pp. 85–98.
Sadek, A. 1977. Early Paleozoic sediments of the Zagros-Taurus ranges. *Geol. Rundschau* **66**(1):263–276.
Saint-Marc, P. 1978. Arabian Peninsula. In Moullade, M., and A. E. M. Nairn, eds., *The Phanerozoic Geology of the World, vol. II. The Mesozoic A.* Amsterdam: Elsevier, pp. 435–462.
Sander, N. J. 1968. Pre-Mesozoic structural evolution of the Mediterranean region. In Barr, F. T., ed., *Geology and Archaeology of North Cyrenaica*. Petrol. Explor. Soc. Libya, Ann. Field Conf. No. 10, 47–70.
Sander, N. J. 1970. Structural evolution of the Mediterranean region during the Mesozoic era. In Alvarez, W., and K. H. A. Gohrbandt, eds., *Geology and History of Sicily*. Petrol. Explor. Soc. Libya, Ann. Field Conf. No. 12, 43–132.
Sborschikov, I. M. 1976. Tectonics of Afghanistan and structural evolution of the Alpine belt (Pamirs—eastern Iran segment). *Geotectonics* **10**(3):189–198.
Schlager, W. 1969. Das Zusammenwirken von Sedimentation und Bruchtektonik in den triadischen Hallstätterkalken der Ostalpen. *Geol. Rundschau* **59**(1): 289–322.
Schlager, W. 1974. Preservation of cephalopod skeletons and carbonate dissolution on ancient Tethyan sea floor. In Hsu, K. J., and H. C. Jenkyns, eds., *Pelagic Sediments: On Land and Under the Sea*. Internat. Assoc. Sedimentology Spec. Pub. 1, pp. 49–70.
Schmidt, G. C. 1964. A review of Permian and Mesozoic formations exposed near the Turkey-Iraq border at Harbol. *Maden Tetkik ve Arama Enst. Bull.* **62**:103–119.
Schmidt-Effing, R. 1977. Das marine Unterjura Mexicos und seine Beziehung zur Entstehung des Golfes von Mexico. *Deutsch. Geol. Gesell. Nachr.* **17**:59–62.
Scholle, P. A. 1970. The Sestri-Voltaggio line: a transform fault induced tectonic boundary between the Alps and Apennines. *Am. Jour. Sci.* **269**(4):343–359.
Schönlaub, H. P. 1970. Die fazielle Entwicklung im Altpaläozoikum der karnischen Alpen. *Deutsch. Geol. Gesell. Zeitschr.* **122**:97–111.
Schütz, K. I. 1979. Die Aptychen-Schichten der Thiersee- und der Karwendelmulde. *Geotektonische Forschungen* **57**:1–84.
Schwinner, R. 1919. Vulkanismus und Gebirgsbildung: Ein Versuch. *Zeitschr. f. Vulkanologie* **5**(4):175–230.
Sdzuy, K. 1972. Das Kambrium der acadobaltischen Faunenprovinz. Gegenwärtiger Kenntnisstand und Probleme. *Zentralbl. Geol. u. Palaeont.* **2**(1): 1–19.
Seilacher, A. 1963. Kaledonischer Unterbau der Irakiden. *Neues Jahrb. Geologie u. Paläontologie Monatsh.* **9**:527–542.
Selley, R. C. 1969. Near-shore marine and continental sediments of the Sirte Basin, Libya. *Geol. Soc. London Quart. Jour.* **124**(496):419–460.
Selley, R. C. 1970. Ichnology of Paleozoic sandstones in the southern desert of Jordan: a study of trace fossils in their sedimentological context. In *Trace Fossils*, Geol. Jour. Spec. Issue No. 3, pp. 477–488.
Selli, R., and A. Fabbri. 1971. Tyrrhenian: a Pliocene deep sea. *Accad. Naz. Lincei Atti, Cl. Sci. Fis., Mat. e Nat. Rend.*, ser. 8, **50**(5):580–592.
Setuhdenia, A. 1978. The Mesozoic sequence in SW Iran and adjacent areas. *Jour. Petroleum Geology* **1**(1):3–42.
Seyfert, H. 1978. Der subbetische Jura von Murcia (SE-Spanien). *Geol. Jahrb.* **29**: 3–201.

Shevchenko, V. I., Ye. K. Stankevich, and I. A. Rezanov. 1974. Pre-Cenozoic magmatism in Caucasus and western Turkmenia in relation to deep-seated structures. *Internat. Geology Rev.* **16**(3):301–309.

Sikosek, R., and W. Medwenitsch. 1965. Neue Daten zur Fazies und Tektonik der Dinariden. *Deutsch. Geol. Gesell. Zeitsch.* **116**(2):342–358.

Smith, A. G. 1971. Alpine deformation and the oceanic areas of the Tethys, Mediterranean and Atlantic. *Geol. Soc. America Bull.* **82**(8):2039–2070.

Sonnenfeld, P. 1974. The Upper Miocene evaporite basins in the Mediterranean region—a study in paleo-oceanography. *Geol. Rundschau* **63**(3):1133–1172.

Sonnenfeld, P. 1975. The significance of Upper Miocene (Messinian) evaporites in the Mediterranean Sea. *Jour. Geology* **83**(3):287–311.

Sonnenfeld, P. 1977. Origins of Messinian sediments in the Mediterranean region—some constraints on their interpretation. *Annales Géol. Pays Helléniques* **27**:160–189.

Sonnenfeld, P. 1978a. Eurasian ophiolites and the Phanerozoic Tethys Sea. *Geotektonische Forschungen* **56**:1–88.

Sonnenfeld, P. 1978b. Effects of a variable sun at the beginning of the Cenozoic era. *Climatic Change* **1**:355–382.

Sonnenfeld, P. 1980. Sources of energy for plume propagation. *Berliner Geowiss. Abh.*, ser. A, **19**:223–224.

Stöcklin, J. 1968. Structural history and tectonics of Iran: a review. *Am. Assoc. Petroleum Geologists Bull.* **52**(7):1228–1258.

Suensilpong, S., C. K. Burton, N. Mantajit, and D. R. Workman. 1978. Geological evolution and igneous activity of Thailand and adjacent areas. *Episodes* **3**:12–17.

Suess, E. 1875. *Die Entstehung der Alpen.* Wien: W. Braümuller, 227pp.

Sugisaki, R., S. Mizutani, H. Hattori, M. Adachi, and T. Tanaka. 1972. Late Paleozoic geosynclinal basalt and tectonics in the Japanese islands. *Tectonophysics,* **12**(5):393–413.

Svoboda, J. et al. 1966–1968. *Regional Geology of Czechoslovakia,* 3 vols. Prague: Czechoslovak Akad. Sci.

Termier, H., and G. Termier. 1974. Distribution des faunes marines dans le sud de la Téthys et sur la bordure septentrionale du Gondwana au cours de Paléozoique supérieur. *Soc. Géol. Belgique Annales* **97**:387–446.

Termier, H., and G. Termier. 1976. Configuration de la Téthys en connexion avec la Gondwanie au Paléozoïque superieur. *Acad. Sci. Comptes Rendus,* ser. D, **283**(2):139–142.

Tollmann, A. 1963. *Ostalpensynthese.* Wien: Deuticke, 265pp.

Tollmann, A. 1969. Die Bruchtektonik in den Ostalpen. *Geol. Rundschau* **59**(1): 287–288.

Tollmann, A. 1970. Die bruchtektonische Zyklenordnung im Orogen am Beispiel der Ostalpen. *Geotektonische Forschungen* **34**:1–90.

Tromp, S. W. 1948. Shallow-water origin of radiolarites in southern Turkey. *Jour. Geology* **56**(3):492–494.

Trümpy, R. 1960. Paleotectonic evolution of the central and western Alps. *Geol. Soc. America Bull.* **71**:843–908.

Trümpy, R. 1965. Zur geosynklinalen Vorgeschichte der Schweizer Alpen. *Umschau* **18**:573–577.

Trümpy, R. 1971. Stratigraphy in mountain belts. *Geol. Soc. London Quart. Jour.* **126**:293–318.

Trunkö, L. 1977. Karpatenbecken und Plattentektonik (Carpathian Basin and Plate Tectonics). *Neues Jahrb. Geologie u. Paläontologie Abh.* **153**(2):218–252.

Van Bemmelen, R. W. 1969. The Alpine loop of the Tethys zone. *Tectonophysics* **8**(2):107–113.

Van Bemmelen, R. W. 1972. *Geodynamic Models, an Evaluation and a Synthesis.* New York: Elsevier, 267pp.

Van Calsteren, P. W. C., and E. Den Tex. 1979. A mantle plume interpretation for the Variscan basement of Galicia (NW Spain). *Padova Univ. Ist. Geologia e Paleontologia Mem. Sci. Geol.* **33**:243–246.

Van Hilten, D. 1964. Evaluation of some geotectonic hypotheses by paleomagnetism. *Tectonophysics* **1**(1):3–71.

Virgili, C. 1977. Le Trias du nord de l'Espagne. *Bur. Recherches Géol. et Miniere Bull. (France), Ser. 4, no. 3,* pp. 205–213.

Voskresenskiy, I. A., K. N. Kravchenko, E. B. Movshovitch, and B. A. Sokolov. 1971. *A Short Description of the Geology of Pakistan* (in Russian). Moscow: Nedra, 168pp.

Vylov, O. S. 1967. Some aspects of the history of the Carpathians. *Geotectonics* **1**(2):251–254.

Wellmann, H. W. 1966. Active wrench faults of Iran, Afghanistan and Pakistan. *Geol. Rundschau,* **55**(2):710–735.

Wolfart, R. 1967. Zur Entwicklung der paläozoischen Tethys in Vorderasien. *Erdöl u. Kohle, Erdgas u. Petrochemie* **20**(3):168–180.

Wolfart, R. 1972. Das Kambrium im mittleren Südasien (Iraq bis Nordindien). *Zentralbl. Geol. u. Palaeont.,* Pt. 1, No. 5-6, pp. 347–376 (227–256).

Wolfart, R., and M. Kursten. 1974. Stratigraphie und Palaogeographie des Kambriums im mittleren Süd-Asien (Iran bis Nordindien). *Geol. Jahrb. Beihefte* **8**:185–234.

Wunderlich, H. G. 1973. Plattentektonik in kritischer Sicht. *Deutsch. Geol. Gesell. Zeitsch.* **124**:309–328.

Zavaritskiy, V. A. 1960. The spilite-keratophyre formation in the region of the Blyava deposit in the Ural Mountains. *Internat. Geology Rev.* **2**(7):551–581, **2**(8):645–687.

Zonenschein, L. P. 1973. The evolution of Central Asiatic geosynclines through sea floor spreading. *Tectonophysics* **19**:213–232.

Part III

THE EPICONTINENTAL PALEOZOIC TETHYS

Editor's Comments on Papers 4 Through 8

To answer questions about the paleoenvironment, how wide the Tethys Sea was originally, how its width oscillated through time, or how its position versus climatic belts changed, it is best to look at the heart, the central part, of the Tethys. The degree of orogenic compression or of block rotation was much less intensive than in the western or eastern margins. It is fortuitous to have now some summaries of the central part of the Paleozoic Tethys available, the Paleotethys of the Near and Middle East. In the western Tethys, in Europe, equivalent sediments are often highly metamorphosed in Caledonian and Variscan orogenies; in the eastern Tethys, in Burma or adjacent China and Thailand, not enough fieldwork has been done in enormously inaccessible mountain chains to gain an authoritative picture.

In his article, R. Wolfart addresses the fate of the pre–Variscan Tethys proper and of contemporaneous waters between the Mediterranean macroisthmus and the shores of the continental platform in the north, the miogeosynclinal ancestral Parathethys. He shows that the Paleotethys was at all times a reasonably shallow sea with epicontinental sedimentation (Paper 4).

On a more specialized scale, V. Havlíček compares the north and south shores of the western Paleotethys, which were both marked by periglacial deposits of Ordovician age (Paper 5). H. Termier and G. Termier are similarly concerned with the north and south shores of the Tethys in the later part of the Paleozoic era, when it became a narrow seaway from the Alps to Tibet and hence into China (Paper 6).

The Tethys Sea never spanned more than one climatic zone in a north-south direction, but almost always did so in an east-west direction, giving it the shape of a long finger. A certain cyclicity can be observed in the paleocurrent regimes.

Toward the end of Precambrian time, much of western Europe was land. This land extended into Eria, alias North America. Cambrian seas invaded Morocco and Atlantic North America and opened a passage deep into northern Europe, the Mid-European Sea with its Baltic connection, and another one deep into southern Europe, the Paleotethys. Gradually, the Atlantic faunas spread eastward and displaced Pacific faunas. Toward the end of the Paleozoic era, the seaway between Paleotethys and Atlantic was severed in response to Variscan movements. Westernmost Europe and Africa became a land barrier between Atlantic and Tethyan faunas.

The Permian Tethys is the transition between pre-Variscan and pre-Alpine expanses of water. I. Argyriadis shows that the conversion of the Tethys from an epicontinental to a geosynclinal sea is post-Permian and that it occurs mostly to the south and west of Hercynian diastrophism (Paper 7). M. Kamen-Kaye charts the western shelf and shore line; the Permian Tethys terminated east of Spain (Paper 8).

4

Reprinted from *Erdol. u. Kohle, Erdgas, Petrochemie* **20**:168–180 (1967)

ZUR ENTWICKLUNG DER PÄLAOZOISCHEN TETHYS IN VORDERASIEN

Reinhard Wolfart*

On the development of the Paleozoic Thetys of the Middle East: The Paleozoic Tethys of the Middle East developed in the unstable shelf adjoining to the Gondwanaland in the north. The southern limit of the Tethys was formed by the Nubian-Arabian Shield, the northern limit by a belt of orthogeosynclines. The Middle East Tethys was composed of a southern zone of relatively strong subsidence — the future Alpidian geosyncline between E-Arabia and SW-Iran — and the Turkish-Afghan archipelago as a northern zone. The Paleozoic history of the Middle East Tethys is divided into three periods which are equivalents of the Postassyntian-Kaledonian, the Postkaledonian-Hercynian and the Posthercynian eras. After the orogenies which affected the Tethys region by uplifting movements the periods began with a continental phase and ended with a marine phase. Young Alpidian structures were already perceptible in the Paleozoic era; presumably they are connected with lineaments of Precambrian age.

Vorbemerkung

Mehrjährige Tätigkeit des Verf. als Geologe in verschiedenen Gebieten Vorderasiens sowie die Beschäftigung mit biostratigraphischen und paläogeographischen Fragen aus dem vorderasiatischen Paläozoikum gaben den Anstoß, die Vorstellungen über die Entwicklung der paläozoischen Tethys in Vorderasien kartographisch aufzuzeichnen. Dabei fanden außer der umfangreichen Literatur auch die z. T. noch unpublizierten Erkenntnisse Verwendung, die von Geologen der Bundesanstalt für Bodenforschung in Afghanistan, Iran, Jordanien, Saudi-Arabien und Syrien gewonnen worden sind. Die vorliegende Arbeit ist ein Entwurf, der sich in zahlreichen Punkten auf lückenhafte Unterlagen stützen muß. Im einzelnen ist daher eine genaue kartographische Darstellung der paläogeographischen Verhältnisse vielfach gar nicht möglich.

1. Einleitung

Seit *Neumayr* und *Suess***) versteht man unter „Tethys" ein W—E erstrecktes Meer, das sich „à travers toute l'Asie centrale et les Alpes jusqu'à la Méditerranée occidentale" hinzieht. *Richter* verwandte den Begriff „Tethys" für paläozoische Meere nur dann, wenn folgende ozeanographische und tiergeographische Voraussetzungen erfüllt waren: 1. Lage des Meeres im Bereich des europäisch-asiatischen Mittelmeeres, 2. offene Verbindung mit dem Pazifik und 3. tiergeographische Einheitlichkeit innerhalb dieses ganzen Gebietes. Von gegensätzlichen Tierwelten belebte Teilbecken im Bereich des späteren Tethys-Gürtels sind nach tektonischen Gesichtspunkten zwar als Vorboten der Tethys, nicht aber als echte Tethys anzusehen. Auf sibirische Meeresstraßen läßt sich der Begriff nicht übertragen.

Entsprechend der oben angeführten Abgrenzung der Tethys, werden im folgenden die Gebiete der östlichen Mittelmeerumrandung, Iran, Afghanistan und Pakistan sowie die Grenzgebiete der südlichen Sowjetunion in die Betrachtungen mit einbezogen. Paläogeographisch bedeckte die Tethys demnach das Gebiet zwischen dem Gondwana-Land im S und den asiatisch-europäischen Festlandsblöcken im N.

*) Anschrift: Regierungsgeologe Dr. R. Wolfart, Bundesanstalt für Bodenforschung, 3 Hannover-Buchholz, Alfred-Bentz-Haus.

**) Literaturverzeichnis am Schluß der vorliegenden Arbeit.

Eine recht ausführliche Übersicht über die voralpidische Zeit in Vorderasien stammt von *Flügel* (1964), der als erster die zahlreichen Beobachtungen zusammenfaßte und den Ablauf der paläozoischen Erdgeschichte in diesem Gebiet skizzierte. Die meisten älteren Publikationen befassen sich mit den geologischen Problemen vergleichsweise kleiner Teilgebiete, ohne aber auf die regionalen Zusammenhänge einzugehen. Wegen sehr lückenhafter Kenntnisse über die paläozoischen Formationen in Vorderasien war dies bis vor wenigen Jahren vorwiegend auch gar nicht möglich. — Es ist das Verdienst *Richters*, auf Grund von Trilobiten pazifischen Gepräges zum ersten Male nachgewiesen zu haben, daß sich bereits im Kambrium ein Vorbote der Tethys bis nach Palästina erstreckte. Seitdem ist diese Kenntnis als fester Bestandteil in unsere paläogeographischen Vorstellungen eingegangen.

Vorliegende Arbeit versucht, durch die Darstellung der paläogeographischen Verhältnisse die erdgeschichtliche Entwicklung des Tethysraumes in Vorderasien vom Kambrium bis zum Perm aufzuhellen und Rückschlüsse auf die strukturelle Gliederung dieses Gebietes im Paläozoikum zu ziehen.

2. Paläogeographie der vorderasiatischen Tethys

Paläozoische Ablagerungen treten in Vorderasien beiderseits des in W-Pakistan, im SW-Iran, im Irak und in Syrien gelegenen mesozoisch-känozoischen Senkungsgebietes zutage. Sie markieren die alten Kerne des präkambrisch konsolidierten Fundamentes, auf dem sich die erdgeschichtliche Entwicklung im Paläozoikum abgespielt hat. Im Kernbereich des Nubisch-Arabischen Schildes sind die kristallinen Gesteinsmassen eine geschlossene Einheit, die vermutlich während des ganzen Paläozoikums mehr oder weniger stark der Erosion ausgesetzt war. Im türkisch-iranisch-afghanischen Raum ragen die höchsten Erhebungen des Fundamentes inselförmig auf; in Anlehnung an *Paffengolz* kann man annehmen, daß sie in den marinen Abschnitten des Paläozoikums von der Türkei bis nach Pakistan einen riesigen Archipel bildeten.

Unsere Kenntnisse von den paläozoischen Ablagerungen in dem mesozoisch-känozoischen Senkungsgebiet sind gering und auf wenige im Verlauf der Erdölexploration abgeteufte Bohrungen sowie auf die Salzstöcke in SW-Persien und im Persischen Golf beschränkt. Immerhin ergibt sich aus der geringen Anzahl von Daten unter Zugrundelegung gewisser geotektonischer Vorstellungen ein recht einleuchtendes Bild von den jeweiligen paläogeographischen Verhältnissen.

2.1. Kambrium (*Abb. 1*)

Die letzten präkambrischen Orogenesen — *Flügel* (1964) schrieb ihre Auswirkungen der assyntischen Faltung zu — konsolidierten die in einem präkambrischen Geosynklinalraum abgelagerten Gesteinsmassen und schufen damit in Vorderasien ein stark verfaltetes und metamorphes, seinem zukünftigen Verhalten nach aber uneinheitliches Fundament. Treffen die Auffassungen von *Bogdanow* und *Ponikarow* et al. zu, dann dürfte die Uneinheitlichkeit des Fundamentes auf Faltungen unterschiedlichen Alters zurückgehen.

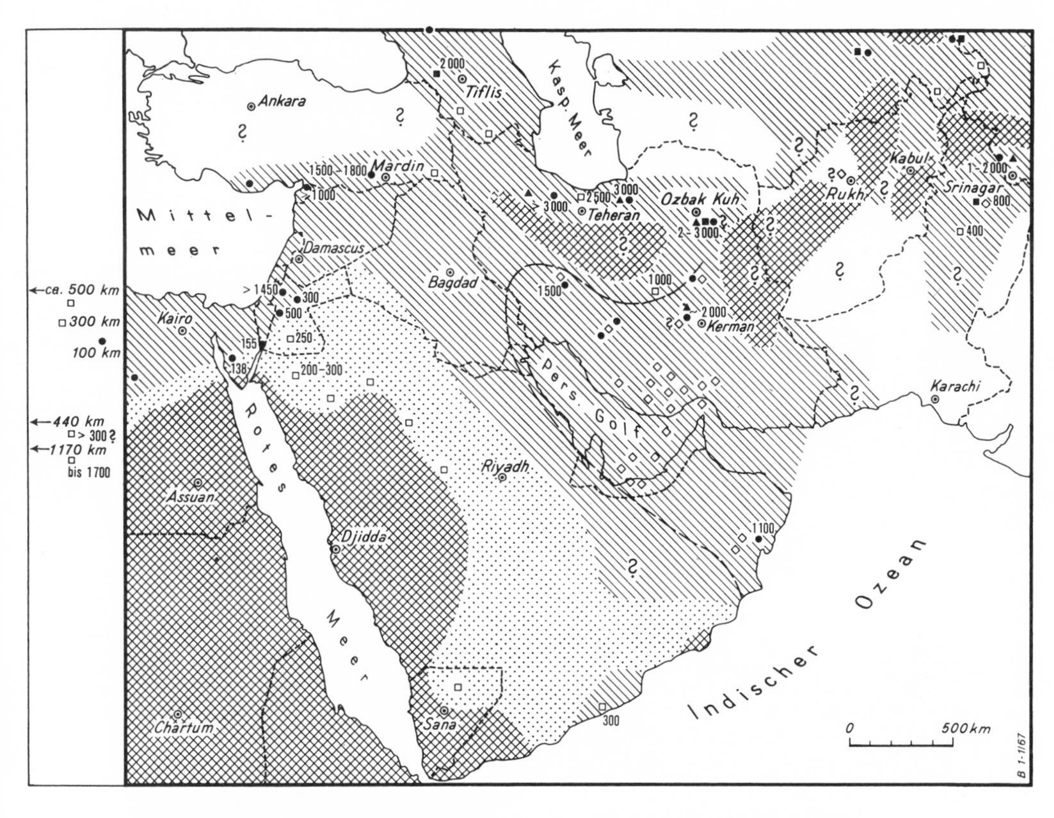

Hochgebiet: Abtragung oder (geringe) kontinentale Sedimentation

kontinentales Kambrium, marine Ingressionen vor allem im Mittel-Kambrium möglich

marines Kambrium über kontinentalem Unter-Kambrium bzw. hohem Präkambrium

Mächtigkeit in Metern

▲ marine Fauna des Ober-Kambriums

● marine Fauna des späten Unter-Kambriums bzw. des frühen Mittel-Kambriums

■ marine Fauna des Unter-Kambriums

□ Fundorte kambrischer Ablagerungen (? kontinental)

◇ Steinsalzbildungen größerer Mächtigkeit im Unter-Kambrium nachgewiesen

Abb. 1. Die Tethys im Kambrium

Im Anschluß an die Konsolidierung wurde unter lang andauernden kontinentalen Verhältnissen das Fundament weitgehend eingeebnet. Schon während der Einebnung zeigte sich die uneinheitliche Beschaffenheit des Fundamentes dadurch, daß neben den kratonischen Festlandskernen im S und den inselförmigen Aufragungen des Sockels im türkisch-afghanischen Raum stark ausgeprägte Senkungsgebiete vorhanden waren, die im späten Präkambrium und im Unter-Kambrium sehr mächtige klastische Sedimente aufgenommen haben. Mächtigkeit und Verbreitung jener unterkambrischen „Old-Red"-Serie sprechen dafür, daß in diesem Raum weithin gleichartige Verhältnisse geherrscht haben, u. zw. sowohl in klimatischer Hinsicht als auch bezüglich der Absenkungsbedingungen. Eine Differenzierung in Becken und Schwellen ist nicht festzustellen. Auffällig ist die besonders große Mächtigkeit der kontinentalen Serie gerade in jenen Gebieten — z. B. bei Mardin und im Elburs-Gebirge —, die mindestens vom jüngeren Ordovizium bis zum älteren Ober-Devon Hochgebiete waren.

Das unterkambrische kontinentale Sedimentationsgebiet erstreckte sich in nördlicher Richtung bis etwa nach Armenien und wenigstens bis zur nordostiranisch-russischen Grenze. Im Dsirula-Massiv, westlich von Tiflis, und in Tadschikistan, nördlich von Afghanistan, ist marines Unter-Kambrium mit Archaeocyathus und Coscinocyathus entwickelt. Möglicherweise erfolgte die unterkambrische Sedimentation zeitweilig auch in anderen Gebieten Vorderasiens unter mariner Beeinflussung, so z. B. bei Ozbakh Kuh und in den Salt Ranges, W-Pakistan. Aus diesen Gebieten wurden Funde von gleichartigen Arthropoden-Spuren gemeldet (*Flügel* u. *Ruttner*; *Schindewolf* u. *Seilacher*), die wenigstens in den Salt Ranges dem marinen Bereich angehören. In den Salt Ranges befindet sich auch die regio typica von Redlichia noetlingi (*Redlich*), einem Vertreter der westpazifischen Provinz.

Eingebettet in den Bereich vorwiegend kontinentaler Sedimentation des Unter-Kambriums sind die Salinarvorkommen von SW-Persien, dem Persischen Golf und den Salt Ranges. In W-Pakistan, hat die Salinarserie eine Mächtigkeit von mehr als 500 m. Im Gebiet des Persischen Golfes ist ihre Mächtigkeit unbekannt; sie dürfte aber ebenfalls ursprünglich recht beträchtlich gewesen sein. Ohne große primäre Mächtigkeit der Salzlager wäre es kaum denkbar, daß eine so große Anzahl von Salzstöcken einen Aufstieg von einigen Kilometern hätte bewältigen können. I. allg. wird die Salinarserie in spätes

Prä-Kambrium bis frühes Unter-Kambrium eingestuft. Als Begründung dafür wird die Überlagerung der Salinarserie durch das fossilführende Unter-Kambrium in W-Pakistan bzw. durch Mittel-Kambrium in SW-Persien herangezogen. Die paläogeographische Entwicklung dürfte somit folgendermaßen verlaufen sein: Zwischen dem Nubisch-Arabischen Schild und dem Türkisch-Afghanischen Archipel senkte sich im späten Prä-Kambrium eine breite, SE—NW streichende Rinne ein, die zunächst mit terrestrischem Schutt angefüllt wurde. Im späten Prä-Kambrium bis frühen Unter-Kambrium führten Meeresvorstöße von E oder SE her zur Salzbildung im Zentrum der Rinne. Lateral verzahnen sich die Steinsalzbildungen mit Gipsen, Carbonaten und sandigen Ablagerungen, wie z. B. im Raume von Kerman. Nördlich des Archipels dürfte es kaum noch nennenswert zur Ablagerung von Salinargesteinen gekommen sein. Über 1000 m mächtige Dolomite werden im Elburs-Gebirge als Randfazies aufgefaßt (*Stöcklin* et al., 1964), die in vorwiegend kontinentale klastische Gesteine eingeschaltet ist. In der Verbreitung der Salzvorkommen deutete sich bereits im Unter-Kambrium eine großräumige Struktur an, die in ihren Umrissen der späteren Tethys-Meeresstraße ähnelte.

Eine Salinarserie unbekannten Alters befindet sich nach *Weippert* u. *Wittekindt* bei Rukh im nördlichen Zentral-Afghanistan. Gewisse Anzeichen, z. B. die Abfolge kristallines Fundament — Konglomerate mit Vulkaniten — rote Serie mit Steinsalz, deuten darauf hin, daß es sich hierbei um unterkambrische Bildungen handeln könnte. Die konkordante Überlagerung der Salinarserie durch fossilführendes Mittel-Devon darf nicht darüber hinwegtäuschen, daß zwischen der möglicherweise kambrischen Serie und dem Mittel-Devon — ebenso wie in anderen Hochgebieten des türkisch-afghanischen Archipels, z. B. im Elburs-Gebirge — eine beträchtliche Schichtlücke bestehen kann.

Der Salinar-Abschnitt wurde im Iran von kontinentalen, in Pakistan von i. w. marinen Verhältnissen abgelöst, bevor sich im späten Unter-Kambrium und im frühen Mittel-Kambrium die erste große marine Transgression des Paläozoikums vollzog. Wie durch Fossilfunde belegt oder durch geotektonische Überlegungen wahrscheinlich gemacht werden kann, wurden in dieser Zeit große Gebiete des Irans, des östlichen Saudi-Arabiens, des Irak und im Bereich der östlichen Mittelmeerumrandung überflutet. Lediglich aus den Salt Ranges könnte sich das Meer bereits im höheren Unter-Kambrium wieder zurückgezogen haben, während im Kleinen Kaukasus mittelkambrische Schichten zwar abgelagert, später aber metamorph umgewandelt worden sind. Die Fossilfunde deuten darauf hin, daß sich Westpazifische und Atlantische Provinz etwa im Raume Palästina — SE-Türkei verzahnen.

Westpazifische Trilobiten treten am Toten Meer auf — Redlichops blanckenhorni *Richter* u. *Richter*, Palaeolenus campbelli (*King*) und Protolenus orientalis *Picard* — sowie im S- und N-Iran. Im S-Iran wurden mit Redlichia cf. chinensis *Walcott* und „Anomocare megalurus" (*Dames*) Trilobiten des Unter-Kambriums und mit Ptychoparia s. l. sowie Alokistocariden solche des Mittel-Kambriums gefunden (*Huckriede* et al.). Da die genannten südiranischen Trilobiten noch nicht genau untersucht worden sind, kann einstweilen nur Redlichia als typischer Vertreter der Westpazifischen Provinz bezeichnet werden. Im N-Iran ist spätes Mittel-Kambrium durch die westpazifischen Trilobiten Lioparella und Anomocarella nachgewiesen. Von Mardin und vom Amanos-Gebirge (S-Türkei) liegen mit Paradoxides cf. mediterraneus *Pomp.* und Pardailhania cf. barthouxi (*Mansuy*) Trilobiten der Atlantischen Provinz vor, zu der auch der nördliche Kaukasus gehören dürfte, wie das Auftreten von Solenopleura andeutet. Im südlichen Tadschikistan kommen nach *Markowskij* et al. neben den westpazifischen Trilobiten Anomocarella und Manchuriella anscheinend auch atlantische Elemente vor (Solenopleura).

Marines Ober-Kambrium wurde im Bereich der vorderasiatischen Tethys bisher lediglich im Elburs-Gebirge, im E-Iran und in Kaschmir faunistisch nachgewiesen. Wie Funde von Kaolishania, Chuangia, Prochuangia, Quadraticephalus und einer Damesellidae im Elburs-Gebirge zeigen, gehörten N- und E-Iran tiergeographisch auch im Ober-Kambrium der Westpazifischen Provinz an. Aus allen anderen Gebieten hatte sich das Meer anscheinend bereits im Mittel-Kambrium wieder zurückgezogen, und kontinentale Bedingungen stellten sich, wenigstens in den Randgebieten des Überflutungsraumes, wieder ein.

Während des ganzen Kambriums bildete sich um das Abtragungsgebiet des Nubisch-Arabischen Schildes herum ein breiter Saum kontinentaler Sedimente. Wie Funde von Cruziana in S-Jordanien und NW-Saudi-Arabien zeigen, muß jedoch damit gerechnet werden, daß einige Meeresvorstöße auch diesen Raum gelegentlich überflutet haben. — Trotz eingehender Untersuchungen ist es wegen ungünstiger Aufschlußverhältnisse bisher nicht gelungen, kambrische Ablagerungen auch im ostafghanischen Trog durch Fossilfunde nachzuweisen. Die Entwicklung in vergleichbaren Gebieten Vorderasiens macht es jedoch wahrscheinlich, daß die Absenkung im Bereich des ostafghanischen Troges nicht erst mit Beginn des Ordoviziums einsetzte, sondern daß der Trog bereits im Kambrium Sedimentationsgebiet war. — Über das Kambrium von Anatolien ist wenig bekannt. Auf Grund der Tatsache, daß die mächtige, klastische, rote Gesteinsserie, die in Vorderasien sonst weit verbreitet ist, in Anatolien bisher noch nicht angetroffen wurde, vermutete *Flügel* (1964), daß dieser Bereich eine andere Entwicklung erlebt habe als die Gebiete im S und SE. Die Entwicklung der paläozoischen Ablagerungen deutet aber darauf hin, daß in der Türkei, ähnlich wie im iranisch-afghanischen Raum, zumindest zeitweise ebenfalls archipelartige Verhältnisse geherrscht haben.

Prä- bis frühkambrische Vulkanite sind von mehreren Lokationen, z. B. aus dem Raume von Mardin, im Hadramaut, im SW-Iran, möglicherweise auch im N-Iran und bei Rukh bekanntgeworden. Sie wurden verschiedentlich als subsequente Vulkanite einer spät-präkambrischen Faltung — assyntische Faltung — zugeordnet.

2.2. Ordovizium (*Abb. 2*)

Im gesamten Tethysraum Vorderasiens liegen die ordovizischen Schichten konkordant über dem Kambrium. Wegen Fossilmangels in den Grenzschichten läßt sich nicht feststellen, ob überhaupt irgendwo in Vorderasien ein lückenloser Übergang vorhanden ist. Die regressive Tendenz des Ober-Kambriums dürfte jedenfalls in einem sehr frühen Stadium des Ordoviziums in eine transgressive Tendenz umgeschlagen sein. Dies beweisen die marinen Faunen vom Westrand des ostafghanischen Troges, bei Kerman, im Elburs-Gebirge, nördlich von Teheran, im Oman, in S-Jordanien und im N-Irak.

In Afghanistan sind Trilobiten des oberen Tremadoc bis unteren Arenig — Euloma (E.) n. sp. A, Harpides? sp., Asaphellus? sp. und Protopliomerops n. sp. A u. a. — gefunden worden*). Ähnliches Alter wie die ostafghanischen Trilobiten hat eine Brachiopoden-Trilobiten-Conodonten-Fauna von Dahu bei Kerman mit Nanorthis, Syntrophina, cf. Asaphellus u. a., während die Trilobiten (Symphysurus, Presbynileus, Ptychopyge cf. cincta *Brögger*, Megalaspis und cf. Ogygites yunnanensis *Reed*) aus dem Elburs-Gebirge vielleicht etwas jünger, u. zw. nach *Glaus* etwa als unteres bis unterstes mittleres Ordovizium einzustufen sind. Schichten des Tremadoc-Arenig sind in den übrigen Gebieten Vorderasiens, z. B. in S-Jordanien und im N-Irak, nur durch Spurenfossilien (Cruziana furcifera *D'Orbigny*) belegt; sie zeigen starke terrestrische Einflüsse. Auffälligerweise sind die unterordovizischen Ablagerungen der nördlichen Grenzbereiche Vorderasiens, nämlich bei Istanbul, im Kaukasus, in Tadschikistan und

*) Eine Publikation über die unterordovizischen Trilobiten E-Afghanistans wird z. Z. von *D. Weippert* u. *R. Wolfart* vorbereitet.

in Kaschmir, kontinentalen Ursprungs oder stark terrestrisch beeinflußt. Funde von Didymograptus bifidus (*Hall*) bei Kerman, im Oman, in NW-Saudi-Arabien und in S-Jordanien deuten darauf hin, daß sich die transgressive Tendenz zu Beginn des Mittel-Ordoviziums verstärkte. Im weiteren Verlauf des Ordoviziums herrschten in den meisten der seit dem Kambrium bekannten Senkungsgebiete Vorderasiens marine Verhältnisse, wie auf Grund zahlreicher Fossilfunde angenommen werden darf.

In E-Afghanistan sind Mittel- und Ober-Ordovizium durch das Auftreten von Pharostoma n. sp. A und Onniella? sp. belegt, bei Kerman und in S-Jordanien durch verschiedene Brachiopoden — (Trematis, Orbiculothyris, Dalmanella sp. aff. horderleyensis (*Whittington*) —, in Libyen durch Graptolithen, Trilobiten und Brachiopoden (*Collomb*), in NE-Syrien durch Diplograptus spinulosus *Sudbury*, Pseudobasilicus cf. nobilis (*Barr.*), Onnia ornata (*Sternb.*), Dalmanitina socialis (*Barr.*) u. a. m. Im N-Irak wurde Selenopeltis buchi gefunden, im nördlichen Giaour Dagh mehrere Arten von Dalmanitina. Aus der Umgebung von Istanbul führen *Baykal* u. *Kaya* mehrere mittelordovizische Arten von Exoconularia an, aus Tadschikistan ist Pseudobasilicus nobilis bekannt. Auch in Armenien sind im Ordovizium mit großer Wahrscheinlichkeit marine Sedimente abgelagert worden; sie sind jedoch metamorph und können daher nicht sicher eingestuft werden.

In einigen Gebieten am Rande der Tethys bahnte sich im mittleren Ordovizium eine Entwicklung an, die im Silur und Unter-Devon zum Rückzug des Meeres aus nahezu dem gesamten Tethysraum Vorderasiens führen sollte. Im Elburs-Gebirge und vermutlich auch im Oman leiteten ausgedehnte Hebungen im Mittel-Ordovizium eine bis zum Ober-Devon bzw. Perm andauernde kontinentale Phase ein. Aus W-Pakistan und aus dem Raume von Mardin sind ordovizische Ablagerungen bisher nicht mit Sicherheit bekanntgeworden.

Nach der Mächtigkeit der Schichten bietet sich im Ordovizium ein von den Verhältnissen im Kambrium abweichendes Bild. Weite Gebiete mit mächtigen kambrischen Sedimenten erlebten im Ordovizium nur noch eine geringfügige Absenkung. Dazu gehören vor allem die Gebiete des türkisch-afghanischen Archipels, u. zw. der Raum von Mardin, das Elburs-Gebirge

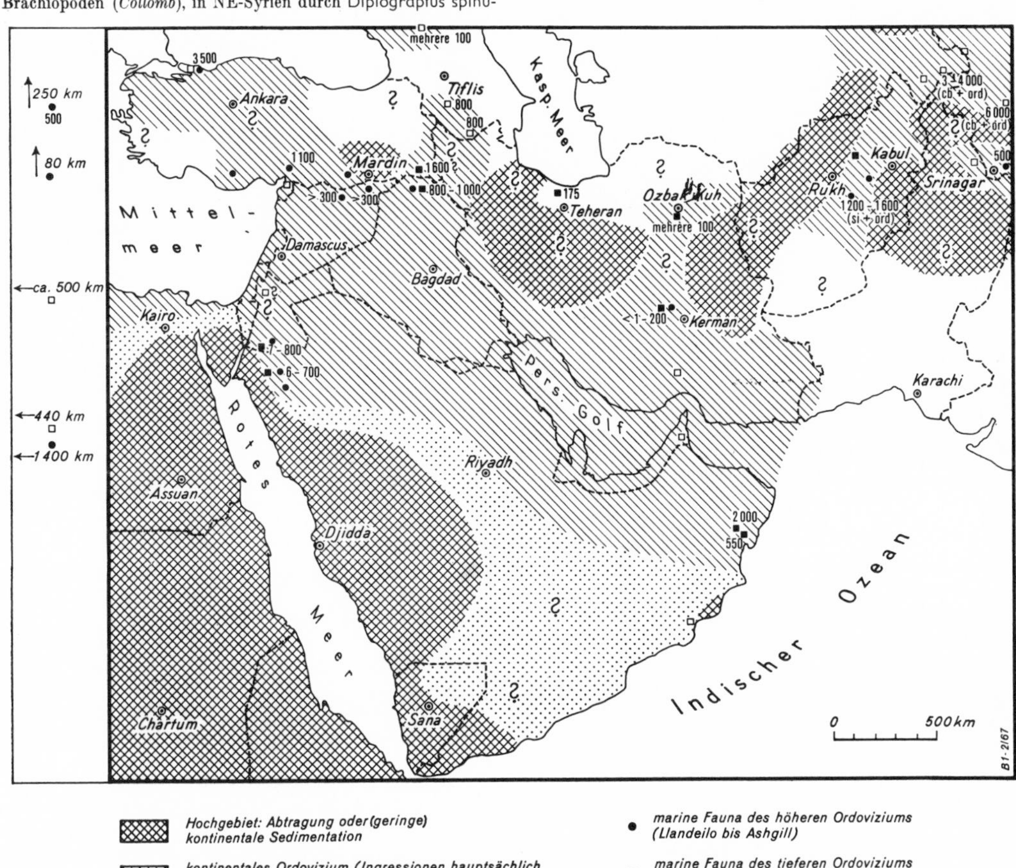

Abb. 2. Die Tethys im Ordovizium

und der S-Iran (Kerman), wo die ordovizischen Schichten nur wenige hundert Meter mächtig sind. Im SW dieses Gebietes, also im arabischen Raum, sind die ordovizischen Schichten bis etwa doppelt bis dreifach so mächtig wie die kambrischen. Große Mächtigkeiten von mehreren 1000 m sind insbesondere in den nördlichen Grenzgebieten des Tethysraumes zu verzeichnen, ferner im ostafghanischen Trog und im N-Irak. In den südlichen Randgebieten, Libyen und Arabien, setzte sich die vorwiegend sandig-tonige Sedimentation des Kambriums beinahe unverändert auch im Ordovizium fort. Klastische Sedimentation herrschte auch in den übrigen Gebieten der vorderasiatischen Tethys vor, wenngleich örtlich, wie bei Kerman und Ozbakh Kuh, ebenso wie in den geosynklinalen

Grenzgebieten im N, wiederholt carbonatische Gesteine auftreten können.

Die paläogeographischen Verhältnisse im Ordovizium Vorderasiens bedeuten i. w. eine Fortsetzung der kambrischen Entwicklungstendenzen. Während der Nubisch-Arabische Festlandskern außer im SE (Oman) keine wesentlichen Veränderungen erfahren haben dürfte, dehnten sich die Festlandsgebiete des Indischen Schildes bis zum ostafghanischen Trog aus, und die Hochgebiete des türkisch-afghanischen Archipels erhielten besonders im Iran und in der SE-Türkei beträchtlichen Zuwachs. Ein wesentliches Ergebnis der ordovizischen Entwicklung ist, daß sich die Tethys-Meeresstraße, die nach S hin bereits im Kambrium klar abgegrenzt war, nun auch im N deutlich von den anschließenden geosynklinalen Gebieten unterscheiden ließ. Es handelt sich dabei um große Gebiete in Tadschikistan und wahrscheinlich auch in N-Anatolien, die sich durch vergleichsweise große Mächtigkeit der Schichten auszeichnen, und um den N-Kaukasus, in dem der hohe Anteil vulkanogener Bestandteile am Geosynklinalkomplex auffällt. Außerdem dürften die Inseln des türkisch-afghanischen Archipels im ausgehenden Ordovizium bereits eine mehr oder weniger geschlossene Landmasse gebildet haben, die die Tethys im N begrenzte.

Aus der unmittelbaren Umgebung der Nubisch-Arabischen Landmasse wurde bisher nichts über vulkanische Erscheinungen im Ordovizium berichtet. Es ist aber möglich, daß, wie bei Kerman und im Elburs-Gebirge, auch in den übrigen Gebieten, in denen Verhältnisse des mobilen Schelfes herrschten, örtlich in etwas verstärktem Maße vulkanische Eruptionen im Ordovizium stattfanden. Der ordovizische Vulkanismus dürfte allerdings in seiner Bedeutung weit hinter dem prä- bis frühkambrischen subsequenten Vulkanismus zurücktreten.

2.3. Silur (*Abb. 3*)

Nachdem sich bereits im mittleren und oberen Ordovizium das Meer aus SE-Arabien (Oman), dem Elburs-Gebirge und vielleicht auch aus großen Teilen Armeniens zurückgezogen hatte, setzte sich die marine Sedimentation zunächst auch im Llandovery in den übrigen aus dem Ordovizium bekannten Ablagerungsräumen der Tethys i. w. in unveränderter Fazies fort. Fossilfunde und Fazies in S-Jordanien lassen vermuten, daß der Übergang vom Ordovizium zum Silur möglicherweise ohne Sedimentationsunterbrechung erfolgt ist.

An wenigen Fundstellen kommen dort nämlich mit Climacograptus scalaris cf. scalaris (*Hisinger*), C. innotatus n. ssp. und Monograptus (Demirastrites) sp. ex. gr. triangulatus *Harkness* Fossilien der gregarius- oder convolutus-Zone (Unter-Llandovery) vor, die sich nach den Erfahrungen in anderen Gebieten unweit der Silur-Basis befindet. Distomodus kentuckyensis zeigt ebenfalls Unter-Llandovery an, während Howellella n. sp. A bisher nur aus dem oberen Llandovery bekannt war.

Ebenso wie in NW-Saudi-Arabien ist auch in Jordanien noch so wenig über die Fossilführung im Grenzbereich Ordovizium/Silur bekannt, daß ergänzende Untersuchungen zur Klärung der Grenzziehung abgewartet werden müssen. Etwas jüngere Graptolithen als in S-Jordanien wurden im Antitaurus gefunden, u. zw. Rastrites carnicus (*Seelmeier*), Spirograptus planus (*Barr.*), Monograptus (M.) lobiferus bulgaricus (*Haberf.*) u. a., die der Zone des Rastrites linnaei angehören. Auch aus Libyen sind Graptolithen-Faunen des Llandovery bekannt (*Hecht* et al.).

Spärlicher sind die Kenntnisse von frühsilurischen Faunen östlich der eben behandelten Fundgebiete. In den Grenzbereich Llandovery/Wenlock stuften *Flügel* u. *Ruttner* eine reiche Korallen-Fauna von Ozbakh Kuh ein, die deutliche tiergeographische Beziehungen zum baltisch-englischen Raum aufweist und bisher das einzige Vorkommen von korallenführendem Silur in Vorderasien darstellt (s. w. u.). Dem älteren bis mittleren Silur gehören nach *Huckriede* et al. möglicherweise auch einige Faunen aus dem Gebiet von Kerman an. Climacograptus cf. scalaris u. a. vom Kuh-i-Gahkun (S-Iran) wurde von *Bulman* (in *Douglas*, l. c.) als Llandovery eingestuft. Im Bereich des ostafghanischen Troges konnte Llandovery faunistisch bisher noch nicht nachgewiesen werden.

Höheres Llandovery und Wenlock sind für die vorderasiatische Tethys eine Zeit der Regression. Über den graptolithen- und tribolitenführenden Sandsteinen des tieferen Llandovery folgen in S-Jordanien und NW-Saudi-Arabien kontinentale oder kont. beeinflußte, fossilleere Ablagerungen. Dieser in S-Jordanien weniger als 100 m mächtige Komplex dürfte etwa den oben angegebenen Zeitabschnitt umfassen. Als Äquivalent können in Libyen Sandsteine mit Arthrophycus und Cruziana angesehen werden, die über graptolithenreichen, tonigen Gesteinen folgen. In den Becken N-Libyens dauerte die marine Sedimentation auch während des Wenlock und Ludlow an; in den Schwellen- und südlichen Randgebieten leitete die Entwicklung dagegen allmählich zu den kontinentalen Verhältnissen des Unter-Devons über. — In NE-Syrien setzte sich die marine Sedimentation auch im Silur zwar noch kräftig fort, doch machte sich der abnehmende marine Einfluß in der großen Fossilarmut der tonig-sandigen Ablagerungen bereits bemerkbar. Das silurische Alter jener 500 m mächtigen Schichten ergibt sich lediglich aus dem Auftreten hochentwickelter Chitinozoen-Formen und der Gattungen Veryhachium, Micrhystridium und Hystrichosphaeridium. — Aus den weiten Gebieten zwischen E-Syrien und dem E-Iran liegen keine Beobachtungen über marines Silur vor. Im E-Iran ist unteres bis mittleres Silur bei Ozbakh Kuh und vielleicht auch bei Kerman entwickelt. Bei Ozbakh Kuh kommen folgende rugose Korallen vor: Cystiphyllum siluriense siluriense *Lonsdale*, Holmophyllum holmi *Wdkd.*, Tryplasma lonsdalei *Eth.* u. a. sowie zahlreiche tabulate Korallen. — In E-Afghanistan konnte mittleres Silur bisher nicht mit Sicherheit nachgewiesen werden. Zwar wurde östlich von Kabul ein Vertreter der im Mittel-Silur häufigen Gattung Leurocycloceras gefunden, doch deuten Conodonten aus denselben Schichten eher auf obersilurisches Alter.

Im Bereich der vorderasiatischen Tethys wurden marine Schichten obersilurischen Alters, außer in den Becken N-Libyens, bisher lediglich in S-Jordanien und E-Afghanistan nachgewiesen. In S-Jordanien tritt Onchus in Sandsteinen auf, im ostafghanischen Trog wurde eine Reihe von Conodonten gefunden — z. B. Kockelella, Spathognathodus primus (*Branson* u. *Mehl*) u. a. —, die als unteres bis mittleres Ludlow gewertet werden. Da in den übrigen Gebieten sichere Hinweise für marines Ludlow fehlen, muß die Frage offen bleiben, ob das Ober-Silur S-Jordaniens auf einen Vorstoß des Meeres aus dem SE oder aus dem W zurückgeht. — In den Geosynklinalen Tadschikistans und in N-Anatolien, bei Istanbul, sind alle drei Stufen des Silurs vertreten. Im N-Kaukasus ist faunistisch belegtes Ober-Silur nur aus dem Flußgebiet der Malka bekannt. In Tadschikistan herrschte im Llandovery und im Wenlock Graptolithen-Fazies vor, während die Schichten des Ludlow eine reiche Brachiopoden-Fauna aufweisen. Aus dem Llandovery sind u. a. Rastrites peregrinus *Barr.* und R. longispinus *Pern.* zu nennen, im Wenlock gesellten sich Vertreter der Atrypiden und Spiriferiden zu den zahlreichen Monograptiden. — Im Gebiet von Istanbul werden Graptolithen-Schiefer des oberen Llandovery von Brachiopoden-Korallen-Kalken des Wenlock und Ludlow überlagert.

Im gesamten Gebiet der vorderasiatischen Tethys übersteigt die Mächtigkeit der silurischen Ablagerungen nur selten 500 m, häufig ist sie noch weit geringer. Lediglich im Bereich der

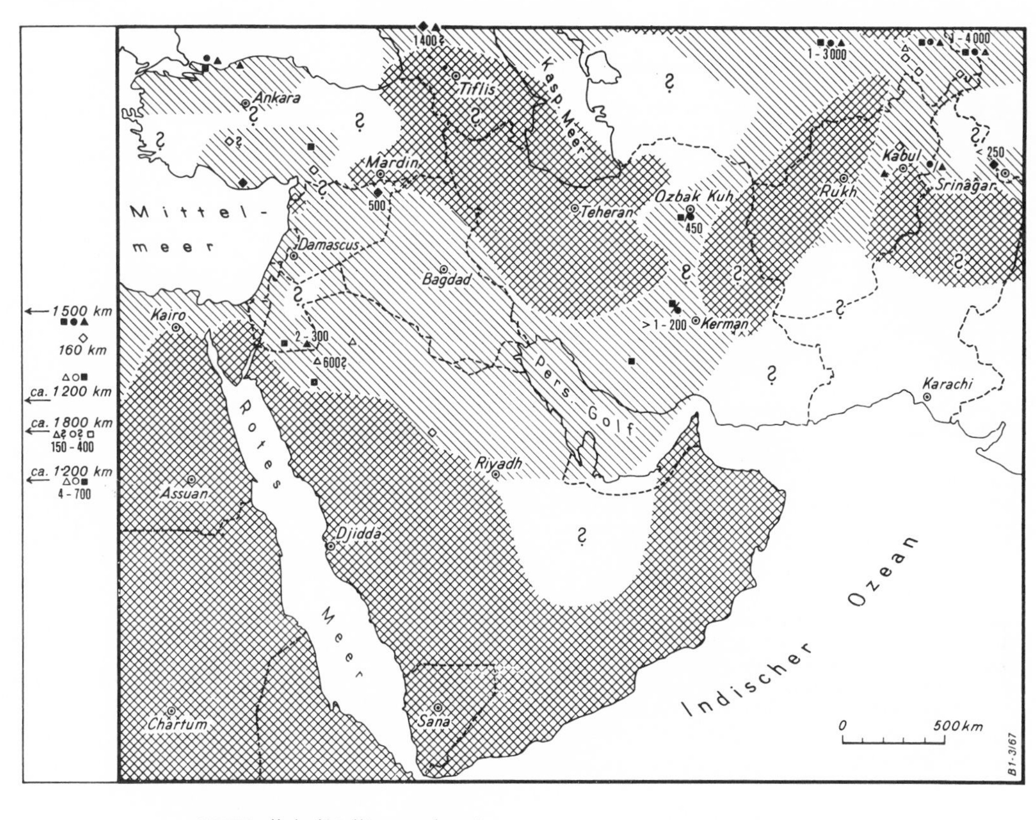

Hochgebiet: Abtragung oder geringe kontinentale Sedimentation

marines Silur

▲ marine Fauna des Ludlow

● marine Fauna des Wenlock

■ marine Fauna des Llandovery

◆ marines Silur

△○□◇ Fundorte silurischer Ablagerungen

Mächtigkeit in Metern

Abb. 3. Die Tethys im Silur

tadschikischen Geosynklinalen wurden über 4000 m mächtige Ablagerungen gebildet.

Im Silur verstärkte sich die gebietsweise bereits seit dem mittleren Ordovizium erkennbare Tendenz zur Einschnürung der vorderasiatischen Tethys-Meeresstraße in zunehmendem Maße. Nach dem in weiter Verbreitung marin entwickelten Abschnitt des Unter-Llandovery mußte sich vom höheren Llandovery bis zum Ausgang des Silurs eine fortschreitende Einengung des überfluteten Gebietes vollzogen haben, nur unterbrochen durch einen kurzen marinen Abschnitt im Ludlow. Über die Ausdehnung des Ludlow-Meeres ist wenig bekannt. Die marine Sedimentation spielte sich während des Silurs in einem vergleichsweise klar begrenzten Raum zwischen dem Nubisch-Arabischen Schild im S und einem großen, vom Schwarzen Meer und Mardin bis weit in den Zentral-Iran sich erstreckenden Hochgebiet, das im höheren Silur möglicherweise mit dem beständigen afghanischen Hochgebiet in Verbindung stand, ab. Im N wurde das iranisch-armenische Hochgebiet von der Kaukasus-Geosynklinale begrenzt, im NW verlief die Küstenlinie im Llandovery irgendwo in der E-Türkei. Im weiteren Verlauf des Silurs zog sich das Meer vorübergehend möglicherweise ganz aus S-Anatolien zurück. — Im Silur dürfte im gesamten Bereich der vorderasiatischen Tethys und der geosynklinalen Grenzgebiete nur geringe vulkanische Tätigkeit geherrscht haben.

Anscheinend im gesamten Bereich der vorderasiatischen Tethys herrscht konkordante Lagerung zwischen den Schichten des älteren Paläozoikums, obwohl sich gebietsweise große Schichtlücken oder rasche vertikale Faziesänderungen nachweisen lassen. Gesicherte Beweise für alpinotype kaledonische Bewegungen gibt es also nicht (*Flügel*, 1964). Die Schichtlücken und Faziesänderungen gehen jedoch auf epirogene Bewegungen zurück, die zeitlich etwa mit den Kaledonischen Faltungsphasen zusammenfallen. Epirogene Hebungen, die der altkaledonischen Faltungsphase zugeordnet werden könnten, machten sich in einzelnen Gebieten bereits vom mittleren Ordovizium an bemerkbar, nämlich im Elburs-Gebirge, im Raum von Mardin und wahrscheinlich auch in SE-Arabien (Oman). In den meisten Gebieten der vorderasiatischen Tethys setzte sich die ordovizische Sedimentation jedoch unverändert bis zum Unter-Silur fort. Epirogene Hebungen der jungkaledonischen Zeit dürften im Tethysbereich schon im Spät-Llandovery begonnen und sich — abgesehen von vergleichsweise kurzfristigen rückläufigen Bewegungen im Ludlow — bis zum Ausgang des Silurs noch verstärkt haben. Diese Hebungen erzeugten schließlich im Unter-Devon in fast der ganzen vorderasiatischen Tethys kontinentale Verhältnisse. — Abweichend davon vollzog sich die Entwicklung in den geosynklinalen Grenzgebieten im N. Im Raume östlich von Istanbul deutet die diskordante Überlagerung ordovizischer durch devonische Schichten auf altkaledonische Bewegungen. Ebenso sollen auch im N-Kaukasus die ersten orogenetischen Bewegungen gegen Ende des Ordoviziums stattgefunden haben. Mit der altkaledonischen Phase des N-Kaukasus wird ein Magmenkomplex aus Granitgneisen und pegmatitischen Gesteinen in Verbindung gebracht (*Paffengolz*). Einen wiederum abweichenden Verlauf nahm die Entwicklung im südlichen Tadschikistan (Pamir usw.). Hier ereigneten sich während der kaledonischen Zeit gebietsweise zwar Hebungen, doch vollzog sich der Übergang von der kaledonischen zur varistischen Ära unter fortwährender Absenkung (*Krestnikow*).

2.4. Devon (*Abb. 4*)

Nachdem sich bereits im Verlaufe des Silurs die regressive Tendenz im größten Teil der vorderasiatischen Tethys immer mehr verstärkt hatte, herrschten während des Unter-Devons arid-kontinentale oder stark kontinental beeinflußte Verhältnisse, unter denen es vor allem zur Ablagerung von roten, schräggeschichteten Sandsteinen, hellen Quarziten, sandigen Tonschiefern und Konglomeraten kam. Aus dem Fehlen von Netzleisten, Steinsalz-Pseudomorphosen, Arthropoden-Spuren und Phyllopoden schlossen *Huckriede* et al. auf marine Beeinflussung des unterdevonischen Sedimentationsraumes von Kerman. Im oberen Teil dieser dem „Old Red" vergleichbaren Gesteinsserie, der auch noch älteres Mittel-Devon umfassen kann, wurde im Raume von Kerman ein Gipshorizont nachgewiesen. In ähnlicher stratigraphischer Position ist auch bei Ozbakh Kuh und im östlichen Elburs-Gebirge ein evaporitischer Horizont entwickelt. Ablagerungen der unterdevonischen Old-Red-Fazies wurden ferner aus dem N-Irak, dem Taurus und in sehr geringer Mächtigkeit auch aus dem Bereich der mittelsyrischen Senke beschrieben. In Syrien dürften diese Ablagerungen jedoch — wie möglicherweise auch in W-Anatolien — das gesamte Devon vertreten. Aus Sinai berichtete *Saad* über fossile Pflanzenreste aus dem „Nubischen Sandstein", die auf Devon hindeuten. In Libyen liegt das Unter-Devon vorwiegend in kontinentaler Entwicklung vor; lediglich in den nördlichen Becken machen sich marine Einflüsse bemerkbar.

Eine eigenartige Stellung nimmt das marine Unter-Devon des Jauf-Gebietes im nördlichen Saudi-Arabien ein, das nach *Cooper* (in *Steineke* et al., l. c.) auf Grund von Anathyris und „Rensselaeria" eingestuft wurde. Es handelt sich dabei um Ton- und Sandsteine mit Gipslagen und fossilführende Kalke. Wie im Ludlow kann auch in diesem Falle noch nicht festgestellt werden, auf welchem Wege das Meer bis nach Jauf vorgedrungen ist. — Im Gegensatz zum vorderasiatischen Tethysbereich herrschten in den geosynklinalen Grenzgebieten des Nordens ganz oder zeitweise marine Sedimentationsbedingungen im Unter-Devon. Dies ist im Raume von Istanbul und in Tadschikistan durch zahlreiche Fossilfunde belegt, während im N-Kaukasus die unterdevonischen Schichten vorwiegend aus vulkanogenen Gesteinen mit Einschaltungen von ?marinen klastischen Gesteinen bestehen. Bei Istanbul ist Emsium in karbonatischer Entwicklung durch das Vorkommen verschiedener Arten von Acastoides, einigen Brachiopoden und tabulaten Korallen nachgewiesen. In Tadschikistan enthalten mächtige Kalke eine reiche unterdevonische Brachiopoden-Fauna. Im ostafghanischen Trog, der vielleicht als ein den tadschikischen Geosynklinalen angegliedertes Senkungsgebiet zu betrachten ist, unterlagern mächtige, fossilarme dolomitische Kalke kalkige Schichten mit einer Fauna von altmitteldevonischem Gepräge. Möglicherweise handelt es sich bei diesen dolomitischen Kalken um Unter-Devon.

Im Verlaufe der mittel- bis oberdevonischen Transgression wurden vor allem im Bereich des türkisch-afghanischen Archipels Gebiete überflutet, die mindestens seit dem jüngeren Kambrium Hochgebiete gewesen waren. Dagegen hatte sich das Meer aus den alten Sedimentationsgebieten am Nordrand des Nubisch-Arabischen Schildes zurückgezogen. Spätestens im Ober-Devon, nachdem auch das Elburs-Gebirge bis auf geringe Reste überflutet war, bestand wieder eine breite Meeresstraße zwischen der Türkei und Afghanistan, die auch tiergeographisch Beziehungen zum pazifischen Raum aufwies. Die altpaläozoische Tethys war somit wieder hergestellt, wenn auch unter Verlagerung nach N. — Im älteren Paläozoikum herrschte im vorderasiatischen Tethysbereich — abweichend von den geosynklinalen Grenzgebieten — klastische Sedimentation vor. Mit der mitteldevonischen Transgression setzte nun auch im Tethysbereich in einem verstärkten Maße kalkige Sedimentation ein, die vielfach erst im Unter-Karbon endete. Außerdem gelangte allerdings in den meisten Gebieten immer wieder auch sandig-toniges Material zur Ablagerung, z. T. mit einem Kalkanteil unterschiedlicher Höhe. Am NW-Rand des Nubisch-Arabischen Schildes, in Libyen, dauerte demgegenüber die klastische Sedimentation des älteren Paläozoikums während des ganzen Devons an.

Die Zusammensetzung der mittel- bis oberdevonischen Fauna des vorderasiatischen Tethysraumes und seiner geosynklinalen Grenzgebiete — die Fauna besteht aus einer individuen- und artenreichen Vergesellschaftung von Korallen und Brachiopoden sowie anderen Tiergruppen — deutet auf ziemlich einheitliche Lebensbedingungen in dem gesamten Gebiet hin. Oberdevonische Schichten sind in großen Gebieten durch massenhaftes Auftreten von Cyrtospiriferiden, Rhynchonelliden und Productelliden gekennzeichnet. In den klastischen Gesteinen Libyens sind hauptsächlich Brachiopoden, Tentaculiten und Spurenfossilien vertreten, während Korallen fehlen.

Nach der Mächtigkeit der Ablagerungen und dem Auftreten eines geosynklinalen Diabas-Vulkanismus unterscheiden sich die geosynklinalen Grenzgebiete deutlich von dem Bereich der vorderasiatischen Tethys. Über basischen bis intermediären Vulkanismus im Unter-Devon wird allerdings auch aus dem Gebiet von Kerman und aus dem Elburs-Gebirge berichtet. Im ostafghanischen Trog sind zwar mächtige Devon-Schichten entwickelt, doch fehlen hier Anzeichen eines Vulkanismus. Im N-Kaukasus dagegen entstand im Unter- und Mittel-Devon eine über 3000 m mächtige vulkanogene Serie aus Diabasen, Porphyriten und Tuffen mit Einschaltungen von Phylliten und Sandsteinen.

Über lebhaften geosynklinalen Diabas-Vulkanismus wird auch aus der NW-anatolischen Kernzone des varistischen Gebirges berichtet (*Brinkmann*), wobei jedoch über das Alter keine genauen Aussagen gemacht werden können. Es wäre verlockend, diesen Vulkanismus als Äquivalent des nordkaukasischen Vulkanismus zu betrachten. Auch aus den Geosynklinalen Tadschikistans ist basischer und porphyritischer Vulkanismus bekannt.

2.5. Karbon (*Abb. 5*)

Die Sedimentationsverhältnisse des Mittel- und Ober-Devons dauerten im Unter-Karbon unverändert an. In den aus dem Devon bekannten marinen Ablagerungsräumen kam es zur Bildung von Kalken und Sandsteinen unterschiedlicher Mächtigkeit. Lediglich im N-Irak, in Syrien und auf der Halbinsel Sinai sind größere Gebiete neu überflutet worden; im übrigen ähnelte die Verteilung von Land und Meer während des Unter-Karbons derjenigen des Devons. Die Zusammensetzung der Fauna spiegelt im marinen Bereich des unterkarbonen Tethys-

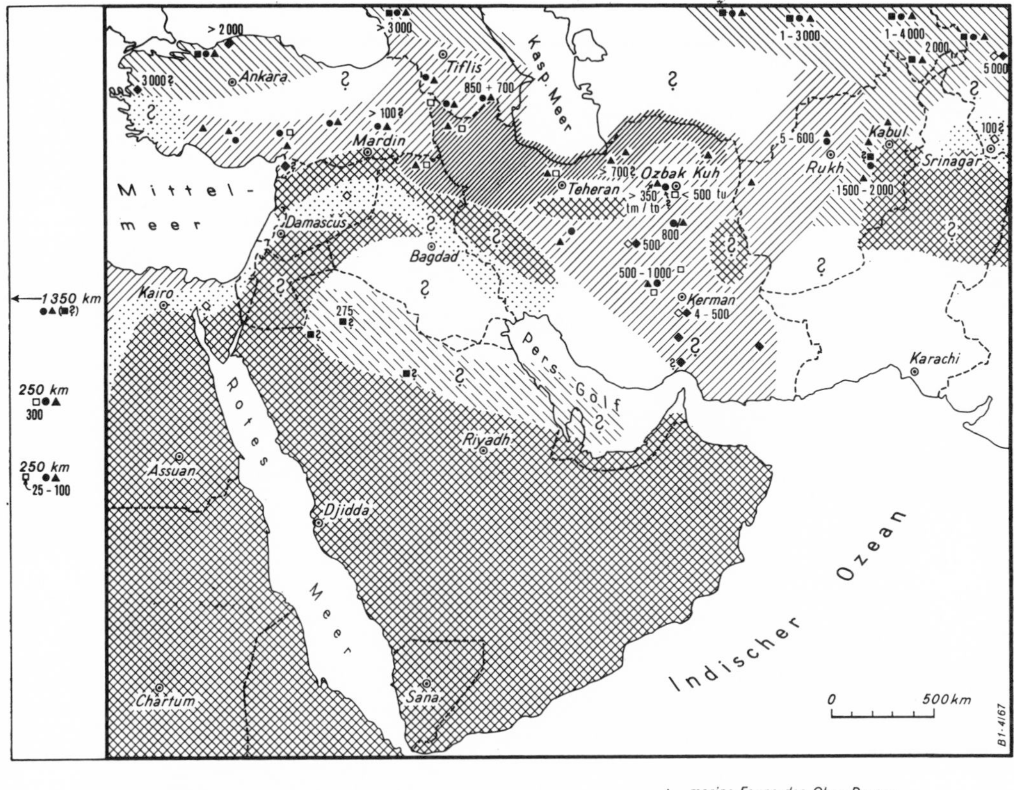

Hochgebiet: Abtragung oder (geringe) kontinentale Sedimentation

kontinentale Sedimentation

kontinentales Unter- und Mittel-Devon, marines Ober-Devon

kontinentales oder stark kontinental beeinflußtes marines Unter-Devon, marines Mittel- und Ober-Devon

marines Unter-Devon, im Mittel- und Ober-Devon kontinentale Verhältnisse

marine Sedimentation während des ganzen Devons

▲ marine Fauna des Ober-Devons

● marine Fauna des Mittel-Devons

■ marine Fauna des Unter-Devons

◆ marines Devon

◇ kontinentales Devon

□ kontinentales Unter-Devon

Mächtigkeit in Metern

Abb. 4. Die Tethys im Devon

raumes Vorderasiens die faziellen Verhältnisse wider. An weit voneinander entfernten Fundpunkten tritt eine Vergesellschaftung von Brachiopoden-Gattungen auf, die eine erstaunliche Einheitlichkeit der tiergeographischen Verhältnisse erkennen läßt. In fast allen der im folgenden genannten Gebiete, nämlich in S-Anatolien, NE-Syrien, Armenien, im Elburs-Gebirge, bei Kerman, in Kaschmir und in Libyen, tritt eine Brachiopoden-Gemeinschaft auf, die für die Stufen Tournai-Visé bezeichnend ist.

Es handelt sich dabei um: Leptagonia analoga (*Phillips*), Schuchertella sp. (z. T. wohl auch als Orthothetes crenistria bezeichnet), Echinoconchus punctatus (*Sow.*), Dictyoclostus, „Camarotoechia", Prospira, „Spirifer tornacensis" *De Kon.*, Syringothyris cuspidata (*Martin*), Athyris lamellòsa (*L'Ev.*), Composita, Dielasma u. a.

Diese Brachiopoden sind auch in Europa, im Moskauer Becken und Australien weit verbreitet. Außerdem sind Korallen (z. B. Zaphrentiden, Cladochonus, Michelinia, Caninophyllum usw.) sowie Bryozoen häufig mit den Brachiopoden vergesellschaftet.

In Zentral- und W-Afghanistan dürfte die Grenze Devon/Karbon etwa über den Schichten mit Phacops accipitrinus (*Phill.*) liegen. Während das aus dickbankigen Kalken bestehende Unter-Karbon dieser Gebiete arm an Makrofossilien ist, weisen gleichalte Schichten des ostafghanischen Troges eine an Brachiopoden, vor allem Productiden, Spiriferiden und Einzelkorallen, reiche Fauna auf. In dieser Fauna sind die oben genannten Formen aus dem vorderasiatischen Tethysbereich nicht enthalten; sie weist eher Beziehungen zu den tadschikischen Geosynklinalen auf. — In den geosynklinalen Grenzgebieten liegt das Unter-Karbon durchweg in mariner, z. T. kalkiger, z. T. sandig-toniger Fazies vor. Hier und da machten sich allerdings frühvaristische Bewegungen bemerkbar, wie im Raume von Istanbul und überhaupt in N-Anatolien sowie gebietsweise auch in SE-Tadschikistan, wodurch es zur Ablagerung von klastischen Sedimenten mit Pflanzenresten und tuffogenen Schichten kam. Bei Istanbul sind allerdings auch Visé-Kalke über jenen kontinentalen Schichten vertreten.

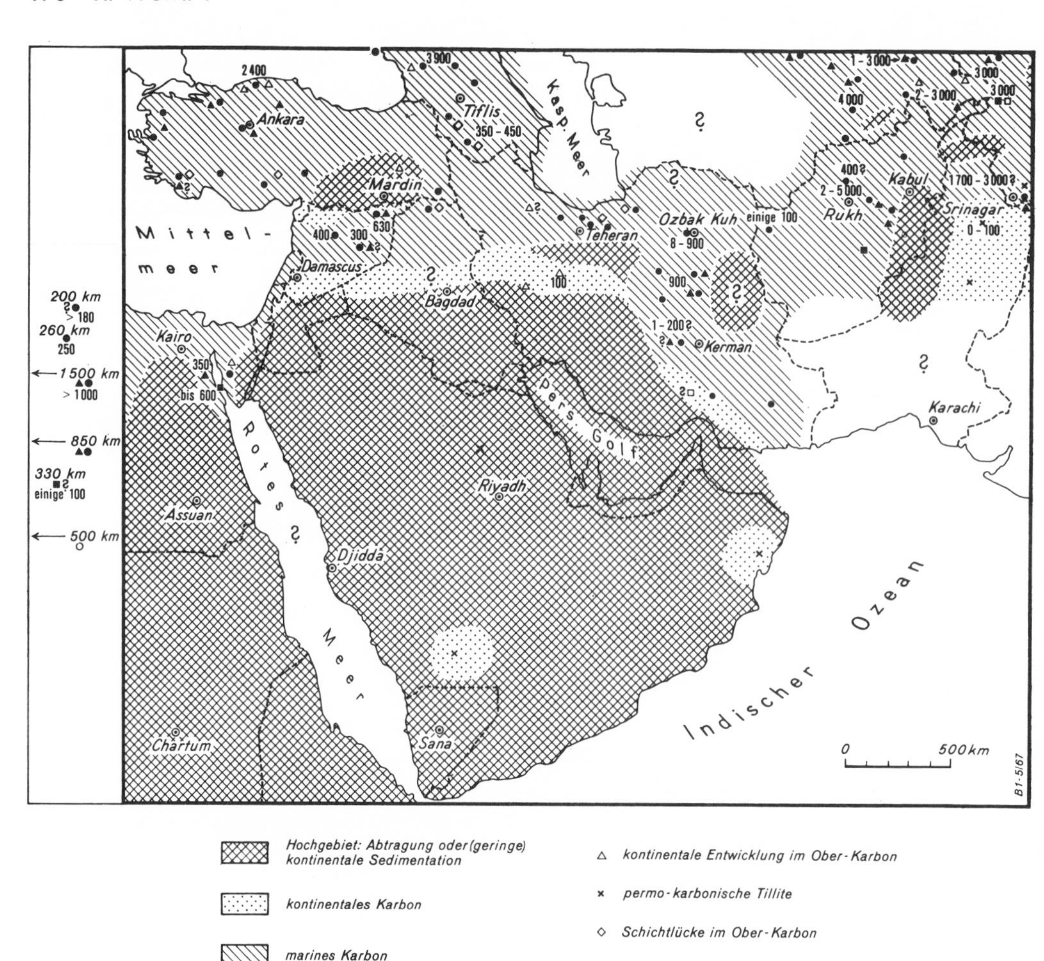

Abb. 5. Die Tethys im Karbon

Es dürfte den Auswirkungen der varistischen Orogenese zuzuschreiben sein, daß vom Grenzbereich zum Ober-Karbon an die weitere Entwicklung unterschiedlich verlief. Im Raume von Ankara liegen über epi- bis mesozonalen Metamorphiten Konglomerate, Sandsteine und Kalke mit einer Visé-Fauna (*Flügel*, 1964), die beweisen würde, daß die von *Brinkmann* angenommene varistische Orogenese im Prä-Visé stattgefunden haben müßte. *Brinkmann* stufte diese Gesteine, die Visé-Kalkschollen enthalten, als Ober-Karbon ein. Im N, bei Zonguldak, sind mächtige marine Karbonate mit terrestrischen Einschaltungen (Asterocalamites, Lepidodendron) entwickelt. Die Fauna der geosynklinalen Grenzbereiche besteht zwar aus einer Vergesellschaftung von Brachiopoden, Korallen usw., die im Tethys-Bereich verbreiteten bezeichnenden Gattungen sind darin anscheinend jedoch nicht enthalten.
Ganz allgemein machte sich bereits im jüngeren Unter-Karbon eine regressive Tendenz bemerkbar, die schließlich im Tethys-Bereich zum Rückzug des Meeres führte, u. zw. aus Anatolien, dem N-Irak, Armenien, dem Elburs-Gebirge und vermutlich auch aus W- und Zentral-Afghanistan (Rukh). Über älteres Ober-Karbon mit Gastrioceras (Branneroceras) branneri *Smith*, Productiden und Spiriferiden berichteten *Stöcklin* et al. (1965) aus dem E-Iran. Die Ablagerungen bestehen überwiegend aus Sand- und Tonsteinen mit Einschaltungen von Kalksteinen. Bei Kerman dürfte Ober-Karbon in dolomitischen, fossilleeren Gesteinen zu sehen sein, die sich nach oben nicht abgrenzen lassen. — Marines fossilführendes Ober-Karbon wurde auch im W des vorderasiatischen Tethys-Bereiches nachgewiesen, u. zw. vor allem in NE-Syrien, in einigen Gebieten der Türkei, im Suez-Gebiet und ferner auch in Libyen. In NE-Syrien besteht das Karbon nach Tiefbohrungen aus über 600 m mächtigen Schiefertonen und Sandsteinen, die nach palynologischen Untersuchungen alle Karbon-Stufen enthalten sollen. Allerdings herrschte offenbar ein lebhafter Wechsel zwischen marinem und kontinentalem Milieu, wie das Auftreten von Pila-Algenkolonien einerseits und Brachiopoden andererseits erweist. Am Hacertun Dagh, SE-Anatolien, ist klastisches kohleführendes Ober-Karbon (West-

fal) in limnischer Entwicklung diskordant über Devon ausgebildet. Auf Grund mikropaläontologischer Untersuchungen stuften *Omara* u. *Kenawy* das marine Ober-Karbon des westlichen Suez-Gebietes als Westfal und Unter-Stephan ein. Es handelt sich dabei um Kalksteine und Schiefertone, die in Sandsteine von Nubischer Fazies eingeschaltet sind. Auch in Libyen macht sich ein Wechsel zwischen marinem und kontinentalem Milieu bemerkbar. Vom Visé bis zum Westfal bildeten sich Sandsteine mit Lepidodendron im Wechsel mit Brachiopoden führenden Ton- und Kalksteinen. Im ausgehenden Ober-Karbon dürfte der kontinentale Einfluß zunehmen. Schichten des Stephans sind bisher nicht beobachtet worden.

In den oberkarbonen Sedimentationsverhältnissen der geosynklinalen Grenzbereiche im nördlichen Anatolien und im Kaukasus spiegelt sich das große Ereignis der varistischen Orogenese. In N-Anatolien, bei Ankara, Bursa usw. liegen wildflyschähnliche oberkarbone Grauwacken mit Visé-Kalkschollen diskordant über epimetamorphen Schiefern des Alt-Paläozoikums (*Brinkmann*). Gleiches gilt für den N-Kaukasus, wo Konglomerate und Sandsteine sowie Effusiva über dem metamorphen altpaläozoischen Komplex lagern. Die Mächtigkeit der kohleführenden Serie des Ober-Karbons erreicht im N-Kaukasus 1100 m. Noch mächtiger (1500 m) ist die oberkarbone flözführende Serie von Zonguldak am Schwarzen Meer, die auf Grund fossiler Pflanzenreste als Namur bis Stephan eingestuft wird. Abweichend vom Kaukasus und der Kernzone des varistischen Gebirges in N-Anatolien folgt das kontinentale Ober-Karbon von Zonguldak konkordant über den nichtmetamorphen Schichten des älteren Paläozoikums. Es entstand nach *Brinkmann* in der nördlichen Saumsenke des varistischen Gebirges. Auch in den tadschikischen Gebieten sprechen viele Anzeichen dafür, daß sich orogenetische Vorgänge etwa an der Wende Unter-/Ober-Karbon abgespielt haben, die jedoch nicht mit einer Metamorphose der älteren Schichten verbunden waren. So wird von Effusiva und Granitintrusionen im nördlichen Pamir berichtet, von Hebungen, Konglomeraten und Molasse, die durch eine Schichtlücke von älteren Sedimenten getrennt sind. Ober-Karbon ist nicht überall in mariner Entwicklung vertreten. Die marinen Schichten führen Productiden und Fusuliniden und sollen, außer in Tadschikistan, auch in NE-Afghanistan und in Kaschmir vorhanden sein. Im ostafghanischen Trog deuten sich varistische Bewegungen in Konglomerathorizonten und einem Fazieswechsel im höheren Unter-Karbon an. Über den fossilreichen Kalken des tieferen Unter-Karbons folgen hier mächtige marine Schiefertone und Sandsteine, die ihrerseits von mächtigen Kalken und Dolomiten überlagert werden. Wenn sich auch die oberkarbonen nicht gegen die permischen Schichten abgrenzen lassen, und wenn auch die bisher vorliegende spärliche Brachiopoden-Fauna nicht als spezifisch für Ober-Karbon anzusehen ist, so dürfte doch — zumal auf Grund der Funde von Humilogriffithides (*Lapparent* et al., 1965) — an dem Vorkommen mariner oberkarboner Ablagerungen im ostafghanischen Trog nicht zu zweifeln sein. Am Nordrand des Gondwana-Landes wurden im Karbon gebietsweise, wie z. B. im Zagros-Gebirge (W-Iran), bis mehrere 100 m mächtige kontinentale Schichten mit Pflanzenresten (Sigillaria persica *Seward*) abgelagert, die konform über dem Kambrium folgen.

Im Bereich des gondwanischen Festlandes, u. zw. in W-Pakistan, Oman und SW-Saudi-Arabien, wurden verschiedentlich erratische Blöcke — „boulder beds" — beobachtet, die mit der permokarbonischen Vereisung in Zusammenhang gebracht werden (*Morton*, *Helal*). Nach *Gansser* transgredieren die Talchir-Tillite in Kaschmir bereits im tieferen Ober-Karbon über marinem Unter-Karbon. Sie werden von Trapp-Vulkaniten und unterpermischen Sandsteinen mit Gangamopteris überlagert.

2.6. Perm (*Abb. 6*)

Mit der devonisch-unterkarbonischen Transgression war das Endstadium der nach N fortschreitenden Verlagerung der vorderasiatischen Tethys erreicht, die während der kaledonischen Ära vorbereitet worden war. Epirogenetische Bewegungen der varistischen Ära bewirkten eine Umkehrung der Verlagerungsrichtung, so daß fast das gesamte Gebiet im SW-Iran und in Arabien, das seit dem höheren Ordovizium schrittweise dem Nubisch-Arabischen Festland angegliedert worden war, im Verlaufe des Perms wieder überflutet wurde. Der im Alt-Paläozoikum so deutliche türkisch-afghanische Archipel muß wohl im Perm — wenn auch in kleinerem Ausmaße — in ähnlicher Weise bestanden haben.

Nachdem große Gebiete der Tethys bis zum höheren Ober-Karbon unter kontinentalen Einfluß geraten waren, kehrte das Meer während des Unter-Perm in den Kernbereich des vorderasiatischen Tethysraumes — Iran, Oman, Armenien, Teile von S-Anatolien — wieder zurück, ohne daß jedoch eine unmittelbare Meeresverbindung zwischen den südöstlichen und den westlichen Teilen des Tethysraumes geschaffen wurde. Im Elburs-Gebirge lagern unterpermische Sand- und Kalksteine mit Pseudoschwagerina, Zellia, Yatsengia, Brachiopoden usw. konkordant über Unter-Karbon-Kalken. Im E-Iran sind Korallenkalke und Dolomite von vermutlich mittelpermischem Alter (Yatsengia, Liangshanophyllum, Schwagerina, Climacammina sowie Neoproetus cf. indicus) durch eine Schichtlücke von älterem Ober-Karbon getrennt. In Armenien sind Unter- und unteres Ober-Perm, wie im Iran, in Korallen-Fusuliniden-Fazies entwickelt. Aus dem S-Oman sind Sandsteine und Kalke mit einer reichen Brachiopoden-Trilobiten-Goniatiten-Fauna der Sakmara-Artinsk-Stufe bekannt. Im südwestlichen Saudi-Arabien kommen bis 300 m mächtige, kontinentale oder kontinental beeinflußte Sandsteine vor, die auf Grund der Überlagerung durch die über 200 m mächtigen oberpermischen Kalke als „Unter-Perm oder älter" bezeichnet werden. — Das Gebiet der Salt Ranges im NW-Vorland des Indischen Schildes erlebte im Perm zum ersten Male seit dem Kambrium wieder eine marine Transgression. Über dem Kambrium folgt hier der nach *Gansser* oberkarbone Talchir-Tillit, darüber der wohl größtenteils oberpermische Productus limestone. Im Unter-Perm dürfte die marine Sedimentation nur vergleichsweise schwach gewesen sein. In Kaschmir ist das Unter-Perm sandig-schiefrig entwickelt und enthält Gondwana-Floren mit Gangamopteris im tieferen und Glossopteris im höheren Teil. — Im westlichen Teil des Tethysraumes ist marines Unter-Perm bisher nur vereinzelt nachgewiesen, wie z. B. Korallenkalke in SW-Anatolien.

Vom Hauptbereich des Tethysraumes abweichend verlief die permische Entwicklung am NW-Rand des Nubisch-Arabischen Schildes sowie in den nordwestlichen geosynklinalen Grenzgebieten. Palynologische Untersuchungen ergaben (*Schuermann* et al.), daß im westlichen Suez-Gebiet oberstes Karbon bis tiefstes Perm in Nubischer Fazies (mit limnischen Einschaltungen) entwickelt ist. In Libyen transgredieren kontinentale Sand- und Tonsteine des ?Unter-Perms über verschiedenalte paläozoische Formationen. Am Schwarzen Meer, bei Zonguldak, gelten rote, z. T. grobklastische Ablagerungen als kontinentale Bildungen des Perms. Im N-Kaukasus wurde im Unter-Perm eine mächtige rote Molasse-Serie gebildet, die, wie auch bei Zonguldak, von heftigen tektonischen Bewegungen während der Sedimentation zeugt. Lediglich in den Geosynklinalen Tadschikistans kam es im Unter-Perm auch weiterhin zur Ablagerung mächtiger mariner Sand- und Tonsteine, untergeordnet auch von Kalksteinen. Im ostafghanischen Trog

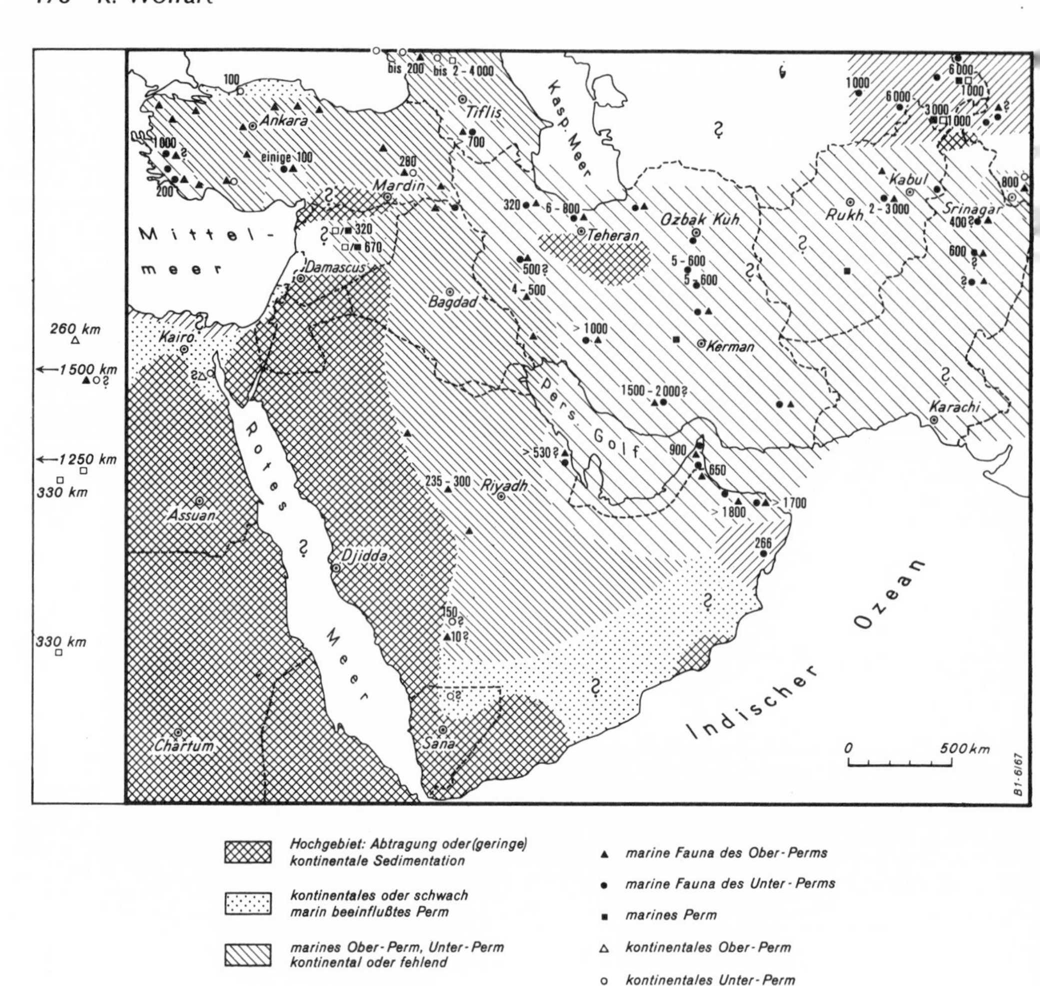

Abb. 6. Die Tethys im Perm

wurden außer mächtigen klastischen Sedimenten auch mächtige Kalke abgelagert, deren Einstufung z. Z. noch Gegenstand verschiedener Untersuchungen ist. *Flügel* (1965) beschrieb eine Korallenfauna — Wentzelellites molengraaffi (*Gerth*), Waagenophyllum indicum indicum (*Waagen* u. *Wentzel*), Iranophyllum —, die in anderen Gebieten anscheinend auf tieferes Perm hindeutet. Die Fauna wird als Bindeglied zwischen den ost- und den vorderasiatischen Korallen-Vorkommen gewertet. Auf unterpermisches Alter weisen Funde von cf. Yatsengia asiatica *Huang* in Kalksteinen östlich von Kabul, bei Jalalabad, hin.

Im Ober-Perm griff das Meer weit nach W und NW, auf Saudi-Arabien, die Türkei und den Kaukasus, über; im E wurde auch das Vorland des Indischen Schildes, W-Pakistan und Kaschmir, weitgehend überflutet, so daß die Tethys schließlich im Ober-Perm ihre größte Ausdehnung während des gesamten Paläozoikums erreichte. Lediglich Jordanien und NW-Saudi-Arabien blieben, abweichend vom Alt-Paläozoikum, Hochgebiet. Sichere oberpermische Kalke und schiefrig-sandige Gesteine mit einer reichen indo-armenischen Brachiopodenfauna (Costiferina, Marginifera, Leptodus, Enteletes u. a. m.) wurden außer in Armenien auch im Elburs-Gebirge nachgewiesen (*Glaus*). Carbonatische Entwicklung des Ober-Perms herrscht auch im W-Iran und im E-Iran. Bei Kerman und Ozbakh Kuh ließ sich Ober-Perm faunistisch jedoch nicht belegen; es hat den Anschein, daß im östlichen Iran mit Schichtlücken im Perm gerechnet werden muß. Im nordwestlichen Vorland des Indischen Schildes besteht der tiefere Teil des hauptsächlich wohl oberpermischen Productus limestone aus massiven Korallenkalken, der obere Teil aus gebankten, Brachiopoden und Ammoniten führenden Kalken. In Kaschmir verzahnt sich der Productus limestone mit Trapp-Vulkaniten. Oberpermische Kalke ohne Fusulinen sind in von W nach E zunehmender Mächtigkeit in der gesamten Osthälfte der Arabischen Halbinsel verbreitet. Fossilarme Schiefertone und Sandsteine sowie Carbonate, die durch Tiefbohrungen in der mittelsyrischen Senke erschlossen wurden, sind mit Vorbehalt dem Perm zuzuordnen; als einzige Fossilien wurden daraus bisher Ammodiscus incerta *D'Orb.* und Hollinella erwähnt. Nordöstlich von Mardin transgredieren über limnischem Ober-

Karbon oberpermische Sandsteine, die Elemente der cathaysianischen *Gigantopteris*-Flora und der gondwanischen *Glossopteris*-Flora enthalten. Darüber folgen marine Kalke des Ober-Perms mit Brachiopoden, *Bellerophon*, Algen usw., die auch in W-Anatolien anzutreffen sind.

In Libyen ist das Ober-Perm vorwiegend kontinental entwickelt, außer in den nördlichen Becken, wo sich mariner Einfluß in dolomitischen Gesteinen, die Elemente der *Streblospira*- und *Hemigordius*-Zone führen, bemerkbar macht.

Während sich bei Zonguldak am Schwarzen Meer Ober-Perm überhaupt nicht nachweisen läßt — es muß in diesem Gebiet, ebenso wie das Unter-Perm, in über 100 m mächtigen roten, klastischen Sedimenten enthalten sein —, zeugen im N-Kaukasus Kalke mit Foraminiferen (*Parafusulina*, *Neoschwagerina* usw.) von mariner Sedimentation. In den tadschikischen Geosynklinalen fand oberpermische tonig-sandige Sedimentation nur noch in Restsenken statt, im ostafghanischen Trog wurden noch mächtige Fusulinenkalke gebildet.

Im Perm sind wiederum deutliche Unterschiede in der Entwicklung des vorderasiatischen Tethysraumes einerseits und der geosynklinalen Grenzgebiete andererseits festzustellen. Der Tethysraum wurde im Verlauf der unter- und der oberpermischen Transgression so weitgehend überflutet, daß die Meeresausdehnung diejenige des Kambriums gebietsweise noch übertraf. Wie die Verteilung der Sedimentmächtigkeit zeigt, gliederte sich das Gesamtgebiet im Perm — ebenso wie im Kambrium und Ordovizium — in den Nubisch-Arabischen Schild im SW, eine SE—NW streichende Zone vergleichsweise starker Absenkung im Bereich zwischen E-Arabien und dem SW-Iran, den türkisch-afghanischen Archipel und die im N anschließenden Grenzgebiete geosynklinalen Gepräges. Die permischen Transgressionen im Tethysraum können der Ausdehnung nach am ehesten mit den kambrisch-ordovizischen verglichen werden. Beide Transgressionen überfluteten Landgebiete, die auf Grund von Hebungen im Gefolge der spätkambrischen bzw. varistischen Orogenese entstanden waren. Die postkaledonische Transgression erlangte demgegenüber nur vergleichsweise geringe Ausdehnung. — Die geosynklinalen Grenzgebiete in Tadschikistan und im N-Kaukasus zeichneten sich zumal im Unter-Perm durch große Absenkungsunterschiede aus. Während im Kaukasus im Ober-Perm eine Beruhigung der vertikalen Bewegung eingetreten war, verband sich die oberpermische Sedimentation im Pamir mit porphyritischen Eruptionen, die zur Bildung von über 3500 m mächtigen Vulkaniten führten. In N-Anatolien herrschten offenbar nach der von *Brinkmann* postulierten varistischen Orogenese im Perm i. w. stabile Verhältnisse.

3. Bemerkungen zur Struktur der paläozoischen Tethys (*Abb. 7*)

Die paläozoische Tethys Vorderasiens entwickelte sich im Bereich des dem Gondwanaland im N vorgelagerten mobilen Schelfes. Nachdem im ganzen nördlichen Vorland des Gondwana-Kontinentes während des späten Prä-Kambriums und tieferen Kambriums, also im Gefolge der assyntischen Orogenese, im wesentlichen gleichartige Verhältnisse geherrscht hatten, stellten sich im südlichen Teil des Vorlandes, in dem Zeitraum zwischen dem mittleren Kambrium bis tieferen Ordovizium, Verhältnisse des mobilen Schelfes ein. Gleichzeitig begann im nördlichen Teil des Vorlandes — N-Anatolien, N-Kaukasus, Tadschikistan — die Entwicklung umfangreicher Geosynklinalen. Dadurch war der Nordrand der vorderasiatischen Tethys, also des dem Gondwanaland zugeordneten mobilen Schelfes, festgelegt. In dem so abgegrenzten Nordsaum des Gondwanalandes überdauerten die Verhältnisse des mobilen Schelfes die kaledonische und varistische Ära. Erst im Mesozoikum, mit Beginn der alpidischen Ära, vollzogen sich in der strukturellen Gliederung des Gondwana-Nordsaumes tiefgreifende Änderungen, indem sich außer dem teilweise rejuvenierten nördlichen Geosynklinal-Gürtel des Paläozoikums auch im S des paläozoischen mobilen Schelfes, u. zw. in einem schon frühzeitig vorgezeichneten Bereich, ein weiterer Geosynklinal-Gürtel bildete.

Die paläozoische Geschichte der Tethys gliedert sich in drei Abschnitte. Im postassyntisch-kaledonischen Abschnitt verlief die Kernzone der Tethys-Meeresstraße zwischen E-Arabien und dem SW-Iran von SE nach NW und bog etwa im Gebiet von Syrien nach W bis SW um. Der türkisch-afghanische Archipel war deutlich entwickelt und erhielt durch Teilregressionen bis zum Ausgange des Silurs immer mehr Landzuwachs. Im postkaledonisch-varistischen Abschnitt verlagerte sich die

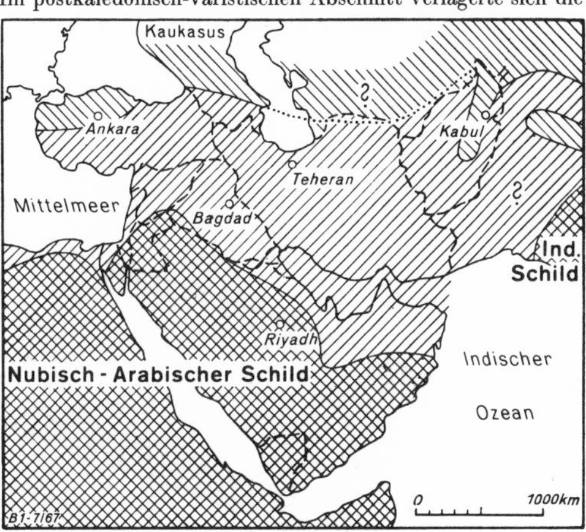

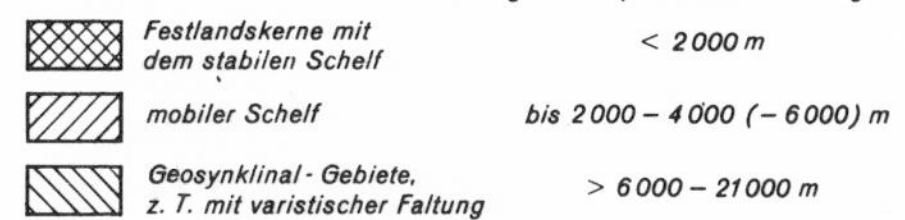

Abb. 7. Strukturelle Gliederung Vorderasiens im Paläozoikum

Tethys-Meeresstraße in den nördlichen Bereich des mobilen Schelfes. Dabei wurden große Teile des altpaläozoischen Inselgürtels, besonders in Afghanistan und im Iran, überflutet, während die bisherige Kernzone der Tethys dem Nubisch-Arabischen Festland angegliedert wurde. Im postvaristischen Abschnitt, von dem hier nur das Perm beschrieben wurde, verlief die Kernzone der Tethys, wie im Alt-Paläozoikum, wieder in dem SE—NW streichenden Bereich zwischen E-Arabien und dem SW-Iran. Es erscheint einleuchtend, daß es sich bei dieser bereits im Kambrium vorgezeichneten Zone um einen Vorläufer der späteren alpidischen Geosynklinale handelte. In ähnlicher Weise sind junge alpidische Strukturen — Senkungszonen und Grabenbrüche — z. B. auch in Syrien und im Suezgebiet schon im Paläozoikum erkennbar. Wie u. a. auch von *Wolfart* (1967) dargelegt wurde, dürften diese Strukturen mit den großen, im präkambrischen Gefüge der Erdkruste verankerten Lineamenten in engem Zusammenhang stehen.

Der Bundesanstalt für Bodenforschung dankt Verf. für den Auftrag, seine Untersuchungsergebnisse zu veröffentlichen. Herzlich gedankt sei den Herren Dr. *H. Kleinsorge*, Dr. *H. Venzlaff* und Dr. *D. Weippert* für Diskussionen und wertvolle Hinweise zum Entwurf der paläogeographischen Karten.

Literatur

Die ältere Literatur ist hauptsächlich in den nachstehend aufgeführten Arbeiten von *Flügel* (1964), *Gansser*, *Hecht* et al., *Said*, *Wirtz* et al. (1964b) zitiert.

Abdallah, A. M., u. *A. El Adindani*, Stratigraphy of Upper Paleozoic rocks western side of the Gulf of Suez. Geol. Surv. Miner. Res. Dept., Pap. 25. Cairo 1965.

Bakr, M. A., u. *R. O. Jackson*, Geological map of Pakistan. Geol. Surv. Pakistan. Rawalpindi 1964.

Baykal, F., u. *O. Kaya*, Note préliminaire sur le Silurien d'Istanbul. Bull. Miner. Res. Inst. Turkey **64**, 1/8 [1965].

Bender, F., et al.: Geologische Karte von Jordanien 1 : 250 000, Bl. Aqaba-Ma'an und Bayir. Bundesanst. Bodenforsch. Hannover 1966.

— —, *V. Stein* u. *R. Wolfart*, Ober-Ordovizium (Caradoc-Ashgill) und Unter-Silur (Unter-Llandovery) in Südjordanien. Geol. Jb. **85** (im Druck). Hannover 1967.

Bogdanow, A., Sur certains problèmes de structure et d'histoire de la plate-forme de l'Europe orientale. Bull. Soc. géol. France **4**, Nr. 7, 898/911 [1962].

Brinkmann, R., Geotektonische Gliederung von Westanatolien. Neues Jb. Geol. Paläontol., Mh., **1966**, Nr. 10, 603/18.

Collomb, G. R., Étude géologique du Jebel Fezzan et de sa bordure paléozoique. Cie. franç. Pétroles, Notes Mém., Nr. 1. Paris 1962.

Dean, W. T., u. *R. Krummenacher*, Cambrian trilobites from the Amanos Mountains, Turkey. Paleontology **4**, Nr. 1, 71/81 [1961].

Douglas, J. A., The Carboniferous and Permian faunas of South Iran and Iranian Baluchistan. Mem. geol. Surv. India, Palaeont. Indica, n. S. **22**, Nr. 7. Calcutta 1950.

Fantini Sestini, N., Brachiopods from Geiroud formation, Member D (Lower Permian). Riv. ital. paleontol. stratigr. **72**, Nr. 1, 9/50 [1966].

Flügel, H., Die Entwicklung des vorderasiatischen Paläozoikums. Geotektonische Forsch., Nr. 18. Stuttgart 1964.

— —, Permian corals from Ruteh limestone. Riv. ital. paleontol. stratigr. **70**, Nr. 3, 403/32 [1964].

— —, Rugosa aus dem Perm Afghanistans. Neues Jb. Geol. Paläontol., Mh. **1965**, Nr. 1, 6/17.

— — u. *A. Ruttner*, Vorbericht über paläontologisch-stratigraphische Untersuchungen im Paläozoikum von Ozbakh-Kuh (NE-Iran). Verh. geol. Bundesanstalt 1962, S. 146/50. Wien 1962.

Gaetani, M., Brachiopods and molluscs from Geiroud Formation, member A (Upper Devonian and Tournaisian). Riv. ital. paleontol. stratigr. **71**, Nr. 3, 679/770 [1965].

Gansser, A., Geology of the Himalayas. London 1964.

Glaus, M., Die Geologie des Gebietes nördl. des Kandevan-Passes (Zentral-Elburs), Iran. Mitt. geol. Inst. Eidgenöss. Hochschule, n. S., Bd. 48. Zürich 1965.

Hecht, F., *M. Fürst* u. *E. Klitzsch*, Zur Geologie von Libyen. Geol. Rdsch. **53**, Nr. 2, 413/70 [1964].

Helal, A. H., On the occurrence and stratigraphic position of Permo-Carboniferous tillites in Saudi-Arabia. Geol. Rdsch. **54**, Nr. 1, 193/207. Stuttgart 1964.

Huckriede, R., *M. Kürsten* u. *H. Venzlaff*, Zur Geologie des Gebietes zwischen Kerman und Sagand (Iran). Beih. geol. Jb., Nr. 51. Hannover 1962.

Inst. Min. Res. Explor. Turkey: Geological Map of Turkey 1 : 500 000. Ankara 1961—1964.

Krestnikow, V. N., History of oscillatory movements in the Pamirs and adjacent regions of Asia. Izdatelstvo Akad. Nauk SSSR. Moskau 1962. (Translated from Russian by Israel Program for Scientific Translations, Jerusalem 1965.)

Lapparent, A. F. de, u. *H. de Lavigne Saint-Suzanne*, Le Carbonifère marin aux environs de Wakak (province de Ghazni, Afghanistan). C. R. hebd. Séances Acad. Sci. **258**, Nr. 20, 5018/19 [1965].

— — u. *M. Lys*, Attribution au Permien supérieur du gisement à Fusulines et Brachiopodes de Kwaja Gar (Bamian, Afghanistan). C. R. hebd. Séances Acad. Sci. **262**, Nr. 20, 2138/40 [1966].

Markowskij, A. P., et al., Geologische Beschreibung der Tadschikischen SSR. Geologie der UdSSR, Bd. 24. Moskau 1959.

Morton, D., The geology of Oman. Proc. 5th Wld. Petroleum Congr., Sect. I, 277/94. New York 1959.

Naliwkin, D. W., et al., The geological map of the UdSSR 1 : 2 500 000. Ministry Geol. UdSSR. Moskau 1956.

— —, Stratigraphie der UdSSR in 14 Bänden. Silur. Moskau 1965.

Neumayr, M., Die geographische Verbreitung der Juraformation. Denkschr. Akad. Wiss., math.-naturwiss. Cl., Nr. **50**, S. 57/142. Wien 1885.

Omara, S., A micropaleontological approach to the stratigraphy of the Carboniferous exposures of the Gulf of Suez region. Neues Jb. Geol. Paläontol., Mh. **1965**, Nr. 7, 409/19 [1965].

— — u. *A. Kenawy*, Upper Carboniferous microfossils from Wadi Araba eastern Desert, Egypt. Neues Jb. Geol. Paläontol., Abh. **124**, Nr. 1, 56/83 [1966].

Paffengolz, K. N., Geologischer Abriß des Kaukasus. Fortschr. sowj. Geol., Bd. 5/6. Berlin 1963.

Ponikarow, W. P., *E. D. Sulidi-Kondrat'ew*, *V. V. Kozlow* u. *V. G. Kaz'min*, Die Tektonik des nördlichen Teiles der Arabischen Tafel. Sowjet-Geol. **1**, 39/48 [1964].

Rahman, H., Geology of petroleum in Pakistan. Proc. 6th Wld. Petroleum Congr., Sect. I, 659/83. Hamburg 1964.

Richter, R. u. *E.*, Das Kambrium am Toten Meer und die älteste Tethys. Abh. senckenberg. naturforsch. Ges., Nr. 460. Frankfurt (Main) 1941.

Saad, S. J., Pollen and spores recently discovered in the coals of Sinai region. Palaeontogr., B **115**, Nr. 4/6, 139/49 [1965].

Said, R., The geology of Egypt. Amsterdam – New York 1962.

Sayer, C., Ordovician conulariids from the Bosphorus area, Turkey. Geol. Mag. **101**, Nr. 3, 193/97 [1964].

Schindewolf, O. H., Über die Faunenwende vom Paläozoikum zum Mesozoikum. Z. dtsch. geol. Ges. **105** (1953), 153/82 [1955].

— — u. *A. Seilacher*, Beiträge zur Kenntnis des Kambriums in der Salt Range (Pakistan). Akad. Wiss. Mainz, math.-naturwiss. Kl., Bd. 10, S. 257/446. Wiesbaden 1955.

Schuermann, H. M. E., *D. Burger* u. *S. J. Dijkstra*, Permian near Wadi Araba, eastern desert of Egypt. Geol. en Mijnbouw **42**, 329/36 [1963].

Steineke, M., *R. Bramkamp* u. *N. J. Sander*, Stratigraphic relations of Arabian Jurassic oil. In: Habitat of Oil, S. 1294/1329. Tulsa 1958.

Stöcklin, J., *J. Eftekhar-Nezhad* u. *A. Hushmand-Zadeh*, Geology of the Shotori Range (Tabas area, East Iran). Geol. Surv. Iran, Rep. 3. Teheran 1965.

— —, *M. Nabavi* u. *M. Samimi*, Geology and mineral resources of the Soltanieh Mountains (Northwest Iran). Geol. Surv. Iran, Rep. 2. Teheran 1965.

— —, *A. Ruttner* u. *M. Nabavi*, New data on the Lower Paleozoic and pre-Cambrian of North Iran. Geol. Surv. Iran, Rep. 1. Teheran 1964.

Suess, E., Note sur l'histoire des océans. C. R. hebd. Séances Acad. Sci. **121**, 1113/16 [1895].

Thiele, O., Zum Alter der Metamorphose in Zentral-Iran. Mitt. geol. Ges. Wien **58**, 87/101 [1966].

U. S. Geological Survey u. Arabian American Oil Company, Geological map of the Arabian Peninsula, 1 : 2 000 000. Washington 1963.

Weippert, D., u. *H. Wittekindt*, Ein Vorkommen von paläozoischem Salz im westlichen Zentralafghanistan. Geol. Jb. **82**, S. 99/102. Hannover 1964.

— — u. *R. Wolfart*, Altordovizische Trilobiten von Kirman bei Panjao im östlichen Zentral-Afghanistan. (Erscheint: 1967/1968).

Wellman, H. W., Active wrench faults of Iran, Afghanistan and Pakistan. Geol. Rdsch. **55**, Nr. 3, 716/35 [1966].

Wirtz, D., et al., Geological map of Afghanistan, central and southern part. Bundesanst. Bodenforsch. Hannover 1964 (a).

— —, Zur Geologie von Nordost- und Zentral-Afghanistan. Beih. geol. Jb., Nr. 70. Hannover 1964 (b).

Wolfart, R., Die Fauna des Karbons vom Jebel Abd-el-Aziz (Nordost-Syrien). Geol. Jb. **83**, S. 277/326. Hannover 1965.

— —, Geologie von Syrien und dem Libanon. In: *H. J. Martini*, Beiträge zur regionalen Geologie der Erde. Berlin 1967.

— —, Tektonik und paläogeographische Entwicklung des mobilen Schelfes im Bereich von Syrien und dem Libanon. Z. dtsch. geol. Ges. (Erscheint 1967/68).

5

Reprinted from *Věstník Ústřed. Úst. Geol.* **49**:343–348 (1974)

Some problems of the Ordovician in the Mediterranean region

(1 text-fig.)

VLADIMÍR HAVLÍČEK[1]

Abstract. The author discusses some problems of the Ordovician of the Mediterranean region, namely the geographic delimitation of the proto-Tethys, the migration of fauna in the proto-Tethys, the relationship between fauna and Upper Ordovician glaciation and the relationship between glaciation and continental drift.

Abstakt. Jsou diskutovány některé problémy ordoviku mediteránní oblastí, a to geografické vymezení Prototethydy, migrace fauny v Prototethydě, vztah fauny a svrchnoordovického zalednění a konečně vztah zalednění a kontinentálního driftu.

Recently, a great number of new data concerning the stratigraphy of the Ordovician, paleogeography, evolution of animal assemblages, alternation of cold and warm periods, paleomagnetism, continental drift, etc. has been acquired. They are used in the present paper to elucidate some problems of the Ordovician in the Mediterranean province.

1) Delimitation of the Mediterranean province: The Mediterranean province is considered to be the area on both sides of the Mediterranean Sea, occupying a substantial part of North Africa, western and central Europe, the Balkans and reaching eastward as far as Turkey, Syria and Jordan, or even farther to the E. Geologically, this province corresponds to the miogeosynclinal and eugeosynclinal regions of the proto-Tethys bordering the northern margin of Gondwana and the adjacent shelves. From the zoogeographic point of view its extent corresponds to the Selenopeltis province as conceived by Whittington and Hughes (1972). On the geological map we may trace the northern boundary of the Mediterranean region along the Odra lineament (sensu Dvořák and Paproth 1969), which separates it from the east-European platform.

Due to regression of the Upper Cambrian sea, a substantial part of the Mediterranean province became dry land. At that period, Gondwana consisted of not only all territories S of the Mediterranean Sea but included also extensive regions in western and central Europe. A marine regression with interrupted sedimentation between the Cambrian and Ordovician has been established (after Sdzuy 1972) in the Montagne Noire, the Frankenwald, in Lusatia, Bohemia and in most of North Africa. Marine Upper Cambrian (including a gradual transition into the Tremadoc) has been proved in Spain and according to Legrand (1973) also in the northern part of the Sahara.

Extensive Mediterranean areas that were dry land in the Upper Cambrian became shallow epicontinental seas in the Lower Tremadoc; during the Ordovician, many of them developed into miogeosynclines and eugeosynclines, in places with strong volcanic activity (Busacco area in Portugal, Spain, Normandy and Bohemia); in the Upper Devonian and Carboniferous they were affected by Hercynian folding that in places was connected with weak to strong regional metamorphism (Krušné hory Mts. and Krkonoše Mts.).

In comparing the picture of the Mediterranean region in the Upper Ordovician with that in the Upper Cambrian we may notice a striking difference. Whereas in the Upper Cambrian dry

---

[1] Ústřední ústav geologický, 118 21 Praha 1, Malostranské nám. 19.

land prevailed, this region was covered in the Upper Ordovician by the extensive sea of the proto-Tethys, which was bordered on the N (i.e. at its boundary with the mid-European sea — fig. 1) by an archipelago with many islands of various size. One of these islands was the Precambrian core of the Bohemian Massif.

2) Even when speaking of the proto-Tethys as a zoogeographic province we are aware of the fact that the faunal assemblage within this province is not uniform. This is especially true in the Lower Ordovician, when almost every region of the proto-Tethys was inhabited by a different assemblage of shelly fauna. Such a scattered distribution of fauna is undoubtedly due to the various modes and speeds of migration of individual animal groups. Whereas the Bohemian Lower Arenig is almost devoid of shelly fauna (except for several species of inarticulate brachiopods) a rich assemblage of trilobites, bivalves, echinoids, etc. has been reported from the Montagne Noire. Thoral (1935) and Dean (1966) have described a great number of genera from this region that appeared in Bohemia much later, most of them as late as the Llanvrin, e.g.: *Bathycheilus, Colpocoryphe, Ormathops, Platycoryphe, Selenopeltis, Ampyx, Pricyclopyge, Symphysurus, Geragnostus, Ribeiria, Ribeirella, Anatifopsis, Mitrocystites, Babinka, Redonia, Synek, Lesueurilla,* and others. The first weak inflow of this new fauna into the Bohemian region is known to have occurred as late as the Upper Arenig, when the trilobite genera *Geragnostus, Symphysurus, Colpocopyphe* and *Ormathops* (mentioned by Dean from the Lower Arenig in the Montagne Noire) appear for the first time in Bohemia. The main inflow of new fauna occurred not earlier than at the beginning of the Llanvirn. At that time, a great number of new genera of trilobites, bivalves, gastropods and echinoderms appeared but the articulate brachiopods arrived much later and thus the brachiopod assemblage of the Llanvirn was very poor; an increased immigration of brachipods into Bohemia occurred as late as the Dobrotiv, the major influx, however, was delayed until the beginning of the Beroun. Some of the brachiopod genera appearing in Bohemia for the first time in the Lower and Middle Beroun have been recorded from the Llanvirn and Llandeilo of Wales, as for instance the genera *Rostricellula* and *Onniella.*

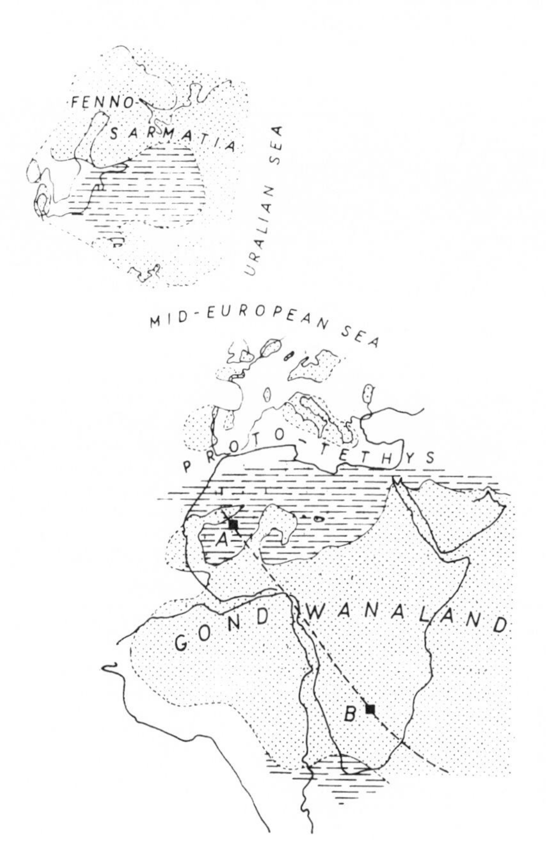

Fig. 1. Proto-Tethys in the Upper Ordovician. Dotted: without marine deposits. Shading: epicontinental sea. Without shading: mio- and eugeosynclinal areas. *A, B* — presumed South pole wander path: *A* — position of pole during Cambrian and Ordovician, *B* — position of pole during Upper Silurian to Lower Carboniferous

The irregular distribution of fauna in the proto-Tethys during the Upper Ordovician may be traced less easily. Thus, for instance, *Kozlowskites* has been always considered to be the index genus of the Královdvor, but recently it has been found in the Beroun of the Carnic Alps (Vai 1971). Another interesting brachiopod in the Beroun of the Carnic Alps is *Plectothyrella* which in the Kosov spread very widely all over the proto-Tethys and partly even beyond its boundaries. An interesting example of the uneven distribution of the assemblages in the proto-Tethys may be also offered by *Philipsinella*: its first occurrence was reported by Dean (1973) from the Turkish Arenig, whereas in Bohemia it appeared as late as the beginning of the Královdvor.

Table 1

Correlation of the British and Bohemian successions

Great Britian	Bohemia (according to Havlíček and Marek)
Series	Series
Ashgill	Kosov
	Králův Dvůr
Caradoc	Beroun
Llandeilo	Dobrotivá
Llanvirn	Llanvirn
Arenig	Arenig
Tremadoc	Tremadoc

The before-mentioned data indicate that faunal migration in the proto-Tethys is incontestable. A generally even and almost undisturbed development of shelly fauna occurred in the miogeosynclinal regions of the proto-Tethys, where the fauna was not exposed to the drastic environmental changes that were manifested to a greater extent in the marginal parts of the proto-Tethys. Far-reaching changes in the configuration of the sea and dry land often resulted in a drastic depletion of the faunal assemblages in the shelf areas. An almost complete extinction of brachiopod assemblages is known to have occurred at the Tremadoc/Arenig and Arenig/Llanvirn boundaries. The vacant areas where then re-occupied by new immigrants all of them beloging to genera unknown in Bohemia before.

Foreign elements also occur in the Ordovician of the proto-Tethys. These are for instance some brachiopods characteristic of the warm sea of the Baltic region such as *Porambonites, Christiania, Strophomena, Leptaena* and *Anisopleurella,* cited from the Beroun of the Carnic Alps by Vai (1971) and Spjeldnaes (1967). They spread within the proto-Tethys exclusively within the miogeosynclinal areas and did not proceed into the platform areas, as for instance into the Sahara.

3) Fauna and climate. The close relationship between Ordovician fauna and climate has been already pointed out by Spjeldnaes (1961), Havlíček and Vaněk (1966), Williams (1969, 1973), Whittington and Hughes (1972, 1973). These authors agreed in stating that the proto-Tethys had been a cold-water region influenced by the position of the pole approximately in the centre of the Sahara. The cold (Upper) Ordovician climate has been also proved by the well established evidence of the extensive glaciation of Gondwana. Beuf, Biju-Duval, Stevaux and Kulbicki (1966) quote examples of both glacial erosion and glacial and periglacial sedimentation on the Sahara. Evidence for Upper Ordovician glaciation has been found now in Morocco (Destombes 1968, 1971), Algeria (Beuf et al. 1966, Whiteman 1973), Libya (Havlíček - Massa 1973), Sierra Leone (Tucker - Reid 1973) and South Africa (Cocks et al. 1970). Also it is possible that the Upper Ordovician sediments of Ethiopia are glacial in origin but their age has not been proved by paleontological evidence (Dow et al. 1971). The Upper Ordovician glaciation has been further established in Spain (Arbey - Tamain 1971) and Normandy (Dangeard - Doré 1971; Doré - Le Gall 1971) and according to Zimmermann (1914) and Katzung (1961) also in Thuringia.

The mutual relationship between glaciation and richness of fauna has been stressed by Whiteman (1973). He states that the distribution of fauna in the Sahara was controlled by glacial and periglacial conditions. The absence of current stenohaline forms from Ordovician rocks indicates deposition under restricted(?) marine shelf or lagoonal conditions. According to this author, marine (even though depleted) fauna existed in Africa mainly in Morocco, i.e. at the periphery of the miogeosynclinal region (Anti-Atlas, Ougarta). The glaciation is thus the reason why Ordovician rocks on such extensive territories are unusually poor in fossils. This has considerably impeded the dating of glacial episodes. The maximum of Ordovician glaciation was incontestably in the Kosov, as evidenced by the presence of brachiopods in the periglacial deposits of Morocco (Havlíček 1971) and Libya (Havlíček - Massa 1973). The Formation du deuxième Bani (Morocco), Unité IV in the sense of Beuf et al. (Algérie), the Mémouniat Formation (Libya) and the Tillite of Feuguerolle (Normandie) are all thus of Kosovian age and also possibly the Lederschiefer (Thuringia). The Kosov Formation corresponding in Bohemia to the Uppermost Ordovician has sofar lacked direct evidence of the glaciation of the Bohemian Massif during that period, but the presence in places of distinctly unsorted material (including pebbles of granites and older Ordovician rocks of paleontologically evidenced Upper and Lower Berounian age) indicates the possibility of deep glacial erosion having occurred at the periphery of the Bohemian sedimentary basin.

Evidence for the older pre-Kosov glaciation has sofar only been found in Libya. Poor fauna of the Melez Chograne Formation, whose material is partly of moraine origin (Collomb 1962) gives this older Ordovician glaciation a Berounian age (Havlíček - Massa 1973).

The influence of the Kosov glaciation on the development of assemblages has been enormous. Sheehan (1973) has compared the extent of this glaciation with Pleistocene continental glaciation and has arrived at the conclusion that in accordance with this younger period the level of the Upper Ordovician seas could have dropped 100 m or more due to the large quantity of water frozen in continental glaciers. This dropping of the sea level drastically interfered with the fauna of the epicontinental seas. In the proto-Tethys it caused the extinction of a large part of the shelly fauna (of the Králodvor type) which was replaced in the Kosov and Lower Silurian by immigrating new forms.

4) Glaciation and drift. Ordovician glaciation is in accordance with the data on paleomagnetism. McElhinny and Briden (1971) suggest that the South Pole was situated approximately in the western part of North Africa during Cambrian and Ordovician times. A long "quasi-static interval" was succeeded by a comparatively short drift period which was manifested by a rapid movement of Gondwana with respect to the Pole. As stated by McElhinny and Briden, there were two such periods of polar wandering with respect to Gondwana during the Paleozoic: an earlier one in the Ordovician and a later one in the Carboniferous. Of great significance for the development of shelly fauna is the Late Ordovician drift episode, when the South Pole moved in the general direction of the territory of South Africa. This means that Gondwana as a whole (including the Mediterranean area) drifted north to northwestwards causing a substantial narrowing of the mid-European sea. There are various reasons why the Ordovician faunal provincialism was replaced by Silurian cosmopolitism. One of the most serious ones is still the movement of Gondwana accompanied by a substantial narrowing of the mid-European sea which together with a probable elimination of longitudinal currents allowed the fauna to cross it. This was perhaps the main reason why an independent Mediterranean province ceased to exist in the Late Ordovician and why a uniform north-European province developed.

The displacement of Gondwana during the Upper Ordovician was manifested also otherwise: together with the general warming up of the climate at the Ordovician/Silurian boundary it co-acted with the melting of the continental ice-sheet on the Saharian region. According to Beuf et al. (1966) this melting was so quick that it can be regarded as a "catastrophic phenomenon" ("phenomène catastrophique") on a continental scale and it was manifested by a striking change of the mineralogical composition of the sediments between the Uppermost Ordovician (Unité IV) and the Lower Silurian graptolite shales. Beuf et al. prove by using geochemical analyses that the Silurian sea in North Africa was a cold post-glacial one with a considerably lowered salinity as a rule, being approximately equal in salinity to the present-day Baltic sea. The boundary between the Ordovician and Silurian in the proto-Tethys has thus been determined primarily by the sudden climatic changes that influenced both the mineralogic composition and the faunal content of the boundary beds.

Another comparatively short drift episode occurred according to McElhinny and Briden (1971) in the Carboniferous. At this period, the northern boundary of Gondwana approached so close to the Fenno-Sarmatian continent as to cause the further narrowing and finally the disappearance of the mid-European sea accompanied by the Hercynian folding.

5) Possible causes of (Upper) Ordovician glaciation: When evaluating the present galactic model Steiner and Grillmair (1973) pointed to the regular interchange of warm periods ("galactic

summers") and comparatively short cold periods ("galactic winters") which have been recorded during the past 3 b.y. These authors claim that the solar system moves on an elliptic path around the center of the galaxy. One complete revolution is called a cosmic year, whose duration is supposed to have decreased from 400 m.y. in the Precambrian to the present duration of 274 m.y. They call the orbital position closest to the galactic centre perigalacticum and the most remote position apogalacticum. The glacial episodes in the Precambrian correspond to the apogalacticum, whereas in the Phanerozoic they correspond to the Perigalacticum. After Steiner and Grillmair the glaciation of the Ordovician/Silurian boundary is in a transitional position between the two before-mentioned types. The causal relationship between the periods of glaciation and the galactic model is evident but cannot be explained as yet.

G. Williams (1972) suggested a new hypothesis for the origin of the solar system ("prominence hypothesis") which explains much better than before the fluctuations of Earth climate including the Upper Ordovician period of glaciation. Secular change in the obliquity of the spin axis of Earth, which is attributed to secular rotation of the plane of the solar system, is responsible for long-period (1000—2000 m.y.) changes of temperature, whereas moderate-period (100—200 m.y.) changes of temperature are due to rotation of the solar system about the galactic center.

Submitted April 10, 1974, received for publication May 2, 1974

References

Arbey F. - Tamain G. (1971): Existence d'une glaciation siluro-ordovicienne en Sierra Morena (Espagne).— C. R. Acad. Sci. Paris, 272, *1721—1723*. Paris.

Beuf S. - Biju-Duval B. - Stevaux J. - Kulbicki G. (1966): Ampleur des glaciations „siluriennes" au Sahara: Leurs influences et leurs conséquences sur la sédimentation. — Rev. Inst. franç. Pétrole, 31, *363—380*. Paris.

Cocks L. R. M. - Brunton C. H. C. - Rowell A. J. - Rust I. C. (1970): The first Lower Palaeozoic fauna proved from South Africa. — Quart. J. Geol. Soc. London, 125, *583—603*. London.

Collomb G. R. (1962): Etude géologique du Dj. Fezzan et de sa bordure paléozoïque. — Not. Mém. (Co. franc. Pétrol.), 1. Paris.

Dangeard L. - Doré F. (1971): Faciès glaciaires de l'Ordovicien supérieur en Normandie. — Mém. Bur. Rech. géol. min. (Paris), 73, *119—128*. Paris.

Dean W. T. (1966): The Lower Ordovician stratigraphy and trilobites of the Landeyran Valley and neighbouring district of Montagne Noire, southwestern France. — Bull. Brit. Mus. natur. Hist., Ser. Geol., 12, *245—353*. London.

— (1973): The Lower Palaeozoic stratigraphy and fauna of the Taurus Mountains near Beysehir, Turkey III. The trilobites of the Sobova Formation (Lower Ordovician). — Bull. Brit. Mus. natur. Hist., Ser. Geol., 24, *281—348*. London.

Destombes J. (1968): Sur la nature glaciaire des sédiments du groupe du 2° Bani, Ashgill supérieur de l'Anti-Atlas (Maroc). — C. R. Acad. Sci. Paris, 267, *565—567*. Paris.

— (1971): L'Ordovicien au Maroc. Essai de synthèse stratigraphique. — Coll. Ordovicien-Silurien. Brest, 1971, Mém. Bur. Rech. géol. min. (Paris), 73, *237—263*. Paris.

Doré F. - Le Gall (1972): Sédimentologie de la „Tillite de Feuguerolles" (Ordovicien supérieur de Normandie). — Bull. Soc. géol. France, 14, *199—211*. Paris.

Dow D. B. - Beyth M. - Tsegaye Hailu (1971): Paleozoic glacial rocks recently discovered in northern Ethiopia. — Geol. Mag., 108, *53—59*. London.

Dvořák J. - Paproth E. (1969): Über die Position und die Tektogenese des Rhenoherzynikums und des Sudetikums in den mitteleuropäischen Varisziden. — Neu. Jb. Geol. Paläont., Mh., *65—88*. Stuttgart.

Havlíček V. (1971): Brachiopodes de l'Ordovicien du Maroc. — Not. Mém. Serv. géol. (Rabat), 230, *135*. Rabat.

Havlíček V. - Marek L. (1973): Bohemian Ordovician and its international correlation. — Čas. Mineral. Geol., 18, *225—232*. Praha.

Havlíček V. - Massa D. (1973): Brachiopodes de l'Ordovicien supérieur de Libye occidentale. Implications stratigraphiques régionales. — Geobios (Lyon), 6, 4, *267—290*. Lyon.

Havlíček V. - Vaněk J. (1966): The biostratigraphy of the Ordovician of Bohemia. — Sbor. Ústř. Úst. geol., Odd. paleont., 8, *7—69*. Praha.

Katzung G. (1961): Die Geröllführung des Lederschiefers (Ordovizium) an der S. O. Flanke des Schwarzburger Sattels (Türingen). — Geologie (Berlin), 10, *778—802*. Berlin.

Léspérance P. J. (1974): The Hirnantian fauna of the Percé area (Quebec) and the Ordovician-Silurian boundary. — Amer. J. Sci., 274, *10—30*. New Haven.

McElhinny M. W. - Briden J. C. (1971): Continental drift during the Palaeozoic. — Earth planet. Sci. Lett. (Amsterdam), 10, *407—416*. Amsterdam.
Schönlaub H. P. (1971): Palaeo-environmental studies at the Ordovician/Silurian boundary in the Carnic Alps. — Coll. Ordovicien-Silurien, Brest 1971, Mém. Bur. Rech. géol. min. (Paris), 73, *367—378*. Paris.
Sdzuy K. (1972): Das Kambrium der acadobaltischen Faunenprovinz. Gegenwärtiger Kenntnisstand und Probleme. — Zbl. Geol. Paläont., 2, *1—91*. Stuttgart.
Sheehan P. M. (1972): The relation of Late Ordovician glaciation to the Ordovician-Silurian changeover in North American brachiopod faunas. — Lethaia (Oslo), 6, *147—154*. Oslo
Spjeldnaes N. (1961): Ordovician climatic zones. — Nor. geol. Tidsskr., 41, *45—77*. Bergen.
— (1967): The palaeogeography of the Tethyan region during the Ordovician. — *In* Aspects of Tethyan Biogeography. Syst. Ass. Pub. 7, *45—57*. London.
Steiner J. - Grillmair E. (1973): Possible galactic causes for periodic and episodic glaciations. — Geol. Soc. Amer. Bull., 84, *1003—1018*. New York.
Thoral M. (1935): Contribution à l'étude paléontologique de l'Ordovicien inférieur de la Montagne Noire et révision sommaire de la faune cambrienne de la Montagne Noire. Montpélier.
Vai F. B. (1971): Ordovicien des Alpes Carniques. — Coll. Ordovicien-Silurien, Brest 1971, Mém. Bur. Rech. géol. min. (Paris), 73, *437—448*. Paris.
Whiteman A. J. (1973): "Cambro-Ordovicien" rocks of Al Jazair (Algeria) — a review. — Amer. Assoc. Petrol. Geologists Bull., 55, *1295—1335*. Tulsa.
Whittington H. B. - Hughes C. P. (1972): Ordovician geography and faunal provinces deduced from trilobite distribution. — Philosoph. Trans. Roy. Soc. Lond. Biol. Sci., 263, 850, *235—278*. London.
— (1973): Ordovician trilobite distribution and geography. — *In* Organisms and continents through time. Spec. Pap. in Palaeontology (London), 12, *235—240*. London.
Williams A. (1969): Ordovician of British Isles. — North Atlantic geology and continental drift, Mem. 12. Amer. Assoc. Petrol. Geologists Bull., *236—264*. Tulsa.
— (1973): Distribution of brachiopod assemblages in relation to Ordovician palaeogeography. — In Organisms and continents through time. Spec. Pap. in Palaeontology (London), 12, *241—269*. London.
Williams G. (1972): Geological evidence relating to the origin and secular rotation of the solar system. - Mod. Geol., 3, 165—181. Amsterdam.
Zimmermann E. (1914): Gerölltonschiefer im Untersilur Thüringens. — Z. Dtsch. geol. Gesell. 66, *269—271*. Hannover.

6

Reprinted from *Acad. Sci. Comptes Rendus*, ser. D, **283**:139–142 (1976)

CONFIGURATION DE LA TÉTHYS EN CONNEXION AVEC LA GONDWANIE AU PALÉOZOIC SUPÉRIEUR

H. Termier and G. Termier

Depuis le Protérozoïque, le cadre de la distribution des continents et des océans est composé par deux supercontinents subégaux (Laurasie et Gondwanie) séparés par une coupure, la *Téthys*. Le schéma de Dietz et Hòlden ([6]) figure la Téthys comme un golfe du Pacifique largement ouvert à l'Est entre l'Australasie et le sud-est asiatique.

SUD DE LA TÉTHYS. — Pendant toute l'ère Primaire, des liens continentaux étroits ont existé entre l'est de l'Amérique du Nord, la Méso-europe et l'Afrique par l'intermédiaire, entre autres, de la Moghrabia ([12]). Aucune trace d'une large entaille océanique téthysienne n'est présente dans cette zone qui devait être seulement rompue (Téthys fracturale, ([11]) le long de grandes familles de linéaments telles que celle du « Guadalquivir et du sud subbé-tique-nord bétique » et celle de l'« accident sud-atlasien » prolongé par les bords de la Grande Syrte et le sud des Taurides. En accord avec les résultats d'Argyriadis ([1]) relatifs à la Méditerranée orientale, on doit considérer le sud de la Téthys occidentale comme une plate-forme continentale jusqu'à la fin de l'ère Primaire. C'est celle-ci qui deviendra, lors de l'orogenèse alpine, l'ensemble massifs intermédiaires-« Dinarides ».

Vers l'Est, les continents gondwans, Arabie. Dekan et Australie, étaient alors bordés au Nord par un glacis de terres satellites, Transcaucasie, Iran, Afghanistan, Tien Shan et Tarim, composant une plate-forme accidentée dont les éléments devaient avoir un comportement analogue à celui des Dinarides durant l'orogenèse alpine.

Contrairement au modèle de Dietz et Holden mais conformément à celui de Craw-ford ([4]) qui situe le Tarim contre l'ouest australien en le rétablissant dans sa position initiale par une rotation sénestre, l'Australasie était reliée à la plate-forme de la Sonde, Indonésie, Indochine ([2]), étant elle-même attenante à la Chine du sud (Cathaysie) : cette zone constitue une charnière analogue à celle de Gibraltar. Dans ce glacis, des linéaments étaient présents au Paléozoïque supérieur, prolongement de la suture de l'Indus vers l'Hindou-Kouch, linéament Owen (ouverture du Mozambique) — Argandeh (Chaman s.l.) — Zagros. La fragmentation des continents, seulement esquissée, laissait s'effectuer au Permien et encore au Trias inférieur des migrations de Vertébrés gondwans jusqu'en Indochine ([7]) et, dans le Tarim, le fameux *Lystrosaurus* ([3]).

L'OCÉAN TÉTHYSIEN ? — Une Téthys à fond océanique au Paléozoïque reste une hypothèse. Il est vraisemblable qu'un couloir océanique assez étroit séparait la plate-forme nord de la Gondwanie de la bordure sud de l'Eurasie (Altaïdes et zone varisque méso-européenne) qui n'était pas sans montrer, elle aussi, les caractères d'une plate-forme accidentée.

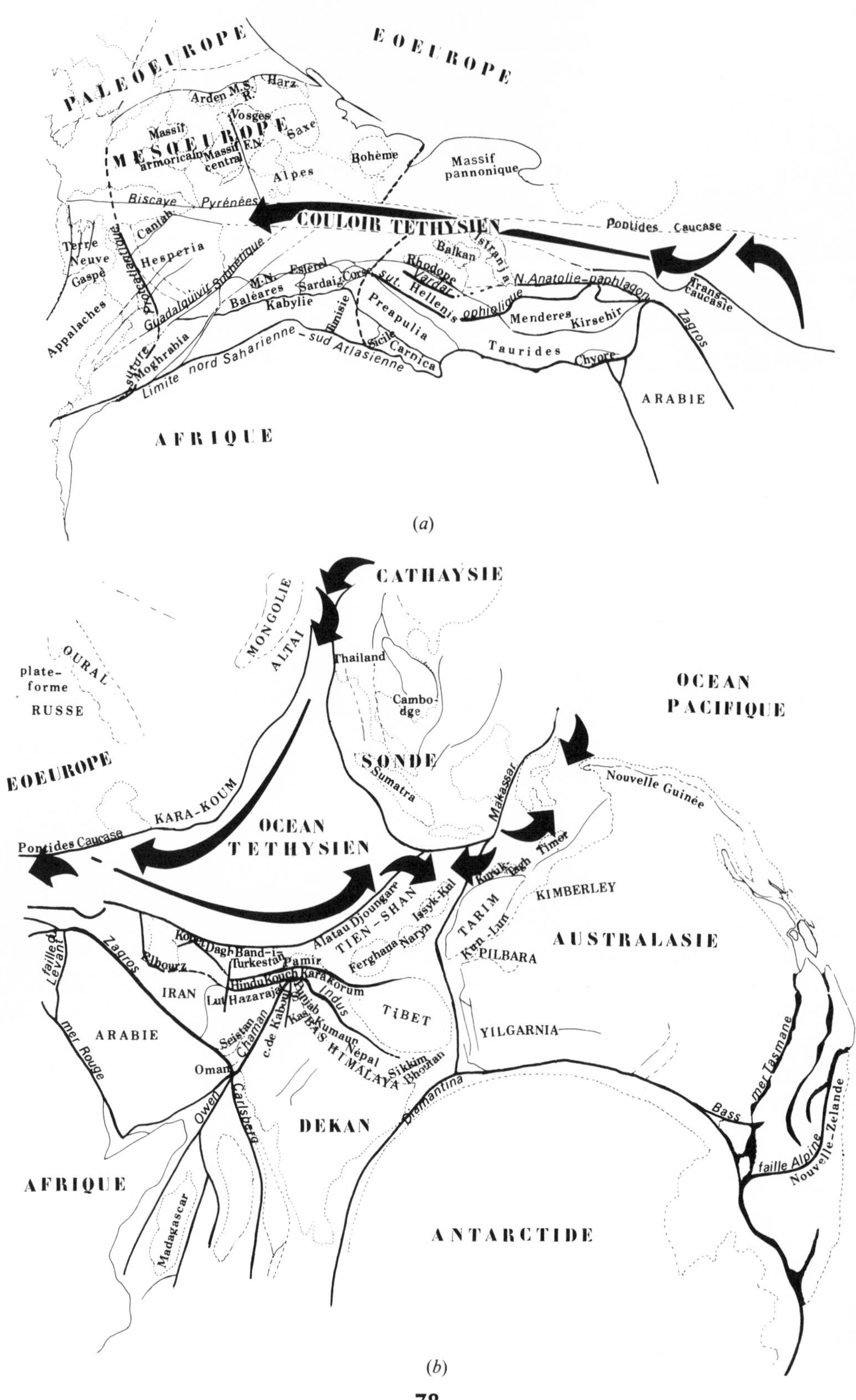

PALEOEUROPE
EOEUROPE
MESOEUROPE
Arden
Harz
Vosges
Massif armoricain
Massif central
Saxe
Bohème
Massif pannonique
Alpes
Biscaye
Pyrénées
COULOIR TETHYSIEN
Pontides
Caucase
Terre Neuve
Gaspé
Hesperia
Balkan
Rhodope
Vardar
Hellenis
N.Anatolie-paphlagon
Trans-caucasie
Appalaches
Baléares
Kabylie
Sardaig
Esterel
Tunisie
Preapulia
Sicile
Carnica
Menderes
Kirsehir
Zagros
Taurides
Chypre
Moghrabia
Limite nord Saharienne - sud Atlasienne
ARABIE
AFRIQUE
CATHAYSIE
MONGOLIE
ALTAI
Thailand
Cambodge
OURAL
plate-forme RUSSE
OCEAN PACIFIQUE
EOEUROPE
SONDE
Sumatra
Makassar
Nouvelle Guinée
KARA-KOUM
OCEAN TETHYSIEN
Pontides Caucase
Timor
KIMBERLEY
Issyk-Kul
TIEN-SHAN
Alatau Djoungarie
TARIM
Kun-Lun
AUSTRALASIE
PILBARA
Ferghana
Naryn
Zagros
Elbourz
Band-i-Turkestan
Pamir
Hindu Kouch
Kara Korum
Indus
TIBET
IRAN
Lut
Hazarajat
YILGARNIA
ARABIE
mer Rouge
Seistan
Chaman
c.de Kabul
Népal
Sikkim
Bhoutan
HIMALAYA
Oman
mer Tasmane
Nouvelle-Zelande
Diamantina
Bass
Owen
Carlsberg
DEKAN
faille Alpine
AFRIQUE
Madagascar
ANTARCTIDE

(a)

(b)

78

LÉGENDE DE LA FIGURE

Tentative d'esquisse de la Téthys au Paléozoïque supérieur. Plusieurs des linéaments figurés ont joué à des périodes diverses, post-paléozoïques. Leur parallélisme avec les sutures anciennes nous a paru l'indice d'une dynamique assez constante. Les grosses flèches noires jalonnent les zones téthysiennes.

(*a*) La Téthys occidentale (zone alpino-méditerranéenne à l'âge varisque). On note une certaine symétrie des plates-formes nord-Gondwana et sud-Europe par rapport au couloir téthysien (futur sillon alpin ?).

(*b*) La Téthys orientale (océan Téthysien et connexions avec le couloir téthysien à l'Ouest, avec l'océan Pacifique à l'Est).

Le « couloir téthysien » préfigurant le sillon alpin et susceptible de fournir les transgressions téthysiennes ([9]) pourrait avoir eu quelque similitude avec une grande « mer Rouge ». Il se terminait à l'Ouest au niveau d'un linéament coïncidant avec celui des Pyrénées et de Biscaye ou parallèle à lui. Une partie de sa bordure nord correspondait avec les Alpes occidentales dont l'essentiel est varisque ou même antérieur ([10]). Ce linéament se prolonge probablement au nord de Terre-Neuve et au nord du Labrador par le détroit d'Hudson. Ce trait majeur recoupe la suture du Protoatlantique, fermé depuis la fin de l'orogenèse calédonienne, et plus tardivement, la chaîne asturienne d'âge varisque (prolongée vers le Sud par les Moghrabides et les Maurétanides?). Au Permien, les mers disparurent à l'ouest de la Tunisie.

Vers l'Est, le couloir téthysien continue en s'élargissant entre le sud de l'Eurasie, que nous limitons au Caucase, au Kara-Koum et à l'Altaï, et la plate-forme nord-gondwane définie plus haut. La superficie est assez grande pour qu'on puisse parler d'un océan téthysien, non largement ouvert, cependant, à l'Est sur le Pacifique car il est bordé par la Terre de la Sonde et communique avec le grand océan par deux voies « géosynclinales » : le géosynclinal Westralien-Timor-Célèbes et le géosynclinal paléo-cathaysien. La première de ces voies n'a pas empêché les passages de Vertébrés terrestres dont nous avons parlé plus haut.

Ouverture de fonds océaniques au Permo-Trias. — A partir de la fin du Permien (Zagros - Hindou-Kouch - suture de l'Indus) et du Trias inférieur (Sud des Taurides), les linéaments téthysiens se sont ouverts, libérant des fonds océaniques. Ces estafilades qui modifient la signification de la Téthys furent concomitantes des premières ouvertures au niveau de l'Atlantique et de l'Est australasien (arc Nouvelle-Zélande - Nouvelle-Guinée). Il s'agit d'un comportement crustal particulier ayant affecté l'ensemble des bordures gondwanes dont le rôle avait été jusque-là celui de plates-formes continentales, enjeux de transgressions marines peu profondes. Lors de cette ouverture précoce, les massifs des « Dinarides » et les satellites du glacis oriental se détachèrent du bloc gondwan en même temps que des fragments beaucoup plus importants comme la Lémurie (Dekan + Madagascar), voire l'Australasie. Plus tard, au Jurassique supérieur, d'autres linéaments joueront dans le même sens, tels ceux dont naquit la suture ophiolitique de la presqu'île hellénique et du sud de l'Anatolie. Le phénomène se continuera au Crétacé par la zone du Vardar, prolongée par l'accident nord-anatolien (paphlagonien) et, encore plus tard, avec le croissant ophiolitique péri-arabe. La Téthys évolue ainsi par jeu et rejeu de fossés subparallèles, individualisant les massifs intermédiaires jusqu'au Crétacé moyen surtout, où l'ampleur des fissurations méridiennes privilégia parmi tous les linéaments ceux d'où devaient issir l'Atlantique et l'océan Indien.

L'orogenèse alpine aboutit à la fermeture de la Téthys par l'affrontement, au niveau même du sillon alpin, de la Laurasie et de la Gondwanie. Mais de cette dernière, seuls ont été concernés la plate-forme fissurée des Dinarides et son homologue du glacis oriental entraînant le Dekan : leurs éléments dissociés par la lente préparation linéamentaire ont été soudés à la Laurasie tandis que des structures parallèles inédites comme la Paratéthys et la Méditerranée s'installaient dans ce nouveau cadre.

CONCLUSIONS. — La Téthys du Paléozoïque supérieur se présente comme une structure différente de tous les autres océans en ce qu'elle fut une vaste zone de fissuration entre Laurasie et Gondwanie, dans laquelle la partie de fond océanique a toujours été restreinte. Son élargissement, pendant l'ère Mésozoïque, résultera de phénomènes comparables aux fossés africains et à la mer Rouge. La différence avec l'expansion océanique qui a donné naissance à de très larges océans n'est pas dans l'origine des fissures mais dans le cadre de leur évolution.

(*) Séance du 10 mai 1976.

(1) I. ARGYRIADIS, *Bull. soc. Géol. Fr.*, (7), 17, 1975, p. 56-67.
(2) M. G. AUDLEY-CHARLES, *Palaeogeogr., Palaeoclim., Palaeocol.*, 1, 1965, p. 297-305.
(3) E. H. COLBERT, Antarctic Gondwana Tetrapods. *2nd Gondwana Symp.*, South Africa, 1970, p. 659-664.
(4) A. R. CRAWFORD, *Geol. Mag.*, 111, (5), 1970, p. 369-480.
(5) J. F. DEWEY, W. C. PITTMAN, W. B. F. RYAN et J. BONNIN, *Geol. Soc. Amer. Bull.*, 84, 1973, p. 3137-3180.
(6) R. C. DIETZ et J. C. HOLDEN, *Nature*, 229, 1971, p. 309-312.
(7) J. PIVETEAU, *Ann. de Pal.*, 27, 1938.
(8) C. Mc A. POWELL et P. J. CONAGHAN, *Earth and Planetary Science Letters*, 20, 1973, p. 1-12 (North-Holland Publ. Cy).
(9) H. TERMIER et G. TERMIER, *Histoire géologique de la Biosphère*, 1952 (Masson éd.).
(10) H. TERMIER et G. TERMIER, *L'évolution de la Lithosphère. II. Orogenèse.* Fasc. 1 et 2, 1956-1957 (Masson éd.).
(11) H. TERMIER et G. TERMIER, *NATO Symp.* Newcastle-u.-Tyne, 4, 3, 1972, p. 477-485.
(12) H. TERMIER et G. TERMIER, *Ann. Scient. Univ. Besançon*, 3, 22, 1974.
(13) H. TERMIER et G. TERMIER, *Ann. Soc. Géol. Belgique*, 97, 1974, p. 387-446.
(14) H. TERMIER et G. TERMIER, *Le Gondwana et l'évolution des faunes marines durant le Carbonifère et le Permien : relations paléogéographiques de l'Afghanistan* [*Liv. Jub. A. F. de Lapparent*, 1976 (sous presse)].

7

Reprinted from *Soc. Géol. France Bull.*, ser. 7, **17**:56–67 (1975)

Mésogée permienne, chaîne hercynienne et cassure téthysienne

par Ion ARGYRIADIS *

Résumé. — Dans la partie du domaine mésogéen comprise entre la Méditerranée moyenne et l'Iran central, la paléogéographie de la fin du Paléozoïque et en particulier celle de la Téthys s'organise autour des restes de la chaîne hercynienne. Aucun hiatus à fond océanique ne sépare, à cette époque, l'Afrique de l'Europe. C'est plus tard, au début du Mésozoïque, que se manifeste un début d'écartement des deux blocs : la « cassure téthysienne ». Celle-ci induit une nouvelle paléogéographie, indépendante de l'héritage hercynien, comme le montre une comparaison des dispositions respectives de la paléogéographie permienne et des ophiolites mésogéennes alpines.

Abstract. — Within the Mesogean domain extending from the central Mediterranean to Iran the paleogeography at the end of the Paleozoic, particularly within the Tethys, was related to remnants of Hercynian chains. At this epoch no depression with oceanic floors separated Africa and Europe. Subsequently, in the beginning of the Mesozoic, the first evidence of separation of the two blocks is indicated by the « cassure téthysienne ». This separation initiated a new paleogeographic style independant of hercynian trends. This independance is indicated by comparison of Permian paleogeography and the distribution of ophiolites in the Alpine Mesogean.

Introduction.

Il est une vieille question de la géologie alpine et, us généralement, de la théorie orogénique : les aînes alpines sont-elles liées génétiquement aux aînes hercyniennes ? Sont-elles des « Altaïdes osthumes » comme le pensait E. Suess, ou le résultat 'orogenèses complètement indépendantes ? Aujour'hui cette question se pose un peu différemment : les chaînes mésogéennes (crétacées et tertiaires) nt le résultat de l'oblitération d'un domaine à nd océanique séparant l'Afrique et l'Eurasie, uels sont les rapports historiques, c'est-à-dire ographiques et génétiques, de ce domaine avec le omaine hercynien plissé ?

Explicitons les données de ce problème. Que ce it dans l'optique de la théorie géosynclinale ou dans lle de la « tectonique globale », il est incontestable ue les chaînes mésogéennes résultent de la déforation violente du domaine qui séparait les blocs ntinentaux gondwanien et eurasiatique, et, pour secteur qui nous concerne, l'Afrique et l'Europe. A la fin du Carbonifère se termine l'histoire ercynienne et commence une nouvelle période qui ous mène jusqu'aux grandes compressions du ésozoïque supérieur et du Tertiaire. Le Permien pparaît donc comme une époque-charnière de première importance pour la compréhension des mécanismes fondamentaux de cette évolution. Le cadre général de celle-ci a fait l'objet de reconstitutions, dans des modèles dérivistes fondés sur des méthodes d'extrapolation retrospective de l'expansion océanique, ou tout simplement sur des concordances géométriques de lignes de rivages ou d'isobathes.

Or toutes les reconstitutions sont à peu près unanimes sur l'allure du domaine mésogéen : à l'Ouest elles placent une masse continentale ininterrompue reliant l'Afrique à l'Europe. Vers l'Est, par contre, elles admettent une Téthys permienne, largement ouverte vers l'océan [J. F. Dewey *et al.*, 1973 ; R. Dietz et J. C. Holden, 1970 ; M. Kamen-Kaye, 1972]. Des mesures de paléomagnétisme effectuées sur des roches permiennes de la Téthys [C. B. Gregor et J. D. A. Zijderveld, 1964 ; D. van Hilten et J. D. A. Zijderveld, 1966] tendent à montrer l'existence d'un décrochement dextre entre Eurasie et Gondwana. Malheureusement, ces mesures sont encore trop rares et ponctuelles pour pouvoir en tirer des conclusions définitives.

* Lab. de géologie historique. Univ. Paris Sud, 91405 Orsay.
Note présentée oralement à la séance du 4 novembre 1974, manuscrit définitif remis le 7 novembre 1974.

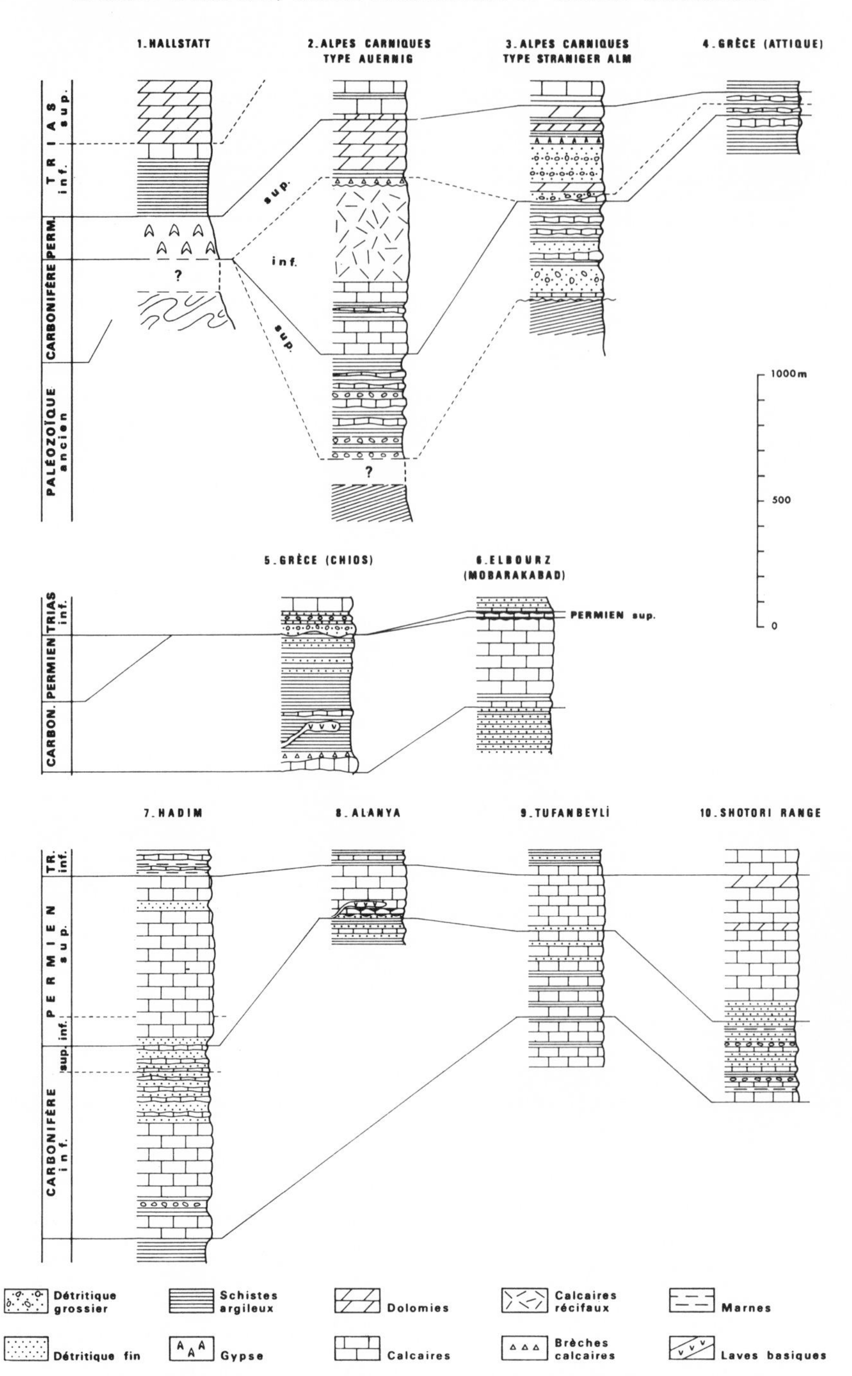

FIG. 1. — Quelques séries stratigraphiques caractéristiques de la Téthys fini-paléozoïque *(Voir légende page suivante).*

A. — La mésogée permienne.

a) *Un problème majeur : les relations de l'Europe [e]t de l'Afrique à la fin du Paléozoïque et à l'aube [d]u Mésozoïque.*

Quelle était la nature du fond de cette mer ? [L]e problème est de taille pour la reconstitution de [l]a dérive des continents, mais aussi, en particulier, [p]our la réponse à la question posée au début de cet [a]rticle.

Si, en effet, ce fond était de nature océanique, [e]t étant donné que des ophiolites assimilées par [b]eaucoup à un fond de cette nature prennent part [s]ystématiquement à la constitution de l'édifice [p]lissé mésogéen, on pourrait admettre un modèle [q]ui ferait de cet édifice le résultat final de la ferme[t]ure d'un « coin » océanique existant déjà depuis [(]et peut-être avant) l'Hercynien. Si, par contre, un tel océan n'existait pas à la fin du Paléozoïque, il faudrait admettre que les ophiolites mésogéennes ont été mises à nu à la faveur d'une distension postérieure à cette époque et au moyen de processus encore discutés, parmi lesquels une océanisation post-permienne n'est pas le moins probable.

La plupart des modèles dérivistes proposés à ce jour admettent la nature océanique de la Téthys permienne. Il me semble pourtant que leur majorité a fait la part trop belle à l'extrapolation à partir de données géométriques et que certains d'entre eux ont cédé à la tentation de l'assimilation pure et simple d'une mer largement ouverte à un océan. Les données stratigraphiques concernant le Paléozoïque et le Trias qu'ils intègrent sont par contre, je pense, insuffisantes.

Il y a, cependant, une façon correcte d'aborder ce problème : c'est celle fondée sur l'analyse des

---

(Légende de la fig. 1, page précédente).

1) Série de la nappe supérieure de Hallstatt (d'après [A]. Tollmann). La discordance hercynienne n'est pas direc[t]ement observable à cause du décollement tectonique géné[r]alisé des Alpes calcaires septentrionales sur la zone des [G]rauwackes. Cependant, les spécialistes s'accordent pour [f]aire de cette dernière le substratum hercynien plissé des [u]nités austro-alpines supérieures.

2) Série d'Auernig dans les Alpes Carniques. D'après [F]. Heritsch, F. Kahler et K. Metz [1934] et I. Argyriadis [1968]. Colonne reconstituée d'après les coupes de terrain [d]u Garnitzenberg pour le Carbonifère supérieur du Trogkofel [e]t de Reppwand-Gartnerkofel pour le Permo-Trias. L'« unité [d]'Auernig » est décollée à sa base [I. Argyriadis, 1970] mais il [e]xiste une différence de plissement certaine entre le Carbonifère [s]upérieur et son substratum (« couches de Hochwipfel »).

3) Série de Straniger Alm dans les Alpes Carniques. D'après [A]. Mariotti (1972) modifiée sur la base de données de [F]. Heritsch [1936] et de mes propres levés de terrain. Entre [l]e Carbonifère supérieur débutant au Kasimovskien et la [t]ransgression du Permien supérieur décrite par A. Mariotti [s]'intercale, notamment au Nord du point côté 1 781 m, [a]u SW de Zöllner See, un petit affleurement de « calcaire à [P]seudoschwagérines inférieur » (Sakmarien). La discordance [a]ngulaire à la base du Carbonifère supérieur est faible. Elle [e]st cependant réelle sur le plan régional, puisque la même [s]érie transgresse sur des termes différents du substratum [(]p. ex. aux environs de Paularo d'Incarojo, près de la casera [P]izzul, où les couches du Carbonifère supérieur reposent en [d]iscordance sur les formations du Carbonifère inférieur du [v]ersant sud du Zermula).

4) Série du Mont Beletsi en Attique, décrite par B. Clément, [C]l. Guernet et M. Lys [1971].

5) Série de l'autochtone de Chios, reconstituée d'après [l]es données de H. Besenecker *et al.* [1968] et mes propres [o]bservations. Cet autochtone relatif supporte une nappe [d]e charriage dont le Paléozoïque a des affinités péri-hercy[n]iennes. Retraçant l'histoire de l'autochtone, H. Besenecker [e]t *al.* argumentent en faveur de l'existence d'une phase her[c]ynienne. Les arguments avancés ne sont, cependant, pas [t]out-à-fait convaincants et ils disent d'ailleurs clairement [e]ux-mêmes (p. 138) qu'une incertitude subsiste à propos de cette question. D'après mes propres observations, à Chios comme en face, à Karaburun, il existe bien un ravinement du substratum par les conglomérats permo-scythiens, mais pas de discordance angulaire, donc de plissement hercynien véritable.

6) Série de Mobarakabad, dans l'Elbourz central, à l'ENE de Téhéran, d'après F. Bozorgnia, [1973]. Grande lacune entre le Carbonifère inférieur et le Permien supérieur mais pas de trace de discordance hercynienne.

7) Série des yaylas de Hadim, dans la « nappe de Hadim » (*sensu* M. Blumenthal). Les formations de cette nappe — dont le Paléozoïque — ont été décrites en 1971 par N. Ozgül, qui lui donne le nom d'« unité du Taurus moyen ». J'ai levé en détail la série figurée, indiquée par N. Ozgül, le long de la piste forestière qui va de Hadim à Muzvadi, par la haute vallée du Göksu. Au Sud de Yariçak yayla au niveau du point culminant de la piste (2 000 m) un plateau d'altitude fait affleurer, en pendage monoclinal, les couches allant du Dévonien au Trias. La série est continue. Les faciès régressifs sont la règle à partir du Viséen et le sommet du Carbonifère est très réduit. La base du Permien, finement détritique avec quelques intercalations de lentilles calcaires à Pseudoschwagérines, montre des niveaux de brève émersion avec oxydation et latéritisation. Le faciès calcaire, transgressif mais peu profond, revient au sommet du Permien inférieur. La biostratigraphie de cette série est en cours d'étude en collaboration avec M. Lys.

8) Série du Paléozoïque métamorphique du massif d'Alanya. D'après I. Argyriadis [1973]. Les épaisseurs, estimées dans des formations épimétamorphiques très plissées, ne peuvent être qu'indicatives.

9) Série de Tufanbeyli dans le Taurus oriental, près de Pinarbashi. D'après N. Ozgül *et al.* [1973]. Cette série est avec celle de Hadim une des plus caractéristiques pour le Paléozoïque du Taurus.

10) Série de Shotori Range (Kuh-e-Shotori) dans la région de Tabas. D'après J. Stöcklin [1968]. Cette série, en plein Iran central, est remarquable par sa convergence manifeste avec celles du Taurus par exemple. Cette stratigraphie me semble indiquer que l'Iran central faisait partie, jusqu'au Trias, de la plate-forme arabe.

tendances des séries stratigraphiques, plus particulièrement de celles du Permien, analyse qui cherche en même temps à mettre en évidence leur substratum : chaîne récemment plissée, socle continental ou fond océanique ? C'est ce que cet article va essayer de résumer, pour le secteur qui s'étend de l'actuelle Méditerranée moyenne à l'Iran central, entre l'Eurasie et le bloc afro-arabique.

Il s'agit d'une tentative de reconstitution paléogéographique d'un type particulier : ce n'est pas la recherche d'un instantané reflétant tant bien que mal les conditions géographiques à un moment donné, mais celle du schéma évolutif d'une vaste région, replaçant les phénomènes géologiques dans la dynamique de leur histoire.

Toute reconstitution paléogéographique nécessite, bien entendu, une analyse structurale préalable destinée à replacer dans leur contexte originel des séries dispersées par des tectoniques successives et violentes. Ce n'est évidemment pas l'affaire de cet article. Je la supposerai faite, sur la base de la connaissance critique des derniers travaux en cette matière, dont certains figurent, d'ailleurs, dans la bibliographie.

b) *La paléogéographie fini-paléozoïque (du Carbonifère sup. au Trias basal).*

De nombreux logs levés en Méditerranée moyenne, orientale et en Moyen-Orient donnent une image de ce qui a pu être la Mésogée permienne. Je me bornerai, pour des raisons d'économie, à donner quelques-uns d'entre eux (cf. fig. 1). L'ensemble des séries étudiées définit un domaine marin situé entre deux bordures continentales, eurasiatique et africaine, divisé lui-même en trois domaines particuliers, caractérisés par des stratigraphies diverses sur des socles divers : du côté européen, un sillon marin cerne vers le sud les reliefs hercyniens ; c'est le domaine *péri-hercynien.* Lui fait suite, vers le large, un domaine à tendances positives constantes depuis le Paléozoïque inférieur, non plissé à l'Hercynien, que j'appellerai *haut-fond intermédiaire* ; de l'autre côté de ce haut-fond se trouvait un nouveau domaine marin, très large, souvent subsident, le *domaine marginal arabo-africain* ; il passe à son tour, insensiblement et sur une très grande étendue, au domaine continental africain, partie du domaine gondwanien, qui sort du cadre de cette étude.

1. *Le domaine péri-hercynien.* Avec des séries comme celle de la nappe de Hallstatt ou des Alpes du Gail en Autriche, des Dolomites dans le Haut-Adige, celle de Gebze et d'Amasra en Turquie, nous sommes encore sur l'orogène hercynien. La discordance est nette sur le substrat plissé, l'influence continentale prédominante, la présence marine épisodique et marginale. Elle s'affirme pourtant vers

FIG. 2. — Localisation actuelle des séries du Paléozoïqu supérieur ayant servi à l'élaboration de cet article. *Carr pleins* : Domaine péri-hercynien (les séries peuvent appa tenir, selon les cas, à différentes parties de ce domaine, d régions épi-hercyniennes aux parties externes du sillon *Triangles* : Haut-fond intermédiaire. *Ronds* : Domaine margin arabo-africain. Les numéros correspondent à ceux des colonn de la figure 1.

la fin du Paléozoïque et le début du Trias, où transgression téthysienne gagne sur l'Europe herc nienne. Ce sont des séries comme celles des Alp Carniques ou du Caucase septentrional qui marquer le véritable passage au sillon péri-hercynien. L flexure semble avoir été active surtout au Carb nifère supérieur et au Permien moyen (décharg conglomératiques importantes et répétées venant d la chaîne hercynienne faiblement marquée). A Permien inférieur la morphologie s'étant quelqu peu stabilisée, des récifs viennent s'installer s cette marge. A la fin de l'époque (Permien supérieu Trias basal) le sillon semble comblé par le détritiqu hercynien ; la mer transgresse sur une région pla et monte en pente douce à l'assaut des dernie reliefs varisques. Des séries comme celles de l'A tique-Eubée et de l'allochtone de Chios en Grèc de Balia-Maden et de Bursa en Asie mineure, o la discordance varisque est faible ou absente ma le détritique hercynien présent, doivent apparten à une région plus centrale du même sillon.

Il faut souligner les caractéristiques commun de toutes ces séries : discordance hercynienne pr sente mais tendant à s'effacer vers le sud, appor détritiques importants et fréquents, mal calibrés, o les couleurs rouges prédominent. Pendant les p riodes de relâchement de l'érosion, établissement d récifs. Séries à caractère assez subsident, mais co blement quasi total du domaine vers la fin d l'époque, et, en même temps, fracturation et ar vées occasionnelles de magma, généralement acid

Je rattache à ce domaine la série de Bükk e Hongrie, les séries des Karawanken et de la Sl vénie, les séries de l'Othrys, de l'Eubée et de l'A tique dans les Hellénides. La série du Vélébit e

;roatie ne lui est certainement pas étrangère. En 'urquie, il faut envisager l'appartenance à ce même omaine du Permien de certaines unités des nappes yciennes (p. ex. Haticeana dağ) ce qui s'accomoerait fort bien avec l'attribution à ces mêmes nités d'une origine interne, « ultra-Menderes ». Le illon péri-hercynien est représenté, dans le Grand aucase, par les séries qui affleurent entre les ivières Belaja et Bolshaja Laba, à Permien supéieur marin transgressif sur des conglomérats et rès, eux-mêmes transgressifs et discordants sur le ubstratum hercynien.

Ce domaine très allongé, de l'Autriche au Caucase, st relativement étroit. Les rares cas montrant des éries lui appartenant loin de la marge hercynienne ont imputables à une tectonique ultérieure, polyhasée, très importante.

2. *Le domaine marginal arabo-africain.* C'est dans ertaines unités des chaînes tauriques, en Turquie, u'on peut, le mieux, détailler les séries caractéistiques de ce domaine. Il en existe une vingtaine e coupes, similaires à celles figurées dans la figure 1. l faut y remarquer l'absence de toute discordance ngulaire depuis le Paléozoïque inférieur. Les dépôts ermo-carbonifères sont très peu profonds, souvent nême à la limite de l'émersion. Cependant, leur paisseur atteint plusieurs centaines de mètres, nontrant ainsi une tendance à la subsidence. Le arbonifère supérieur peut être régressif, avec de etites lacunes, notamment de l'Ouralien. Des calaires bio-détritiques y alternent avec des quartites clairs, fins, d'une granulométrie très homogène. u Permien inférieur les quartzites deviennent rédominants, la profondeur de dépôt diminue ncore, et les niveaux de franche émersion, avec aléosols latéritiques, n'y sont pas rares. Cepenant, des intercalations locales de bancs calcaires ouvent teintés de rouge, montrent des pisolites l'origine algaire très caractéristiques, et une très iche faune de Fusulinidés (Pseudoschwagérines). u Permien supérieur — à partir du Murghabien — a sédimentation change de caractère et devient ranchement calcaire ou marneuse. Les calcaires ont souvent argileux, très bitumineux, noirs, enahis par une faune et une flore peu profondes, xtrêmement riches en Fusulinidés, Polypiers, 3rachiopodes et Algues. Vers le haut apparaissent ocalement des dépôts dolomitiques. A partir du Permien moyen commence à se manifester une racturation et une montée de magma à prédomiance basique. Le passage au Trias basal se fait rogressivement, mais le caractère de la sédimenation change à la limite des deux périodes. Le Scythien a le faciès « Seis » des Alpes méridionales, vec des schistes argileux et calcaires fins en plauettes, à faune très appauvrie composée de quelques ares Lamellibranches (Claraia, Unionites) [O. Monod *et al.*, 1974, p. 118] [1].

Des variantes locales (tendance plus détritique ou, au contraire, établissement de récifs) peuvent, bien entendu se manifester çà et là. Je rattache à ce domaine les séries de Kirchaou et Casbah Leguine en Tunisie méridionale, la série des Talea Ori en Crète, celle du Karadağ dans les nappes lyciennes (Turquie), les séries des nappes d'Antalya, celles du Taurus occidental et oriental. Lui appartiennent également les séries permo-carbonifères des massifs métamorphiques du Taurus : Alanya, Bolkar dağ, Keban. A l'Est de la Turquie, les séries de Hazro et celle de Harbol près de la frontière irakienne sont des témoins du domaine marginal araboafricain, comme celle de Djoulfa en Transcaucasie. Enfin, plus loin, les séries du Zagros, certaines séries de l'Elbourz (Dozdehband, Nessen) et celles de l'Oman (autochtones et allochtones) sont attribuables au même ensemble.

De toutes ces séries, il faut retenir les caractéristiques communes, à savoir : absence de discordance angulaire d'âge hercynien, faible profondeur (à la limite de l'émersion) mais tendance subsidente, apports détritiques fréquents et abondants, exclusivement quartzeux, très fins et homogènes, caractères qui définissent un détritique d'origine « africaine », par opposition à celui grossier, rouge et mal calibré d'origine hercynienne. La fracturation de ce domaine à la fin du Paléozoïque et la montée de magma basique en est également une caractéristique importante.

3. *Le haut-fond intermédiaire.* Nous avons déjà examiné deux domaines marins — l'un du côté européen, l'autre du côté africain — dont le caractère épicontinental est hors de doute, leur substratum étant connu. Le domaine qui les sépare n'a rien à faire, non plus, avec un océan. Il existe un certain nombre de séries comme celles (Chios, Elbourz) de la figure 1, qui ne montrent pas de discordance hercynienne mais qui sont marquées par des lacunes fréquentes, réparties chronologiquement d'une manière irrégulière d'une coupe à l'autre. Tantôt c'est le Permien terminal qui transgresse sur le Carbonifère inférieur, tantôt le Trias basal repose à plat sur le Carbonifère supérieur, lui-même en concordance avec son substratum. Tout se passe comme si une région élevée de la Téthys était épisodiquement « léchée » par des transgressions marines hésitantes et éphémères.

Je rattache à ce domaine entre autres, outre les séries figurées, la série de la Bukovica (Cokotin) dans les Dinarides [J. P. Rampnoux, 1968], la série de Molaï (?) dans le Péloponèse [C. Renz, 1955], Karaburun en face de Chios, plusieurs séries de l'Elbourz central (Kalariz, Khoshyeilagh [F. Bozorgnia, 1973]).

Ce domaine appartient, par l'absence de discordances hercyniennes, au bâti africain, stable depuis l'aurore du Paléozoïque. Mais les apports détritiques d'origine européenne n'y sont pas tout à fait absents, quoique rares et tardifs (conglomérat werfénien de Cokotin, conglomérat de même â[ge] dans l'autochtone de Chios). L'âge tardif de c[es] détritiques se justifie par l'emplacement de ce d[o]maine au-delà du sillon péri-hercynien : le « piè[ge]

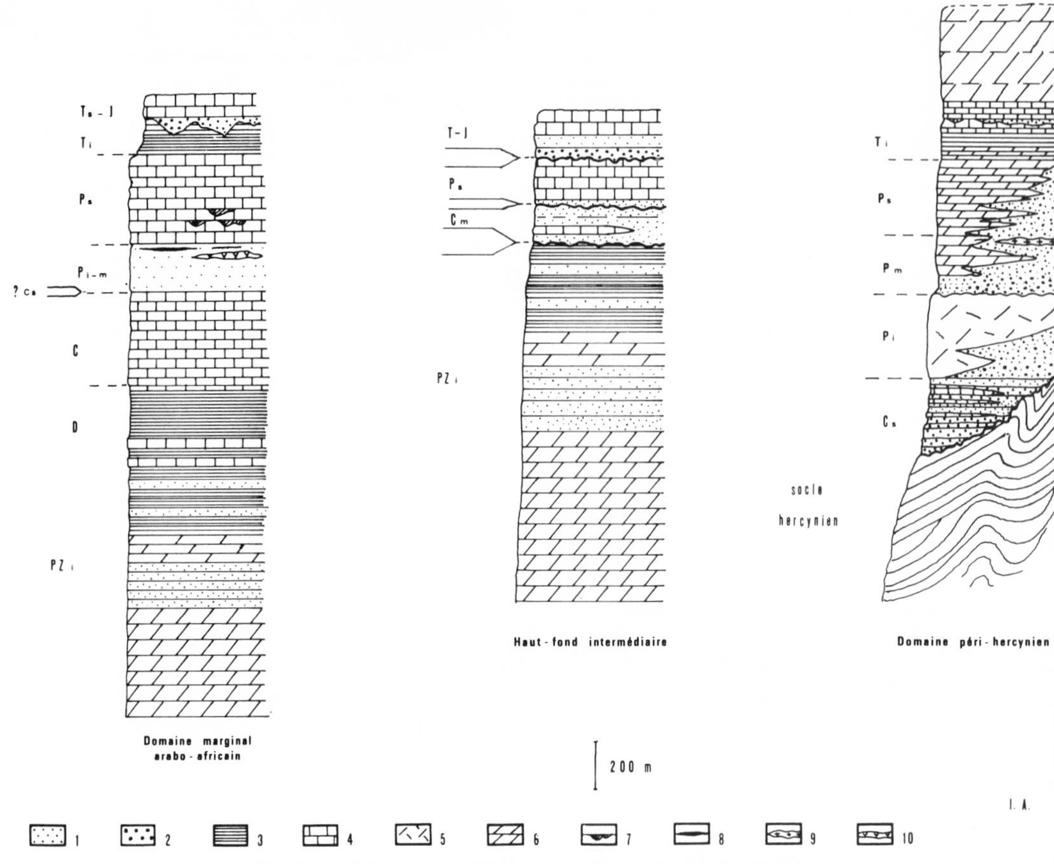

FIG. 3. — Colonnes synthétiques caractérisant chacun des domaines.
1 : détritique fin ; 2 : détritique grossier ; 3 : schistes argileux et pélites ; 4 : calcaires lités ; 5 : calcaires récifaux; 6 : dolomies ; 7 : poches de bauxite ; 8 : charbon ; 9 : volcanisme acide ; 10 : volcanisme basique.

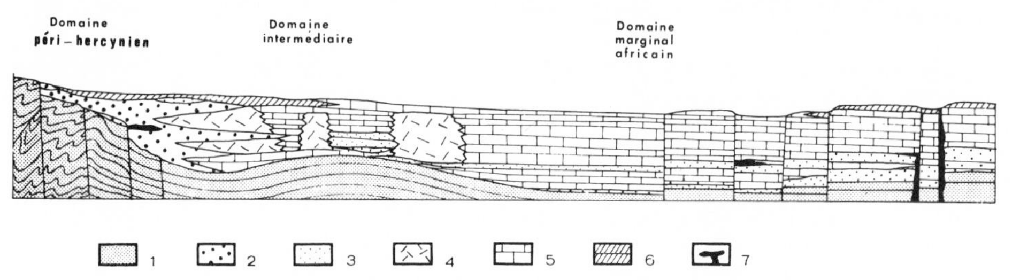

FIG. 4. — Coupe schématique à travers la Téthys permienne.
1 : socle ; 2 : détritique grossier ; 3 : détritique fin ; 4 : calcaires récifaux ; 5 : calcaires lités ; 6 : dolomies ; 7 : laves.

n creux » que constituait ce dernier a cessé de onctionner à la suite de son comblement à la fin u Permien. Le haut-fond intermédiaire est, nous voyons, le lieu où les influences hercyniennes et ondwaniennes se rencontrent et s'interpénètrent, trait d'union entre les deux grands blocs ontinentaux (fig. 3 et 4).

En conclusion de ce chapitre, force nous est de consater, après l'examen rapide des trois domaines, ue la paléogéographie fini-paléozoïque s'organise utour des restes de la chaîne hercynienne, est onditionnée par sa disposition, en un mot est inuite par cette dernière.

Le vaste domaine téthysien est affecté par une racturation marquée à la fin du Permien, qui 'exaspère à partir du Trias : brèches à blocs énornes dans le Scythien de plusieurs unités du Taurus, iversification brusque des faciès au-dessus de ce nême Scythien, manifestations volcaniques fréuentes à partir de l'Aniso-ladinien depuis les Alpes rientales jusqu'en Iran [J. H. Brunn *et al.*, 1970 ; K. W. Glennie *et al.*, 1973 ; J. Stöcklin, 1968 ; . Tollmann, 1963, etc.] Mais nous sommes déjà u-delà de la période envisagée.

B. — Chaîne hercynienne et ophiolites mésogéennes.

Ayant défini cette paléogéographie et ses principales lignes directrices et ayant constaté l'absence d'un océan entre l'Afrique et l'Europe au Permien, il est temps de jeter un coup d'œil sur la distension mésozoïque de la Téthys — préhistoire de la formation des chaînes crétacées et tertiaires — et d'essayer d'apporter une réponse, par cela même, à la question déjà posée à propos de ces dernières : « Altaïdes posthumes » ou chaînes indépendantes ?

Pour ce faire, nous disposons d'un bon « marqueur » : la répartition des ophiolites mésogéennes alpines. Il s'agit de voir comment se sont disposées mutuellement, pendant le Mésozoïque, la cassure de la croûte continentale qui a mis à nu le matériel ophiolitique et la vieille ossature hercynienne.

Nous allons donc essayer de suivre, autant que possible, ces deux ensembles-clés à travers l'espace téthysien (fig. 5).

a) *La chaîne hercynienne.* Il est assez facile de suivre la trace de sa bordure externe — la seule

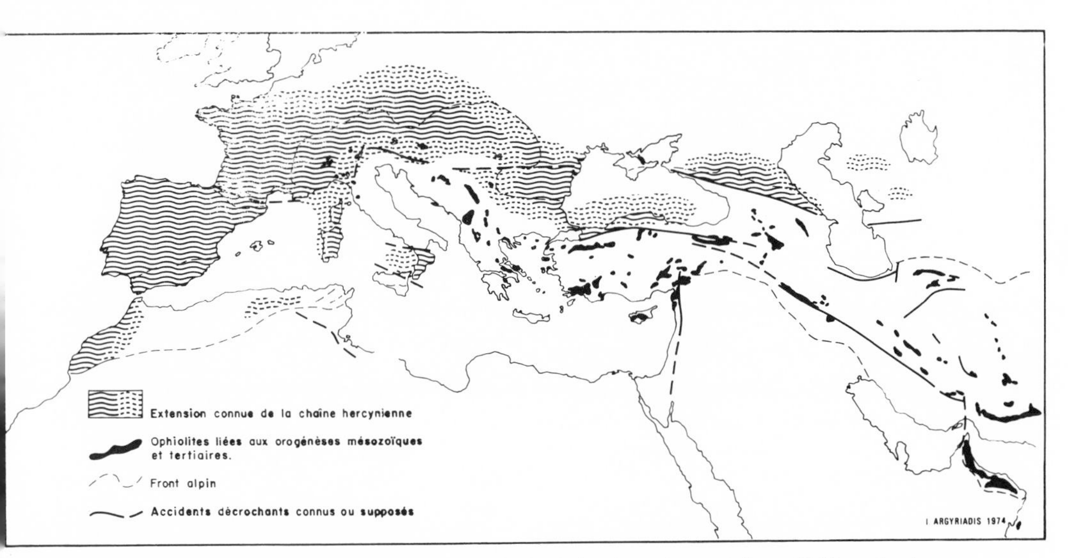

Fig. 5. — Disposition relative actuelle des restes de la chaîne hercynienne et des ophiolites mésogéennes.

ui nous intéresse ici — depuis la Méditerranée noyenne. On la trouve dans les chaînes d'Afrique lu Nord (p. ex. Djurdjura) puis en Calabre dans les nassifs d'Aspromonte et de la Sila, qu'il faut cependant replacer loin vers l'Ouest, puisqu'ils doivent eur emplacement actuel à d'importants charriages alpins. Ensuite, vers le Nord, l'autochtone toscan est épi-hercynien, les Alpes Lombardes et les Dolomites aussi. Avec les Alpes Septentrionales (charriées), les Alpes du Gail, les Alpes Carniques et les Karawanken nous touchons le rivage téthysien de cette chaîne. Ce rivage continue vers le sud, probablement à travers la Serbie, puisque la Serbie orientale — et les Balkans — sont épi-hercyniens et que, par contre, la présence de vraie discordance varisque dans les zones plus externes est très incertaine.

Dans les Hellénides l'investigation est délicate. Si les zones très internes, p. ex. le massif serbo-macédonien, sont incontestablement épi-hercyniennes, les données sont moins sûres pour les zones plus externes. Toute reconstitution paléogéographique est ici compliquée par la discussion structurale qui bat son plein et qui n'a certainement pas dit son dernier mot [J. Aubouin, 1973]. Nous pouvons, cependant, préciser indirectement et approximativement l'emplacement des rivages hercyniens : dans les zones tout-à-fait externes, les rares témoins du Paléozoïque supérieur connus à l'affleurement (Talea Ori en Crète p. ex.) sont similaires, par leur stratigraphie, aux séries du domaine arabo-africain, connues en Turquie dans les Taurides. Plus à l'intérieur, en Attique par exemple, les unités charriées (Mont Parnès, Pateras) sous-jacentes aux ophiolites, ont un caractère certainement péri-hercynien. La même constatation vaut pour les séries de l'Othrys [G. Marinos ; 1960] et de l'Eubée. Je pense donc qu'il est légitime d'admettre qu'en Grèce les reliques de la chaîne hercynienne se cantonnent dans un domaine plus interne que les racines des unités sédimentaires charriées d'Attique et de Béotie, elles-mêmes plus internes que le fameux « massif » d'Attique-Cyclades-Eubée [I. Argyriadis, 1974 ; J. Aubouin, 1973].

En Turquie les choses sont plus claires parce que le Paléozoïque des zones externes comme celui des zones internes affleure largement. Il faut, toutefois, se débarasser une fois pour toutes du concept tenacement ancré de « vieux » massifs métamorphiques tels le massif du Menderes. Pas plus que son homologue grec, le métamorphique de l'Eubée, où le métamorphisme affecte au moins le Crétacé daté, le massif du Menderes n'est un « vieux » massif. Ses formations métamorphiques comprennent du Permien et du Trias [P. Ch. de Graciansky, 1972 ; J. Wippern, 1964] et montent jusqu'au Crétacé sup. [S. Dürr, rens. or.]. Nulle discordance *d'âge hercynien* n'y est discernable. Il existe, bien sûr, en Turquie, des massifs métamorphiques d'âge varisque : ils sont tous cantonnés, d'après les données de terrain et celles qui proviennent de l'étude critique de la bibliographie, au Nord d'une ligne qui coïncide approximativement avec l'accident paphlagonien (nord-anatolien). Mais cet accident récent n'est ici qu'une référence géographique et n'a aucun rapport causal avec notre distribution paléogéographique. Dans l'ensemble de l'Anatolie, la totalité de la chaîne hercynienne reste au Nord de cet accident ou, plus précisément, d'une zone sinueuse qui passe par le golfe d'Edremit (Egée), Eskisehir, le Nord d'Ankara, et qui, comprenant des ophiolites et très fortement tectonisée va, par Erzincan et Erzerum, vers le nœud arménien. Plus à l'Est, le rivage méridional du continent hercynien apparaît dans le Grand Caucase [J. M. Mstislavskiy *et al.*, 197[illegible]] avant de traverser la Caspienne au niveau de son étranglement pour s'en aller vers le Nord sous les plaines du Kazakhstan. Au Nord de cette limite esquissée en Anatolie l'orogenèse hercynienne est mise en évidence et le Permo-carbonifère a le faciès houiller (p. ex. bassin d'Eregli-Zonguldak). C'est le domaine pontique des géologues. Au Sud, par contre, aucune preuve sérieuse de l'existence d'une phase varisque n'a été avancée. Sur le terrain toutes les séries étudiées se sont révélées sans exception exemptes de toute discordance angulaire de cet âge.

b) *Les ophiolites*. Essayons, maintenant de placer les ophiolites mésogéennes alpines par rapport au domaine hercynien : une difficulté majeure est liée au choix de l'interprétation structurale et la détermination en fonction de cette interprétation de la ou des zones de racines des ophiolites. Cette difficulté n'est pourtant pas insurmontable pour ce qui nous intéresse ici. Nous avancerons progressivement, en envisageant toutes les interprétations vraisemblables qui ont été proposées jusqu'ici.

En Méditerranée moyenne la situation est claire dans l'Apennin et dans les Alpes Orientales les principales zones à ophiolites (allochtone Ligure et zone Sud-pennique-faciès du Glockner) sont, en effet, incontestablement enracinées entre deux morceaux hercyniens (respectivement socle de l'île d'Elbe et autochtone toscan, Pennique des Hohe Tauern et dalle de Silvretta-Oetztal). Dans les Dinarides, la zone ophiolitique est encore unique et contiguë à la chaîne hercynienne (zone du Vardar et massif Serbo-macédonien).

Dans les Hellénides il y a deux zones principales à ophiolites (zone du Vardar et zone Sub-pélagonienne) — et deux interprétations : l'une consiste à admettre que la zone la plus interne (Vardar) est la vraie zone des racines de toutes les unités à ophiolites ; celles-ci se seraient avancées à la fin du Mésozoïque et au début du Tertiaire, à la faveur d'une tectonique polyphasée, par dessus les zones plus externes comme par exemple les unités béotiennes et la zone pélagonienne. La seconde hypothèse consiste à admettre deux zones d'enracinement des ophiolites, celle du Vardar et celle subpélagonienne qui serait enracinée en avant de la zone pélagonienne. Je dois dire que c'est à la première des deux hypothèses que va ma préférence [I. Argyriadis, 1974].

Voyons maintenant notre schéma dans chacune des deux hypothèses :

— Dans l'hypothèse de la zone unique de racines cette dernière (Vardar) est adossée, vers l'intérieur u bâti hercynien. Vers l'extérieur nous ne connais-ons pas d'indice sûr de l'existence de la chaîne varisque, mais nous trouvons des séries typiquement éri-hercyniennes (Attique etc.). Nous pouvons dire que la « cassure ophiolitique » a affecté, ici, le talus ercynien, du côté de l'avant fosse.

— Dans l'hypothèse des deux zones de racines, ne autre zone ophiolitique serait enracinée à l'exté-ieur de la précédente : cette zone serait dans ce as, *a fortiori*, encore plus loin des rivages hercyniens.

Dans les deux cas nous constatons que, dans les Hellénides, la ou les cassures ophiolitiques ne coupent lus à travers le continent hercynien, mais lui estent très proches.

Le schéma est le même en Asie Mineure, sauf u'ici il y a non plus deux mais trois zones principales l'affleurement des roches vertes — et autant d'inter-rétations structurales.

Il est inutile de les examiner en détail, les unes près les autres. Le raisonnement reste fondamen-alement le même que pour les Hellénides, fondé, ci, sur des données de terrain mieux exposées à 'observation. Etant donné que les traces des plisse-nents varisques restent résolument au Nord de toute zone possible d'enracinement des ophiolites, la preuve est faite que la ou les cassures sont restées à l'extérieur du continent hercynien. Qui plus est, à l'Est du méridien d'Ankara on ne rencontre plus, au Sud de la zone ophiolitique la plus septentrionale, que des séries appartenant au domaine marginal arabo-africain, ce qui tend à démontrer que la cas-sure s'éloignait de l'orogène hercynien.

Cette dernière constatation devient flagrante plus à l'Est. Dans la région de l'Araxe, Djoulfa est une coupe qui appartient, au Permien, au domaine arabo-africain. De même l'Elbourz, chaîne tecto-niquement complexe, comprend des unités à Paléozoïque supérieur appartenant au « haut-fond intermédiaire », d'autres appartenant à la marge arabo-africaine. Or, la cassure ophiolitique la plus septentrionale possible passe bien au Sud de cette chaîne. Nous nous trouvons donc, déjà, dans des régions où la cassure ophiolitique sépare deux mor-ceaux de la plate-forme arabo-africaine.

C. — Conclusions.

Les conclusions de ce travail se laissent classer en deux catégories :

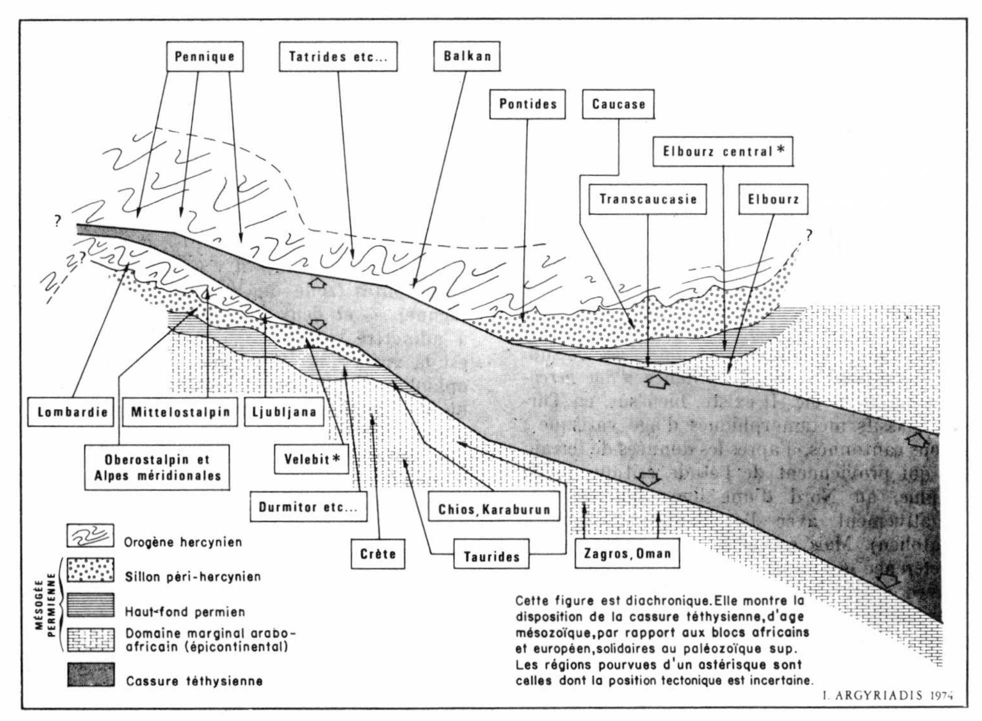

Fig. 6. — Les relations entre l'orogène hercynien, la paléogéographie du Permien et la cassure téthysienne.

La première concerne l'histoire fini-paléozoïque proprement dite. On constate que la paléogéographie du Permo-carbonifère s'organise en étroite liaison avec les reliefs de la chaîne hercynienne. C'est une paléogéographie induite par cette orogenèse.

Nous avons vu, par ailleurs, que l'existence d'un domaine à fond océanique dans la région étudiée (Méditerranée moyenne-Iran central) est contredite par toutes les données de terrain, stratigraphiques et retrotectoniques, et par la synthèse qui en découle : *Les blocs européen et arabo-africain étaient soudés depuis le choc hercynien.*

La seconde catégorie de conclusions concerne justement la création de ce fond de type océanique. Nous savons que les ophiolites mésogéennes étaient mises à nu et ont participé aux bouleversements tectoniques de la fin du Mésozoïque comme à ceux du Tertiaire. Puisque un fond océanique — ou un « plancher » ophiolitique quelconque — n'existait pas au Permien, nous sommes obligés d'admettre que plus tard, c'est-à-dire au Mésozoïque, une cassure a révélé ces ophiolites dans le domaine téthysien, en écartant les blocs continentaux qui l'encadrent. Nous l'appellerons *cassure téthysienne* (fig. 6).

Son parcours est déduit des constatations faites plus haut dans cet article. Il est oblique *et* par rapport aux contours hercyniens *et* par rapport à la paléogéographie permienne. *C'est une nouvelle organisatio de l'espace mésogéen qui est annoncée avec le débi de la cassure téthysienne.*

Elle induit, en particulier, une nouvelle paléc géographie méditerranéenne, et est certainement l'origine de plusieurs de ces zones isopiques fam lières aux géologues alpins : zones à volcanism initial, à radiolarites, etc.

Enfin, son parcours traverse nécessairement, a moins une fois, le front hercynien méridional, e ceci impose une nouvelle remarque : dans cett distension du début du Mésozoïque, la cicatric hercynienne ne se comporte absolument pas comm une zone faible de l'écorce, malgré les indiscutable indices de fracturation qu'on y constate à la fi du Paléozoïque (accidents tardi-hercyniens, volca nisme). Je crois que ceci est dû au fait que les deu phénomènes sont d'un ordre différent : *Le méca nisme qui provoque l'écartement semble apparten à un niveau d'organisation de la matière terrestr bien supérieur à celui des accidents, même profond de la partie supérieure de la lithosphère.*

1. De cet article je retiens la description stratigraphiqu du Trias mais, comme l'on sait, je suis en désaccord av l'interprétation structurale et les corrélations qui y sor proposées [cf. L. E. Ricou *et al.*, 1973].

Bibliographie

Argand E. (1922). — Tectonique de l'Asie. XIIIe Congrès géol. intern., fasc. 1, p. 171-372, Bruxelles.

Argyriadis I. (1968). — Le Permo-Carbonifère marin des Alpes carniques, jalon allochtone entre Nord-alpin et Sud-alpin. D.E.S., Travaux du Laboratoire de géologie historique, Fac. Sc. Orsay, 156 p.

Argyriadis I. (1970). — La position des Alpes carniques dans l'orogène alpin et le problème de la limite alpino-dinarique. *Bull. Soc. géol. Fr.*, (7), XII, n° 3, p. 473-480.

Argyriadis I. (1973). — Le Paléozoïque supérieur métamorphique du massif d'Alanya (Turquie méridionale). Description, corrélations et position structurale. *Bull. Soc. géol. Fr.*, (7), XVI, n° 2, p. 112-115.

Argyriadis I. (1974). — Sur l'orogenèse mésogéenne des temps crétacés. *Rev. Géogr. phys. et Géol. dyn.*, (2), vol. XVI, fasc. 1, p. 23-60.

Assereto R. (1972). — Notes on the Anisian biostratigraphy of the Gebze area (Kocaeli peninsula, Turkey). *Zeit. Deutsch. geol. Ges.*, Bd. 123, p. 435-444.

Aubouin J. (1973). — Des tectoniques superposées et de leur signification par rapport aux modèles géophysiques : l'exemple des Dinarides ; paléotectonique, tectonique, tarditectonique, néotectonique. *Bull. Soc. géol. Fr.*, (7), XV, p. 426-460.

Balogh K. (1964). — Die geologischen Bildungen des Bükk Gebirges. *Ann. Inst. géol. Hongrie.* Budapest.

Belov A. A. (1973). — Paleozoïc tectonics of the Wester and Central Taurus (Turkey). Academy of Science of the U.S.S.R. *Geotectonics*, n° 1, p. 31-38.

Besenecker H., Dürr S. *et al.* (1968). — Geologie vo Chios (Ag iis). Ein Überblick. *Geologica et Paleonto logica*, Marburg, 2, p. 121-150.

Blumenthal M. (1944). — Contribution à la connaissanc du Permo-carbonifère du Taurus entre Kayser Malatya. *Bulletin M.T.A.*, n° 1/31.

Blumenthal M. (1960/1963). — Le système structural d Taurus sud-anatolien. *In* Livre à la mémoir P. Fallot, *Mém. h.s. Soc. géol. Fr.*, t. II, p. 611-66

Bonneau M. (1973). — Sur les affinités ioniennes des « calcaire en plaquettes » de la Crète, le charriage de la séri de Gavrovo-Tripolitza et la structure de l'arc égéen *C.R. Ac. Sc.*, t. 277, p. 2453-2456.

Bozorgnia F. (1973). — Paleozoïc foraminiferal biostrati graphy of Central and East Alborz mountains. Ira *National Iran Oil Company, Geol. Lab. Publicatior* Teheran, n° 4.

Brinkmann R. (1971). — Jungpaläozoïkum und ältere Mesozoïkum in Nordwest Anatolien. *Bull. M.T A.* n° 76, p. 55-67.

Brunn J. H. (1964). — Sur la disposition originelle du systèm alpin en trois rameaux. Application de cette notio à l'analyse de grandes structures charriées : Alpe orientales ; Carpathes. *C.R. Ac. Sc.*, t. 25 p. 4739-4741.

BRUNN J. H. (1967). — Recherche des éléments majeurs du système alpin. *Rev. Géogr. phys. Géol. dyn.*, (2), vol. IX, fasc. 1, p. 17-34.

BRUNN J. H. *et al.* (1970). — Structures majeures et corrélations stratigraphiques dans les Taurides occidentales. *Bull. Soc. géol. Fr.*, (7), XII, n° 3, p. 515-556.

BUBNOFF S. V. (1931). — Über die permische Formation und über geologische Zeitwenden. *Naturwissenschaften*, 19, 29, p. 634-639.

CANUTI P., MARCUCCI M. et RADRIZZANI PIRINI C. (1970). — Microfacies e microfaune nelle formazioni paleozoiche dell'anticlinale di Hazro (Anatolia sud-orientale, Turchia). Consiglio Nazionale delle Ricerche. *Boll. Soc. geol. It.*, 89, p. 21-40.

CLÉMENT B. (1970). — Remarques sur le Permien du Pateras (Attique, Grèce). *C.R. somm. S.G.F.*, fasc. 2, p. 40-41.

CLÉMENT B., GUERNET C. et LYS M. (1971). — Données nouvelles sur le Carbonifère et le Permien du mont Beletsi, en Attique (Grèce). *Bull. Soc. géol. Fr.*, (7), XII, n° 1, p. 88-91.

DEWEY J. F. *et al.* (1973). — Plate tectonics and the evolution of the alpine system. *Geol. Soc. of America Bull.*, v. 84, p. 3137-3180.

DIETZ R. S. et HOLDEN J. C. (1970). — Reconstruction of Pangea : breakup and dispersion of continents Permian to Present. *J. geophys. Res.*, v. 75, n° 26, p. 4939-4956.

DONZELLI G. et CRESCENTI U. (1970). — Segnalazione di una microbiofacies permiana, probabilmente rimaneggiata, nella formazione di M. Facito (Lucania occidentale). *Boll. Soc. Natur. Napoli*, v. LXXIX, p. 3-19.

DOUVILLÉ H., SOLIGNAC M. et BERKALOFF E. (1933). — Découverte du Permien marin au Djebel Tebaga (extrême Sud tunisien). *C.R. Ac. Sc.*, t. 196, p. 21.

EPTING M. *et al.* (1972). — Geologie der Talea Ori/Kreta. *N. Jb. Geol. Pal. Abh.*, 141, 3, p. 259-285.

FLÜGEL H. (1964). — Die Entwicklung des vorderasiatischen Paläozoikums. *Geotektonische Forsch.*, 18, I-II, p. 1-68.

GLENNIE K. W. *et al.* (1973). — Late Cretaceous Nappes in Oman Mountains and their geologic evolution. *The American Association of Petroleum Geologists Bull.*, v. 57, n° 1, p. 5-27.

GLINTZBŒCKEL Ch. et RABATÉ J. (1964). — Microfaune et microfaciès du Permo-carbonifère du Sud tunisien. (International Sedimentary Petrographical Series, v. 7). Leiden, E. J. Brill, 36 p.

GRACIANSKY P. Ch. de (1972). — Recherches géologiques dans le Taurus lycien. Thèse, Orsay, série A, n° 896.

GREGOR C. B. et ZIJDERVELD J. D. A. (1964). — Paleomagnetism and the alpine tectonics of Eurasia. I. — The magnetism of some Permian red sandstones from northwestern Turkey. *Tectonophysics*, 1 (4), p. 289-306.

GÜVENÇ T. (1966). — Description de quelques espèces d'Algues calcaires du Carbonifère des Taurus occidentaux (Turquie). *Rev. Micropal.*, 9 (2), p. 94-103.

HERITSCH F. (1936). — Die Karnischen Alpen. Graz, 205 p.

HERITSCH F. (1940). — Das Mittelmeer und die Krustenbewegungen des Perms. *Wiss. Jb. Univ. Graz*, p. 305-338.

KAHLER F. (1970). — Die Überlagerung des variszischen Gebirgskörpers der Ost- und Südalpen durch jungpaläozoische Sedimente. *Z. Dtsch. geol. Ges. Dksch.*, v. 122, p. 137-143.

KAMEN-KAYE M. (1972). — Permian Tethys and Indian Ocean. *The American Association of Petroleum Geologists Bull.*, v. 56, n° 10, p. 1984-1999.

LAUBSCHER H. P. (1970). — Das Alpen-Dinariden-Problem und die Palinspastik der südlichen Tethys. *Geol. Rdsch.*, 53, Bd. 60, p. 813-833.

MARINOS G. (1960). — (Palaeontologic and stratigraphic investigations on eastern continental Greece). En grec. *Bull. Soc. géol. Grèce*, v. IV, fasc. 1, p. 14-28.

MARIOTTI A. (1972). — Étude stratigraphique et structurale d'une nouvelle série permocarbonifère dans les Alpes Carniques : l'unité de Straniger Alm (Autriche). Conséquences pour la tectonique régionale, varisque et alpine. *Bull. Soc. géol. Fr.*, (7), XIV, p. 25-33.

MASCLE G. (1973). — Étude géologique des Monts Sicani (Sicile). Thèse, Paris, n° C.N.R.S. AO 8294.

MEYERHOFF A. A. et TEICHERT C. (1971). — Continental drift, III : late paleozoic glacial centers, and devonian-eocene coal distribution. *Journ. of Geol.*, v. 79, 3, p. 285-321.

MIKLUKHO-MAKLAI K. V. (1954). — Foraminifères du Permien sup. du Caucase septentrional. V. NAUK. Issl. Geol. Inst. USEGEI. Minist. Geol. i Okhrany Nedr. Moscou.

MONOD O., MARCOUX J. *et al.* (1974). — Le domaine d'Antalya, témoin de la fracturation de la plate-forme africaine au cours du Trias. *Bull. Soc. géol. Fr.*, (7), XVI, n° 2, p. 116-127.

MSTISLAVSKIY J. M., ZUBREV I. N. *et al.* (1971). — Limite sud de la plate-forme épihercynienne scythe et étape cimmérienne de l'histoire du Caucase central. *Dok. Akad. Nauk. S.S.S.R.*, t. 198, n° 3, p. 680-683.

OZGÜL N. (1971). — (The importance of block mouvements in structural evolution of the northern part of central Taurus). En turc. *Türk jeol. kurumu Bült.*, vol. XIV, n° 1, p. 85-101.

OZGÜL N., METIN S. *et al.* (1973). — Tufanbeyli dolayinin Kambriyen ve Tersiyer kayalari. *Türk Jeol. Kurumu Bült.*, XVI, 1, p. 82-100.

RADELLI L. (1970). — La nappe de Balya, la zone des plis égéens et l'extension de la zone du Vardar en Turquie occidentale. *Géologie alpine*, t. 46, p. 169-175.

RAMPNOUX J. P. (1968). — Sur le problème du passage du Paléozoïque au Trias dans les Dinarides yougoslaves (secteur de Serbie centrale et du Monténegro oriental, Yougoslavie). *C.R. Ac. Sc.*, t. 267, sér. D, p. 1087-1090.

RENZ C. (1955). — Stratigraphie Griechenlands. Athens Inst. Geol. and Subsurf. Research.

RICOU L. E. (1971). — Le croissant ophiolitique péri-arabe : une ceinture de nappes mises en place au Crétacé supérieur. *Rev. Géogr. phys. Géol. dynam.*, t. 13, 4, p. 327-349.

RICOU L. E., ARGYRIADIS I. et LEFÈVRE R. (1973). — Proposition d'une origine interne pour les nappes d'Antalya (Taurides occidentales). *Bull. Soc. géol. Fr.*, (7), XVI, n° 2, 1974, p. 107-111.

SCHMIDT G. C. (1964). — A review of Permian and Mesozoïc formations exposed near the Turkey/Iran border at Harbol. *Bull. M.T.A.*, Ankara, n° 62, p. 103-119.

SIMIÇ V. (1938). — Über die Jungpaläozoïschen Fazies im Westserbien. *Bull. Inst. geol. Yougosl.*, Belgrade, VI.

STEPANOV D. L. (1967). — Carboniferous stratigraphy of Iran. Congr. Carbonif. Sheffield.

STÖCKLIN J. (1968). — Structural history and tectonics of Iran : a review. *The American Association Petr. Geol. Bull.*, v. 52, n° 7, p. 1229-1258.

STÖCKLIN J. *et al.* (1965). — Geology of the Shotori Range (Tabas area, East Iran). Geol. Survey of Iran, rapport n° 3.

TARAZ H. (1973). — Correlation of Uppermost Permian in Iran, Central Asia and South China. *Amer. Assoc. Petr. Geol. Bull.*, v. 57, n° 6, p. 1117-1133.

TOLLMANN A. (1963). — Ostalpensynthese. Deuticke, Wien, 256 p.

TOLLMANN A. (1972). — Die Neuergebnisse über die Trias-Stratigraphie der Ostalpen. *Mitt. Geol. Bergbaustud.*, v. 21, p. 65-113.

VAN HILTEN D. et ZIJDERVELD J. D. A. (1966). — The magnetism of the Permian porphyries near Lugano (northern Italy, Switzerland). *Tectonophysics*, v. 3, p. 429-446.

WEGENER A. (1912). — Die Entstehung der Kontinente. *Petermanns Mitteilungen,* v. 58, n° 4, p. 185-195, n° 5, p. 253-256 ; n° 6, p. 305-309 ; *Geol. Rundschau,* v. 3, n° 4, p. 276-292.

WIPPERN J. (1964). — Die Stellung des Menderes Massives in der Alpidischen Gebirgsbildung. *Bull. M.T.A* n° 62, p. 74-82, Ankara.

WOLFART R. (1967). — Zur Entwicklung der paläozoische Tethys in Vorderasien. *Erdöl u. Kohle-Erdga Petrochemie,* Jhrg. 20, fasc. 3, p. 168-180.

8

Reprinted from *Am. Assoc. Petroleum Geologists Bull.* **60**:623–626 (1976)

MEDITERRANEAN PERMIAN TETHYS

Maurice Kamen-Kaye
Cambridge, Massachusetts 02138

Abstract Two localities in the Paleozoic biogeographic province of the western Mediterranean previously listed as Permian, have been found to be medial Carboniferous. The sequence, lithofacies, and presumed diastrophic history of a marine Triassic column in southernmost Spain indicates the possibility of marine Permian strata. Available evidence does not demand a direct connection between the Permian Tethys of the Old World and that of the New World. However, there remains the possibility that southernmost Spain is a portal leading from Permian waters of the western Mediterranean into Permian waters of an Atlantic or proto-Atlantic "ocean."

Western Mediterranean

Previous study of Permian Tethys (Kamen-Kaye, 1972) showed two Permian localities in the Paleozoic biogeographic province of the western Mediterranean. Both localities are west of those usually cited by the major paleogeographers, hence in my 1972 study were considered to have been overlooked. However, information received since writing the original paper shows that a Permian age for these two localities almost certainly is incorrect. The present consensus is that the age is near the boundary between the Early and Late Carboniferous at both localities. The rocks thus could be called "middle Carboniferous." The two localities, marked by crosses in Figure 1, are (1) St. Girons, on the French side of the Pyrenees; and (2) Menorca (Minorca), an island at the eastern end of the Balearic group, that is, east of Spain.

Both Caralp (1903) and Schmidt (1931) concluded that the marine fauna of St. Girons is Permian in age. Apparently both authors were mistaken according to A. F. de Lapparent and M. Lys (written commun., 1973), who cited Delépine (1931) as their authority for placing the St. Girons fauna in the Early Carboniferous ("limit between Visean and Namurian"). This is considerably different from the Permian age of Caralp and of Schmidt. However, the Early Carboniferous age is supported by W. M. Furnish and B. K. Glenister (written commun., 1973), who note that a combination of genera, actually families, given by Casteras (1933) is found only within the "upper Visean" of the Lower Carboniferous. It should be noted that the authority for Casteras is again Delépine.

Bourrouilh (1970; written commun., 1973) has made a careful study of the age of late Paleozoic rocks on the island of Menorca. He pointed out that the impression of a Permian age stems from Schindewolf (1934). However, Schindewolf (1958) reversed his earlier findings, stating that his "Permian" is in reality Namurian, which is to say the base of the Upper Carboniferous. Bourrouilh confirmed this latter finding. His fossils from Menorca indicate the Namurian Stage, Zone "R 2."

The critical taxa of late Paleozoic faunas in the western Mediterranean are the ammonoids, and it is in these forms that paleontologic analysis has been at fault. Intensive study of late Paleozoic ammonoids since 1930 has revealed that sophisticated analysis of septal architecture is necessary to define reliable "genera." The rather simple distinction of earlier years between goniatitic and ceratitic sutures no longer is sufficient. If the ammonoids at St. Girons and those on Menorca have been analyzed correctly the upper Paleozoic rocks at both localities must be called "middle Carboniferous."

With St. Girons and Menorca removed for the present from the roster of Permian localities in the Mediterranean biogeographic province, only Sosio on Sicily and Tobaga in Tunisia remain as the main outposts of Permian Tethys (Fig. 1). West of these outposts we know only the Permian locations of Venezuela and Central America, both near the Caribbean approaches to the western Atlantic Ocean.

Diastrophism

Colom (1950) described how nonmarine rocks, assigned to the Triassic, ride ("*cabalgan*") across folded middle Carboniferous marine rocks on the island of Menorca. Unconformity is manifested clearly. On this basis it is tempting to fill the whole of the interval between middle Carboniferous and Triassic time with a period of orogeny. However, Colom described a very different situation on another part of the island where the rocks are Carboniferous and presumed Triassic, but now are folded together. Colom did not suggest the possibility of two periods of folding. He suggested only that the folding in the second locality must be younger than in the first. Evidently there

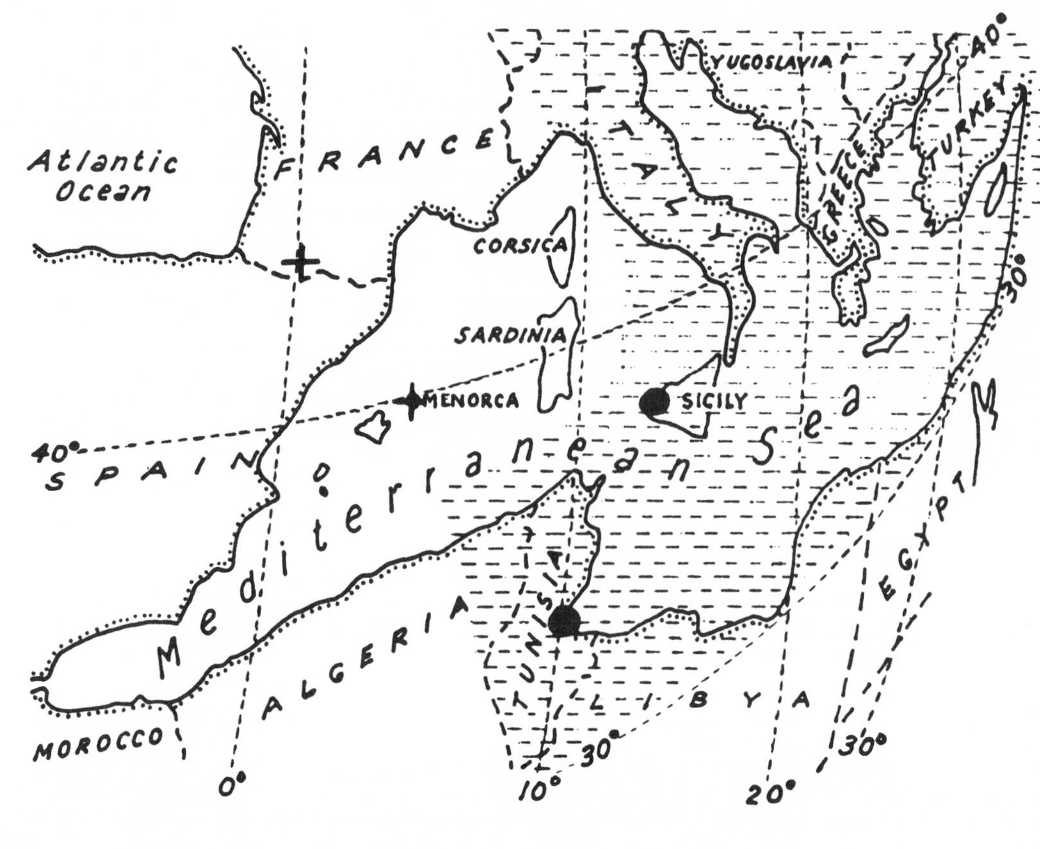

FIG. 1—Western extent of marine Permian in Mediterranean region.

is a lack of uniformity in the duration of orogeny on the island of Menorca.

The classic tectogenesis of the western Mediterranean is the Hercynian orogeny. As briefly reviewed by Querol (1969), this orogeny affects the entire Iberian Peninsula (Spain and Portugal). Rocks ranging in age from Cambrian to the Namurian Stage of the Early Carboniferous are folded and faulted strongly. This complex of deformed rocks is succeeded by much less disturbed "post-orogenic" clastic rocks which contain coal measures belonging to the Westphalian and Stephanian Stages of the Late Carboniferous. Despite the broad extent of the orogenesis and the intensity of its folding, the Hercynian episode spanned less than a single chronostratigraphic stage, and was completed nearly two stages before the beginning of Permian time.

It is difficult to know whether there were discrete episodes of diastrophism in southern Spain during Permian and Triassic times. Central Spain was a topographic high when Lower Triassic sandstones were deposited against it, but metamorphism and diapirism complicate the geologic picture farther south. If fauna is present at all in the metamorphic cores it will be difficult to find. Without such fauna Querol (1969) speculated that carbonate rocks in a metamorphic mass of the Betic cordillera of southernmost Spain are of "Permo-Triassic" age. Another part of the cordillera consists of

> a metamorphic complex of probable Paleozoic age, overlain by a complete Triassic section, several hundred metres thick, and nearly entirely composed of carbonates with marine fossils (Alpine-type Triassic).

Querol referred also to "Alpine Triassic, mainly composed of carbonates" in the legend of a map showing the distribution of Triassic strata in Spain. These Alpine Triassic strata extend from Malaga east-northeast along the southern coast for more than 400 km. The choice of the adjective "Alpine" for these strata is significant insofar as there are localities in the Alps where marine Late Permian strata are overlain by marine strata of Early Triassic age (Brinkmann, 1960). The passage from one to the other may not be unbroken, but there is little reason to believe that diastrophism is important at the Permian-Triassic boundary in Alpine Tethys. Similarly, diastrophism may not be important at the Permian-Triassic boundary in southern Spain, assuming that such a boundary actually exists within the mainly marine rocks of the area.

Paleogeography

Gobbett (1973) dealt with the distribution of Permian fusulinids around the world. In the European Tethys he noted the presence of these Foraminifera as far north as the Alpine province of Italy. The writer draws a corresponding pattern for Permian marine waters in the present Mediterranean area in Figures 1 and 2. As already discussed, no Permian waters can be documented within the Mediterranean west of the islands of Corsica and Sardinia. When Permian marine waters are documented again they appear in the Paleozoic Cordilleran biogeographic province of western North America, and in the Caribbean Paleozoic approaches to the western Altantic as shown in Figure 2.

A reviewer of this paper pointed out the correspondence between sessile brachiopods in the Lower Permian of Sicily and West Texas. Hill (1958) drew attention to a correspondence between whole faunas of Early Permian age in Tethys and North America, and speculated on the possibility of an ancestral marine connection between the two. Gobbett (1973), in turn, noted an identity of fusuline genera for Texas in the west and "Old World Tethys" in the east during Early Permian time. He noted, furthermore, that although this identity of genera could have been the result of dispersal across the Pacific, it could have taken place alternatively between Texas and the western end of Tethys "on the assumption that, at that time, North America occupied the geographical position of the present North Atlantic Ocean."

Gobbett's condition for the dispersal of fusulinids between Texas and Old World Tethys seems arbitrary. In theory at least, a broad sea or ocean between the two would not affect the dispersal. A similar dispersal could hold for the line between the Caribbean approaches to the western Atlantic and Old World Tethys.

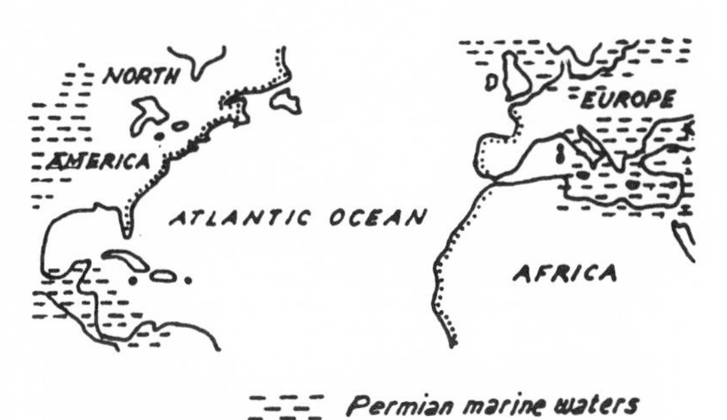

FIG. 2—Permian paleogeography of Atlantic Ocean borders.

Conclusions

Regardless of the relative positions of the Caribbean and the western Mediterranean provinces in Permian time, marine waters of the Permian Caribbean would separate the two. If continental drift is invoked, such marine waters would be confined to an initial rift in the primordial continent. If continental drift is not invoked, such marine waters could exist in stretches broad enough to qualify as an ancestral Atlantic Ocean, although not necessarily with oceanic depths. However, none of these considerations affect the possibility that land stood in the area of the western Mediterranean Sea during most or all of Permian time. The identity of marine invertebrate genera in Tunisia and Central America could be the result of faunal dispersal through the Pacific route. The distance via the Pacific route is great, but is by no means too great for nektonic taxa. The only consideration which can affect Permian paleogeography in the western Mediterranean is the geology of southernmost Spain.

Early marine Triassic strata, perhaps even earliest marine Triassic strata, are present in southernmost Spain. Permian marine strata in this same area are a distinct possibility, especially if diastrophism need not be long-lived during Permian time. The metamorphic or partly metamorphic complexes yet may yield Permian fossils on intensive search, especially if the carbonate rocks are investigated thoroughly in thin section. Alternatively, ultradeep drilling in the Betic trough north of the metamorphosed cordillera might provide important clues. Drilling deep enough to penetrate the diapiric Triassic folds might reveal the presence of Permian strata in littoral or coastal facies. In one way or another evidence might be obtained to suggest that a marine portal opens from western Mediterranean Permian Tethys into a Permian Atlantic "Ocean" or into a Permian proto-Atlantic. The possibility should not be discounted completely.

References Cited

Bourrouilh, R., 1970, Le probleme de Minorque et des Sierras de Levante de Majorque: Lille, France, Soc. Geol. Nord. Ann., v. 90, p. 363-380.

Brinkmann, R., 1960, Geologic evolution of Europe: New York, Hafner Pub. Co., 161 p.; translation and condensation by J. E. Sanders of v. 2, 8th ed., of Brinkmann's *Abriss der Geologie.*

Caralp, J., 1903, Le Permien de L'Ariege, ses divers facies, sa faune marine: Soc. Geol. France, ser. 4, v. 3, p. 635-640.

Casteras, M., 1933, Recherches sur la structure du versant nord des Pyrenees centrales et orientales: France, Serv. Carte Geol. Bull., v. 37, no. 189, 515 p.

Colom, G., 1950, Mas alla de la prehistoria; una geologia elemental de las Baleares: Madrid, Consejo Superior Inves. Cient., Inst. "San José de Calasana" Pedagogia, 285 p.

Delépine, G., 1931, L'age des schistes de Mondette (Ariege): Soc. Geol. France Compte Rendus, f. 12, p. 157-158.

Gobbett, D. J., 1973, Permian Fusulinacea, *in* A. Hallam, ed., Atlas of palaeobiogeography: New York, Elsevier, p. 151-158.

Hill, D., 1958, Sakmarian geography: Geol. Rundschau, v. 47, p. 590-629.

Kamen-Kaye, M., 1972, Permian Tethys and Indian Ocean: AAPG Bull., v. 56, p. 1984-1999.

Querol, R., 1969, Petroleum exploration in Spain, *in* P. Hepple, ed., The exploration for petroleum in Europe and North Africa: London, Inst. Petroleum, p. 49-71.

Schindewolf, O. H., 1934, Ueber zwei jungpalaeozoische Cephalopodenfaunen von Menorca: Gesell. Wiss. Gottingen, Abh., math.-phys. Kl., F.3, H.11, p. 155-191 (1437-1474).

——— 1958, Ueber eine Namur-Fauna von Menorca: Neues Jahrb. Geol. u. Palaontologie Monatsh., H.1, p. 1-8.

Schmidt, H., 1931, Das Palaozoikum der spanischen Pyrenaen: Gesell. Wiss. Gottingen, Abh., math.-phys. Kl., F.3, H.5, p. 1-58.

Part IV

THE GEOSYNCLINAL MESOZOIC TETHYS

Editor's Comments on Papers 9 Through 14

9 SANDER
Structural Evolution of the Mediterranean Region During the Mesozoic Era

10 SCANDONE
Triassic Seaways and the Jurassic Tethys Ocean in the Central Mediterranean Area

11 TERMIER and TERMIER
Une hypothèse sur le rôle des courants marins dans le dépôt de certains calcaires jurassiques et crétacés

12 TRÜMPY
Stratigraphy in Mountain Belts

13 RUKHIN
Paleogeography of the Asiatic Land Mass in the Mesozoic

14 CRAWFORD
The Indus Suture Line, the Himalaya, Tibet and Gondwanaland

The Mesozoic Tethys represents the second cycle in the development of a Mid-Eurasian sea, somewhat different from the Paleotethys with its vast expanses of uniform epicontinental sediments. Block faulting broke up the Tethys area into individual basins separated by submarine or by emergent ridges. This chain of subsiding basins becomes younger from west to east. By Late Triassic time, the Atlantic waters broke through again, thereby terminating a period of extensive evaporite deposition. N. J. Sander's detailed account best summarizes the depositional history of the Mesozoic Tethys. The account is included in its entirety because it is very difficult for most libraries to obtain; even the Library of Congress had trouble locating the entry (Paper 9).

The nature of Mesozoic seaways and their currents are covered by P. Scandone (Paper 10) and H. Termier and G. Termier (Paper 11).

The Jurassic history of the Alps is evidently the key to understanding the demise of the Tethys Sea. Trumpy reviews the sedimentary history of the Alps to show that detailed stratigraphic investigations are an essential prerequisite to explain the complexities of the Alpine mountain chain. He weighs various mechanisms that may have been responsible for the crustal shortening measurable in deformed Mesozoic supracrustal rocks in the Alps. In doing so, he rejects extreme values for the amount of movement in this realignment of Africa and Europe (Paper 12).

One can study the Tethys by examining its sediments; this is the conventional approach. A different method is to study the Tethys by looking at the provenance of its drainage, the source of its sediment supply. The "roof of Asia" has been a source of clastics for most of Phanerozoic history. Central Asian platforms that had been doming up in Paleozoic times after late Proterozoic, Caledonian, and Variscan orogenies were again acting as provenance areas of sediments pouring into Triassic, Jurassic, and Cretaceous seas. The "roof of Asia" as an elevated highland area is thus not a late Cenozoic new addition to the landscape; highland areas of varying extent have been located in Central Asia in many periods of Phanerozoic earth history. The late R. B. Rukhin was a chief propagator of the rather uncommon but very profitable approach of studying the Tethys and its marginal seas from the vantage point of the provenance areas of their sediments (Paper 13).

The Himalayas are younger than the Alps, and the Late Cretaceous to early Cenozoic history is the key period. Crawford shows in his paper that here, too, one has to be careful before interjecting vast distances between lands to the south of the Himalayas and lands to the north of them. Rocks in Central Asia, India, and Australia suggest that the crustal shortening in the orogeny seems to have had more modest dimensions than previously envisaged. This paper (Paper 14) complements rather nicely the findings of the immediately preceding paper.

9

Reprinted from pages 43–132 of *Geology and History of Sicily*, Annual Field Conference No. 12, W. Alvarez and K. H. Gohrbandt, eds., Petrol. Explor. Soc. Libya (now ESSL), 1970, 291pp.

STRUCTURAL EVOLUTION OF THE MEDITERRANEAN REGION DURING THE MESOZOIC ERA

by

N. J. SANDER

American International Oil Company, Chicago, Illinois

TABLE OF CONTENTS

INTRODUCTION

This essay carries through the Mesozoic
;ra the review of tectonic events in southern
;urope and northern Africa begun under
ıe title, "The Premesozoic Structural
;volution of the Mediterranean Region", in
ıe guidebook of the Tenth Annual Field
;onference of the Petroleum Exploration
ociety of Libya. The principal conclusions
f this 1968 paper, serving as points of
eparture for the present discussion, are as
ɔllows:

1. During Precambrian and Paleozoic
.mes cratons developed north, south, and
/est of the site occupied by the present-
ay Mediterranean. Some of them formed
ı troughs flanking two ancient nuclei —
ne in Finland, the other in Africa — and
ɔnsequently increased the size of these
hields, rendered monolithic by granitic
ıtrusion and metamorphism. Other cratons,
ɔmmonly of smaller size, formed in tracts
iscrete from these shields. All, however,
ppear to have developed in areas where
rolonged subsidence accompanied by the
ccumulation of great thicknesses of sed-
nents was followed by orogeny on a
ırge scale.

2. Once formed, these rigid massifs acted
s relatively stable, predominantly positive
tructural elements, so that the development
f new troughs was confined largely to the
nore mobile tracts between them. Thus, any
ıcrease in the number or size of massifs
estricted the area in which subsequent
rogeny and cratonization could occur. At
ıe beginning of Mesozoic times most of the
ıassifs now in existence in the Mediterra-
ean region had already assumed an
pproximation of their current configuration.
ome small metamorphosed and granite-
ıtruded blocks of Paleozoic age have indeed
een caught up in Alpine (Tertiary) orogeny
.ee fig. 1), but they preserve some of their
riginal characteristics in spite of the sub-
equent stresses to which they have been
ıbjected. In any case, the existence in
Iesozoic and Tertiary times of stable, mon-
lithic tectonic elements of Precambrian
nd Paleozoic age influenced profoundly the
placement of younger structures, and, to a lesser degree, their evolution.

Because these massifs bound the Mediterranean on three sides, the area that need be examined in detail here can be much smaller than that covered in the 1968 review of the Premesozoic, for these metamorphosed and intruded blocks were stable enough to justify treatment of tectonic events within the area they circumscribe independently from those in the remainder of Europe and Africa. Obviously, the effects of the principal orogenic pulses were not restricted to this massif-bounded region, but within it the expression of the several orogenies is sufficiently individual to permit discussion of the Mediterranean area as a discrete entity.

MESOZOIC SEDIMENTARY PROVINCES

A brief review of the more characteristic rocks representing the principal environments of deposition in the Mediterranean region during the Mesozoic Era is in order before undertaking the description and analysis of the structural elements that came into existence during these times, for recognition of facies and their significance in terms of tectonism aids greatly in defining structural entities. Facies discrimination is particularly useful for this purpose when, as in the early and medial Mesozoic, strong orogeny was both rare and restricted areally.

This period of relative stability in the Mediterranean area began late in the Paleozoic with the end of the major phases of Hercynian orogeny and persisted in many localities throughout the Mesozoic. From medial Triassic through early Jurassic times tensional tectonism predominated (Glangeaud, 1951, p. 749; 1962, p. 746), but from the Dogger onward compressional phenomena were common locally. The disturbances that can truly be classed as of orogenic proportions appear to have been confined to portions of what is now the Alpine chain; elsewhere movements were mainly epeirogenic.

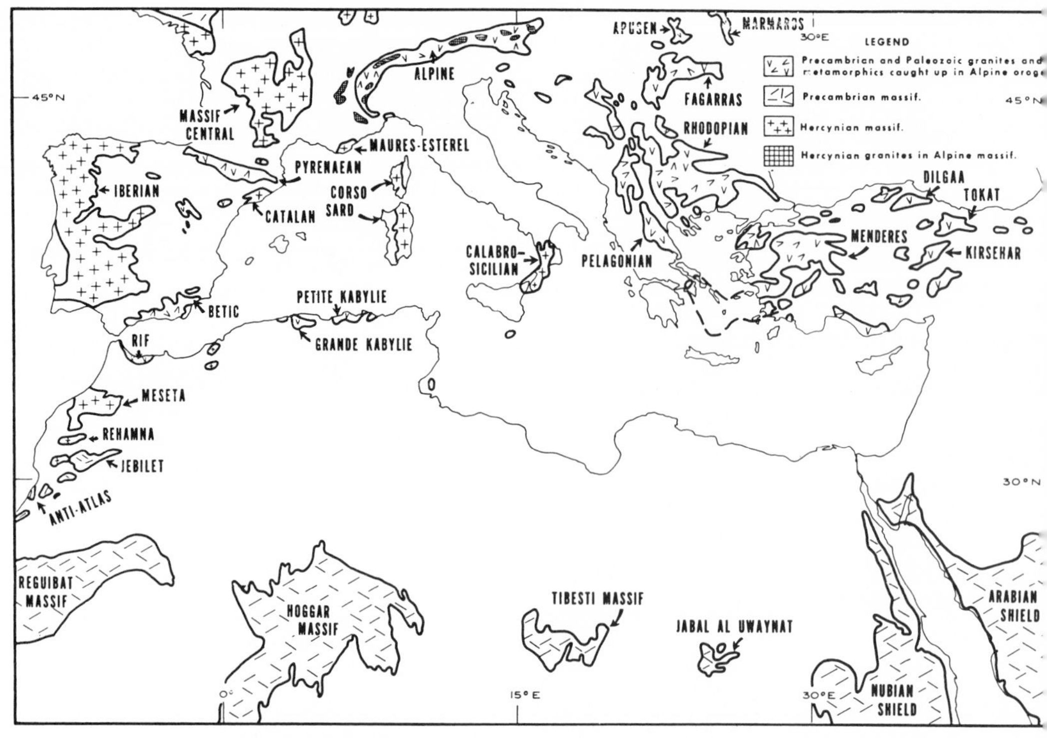

Fig. 1. Mediterranean region. Precambrian and Paleozoic massifs.

The time of relative quiescence early in the Mesozoic was accompanied by destruction of the mountain ranges of the late Paleozoic. The roots of these mountains are the present Hercynian massifs. At the close of the Paleozoic many of the chains had not been leveled completely, and in some places regional uplift (Palatine phase) occurred at this time, so that in much of western Europe and western North Africa strata of Triassic age are for the most part continental, littoral, or epineritic, whereas on the northern shores of the Mediterranean east of the Italian peninsula the Triassic section is made up principally of carbonates laid down in an open sea (Brinkmann, 1960, p. 76).

TRIASSIC FACIES

The Triassic shallow water sequence in the western Mediterranean is known as the "Germanic" facies, and the eastern open-sea succession as the "Alpine" facies. The names are taken from areas in which tl development of the two facies is consider typical.

The Germanic facies consists of thr principal units: a lower, coarse-gra continental sandstone, the *Buntsandstein*; marine limestone with abundant littor molluscs, the *Muschelkalk;* and an upp variegated marl or calcareous shale wi some evaporite and limestone interbeds, tl *Keuper*. The Muschelkalk sea did not rea the coast of northwestern Europe, so th in Great Britain and the Netherlands on the Buntsandstein or "Bunter" and tl Keuper represent the Triassic.

In the Alps of Italy and southweste Austria the Triassic is almost entire marine, consisting predominantly of ca bonates ranging up to more than fifte thousand feet in thickness, which constitu

e type "Alpine" facies. These strata rest
ı marine Permian, or on older Paleozoic.
Lombardy, basal Triassic (Werfenian)
arine sandstones grade eastward to
nestones and shales. With the exception
varicolored shales and thin sandstones in
e Raibl Formation (Carnian) the rocks
nstituting the Triassic section in the
alian Alps are limestones and dolomites,
me of which are biostromal. In some
aces Ladinian and Carnian carbonates are
tercalated with tuffs and flows, and solu-
on phenomena (cargneules) and minor
nounts of gypsum in certain formations
ggest an original content of evaporites.
The great difference between the German-
and Alpine facies led to the assumption
at they must have been separated by a
ıysical barrier. This "Vindelician sill" was
ostulated to run northward from Sardinia
rough Corsica and into southeastern
rance, swinging eastward in the Molasse
asin on the northern side of the existing
lps, and joining the Bohemian massif.
[odern concepts of facies require no such
arrier, but its presumed course serves to
elimit the boundary between the two facies,
ıd indeed, rocks of Triassic age are not
esent in the southern part of the Molasse
ough (Schmidt-Thomé, 1963).
In both western and eastern Europe the
me-rock unit now commonly considered as
ppermost Triassic, the Rhaetian Stage,
nsists of leached limestones (cargneules)
black shales, heralding the return of a
ormal marine environment in the west, and
e resumption of terrigenous clastic deposi-
on in the Alpine geosyncline in the east.
/here marine carbonates were laid down
ithout interruption during the late Tri-
sic, the Rhaetian Stage cannot be rec-
gnized.
As mentioned above, in the Mediterranean
gion the Germanic facies is found mainly
est of the Italian peninsula, and the Alpine
cies east of it. The existence of these two
dimentary provinces shows that the
dividualization of the western and eastern
[editerranean that began late in the
aleozoic continued without interruption
to the Mesozoic. However, the sedimentary sequences characteristic of the two facies are in some places replaced by atypical lithologic successions. These departures from the norm are of much interest for they show that some major structural entities later involved in Alpine orogeny were already labile in Triassic times.

TRIASSIC STRATIGRAPHY

The following summary of Triassic stratigraphy in the vicinity of the existing Mediterranean is intended only as a means of cataloguing the several sedimentary environments present. The location of the structural and geographic entities mentioned can be found on figure 2. An interpretation of land-sea relationships during the late Triassic as well as the gross facies present during that time as inferred from these stratigraphic data is shown on figure 3. The discussion begins with the Iberian peninsula and continues counterclockwise around the Mediterranean.

Iberian Peninsula. In most of the Iberian peninsula the Triassic sections begin with red sandstone, either resting unconformably on strongly tilted Carboniferous or older strata, or continuing unbroken a clastic sequence started in Permian times. The Triassic red bed sequence, very similar to the typical Buntsandstein, covers much of central and eastern Spain (Virgili, 1962, p. 306) as well as the Lusitanian and Algarve basins of Portugal. In the Andalusian portion of the Betic mountains north of the main mountain massif, the "Subbetic" of southern Spain, the early Triassic consists predominantly of red shales intercalated with evaporites, whereas the Muschelkalk and Keuper are marine, with some evidences of an original evaporite content in the Keuper. This partially hypersaline development has been called, "Germano-Andalusian" facies (Dürr, *et al.*, 1962, p. 217; Termier and Termier, 1957, p. 659). A red bed sequence similar to that of the early Triassic in Andalusia occurs at the upper limit of the Buntsandstein as it is commonly found in Spain. This unit is known as the "Röt", the name given a like occurrence in Germany.

The Muschelkalk is present in much of eastern and northern Spain, its lithology and fauna resembling those of the same unit in northern Europe. An intra-Muschelkalk regression of the sea reported by Virgili (1962, p. 302-303) occurs also in eastern France and Germany. The Muschelkalk of portions of the Betic mountains contains a few of the ammonites found in the Alpine Anisian and Ladinian, along with the characteristic littoral fauna. These ammonites have also been found in the Balearic islands, suggesting by their presence intercommunication between Spain and the

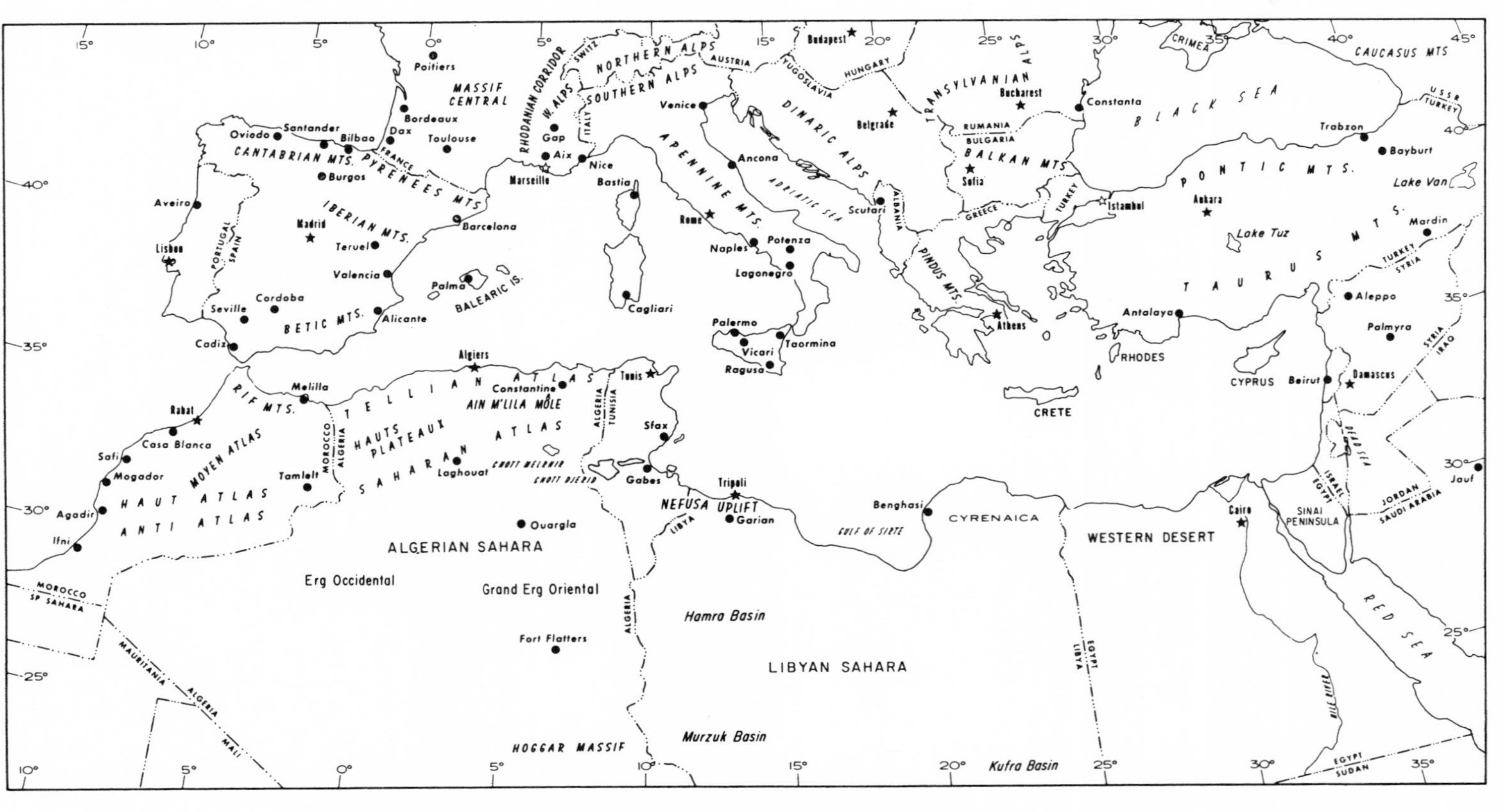

Fig. 2. Situation of some localities mentioned in text.

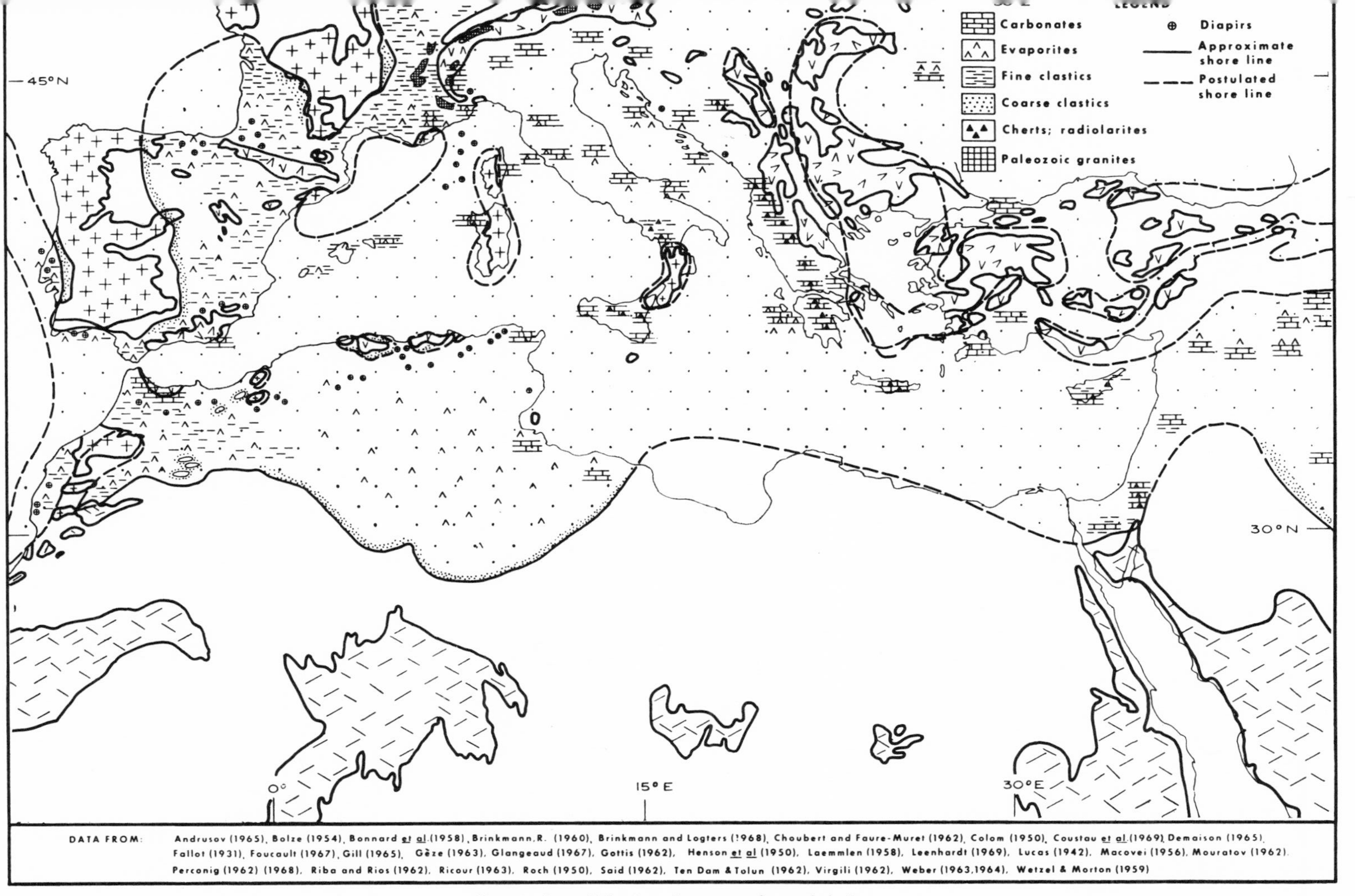

Fig. 3. Late Triassic paleogeography and facies.

eastern Mediterranean during medial Triassic times. (Perconig, 1968, p. 11).

On much of the Iberian peninsula the Keuper consists largely of varicolored calcareous shales and marls intercalated with evaporites. Over extensive tracts evaporites predominate. Where salt is present in considerable thickness in areas later covered by a massive overburden of younger strata, diapirism is common (Lusitanian and Algarve basins of Portugal; Cantabrian and Betic mountains of Spain). Only in a portion of the Betic mountains near the southeastern coast (southern Sierra Nevada) is the post-Werfenian (Anisian-Rhaetian) entirely in a "normal" marine facies. Here more than a thousand feet of limestones with characteristic Muschelkalk and Keuper . fossils supplemented by a few "Alpine" ammonites, rest conformably on Werfenian sandstones, the entire section being highly faulted and disturbed. (Llopis Llado, 1953, p. 11; Fallot, 1941, p. 14).

North Africa. A Triassic section made up predominantly of carbonates is known only in the coastal portion (*Chaine calcaire*) of the Rif mountains of northwestern Morocco where more than four thousand feet of carbonates resting on Werfenian continental sandstones are dated as Anisian-Rhaetian (Roch 1950, p. 224; Fallot and Marin, 1952, p. 8). Immediately to the south, in the Rif complex of southward-directed gravity slides, evaporites, including large blocks of salt, are reported to represent much of the Triassic where the Mesozoic has ridden over younger formations (Lévy and Tilloy, 1952, p. 14). Except in the Rif mountains, the Triassic of Morocco is predominantly varicolored clay shales intercalated with anhydrite or gypsum, locally reaching thicknesses of five thousand feet (*idem*, p. 9). Thick conglomerates and sandstones in discrete, commonly fault-bounded depressions on the Paleozoic erosional surface of southern Morocco, then still in the course of peneplanation, are considered older than the red clay-evaporite sequence which in some localities rests on these older clastics (Choubert and Faure-Muret, 1962, p. 450).

Along the coast of northeastern Morocco and adjacent portions of Algeria marine limestones are interbedded in the lower portion of the evaporites and shales, and, as is common throughout eastern and southern Morocco, many basalt flows are intercalated in the section. In northwestern Algeria on the flanks of the Ghar Rouban and Saida Paleozoic massifs the marine limestones contain a Muschelkalk fauna (Kieken, 1962, p. 547). They rest on sandstones and conglomerates lying on the metamorphosed Paleozoic of the massifs. Dated Keuper has not been found in these outcrops in western Algeria, but here and farther east red unfossilferous sandstones above the Muschelkalk have been referred to the Keuper (Djurdjura; Durand Delga, 1967, p. 66).

Autochthonous Triassic is exposed also in the *Dorsale calcaire,* the narrow tectonic unit on the south side of the Chenoua and Petite and Grande Kabylie massifs where Permo-Triassic sandstones are capped by limestones, dolomites, and gypsum beds. Durand Delga (1967, p. 66) refers the gypsiferous sequence to the lower Triassic; Kieken (1962, p. 547) suggests that is more probably Keuper.

Elsewhere in Algeria — in the Tellian Atlas and the Hauts Plateaux and Saharan Atlas paralleling to the south — the Triassic crops out extensively, b only in the form of diapirs or along thrust plan Triassic strata are consequently in a chaotic assembl with ophiolites, dolomites, and arkosic sandstones. T sequence of the Triassic strata underlying the Telli and Atlas ranges is probably like that exposed alo the coast of southern Tunisia and in northweste Libya. Here, the Triassic, resting on Permian or ol Paleozoic, begins with a basal cross-bedded sandsto and plant-bearing red shale unit ranging up to m than three thousand feet in thickness. This clas sequence represents the Buntsandestein and Muschelkalk, for carbonates with a Muschelkalk fau are intercalated in the sandstones which are capped Carnian dolomite. In northwestern Libya, a sandsto (Bu Sceba Formation) overlies these carbonates, a is succeeded by evaporites or dolomites of Keuper a In southern Tunisia the Carnian carbonate marks upper limit of the terrigenous clastic portion of section which then grades to anhydrite and salt, evaporites attaining a thickness of almost two thousa feet (Busson, 1967, p. 131-135, 143). The Triassic-L boundary lies within the evaporite unit.

Bolze (1954) describes a two thousand-foot secti of slightly metamorphosed ripple marked shales a dolomites in northeastern Tunisia near Jabal Ichke fifteen miles southwest of Bizerte. The lower sevente hundred feet is considered late Triassic, presumal Carnian according to meager faunal evidence, and t remainder, mainly dolomite, is probably Norian a Rhaetian. This sequence, although laid down in shall water for the most part, differs greatly from that fou farther south.

In the southern desert of both Algeria and Lib the Triassic is represented entirely by continen deposits with occasional lacustrine limestone interbe (Lefranc, 1958). The upper portion of this sequer is replaced northward by evaporites, for in deep we in the northern Algerian Sahara, the Triassic consi of a basal continental sandstone ("T1" Series of t Algerian Sahara, equivalent to the Ouled Chet Formation of Libya, probably representing the Bu sandstein and lower Muschelkalk), overlain by a mari sandstone of late Muschelkalk (Ladinian) and ea Carnian age, the "T2" Series (Demaison, 1965). Resti on this clastic unit in the central portion of the Tri sic basin of the northern Sahara, but extending beyond it to the west and southwest and a less distance to the east, is a late Carnian-medial L evaporite section, including both anhydrite and salt a ranging up to more than five thousand feet in thickne Thin dolomites and shales are intercalated in the s tion. The Keuper evaporites present in outcrop southern Tunisia and northwestern Libya represe deposits in the same basin, as does the salt formi the diapirs of the Saharan and Tellian ranges. Lit evidence of strong salt movement has been found that part of the basin now on the Saharan platfor ıt anomalous variations in the thickness of salt found wells in the Algerian Sahara may have been a result flowage.

In Egypt, marine Triassic crops out on the Sinai ninsula and is reported in subsurface at the northern d of the Gulf of Suez (Kostandi, 1959), but none s been encountered in the deep wells of the Western esert. A six hundred and sixty foot outcrop on the nai peninsula was described by Eicher (1946), as nsisting of alternations of greenish-gray shales and ay to brown limestones with a meager fauna showing inities to that of the typical Muschelkalk, but with few ammonoids that suggest open sea influence. The una also includes conodonts, the first ever reported strata of Mesozoic age. Wells drilled in the same ea confirm the existence of one hundred to two ındred and fifty feet of Triassic strata, predominantly rrigenous clastics. Said (1962, p. 245), using informa- on from Kostandi, suggests that carbonates are more oundant seaward, but data are scanty.

The Levant. In Jordan, two incomplete sections of riassic age were described by Wetzel and Morton 959) near the northern end of the Dead Sea. Together ey suggest a Mediterranean expression of the German- facies. One four hundred foot section consists of a oss-bedded sandstone overlain by dolomites and lime- ones with a Muschelkalk fauna, and the other begins ith a few feet of similar limestone capped by three ındred and seventy-five feet of gypsum with interbeds crystalline limestone, succeeded by dark colored ales and sandstones with a Rhaetian-Lias flora. As in gypt, rare ammonites show that the Muschelkalk uivalent was subject to open sea influences. Cinerite nses in the section indicate some volcanic activity, ıt no flows of importance were reported.

Farther north and a short distance inland from the editerranean coast, Shaw (1947) describes a sixteen- ındred foot section of Triassic in southern Palestine. he sequence is like that near the Dead Sea, but each the three units is thicker, the Muschelkalk measur- g more than six hundred and fifty feet, and the euper nearly six hundred feet. Thin flows of basalt cur in the lower portion.

In Lebanon, rocks of Triassic age neither crop out or have been found in wells (Beydoun, 1965, p. 11-12), ıt in Palestine and Syria a number of deep tests have ached or penetrated marine Triassic in widely sep- ated localities. These occurrences, together with similar nds in wells in Iraq, supplemented by outcrops Dunnington, *et al.*, 1959), make it possible to delimit e area occupied by sea during the Triassic in the road easterly continuation of the Tethys between Ara- a and Turkey. The rocks of Triassic age laid down this seaway were predominantly carbonates, with latively minor amounts of terrigenous clastics and aporites near the top of the section. (Weber, 1963). nhydrite forms thick beds locally (*e.g.* Souedie No. 3, amichlie No. 1, El Barde-1 *etc.*, Syria). Triassic shales st conformably on Permian limestones in outcrop in orthern Iraq and southeastern Turkey, but equivalent riassic strata are transgressive elsewhere, both on the nks of the basin and over old highs within it. The upper Triassic may be absent in part of northern Syria (Weber, 1963, fig. 3; 1964, figs. 3, 5) as well as in northwestern Iraq.

In western Iraq the Triassic thins, grades to sand- stone, and wedges out against the Jauf-Mardin high (Sander, 1968, p. 58, 61), which was then a prom- ontory extending northward from the Arabian peninsula into the Tethys (fig. 3). Similar wedge-outs presumably exist on the Mediterranean side of the high, which on that flank may be less highly faulted and slope more gently into the basin. In northern Iraq, the Triassic outcrop attains a thickness of more than fifty-seven hundred feet, with a basal shale-sandstone unit making up less than one-fifth of the total section, which is predominantly carbonates. The Triassic found in wells appears to be considerably thinner.

North of the Tethys in Turkey, the Paleozoic massifs formed during Hercynian orogeny remained for the most part emergent during the Triassic, although the sea advanced over the Anatolian plateau between them (Ten Dam and Tolun, 1962). Most of the strata laid down are calcareous with anhydrite present in the basin of southeastern Turkey (*ibid.*, p. 56). Near the shoreline the limestones are interbedded with or replac- ed by shales. In general the basal unit of the trans- gressive sequence is conglomeratic or psammitic, but in some localities the Triassic has not been differenti- ated in a comprehensive Mesozoic sequence (Erentöz, 1956, p. 45).

Cyprus. Strata of Triassic age in Cyprus consist of rather strongly disturbed shales and radiolarites grad- ing up to oölitic limestones overlain by cherty silts and sandstones. A second occurrence of limestone in a reefal facies may have formed a cap over these older strata. Both are of late Triassic (Carnian-Norian) age (Henson *et al.*, 1952, p. 8-10). Vulcanism accompanied by widespread ejection of ash was common in much of the Mediterranean region during Triassic times, and was presumably responsible for the development of radiolarian cherts in Cyprus and in parts of Greece.

Balkan Peninsula. Clement (1968) describes Triassic sections near Athens in which Werfenian coarse-grade sandstones of undetermined thickness are overlain by Anisian nodular and cherty limestones with red shale and tuff intercalations. A thin radiolarite bed marks the base of the Ladinian which is represented by dolomitic limestones and siliceous limestones with abundant chert nodules, intercalated with fine-grade sandstones and tuffs in the lower part. The Carnian- Norian succession consists of three hundred feet of partially dolomitized limestones with *Gyroporella,* and the complete section exposed is about five hundred and fifty feet thick. This outcrop of Alpine Triassic is in an area where basement is covered by a relatively thin sedimentary cover in an orogenic belt known as an "internal eugeanticlinal ridge" (Aubouin, 1965, p. 44). During Triassic times, only the linear north northwest - south southeast trending depression adjacent to this "Pelagonian massif", called the "Pindus" trough, included somewhat deeper water sediments, as shown by the abundance of platy siliceous limestones containing the pelagic lamellibranch *Halobia,* and radiolarian cherts.

Farther west in Greece, along the Ionian sea, carbonates of Carnian-Norian age are associated with evaporites, the whole attaining thicknesses of several thousands of feet, of which only the upper portion crops out. Deep wells have not, insofar as is known, penetrated through the evaporite section (Jones, 1968).

In Yugoslavia outcrops of Triassic age are divided into two groups by Ciric (1963, p. 567) in accordance with his tectonic scheme of an "Inner" Dinaride belt, and an "Outer" Dinaride belt. The principal difference between the two belts is the existence of volcanic products in the Anisian-Ladinian of the "Inner" one. These consist of green tuffs and andesitic or basaltic flows which are interbedded with limestones and radiolarian cherts. The Werfenian in both belts is made up of sandstones, variegated shales, and some banded limestones, and the remainder of the Triassic section consists of considerable thicknesses of limestones and dolomites, which occur in reefal associations over extensive areas. The evaporites found along the west coast of Greece occur also in coastal Yugoslavia (Sikosek and Medwenitsch, 1965, p. 346), and a similar situation exists in Albania (*ibidem,* fig. 5). In Montenegro near the Albanian frontier some orogenic movement during Anisian times is reported, which is said to have resulted in the development of thick conglomerates and other terrigenous clastics (Ciric, 1963, p. 568). Other deposits of this type occur north of Trieste and along the Albanian coast (Ciric, 1966, p. 500). Epeirogeny in Ladinian times caused brief emergence in parts of coastal Yugoslavia where bauxites occur in the carbonate section.

Little information is available concerning the Triassic of the Balkan ranges north of the Rhodopian massif, but in the Carpathians of Poland and northern Rumania the Triassic is in a more or less typical Germanic facies (Ksiazkiewicz, 1963, p. 534). Farther south, in eastern Rumania, the variegated marls of the Keuper are replaced by dolomites, so that the section takes on Alpine aspect, like that found in northern Turkey near Istanbul (Furon, 1953, p. 61). According to Beloussov (1962, p. 713), the Buntsandstein in Bulgaria is capped by Muschelkalk limestones, and presumably Carnian and Norian carbonates succeed it. Locally, west of Varna, the Buntsandstein is replaced by red shales and evaporites.

The Alps. The geology of the European Alps has been studied as intensively as that of any area in the world, and the literature concerning these mountains is monumental in bulk, if of varying quality. In the summary presented here, only the briefest outline of Alpine stratigraphy is possible. As mentioned in the preceding paper in this series, marine Permian is present in the eastern portion of the Italian Alps, indicating by its presence the first stages of the development of the Alpine geosyncline. The Permian sea was bounded to the north by a Paleozoic massif which remained emergent during early Triassic times, but was covered more or less completely by the Anisian sea. However, the massif remained as a posi although partially submerged structural eler throughout the period. Consequently, the Triassic tion on the massif is thinner and somewhat complete than that present both south and nortl it, and minor differences exist between the nortl and southern Alps in the facies of the several st of the Triassic.

North of the Paleozoic massif the basal Tri clastics of the Werfenian rest on basement. This s consists entirely of conglomerates, sandstones, shales which are continental in the west, but marine to the east. South of the barrier ridge, lower portion of the basal Triassic also consists sandstone and conglomerate in the west (Servino S stone), but in this area it is marine and lies confc ably on Permian continental sandstones known as " rucano". Cavernous dolomites, "Zellendolomit", f the upper part of this Werfenian equivalent. Far east, the Servino Sandstone and the Zellendolomit way to predominantly calcareous strata, the "Sei Limestones and the "Campiler" Beds which incl varicolored marls and limestones with a littoral fa and a few sandstone interbeds. The Seiser Limestc are conformable with the late Permian *Bellerop* Limestones.

Marine limestones and dolomites make up the b of the Anisian south of the barrier. They include abundant fauna of ammonites and thin-shelled lam branchs suggesting open sea conditions. In the nortl Alps the Anisian consists mainly of bedded limest containing many ammonites, presumably laid down a deep-water, locally somewhat euxinic, environm Thus the difference between the facies of the Ani north and south of the intermediate massifs is sm During the Ladinian the sea shallowed in the areas, for reefs made up of corals and calcareous a are found in both (Brinkmann, 1960, p. 76). Howe the sedimentary environment differs, for igneous ac ity, represented by thick tuffs and some andesitic la occurs there with concomitant cherts and cherty li stones. However, in other localities not far from sites of this igneous activity the entire Ladinian made up of crystalline limestone or saccharoidal omite ("Esino", western Italian Alps; "Schlern" east Italian Alps), formed by the alteration of biocla limestones and algal biostromes. These carbonate bu ups are matched by the Wetterstein Limestone Ramsau Dolomite of the northern Alps.

At the end of Ladinian times, emergence was gen in both the northern and southern Alps, with kar phenomena developed in the strata subjected weathering, but the sea returned after a brief inter laying down the transgressive "Raibl Formation" wh represents the Carnian stage. A westward prolongat of the Bohemian island massif, "Vindelicia", then fo ed the northern shore of the Alpine geosyncl Near it were deposited conglomerates, sandstones, shales, but elsewhere the basal unit of the R Formation consists of locally oölitic limestones marls which grade up to cavernous dolomites. remainder of the formations is mainly varicolored sh erbedded with limestones containing a shallow water ına, along with lesser amounts of leached dolomites l limestones (Rauchwacken). Leonardi, (1963, p. 237), licated that many reefs exist in the Raibl Forman, but other authorities do not discuss them.

Nearly everywhere in the Alps the Norian is presented by thick carbonates that range up to more n thirty-five hundred feet in thickness. This seence, called the "Hauptdolomit" and the "Dachinkalk", is characterized by a fauna of thick-shelled nellibranchs, as well as corals. The Rhaetian, where stinguishable, consists of black shales and limestones th *Avicula contorta,* on which lie the dark-colored nestones and dolomites of the "Conchodon Zone".

The total thickness of the Triassic in the Alps ranges to more than fifteen thousand feet, but over the rrier between northern and southern Alps it decreasto about three thousand feet.

Italian Peninsula. The Italian peninsula rose above a level as a result of Hercynian orogeny, and remained nergent until Triassic times. Trevisan (1963) states at in the northwestern part of the peninsula in uscany, Carnian sandstones with brackish or lacustrine olluscs, and at one locality reptile tracks (von Huene, 42), lie with little or no discordance on lithologically nilar but finer-grade sandstones of early Permian e. In the Apuan Alps these sandstones are capped gray or pink bedded limestones and dolomitic nestones, the so-called "grezzoni", with *Worthenia cheri,* a Norian guide-fossil. Farther south, at Monte sano and in other outcrops in southern Tuscany, e Carnian sandstones are overlain by black dolomites, rmerly intergrown and interbedded with evaporites hich have been leached away, leaving "calcare caverso", or *cargneules.* Fossils are rare, but characteristic the Norian and Rhaetian. Farther east, in Umbria, e black dolomites are replaced by light-colored dolnites and dolomitic limestones in outcrop, but evapites are still present at depth as shown by their currence in a number of deep wells, in particular oresta Umbra No. 1 on the Gargano peninsula, hich at 19,391 feet had penetrated at least eight ousand feet of Triassic limestones and evaporites ithout reaching the base of the succession. Trevisan nsiders these evaporites to be of Carnian age, and is age assignment seems reasonable, owing to the gional uplift at the end of Ladinian times in the scent Alpine geosyncline. Furthermore, Carnian fossils cur in a very small outcrop of black shales and nestones in a structurally extremely complex, possibly apiric, situation at a site known as Punte delle etre Nere located at the inner end of the Gargano ninsula.

The Triassic crops out extensively along the Tyrrhean coast south of Naples in the province of Lucania. lany papers have been published on this region hich is highly complex structurally owing to thrustg and gravity tectonics. Grandjacquet (1962, p. 613) views the stratigraphy, indicating that the Triassic the Lagonegro-Potenza region consists of not more an eight hundred feet of cherty, nodular, siliceous mestones, with Ladinian fossils at the base (De Lorenzo, 1894). This limestone is the lower portion of a continous Mesozoic section which according to Grandjacquet (1962, p. 613) is allochthonous, but not far removed from its site of deposition. The autochthonous, Triassic sequence in the area consists predominantly of limestones and dolomites laid down in relatively shallow water and now partially dolomitized. According to both Ippolito (1949) and Quitzow (1935), on the flanks of the Calabro-Sicilian massif unmetamorphosed Triassic is represented only by Carnian and younger strata in a leached dolomite-limestone facies like that in Tuscany: Bousquet (1961) reports a much more complex stratigraphic succession on this massif. In any event, a subsident basin probably existed north and perhaps northwest of the Calabro-Sicilian massif during most of the Triassic, as pointed out by Gill (1965). It was flanked by tracts in which neritic and biostromal limestones were the predominant deposits, for according to Grandjacquet (1962) the siliceous limestones at Lagonegro have been thrust over biostromal carbonates.

In short, the Triassic sea began to cover portions of the southern Italian peninsula at least as early as the Ladinian, whereas the northern portion of the peninsula remained emergent until Carnian or Norian times. Facies interpretation suggest that some kind of persistent structural depression may have been associated with the uninterrupted but relatively thin open-sea Mesozoic section at Lagonegro, yet continous subsidence was also necessary for the development of the thick shallow-water limestones and evaporites present elsewhere on the peninsula.

Sicily. Study of the stratigraphy of Sicily is made difficult by overthrusting and gravity sliding consequent on Alpine orogeny, but the lithologic succession can be determined by careful investigation because the allochthonous sequence forms a continous cover only in the Caltanisetta Tertiary basin, leaving exposed the structurally complex Sicilian Apennines in the north and the stable Iblean plateau in the south.

On the Iblean plateau, the stable southern platform of Sicily, the Triassic section consists of a lower unit of dolomites and dolomitic limestones commonly lying at considerable depths, and presumably very thick, for an incomplete section of more than five thousand feet has been measured in wells. These dolomites represent an altered bioclastic limestone in which algae were a principal source of detritus (Kafka and Kirkbride, 1959; Rigo and Barbieri, 1959). They are capped by black shales containing *Estheria* that were probably laid down in a reducing environment. The shale unit ranges up into the Lias, and the Triassic-Jurassic boundary is therefore indefinite.

In eastern Sicily a section of Triassic is known in the vicinity of Monte Judica west of the Catanian plain. Although the exposures are strongly faulted and thrusted, the section was established by Schmidt di Friedberg and Trovo (1962), who, aided by well data, determined the sequence and thickness of the several stratigraphic units. The Triassic is represented by a basal unit of gray clay shales with thin limestone intercalations together more than two thousand feet thick (base not reached in shallow well) on which lie some eight hundred seventy-five feet of well-bedded limestones with chert nodules and thin marl intercalations. These two units, assigned a late Triassic age, are characteristic of a part of the northern geologic province of Sicily in the Madonie, Termini Imerese, and Palermo ranges. The broad distribution of the cherty limestones is confirmed by Ogniben (1963, p. 206), who gives their thickness as over fifteen hundred feet in the Madonie mountains.

Broquet *et al.* (1966, p. 997) describe the complex structure of western Sicily. They mention a Carnian "flysch" consisting of dark-colored marls and shales with thin sandstone intercalations and interbedded with blackish fine-grained, oölitic limestones, which, together with a Permian sandstone-shale sequence may have served as a gliding surface. This "flysch" must have been formed in relatively shallow water, because of the presence in it of oölites and pseudoöolites. It appears, however, to be confined to the northern part of Sicily where it underlies nearly all the thrust sheets involving the Mesozoic, and its absence farther south may explain the decrease in the allochthony of some Mesozoic units in that direction. The remainder of the Triassic (Norian and Rhaetian) is found in several discrete facies in the several thrust sheets. In one, it forms the lower portion of a six thousand foot massive white limestone; in another, it is represented in a thousand-foot formation of thinly bedded limestones and dolomites. In a third, the Triassic is again massive limestone, but only six hundred feet thick; and in a fourth, it consists of some sixteen hundred feet of Carnian bedded limestone on which lie Norian-Infralias massive dolomites and dolomitic limestones.

Although there are differences in the lithology of the several thrust sheets, the Norian-Rhaetian in western Sicily consists mainly of carbonates, presumably formed in relatively shallow water. Note that terrigenous clastic elements are lacking. Their absence suggests that no land area existed nearby. In general, the thickness of the Triassic section found in the thrust sheets decreases southward.

Triassic formations older than Carnian are not referred to in recent papers on Sicily, but Ladinian along the north coast is mentioned in some older works (Gemellaro, 1921). Probably the dark gray, red, brown, or green clay shales of the Sicani mountains, intercalated with thin beds of platy and cherty limestone, and more rarely with compact sandstones, are the oldest datable formation of Mesozoic age in Sicily (Fabiani and Trevisan, 1937; Mascle, 1968). The fauna includes *Halobia* and other fossils characteristic of the Carnian, and the lithology is similar to that of the Carnian "flysch" of Broquet *et al.* (1966).

In summary, the Triassic in Sicily is represented mainly by carbonates, nearly all of which were laid down in shallow water, although the siliceous nature of the section near Catania described by Schmidt di Friedberg suggests a somewhat more open sea environment than do the algal limestones of the Iblean plateau. Sicily then, at least from Carnian times on, was probably largely below sea level and Triassic deposits were mainly carbonates with an Alpine aspect. Schmidt Friedberg (1962, p. 654) postulates that an east-w trending reef complex of Carnian age exists in cent Sicily, bounded north and south by basins in whi shales predominated. During the Norian the northe shale basin was uplifted so that limestones were la down everywhere north of the Iblean platfo throughout the remainder of Triassic times.

Sardinia. Marine strata of Triassic age are fou only in northwestern Sardinia. They are in a Germa facies. The Buntsandstein consists of plant-bearing sar stones and red and violet shales and marls, with so gypsum in the upper part of the eighty foot sectic The Muschelkalk, blue or gray limestones and dolomit although nearly five hundred feet thick, contains fauna essentially devoid of echinoids, corals or spong and with only a few brachiopods and crinoids. In a dition to the molluscs typical of the Germanic faci the Muschelkalk limestone does, however, include t ammonite species that occur in the Ladinian of t Alpine facies, thus suggesting some connection wi the open sea. The Keuper is represented by varicolor gypsiferous marls capped by two hundred feet of lea ed, cavernous, dolomite. The total thickness of t Triassic in northwestern Sardinia is less than ni hundred feet.

Corsica. In northeastern Corsica limestones referr to the Triassic are involved in intricate thrusti (Brouwer, 1963, p. 279). Nothing is available conce ing the aspect or faunal content of these limeston

Southeastern France. In Provence and in the depr sion lying between the *Massif Central* and the "Rho nian corridor", formations of Triassic age crop c extensively, particularly near the Hercynian mass bordering the French Alps. The Buntsandstein lies u conformably on Permian or on granitic intrusives. Acco ing to Ricour (1963, p. 400), whose discussion is t basis for much of what follows, no Lower Buntsandste is present in the area, and the facies equivalent of t Upper Buntsandstein is of late Muschelkalk age on t several massifs flanking the French Alps.

East of these massifs in the French portion of t nascent Alpine geosyncline, the Werfenian, lyi conformably on Permian sandstones, is more th sixteen hundred feet of azoic white, fine-grained sar stone. In the western Italian Alps, as indicated abo the Servino Formation, which is of the same age, red or brown in color, and contains a few mollus The upper Werfenian in both the French and Itali Alps is represented by leached limestone and dolomi "cargneules" and "Zellendolomit", laid down in hypersaline environment.

A marine incursion into the Rhodanian corridor a Provence from the north and south eventually cov ed all of southeastern France during the Muschelka leaving the several massifs paralleling the Alps em gent until late in the Ladinian. The encroaching s laid down basal sandstones in those areas not previou submerged. Its deposits are, however, predominan dolomites, evaporites, and variegated marls. These gra upward to dark gray and black limestones with b iopods and pelecypods. No emergent barrier existed te in Muschelkalk times to separate this Germanic cies in Provence and the Rhodanian corridor from e Alpine trough, where deposits of the same age e Anisian and Ladinian limestones with ammonites, rals and abundant algae, but the two faunas are ntirely discrete. A late Muschelkalk regression west the Alps is matched by the late Ladinian regression the Alpine geosyncline.

In Provence and the Rhodanian corridor the Keuper egins with gypsum and black shales with *Estheria,* hich were deposited over extensive tracts and followed a brief period of emergence. A thin glauconitic sand-one with remains of plants partly in the form of al and intercalated with shales containing *Equisetum,* es on the very slightly eroded surface of the gypsum eposit. This sandstone, the *Grés à roseaux,* crops out om Austria to the Maritime Alps, and is remarkably niform in lithology. Thin red and green clay shales e on the glauconitic sandstone and are capped by ellow-ocher dolomites that to judge from their meager, npoverished fauna were probably laid down in a ypersaline sea. The uppermost Keuper consists of ricolored marls with some gypsum lentils and car-neules. Near the *Massif Central* it is represented by ndstones.

In the French Alps strata of Carnian age have not een identified with certainty owing to the absence of naracteristic fossils, but may be either evaporites or lack shales and sandstones. The Norian is in an Al-ine facies (light colored dolomites) in the Briançonnais gion only. Elsewhere it has a Germanic aspect.

The Rhaetian in Provence and in the Rhodanian orridor begins with dark gray limestones on which lie laty limestones and greenish shales with a fauna of ttoral lamellibranchs, *Avicula, Mytilus, Cardita.* It has ot been recognized in the French Alps.

This discussion of southeastern France is somewhat ore detailed than that of other areas, first because ore or less precise information concerning the region to hand, and second, because the Triassic sequences emonstrate that maintenance of discrete facies in ontiguous tracts does not require a land barrier, but nly some difference in temperature, salinity, turbidity, epth, or other physical or chemical property of the sea hat serves as an obstacle to migration of faunas as ell as causing differences in the kinds of sediments eposited in it.

Southwestern France. The broad folded and faulted elt now making up the Pyrenees was mainly below ea level during Muschelkalk and Keuper times. The euper in these mountains includes evaporites and hales that facilitated the subsequent development of liding tectonics. The Buntsandstein and Muschelkalk re in a Germanic facies, but the Muschelkalk is absent very much reduced locally, and the Buntsandstein ifficult to distinguish from the Permian on which it ests conformably.

The same sequence underlies nearly all of the Aqui-aine basin of southwestern France, although evaporites ay be present in the Buntsandstein as well as in the euper (Bonnard *et al.,* 1958, p. 1098). The occurrence of salt and anhydrite in the lower Triassic is known also in Germany and in southern Spain, in the "Röt" and its equivalents. Figures 10 and 11 of a recent paper by Coustau *et al.* (1969) show Triassic strata present throughout the basin north of the Pyrenean thrusts, resting unconformably on metamorphosed Paleozoic and certainly including salt and anhydrite, for salt swells are depicted in the figures and mentioned in the text (p. 7). Furthermore, Cuvillier *et al.* (1963) discuss the importance of diapirism in the southwestern part of the basin and estimate the thickness of the Triassic involved in it at nearly thirty-three hundred feet. According to Coustau *et al.* (1969, fig. 2), the thickness of the Triassic is sharply reduced north of a line from Arcachon to Toulouse, but in a well twenty-five miles north of Boudeaux, about one thousand feet of Triassic shales and evaporites rest on basement, so that the Triassic sea must have covered most of the present-day Aquitaine basin (Bonnard, *et al.,* 1958, p. 1098).

TRIASSIC VULCANISM, DIASTROPHISM, AND PALEOGEOGRAPHY

In the immediately preceding review of the Triassic around the Mediterranean some mention was made of the igneous rocks intercalated in a predominantly sedimentary section. In fact, flows and dikes of basalt, andesite, and related intrusives and extrusives are very common, and intrusions of ophiolite (basic igneous rocks altered to rocks rich in serpentine, chlorite, epidote, and albite) are nearly ubiquitous in the evaporitic Keuper of the western Mediterranean. Tuffs are widely distributed in the Alpine geosyncline. They are most common in the Ladinian, but also are found at other levels. The presence of radiolarites and chert in the early Mesozoic sections of the eastern Mediterranean shows that the Triassic sea of the Alpine facies was rich in silica at times, presumably when volcanic eruptions were most numerous.

These igneous phenomena were part of a post-Hercynian tectonic phase in which the long period of quiescence after Asturian orogeny permitted extensive magmatic differentiation. For example, Morre and Thiébault (1963, p. 545) point out that the Permian dacites and andesites on the south side of the Pyrenees differ greatly in composition from the Buntsandstein spilites and albitophyres on the north side.

Most diastrophism during Triassic times seems to have been epeirogenic, with block faulting being the principal mechanism. An exception may be coastal Yugoslavia where thick sandstones and beds and lenses of polygenic conglomerates are intercalated with tuffs and thin limestones in Montenegro, Dalmatia, and northeast of Trieste (Ciric, 1966, fig. 1). This coarse clastic sequence contains a meager Anisian fauna and lies between Werfenian sandstones or limestones and well-dated Anisian limestones. The three conglomeratic sequences, widely separated geographically, were evidently laid down near emergent areas. Ciric has gone so far as to suggest that they are the result of uplift in the west, the first event in an orogenic wave that migrated eastward until Cenomanian times and then reversed its direction. However, because of the lithologic similarity of these clastics to those of the Werfenian, they probably represent the erosion products from large fault blocks brought above sea level during the Carboniferous phases of Hercynian orogeny and rejuvenated by late Palatine epeirogeny. This interpretation is more compatible with the regional transgression observed throughout the Mediterranean in the Anisian and Muschelkalk than an orogeny localized off the coast of Yugoslavia.

Local uplift during the Triassic is reported at Tebaga de Medenine, southeast of Gabes in southern Tunisia, where tilted marine Permian and continental Buntsandstein are covered unconformably by flat-lying Aptian limestones. (Busson, 1967, p. 80). Successively younger strata wedge out against the Buntsandstein on the south flank of the structure which apparently continued to grow slowly throughout the Jurassic and early Cretaceous but became less prominent with time, as shown by onlap relationships. The precise age of the first relatively sharp movement at Tebaga de Medenine cannot be determined, but it is probably pre-Carnian, possibly Ladinian, for dolomitic limestones rest unconformably on the red sandstones, and on regional grounds these carbonates are presumed to be Carnian in age.

In all of Europe, where peneplanati[c] had reached an advanced stage by the e[nd] of the Permian and was continued duri[ng] the early Triassic, the transgression of t[he] Muschelkalk sea appears to have been esse[n]tially simultaneous, geologically speakin[g]. However, this general incursion had be[en] preceded late in Werfenian times by [a] premonitory advance that laid down r[ed] beds and evaporites (Röt) that are conside[r]ed to be a part of the Buntsandstei[n]. Nevertheless, parts of the Alpine trough, [in] southern Turkey and northern Iraq as w[ell] as in Italy and Yugoslavia were continous below sea level throughout most of Permi[an] and all of Triassic times. It is highly unlike[ly] that a tract four thousand miles long a[nd] three hundred miles wide would sink esse[n]tially as a unit, so that a change in sea lev[el] is a more reasonable explanation for t[he] Muschelkalk (Anisian-Ladinian) transgre[s]sion, which is recognizable from the Iberi[an] Peninsula to northern Iraq.

In western Europe this sea becam[e] progressively deeper, except for a bri[ef] regression early in the Ladinian. Howeve[r,] the larger Paleozoic massifs, the Iberia[n,] Armorican, Central, Bohemian, Rhodopia[n] and Menderes blocks, remained for the mo[st] part emergent, and the smaller massi[fs] flanking the Alpine geosyncline; Aar, M[t.] Blanc, Belledonne, Pelvoux, and Mercantou[r] were covered only during the course of th[e] transgression, not in its early stages. In ad[di]tion, "Vindelicia", a peninsula extendi[ng] west along the northern edge of the Alpi[ne] trough from the Bohemian massif, remain[ed] above sea level throughout Triassic times.

In the central Mediterranean, the Calabr[o-]Sicilian massif was emergent throughout t[he] Triassic, but northern Italy and Elba we[re] covered by the sea in Carno-Norian time[s]. It is possible that a Corso-Sard and a Catal[an] massif (see fig. 14) existed during the Tria[s]sic as islands in the Tethys, but evidence f[or] their presence is not conclusive.

In North Africa, the Paleozoic massi[fs] along the coast of Algeria may have remai[n]ed above sea level for some time duri[ng] the early Muschelkalk, or have emerg[ed] during the late Triassic, if Durand Delga terpretation of the Triassic sequence in e *Dorsale calcaire* is correct.

In any event, the Muschelkalk sea in orth Africa did not encroach far inland xcept in Algeria and Tunisia where the ssils found in its deposits are not suffi- ently restricted in range to determine recisely the time of the incursion with rela- on to that in Europe. According to emaison (1965, p. 93) the advance south- ard in Algeria did not begin until Ladinian mes. If so, it coincides with the apogee of uschelkalk transgression in Spain and rance after the intra-Muschelkalk regression entioned by Virgili (1962, p. 303) and icour (1963, p. 401).

A regression occurred at the end of adinian (Muschelkalk) times everywhere in urope and in the trough separating Turkey om the Arabian peninsula, as documented y Dunnington (1959, p. 11) and Weber 963). Brinkmann (1960, p. 78) reports allowing in the southern Alps at this time at caused local emergence and the devel- pment of bauxite. A similar phenomenon curred in Yugoslavia.

Transgression during the late Triassic has een demonstrated in much of the Alpine ough and on its flanks. For example, orthern Italy and Elba were covered by e sea only after Carnian times. In western urope, the lower Keuper includes thick aporites in many localities, sure evidence a recurrent supply of sea water to esiccating tracts in very extensive portions southern France, Spain, and along the eriphery of the western Alpine trough, on e Italian peninsula, in Yugoslavia and in reece. In Algeria, evaporites of Keuper ge are found in tracts that had been above a level since the Paleozoic, although the Vestern Desert of Egypt, central and eastern ibya, and portions of Morocco were not overed by the late Triassic transgression. Iraq, Syria, and probably in the Lebanon, e late Triassic sea returned to tracts that ad been long emergent, as well as to those nly briefly exposed to erosion by Ladinian gression.

No ready explanation exists for the eposition of salt and anhydrite over such very large portions of the Mediterranean region during the late Triassic although worldwide climatic conditions may have been favorable to desiccation. The stratigra- phic section in the Alpine trough itself is not completely devoid of evidences of hypersalinity (Rauchwacken), and thicknesses of five thousand feet or more of evaporites in many localities in the vicinity of this trough suggest prolonged slow subsidence around it as well as in it. The lithologic and faunal characteristics of carbonates associated with these evaporites are, however, indicative of littoral or neritic environments for the most part. Certainly salt and anhydrite are not being formed today in any such quantity over such broad areas.

The differentiation between Germanic and Alpine facies mentioned in the introduction to this review of Triassic sedimentary environments loses much of its cogency when both are evaporitic. However, the varicolored marls and shales associated with the evaporites in areas of Germanic facies are commonly not present in tracts consider- ed representative of Alpine facies. In these localities, carbonates which may be oölitic and detrital are commonly interbedded with the salt and anhydrite. The difference between the two facies is probably depen- dent on the presence or absence of a source of terrigenous clastics in the vicinity of the desiccating basin.

JURASSIC FACIES

The tectonic calm that prevailed in the Mediterranean region during the Triassic was broken in Jurassic times by an increase in the number and strength of epeirogenic movements, and by minor orogeny. The sediments laid down in the seas and those deposited in the newly formed basins on land reflect in their diversity and in their restricted vertical and geographical distribu- tion this moderate acceleration in the tempo and augmentation of the intensity of dias- trophism. The Lias was, however, a period of transition during which the conditions that obtained in Keuper times changed only gradually. Some regions were not affected by crustal movement in Lias times. For example, the hypersaline sea that covered the Algerian Sahara during the Keuper remained undisturbed until late in the Bathonian (Busson, 1967, p. 176).

In the Mediterranean portion of western Europe the cargneules with *Avicula contorta* which represent the Germanic facies of the Rhaetian are commonly grouped with overlying strata of the same type as the "Infra-Lias". No cephalopods, corals, or brachiopods occur in this sequence which is transgressive with respect to the underlying Keuper. The remainder of the early Lias (upper Hettangian-Sinemurian) is calcareous, predominantly dolomites and banded limestones, whereas formations referred to the medial and late Lias are in many areas made up mainly of dark colored calcareous shales and marls with intercalations of limestone, the whole rich in ammonites. In other areas, shales are less common, and light-colored cherty limestones make up much of the section.

Dogger strata are more varied lithologically than those of the Lias because of an increase in both local and regional diastrophism during medial Jurassic times. In some places these movements were accompanied by vulcanism. Dogger epeirogeny is well documented in Spain, Portugal, Minorca, Morocco, western Algeria, Sicily, eastern Syria, and western Iraq. In many other localities shallowing or deepening during the medial Jurassic is indicated by changes in the lithology and fauna of the several predominantly calcareous formations.

Regional uplift during the Bathonian-early Malm was strongest in North Africa where it was accompanied by erosion of the Precambrian and late Paleozoic shields. The relief developed by epeirogeny must have been low and worn down rapidly, for the large tracts on the African coast in which clean calcilutites with tintinnids represent the Tithonian-Berriasian could not have existed had prolific sources of terrigenous clastics been numerous in their vicinity.

The effects of Nevadan and late Cimmerian tectonism are most apparent in Europe where strata assigned to the Malm are as varied as those of the Dogger. Sandstones indicati of uplift during the late Jurassic crop out the Pyrenees and Cantabria, and bauxites a common in Yugoslavia and Greece. In t French and Italian Alps ophiolite flows a interbedded with radiolarian cherts a capped by sublithographic tintinnid-beari limestones. In the Dinaric Alps tuffs a submarine diabase flows are intercalated the late Jurassic limestones that were loca subject to penecontemporaneous erosic However, in some localities Tithonian su lithographic limestones rest on the weathered beds.

In the Mediterranean region the seve manifestations of tectonism were not stro enough to cause long-lasting emergence geosynclinal troughs. However, Kimeridgia Purbeckian epeirogenic uplift was sufficien ubiquitous in Europe to cause widespre regression on shelves and platforms a extensive deposition of continental san stones, mainly of Wealden age. Whe emergence did not occur, changes in t depths of water are reflected in the litholog sequence over tens or hundreds of squa miles.

In southern Europe true orogeny Jurassic times occurred only in the Caucas and Crimea where several sharp upli followed by erosion and transgression a known (Arkell, 1956, p. 636-637). The were accompanied by vulcanism, particula strong in the Bajocian. Although the Crim and the Caucasus are not contiguous to t Mediterranean, tectonic events there we presumably not unrelated to diastroph occurrences in the Tethys, and some a certainly contemporaneous.

A characteristic of the sequence of calc eous strata of Jurassic age throughout t Mediterranean is the presence of red a green nodular limestones with ammonit the *ammonitico rosso.* Aubouin (1965, a 1965a, p. 127) discusses these limestones detail. The characteristic nodular limeston of the true *ammonitico rosso* occur at seve levels in the Jurassic of the weste Mediterranean, and also in the Triassic Greece and Yugoslavia. They are alwa associated with condensation of the sectio that long periods of time are represented only a few feet of rock, and these ndensations are commonly accompanied by rd grounds. Obviously, such an association uld develop only in an area of long-ntinued stability where few terrigenous stics were being deposited. Aubouin lates this stability to the period of orogenic lm following the development of a trough fore it is filled with terrigenous sediments. nd-derived clastics are absent because no nergent areas exist in the region. However, the *ammonitico rosso* and the hard ounds must have formed in relatively allow water, and indeed, are sometimes und near bauxite deposits and in association with phosphatic nodules. Aubouin ates the occurrence of these condensed ata to the existence of submarine ridges rmed during the early stages of the evolun of a geosyncline.

RASSIC STRATIGRAPHY

Like the discussion of Triassic stratigphy, this review of the Jurassic found in ll and outcrop in the Mediterranean region gins with the Iberian peninsula and ntinues counterclockwise around the ores of the existing sea. Because of the eater diversity of facies in Jurassic times, descriptions are even more cursory and s complete than those of the Triassic ccession. However, the relationship tween facies and major structural elnents, already apparent in Triassic rock its, can be traced yet more confidently in rassic formations.

Iberian Peninsula. On both flanks of the Pyrenees assic strata crop out in narrow and discontinuous ts. The lack of coarse terrigenous clastics in the tions, and the similarity of the lithologic succession both belts suggest that they were laid down by a which covered the site of the existing ranges. Their ies indicate that this sea was somewhat deeper in west than in the eastern Pyrenean area. No evidence a geosynclinal development can be found in the ology of these strata, although basalt flows and fs are intercalated in the Hettangian dolomites of northern belt. The Sinemurian also is represented dolomites, but the remainder of the Lias consists shales and limestones, dark-colored in the lower tion, lighter and with fewer shale interbeds near top. All of the post-Sinemurian Lias contains an abundant fauna of molluscs, including ammonites. The Dogger is made up predominantly of limestones, but is somewhat less fossiliferous than the Lias.

Beginning in medial or late Callovian times (Mangin and Rat, 1962, p. 340), uplift occurred in the Santander region of Asturias west of the Pyrenees. Tracts east and west of the uplifted area, in the western Pyrenees and in Oviedo province, were not at first involved in this uplift, for marine limestones assigned to the Oxfordian are reported in both. Uplift continued however, and larger areas became subject to erosion, so that the strata of late Oxfordian age in the western Pyrenees are black pebbly fossiliferous limestones which grade upward to marine sandstones. Kimeridgian sandstones with ammonites crop out along the Oviedo coast where they mark the end of the marine sequence in that region. In the Spanish Basque region at the western end of the Pyrenees no marine strata of Kimeridgian or Portlandian age are known. The late Jurassic is represented in only a small portion of the central Pyrenees by lacustrine limestones of late Kimeridgian or Portlandian age. Consequently, the entire range must have risen above sea level before this time.

Late Dogger uplift also affected the Iberian ranges, where the Jurassic section is somewhat thicker than that of the Pyrenees but similar in aspect. Near the western end of these ranges lower Callovian limestones are covered unconformably by the continental sandstones of the Cretaceous *Utrillas* Formation, but toward the Mediterranean coast successively younger stages of the Malm are found. The net result of the uplift in the Pyrenees and the Iberian ranges, as well as in the Burgos and Santander regions was the formation of an "Aragon Gulf" which during the late Jurassic occupied an area approximately that of the present-day Ebro basin. The sea presumably retreated from this gulf before the end of Kimeridgian times, and nearly all of Spain except the Betic trough rose above sea level.

South of the Iberian ranges in the Betic trough the Jurassic section is much different from that found in Valencia and Catalonia. The Lias is an unbroken sequence of limestones, the lower part of which is similar to that in the rest of Spain, but cherts are common in the Middle Lias along with numerous ammonites like those found in Italy (Fallot, 1932, p. 57). With the Toarcian, the first Jurassic *ammonitico rosso* appears, and indicates, if Aubouin's interpretation is correct, a prolonged period of stability on a submarine ridge. The fauna of the Toarcian-Aalenian is, however, more like that in the remainder of Spain than that in Italy and the Alps.

The Dogger is not well represented in the Betic mountains where most outcrops are measured in tens of feet, but the section may be complete in a limestone sequence locally in an *ammonitico rosso* facies, but more commonly gray and cherty. The fauna is essentially that found in western Europe.

The Malm, on the other hand, has an "Alpine" aspect. It consists of thin beds of limestones and marls, mainly reddish in color, with many outcrops of *"fausses-breches"*, pseudo-breccia, or *"Knollenkalk"*, that may have been formed by disturbance of lime muds in submarine slides before their final consolidation. Rich ammonite faunas occur in this sequence. In some localities the late Tithonian, possibly transgressive, is a very fine-grained, dense limestone with *Calpionella.* The entire Malm ranges from one hundred fifty to two thousand feet in thickness.

The persistence of an open-sea environment in the Betic trough when the remainder of Spain rose above sea level late in Jurassic times emphasizes its individuality as a structural entity. It will be recalled that the discrete nature of the Betic trough had been manifested even earlier by the unique characteristics of the Triassic strata therein.

In the Lusitanian basin of Portugal on the west flank of the Iberian massif the evaporites of the late Triassic seen in wells and inferred from diapirs are succeeded by dolomites, limestones, and calcareous shales representing the Liassic that thicken northward to a maximum of about twenty-five hundred feet. All of the stages of the Lias except the Hettangian (mainly dolomite in wells and sandstone in outcrop) are recognizable in an excellent suite of ammonites and brachiopods. Dogger limestones are less fossiliferous, and for the most part were laid down in shallow water as shown by the abundance of detrital and oölitic levels throughout, particularly in the north where Callovian uplift was followed by erosion. At this same time, diapiric structures that had been growing since late in Liassic times were also subject to erosion, whereas in the basins between them sedimentation was continuous into the Oxfordian.

This regional uplift at the end of the Callovian cut off the open sea to the west. The "Lusitanian" (Upper Oxfordian-Lower Kimeridgian) which ranges up to more than thirty-five hundred feet in thickness was laid down in a trough between an emergent massif in what is now the offshore, and the Iberian shield. It consists predominantly of sandstones and shales in the west, and detrital and oölitic limestones in the east, with a lagoonal facies of bituminous limestones near the eastern shore. The succeeding "Abadia" Formation, included in the "Lusitanian", but of early Kimeridgian age, is marked by the widespread occurrence of sandstones and arkose derived from renewed uplift of the basement, but includes also oölites, and some limestones and marls in the central portion of the basin. Terrigenous clastics are less common in the upper part of the formation, which is terminated by a characteristic oölitic or pisolitic limestone. Reefs exist on the flanks of this late Jurassic trough in strata equivalent in age to the Abadia Formation.

During late Kimeridgian times renewed uplift on the massifs surrounding the Lusitanian basin caused the deposition of calcareous sandstones over extensive tracts. In the southern part of the basin marine limestones are intercalated in the clastic section, but farther north the entire Upper Kimeridgian-Portlandian is represented by sandstones with plant remains and freshwater molluscs. These become finer in grade and more calcareous upward.

The strata of Lias and Dogger age in the Algarve basin of southern Portugal are similar to those in the Lusitanian basin, but even more calcareous. On t other hand, the Malm differs greatly in that clasti are far less abundant in the Upper Jurassic of the sou coast. Nevertheless, uplift during Oxfordian tim occurred throughout the Algarve, for in the east t lithology and fauna of Oxfordian limestones sugge shallowing upward, and in the west strata representi the Oxfordian are absent. There, Kimeridgian detri and oölitic limestones rest in slight unconformity Callovian beds. However, in the east, Kimeridgian a Oxfordian strata are conformable, the Kimeridgian le clastic than in the west. In both areas the remaind of the Jurassic, the late Tithonian, consists of mari limestones that are followed by Cretaceous limeston in an uninterrupted sequence.

Balearic Islands. On Majorca and Minorca Keup shales and limestones grade up to a Rhaetian-Infra-Li represented by cargneules. The overlying Jurassic thickest in what was presumably a structural depressi in the northern part of these islands where a thousa foot dolomite with a meager foraminiferal fau represents the lower Lias. The lower Pliensbachian this trough is distinguished by interbeds in simil dolomites of gray and yellow calcareous shales wi lamellibranchs and brachiopods, and the upp Pliensbachian is gray, calcareous sandstones with Pal ozoic shale pebbles. Northern Majorca was subject to a short episode of subaerial erosion thereafter, a Minorca rose above sea level, remaining emerge throughout the remainder of the Jurassic.

The Lias section in other parts of Majorca is th in comparison to that in the north and includes lim stones with charophytes like those found commonly fresh water. However, as the trough in northe Majorca shallowed, limestones with thin-shelled a monites were laid down on Cabrera island off t south coast and this differentiation of facies fro north to south can be noted on Majorca proper the strata dated as late Lias (Domerian-Aalenian).

Pelagic and neritic environments were widespre in Dogger and Malm times on Ibiza (emergent duri the Lias) and on Majorca, as shown by the numero outcrops of sublithographic limestones which inclu abundant globigerinids, and in the upper portio *Calpionella.* Coccoliths are so numerous at some leve as to comprise the principal constituent of these lim stones (Colom and Escandell, 1962, p. 126).

Toarcian *ammonitico rosso* and radiolarites on M jorca are like those found in the Betic trough, Sicil and Italy, and tintinnid-bearing fine-grade limeston of Malm age are also found in all of these localitie Colom and Escandell postulate that north of the Balea ic islands during Lias times a "Catalan massif" w present which provided the Pliensbachian clastics the northern part of Majorca and Minorca. This mass became smaller and retreated from the immediate vici ity of Majorca and Ibiza during the Dogger-Mal This hypothesis is questionable but the existence Pliensbachian sandstones in northern Majorca and M norca and the prolonged emergence of Minorca the after are undeniable.

Morocco. The *Dorsale calcaire* of the Rif mountains
: coastal Morocco includes a complete Lias section in
limestone facies similar to that of the Betic mountains
: Spain, with common brachiopods, and *ammonitico*
sso present locally. In some places siliceous limestones
ith chert nodules predominate (Durand Delga *et al.*,
)62, p. 407). At many levels ammonites are numerous;
: these, most species are like those found in the Alps
d Sicily. Dogger and Malm have not been recognized
ith certainty in the western portion of the *Dorsale*
lcaire. Malm in the form of sublithographic limestones
ith *Calpionella* is present in the Bokoya, the eastward
tension of the *Dorsale* (Roch, 1950).

The Rif mountains south of the *Dorsale calcaire* are
highly complex mass of olistostromes in which rec-
gnizable Jurassic crops out in isolated exposures. The
uthern portion of these gravity slides is known as
e "Pre-Rif". There, although individual outcrops
clude only a part of the section, the marine Jurassic
complete, made up of Lias and Dogger limestones,
part oölitic and detrital, which grade up to Malm
ales and varicolored limestones with occasional sand-
one and radiolarian chert interbeds. The complex
ratigraphy of this allochthonous sequence is discussed
several publications (Roch, 1950; Lévy and Tilloy,
)52; Durand Delga *et al.*, 1962). Farther north in the
if proper, all or most of the Dogger and Malm is
presented in the very thick, dark-colored calcareous
ales of the *"Zone marnoschisteuse"* (Durand Delga
al., 1962, p. 407). According to Durand Delga, these
ales were derived from a basin lying northeast of
e *Dorsale calcaire.* They could as well have been a
ver on the *Dorsale calcaire.* In any event, the shale
asin was delimited to the south and east by an area in
hich terrigenous clastics were replaced by carbonates.
he area occupied by the autochthonous Jurassic
mestones at the southern edge of this "Rif complex"
as been determined from well information which
rves to define the limits of the Jurassic sea in north-
estern Morocco. Paleogeographic relationships during
e late Malm are shown on figure 6. The sea occupied
ly a slightly larger area to the west and south during
e Lias and Dogger, for the Moroccan Meseta remain-
l emergent throughout the Jurassic period (Roch
)50, p. 292).

East of the Rif complex and the Moroccan Meseta,
e sea advanced southward early in Lias times to the
ont of what are now the Anti-Atlas ranges, thus
vering all of eastern Morocco. It then retreated to
vo great linear subsident belts which are now occupied
y the Moyen and Haut Atlas mountains. These depres-
ons are called, therefore, the Moyen and Haut Atlas
oughs. Their location is shown on figure 13. The
aut Atlas trough continued eastward in Algeria south
: the Hauts Plateaux, where the Saharan Atlas now
es, so this portion of the depression at the edge of
e African shield is known as the Saharan Atlas
ough. The Tamlelt high near the Moroccan-Algerian
ontier acted as a partial barrier to free communica-
on of the sea between the two segments of this linear
bsident zone, but during the early Lias and again
uring the early Dogger when a second marine transgression covered almost the same area as that of the first
one, the seaway of the Haut Atlas and Saharan Atlas
was a single entity.

During Lias and Dogger (to medial Bathonian)
times, great thicknesses of carbonates were laid down
in the Haut Atlas and Moyen Atlas troughs, not only
as reefs and biostromes, but also as calcareous muds,
some of which are now dolomitized. Oölitic limestones,
sandstones, and varicolored shales are intercalated in
the massive carbonates, particularly on the flanks of
these linear depressions. The maximum thickness of
the Lias and Dogger in the troughs is more than
twelve thousand feet.

In the area north of the Haut Atlas trough and
east of the Moyen Atlas trough, known as the "Eastern
Meseta", Lias and Dogger formations are much thinner,
and somewhat less complete. The fauna in the limestones which comprise most of the Jurassic on this
relatively stable platform is made up principally of
lamellibranchs, gastropods, and brachiopods, although
ammonites are sufficiently common to permit recognition of the several stages, and lacunae resulting from
temporary emergence.

During the medial Bathonian the sea left the Haut
Atlas trough and the Eastern Meseta, but remained in
the coastal portion of the Moyen Atlas trough until
Callovian times, its deposits being predominantly sandstones and calcareous gypsiferous shales. The Malm
sea was confined to a narrow coastal belt in which
thin-bedded shales and sandstones were deposited near
the Rif, and thick limestones and dolomites laid down
to the south and east, thinning over the Paleozoic
massifs near the Algerian frontier and in the vicinity
of the southern shore of this seaway.

Algeria. An extension of the Rif trough known as
the Tellian trough continues the length of Algeria
parallel to the coast. Durand Delga (1967, fig. 1)
considers the two to be separated by a spur of foreland
near Melilla. Strata of Jurassic age crop out only rarely
in the Tellian trough west of Algiers, and all exposures
reveal an incomplete section, with conglomerates intercalated at several levels in the Lias and Dogger of the
principal outcrops — the Grand Pic of the Ouarsenis,
and the Ténes and Chenoua massifs. The remainder
of the sequence consists predominantly of carbonates
laid down in the open sea. Deepening in Malm times
is suggested by the presence of *Calpionella*-bearing
limestones. In some localities, particularly along the
Dorsale calcaire, formations of Dogger age are not
present or are very thin. During the late Jurassic in
the vicinity of the Chenoua massif the northern edge
of the trough must have been near the existing coast
where an allochthonous Tithonian "flysch" of interbedded sandstones and shales with thin intercalations
of fine-grained limestones with tintinnids is present.
These terrigenous clastics almost certainly came from
the north.

East of Algiers, exposures of Jurassic rocks in the
Tellian trough are larger and more numerous than
those west of the city, but lithologically and faunally
outcrops in the two areas are similar. In both regions
the Dogger is absent on the *Dorsale calcaire,* but is

present elsewhere, for example, in the western Numidian ranges where the twenty-five-hundred foot Dogger-Malm section consists of siliceous limestones and calcareous shales with a characteristic but meager fauna. However, in the eastern Numidian ranges near the Petite Kabylie massif the Dogger is absent, and the Malm is very thin and incomplete.

North of the Numidian ranges the Jurassic crops out in several nappes, all thrust southward. In the northernmost of these the Tithonian consists of sandstones and shales, and a transition southward to limestones can be followed. Thus, east of Algiers, as in the Chenoua massif west of it, a source of terrigenous clastics must have existed north of the existing coast during Malm times. A portion of this emergent tract may be represented in the Grande and Petite Kabylie massifs.

In western Algeria the Tellian trough is bounded to the south by the Hauts Plateaux or "Hauts Plaines", an eastern extension of the Eastern Meseta of Morocco. Jurassic strata on the Hauts Plateaux are commonly thinner than in the Tellian trough to the north and in the Saharan Atlas trough to the south. The Hauts Plateaux were, however, under progressively deeper water eastward to their eastern limit in what is now the Hodna basin. The Lias on the Hauts Plateaux, and indeed in most of Algeria, consists mainly of dolomitic limestones and dolomites with some intercalations of oölitic and reefal limestones. Its average thickness on the Hauts Plateaux is less than two hundred feet as compared to thirty-five hundred feet in the Saharan Atlas trough, and nearly two thousand feet in the Tellian trough (Kieken, 1962, p. 548). The Dogger is represented by some two hundred feet of oölitic limestones and dolomites in the western Hauts Plateaux, whereas in the Saharan Atlas trough strata of the same age are some eighteen hundred feet of dolomites and limestones with *Cancellophycus,* and in the Tellian trough, where local uplift was common, the Dogger attains a maximum of about one thousand feet, predominantly dolomites and limestones with *Cancellophycus.*

Strata of Callovian and early Oxfordian age on the Hauts Plateaux are for the most part ammonite-bearing calcareous shales with thin sandstone intercalations, their average thickness being nearly one thousand feet. Locally, the Callovian is transgressive. The late Oxfordian consists of about two thousand feet of deltaic sandstones and shales, with some reefal limestones in the northern portion. With the Kimeridgian, terrigenous clastics were no longer supplied to the shallow sea, so that the remainder of the Jurassic is represented by biostromal and reefal limestones, with a maximum thickness of about fifteen hundred feet.

The major faults and other accidents in the Hauts Plateaux are aligned in two principal trends, according to Luças (1952, p. 129). These are N. 55° E. and N. 85° E. Vertical movements along these two trends influenced the facies of Triassic strata, and to a lesser extent the pattern of Jurassic sedimentation. The effect of these movements is particularly noticeable where accidents on each of the trends intersect.

The Hauts Plateaux during Jurassic times we separated from a similar positive block in eastern geria by a depression in which the strata deposited more like those in the Tellian Atlas trough than th on the Hauts Plateaux or its eastern continuation, t "Ain M'lila môle" of Kieken (1962, p. 553). In t eastern portion of this "Hodna" depression where t trough was deepest, the Jurassic is entirely calcareou the Lias is fifteen hundred feet of dolomites; t Dogger three hundred feet of siliceous limestones wi *Cancellophycus;* the Callovian-Kimeridgian three th sand feet of nodular limestones and cherts in an *a monitico rosso* facies (this facies if it really occup three thousand feet, would render Aubouin's interpre tion of the significance of *ammonitico rosso* invali and the Portlandian, seven hundred feet of sublit graphic limestones with *Calpionella.* Thus, in t Hodna mountains the calcareous Jurassic is over fif five hundred feet thick. Elsewhere in the depressi between the Hauts Plateaux and the Ain M'lila mô the Jurassic consists of relatively thin carbona deposited in an epineritic environment. The section incomplete locally because of minor uplift.

The eastern limit of the Ain M'lila môle is n yet defined, but it is present at least to the longitu of Constantine. No terrigenous clastics occur in Jurassic section found on it, which thins northwa from the Sahara Atlas from at least five thousand less than fifteen hundred feet of biostromal limesto and dolomites.

The portion of the Saharan Atlas which lies sou of the Hauts Plateaux is discussed in detail by Corr (1952) and SN REPAL (1952). Successively young Mesozoic strata are exposed eastward in these rang Near the western end, in the Ksour mountains, the L consists of some four thousand feet of dolomites, lin stones, and calcareous shales, with dark colored d omites common at the base, and lighter-colored lin stones and calcareous shales predominant in Domeri and younger strata. Ammonites are common in so localities. The Bajocian is represented by thirty-fi hundred feet of gray and black limestones with abundant fauna of characteristic ammonites, the Bath ian by more than two thousand feet of dark gr sandstones and littoral limestones with abundant la ellibranchs, locally replaced by dolomites, and t Callovian through Oxfordian by nearly four thousa feet of shaly sandstone intercalated with five to twe foot beds of dolomitic limestones containing a mea littoral fauna. Near the Tamlelt basement high on t Moroccan border these sandstones are distinguished fr those of the Kimeridgian with difficulty, but east of Ja Djibissa, a five hundred foot dark gray-blue limesto with common echinoids and lamellibranchs represe the Kimeridgian in a near-shore facies. The remainder the Jurassic consists of thin, ripple-marked sandsto intercalated with and capped by lighter-colored lin stones with oysters and other littoral molluscs.

At the extreme west end of the Saharan Atlas no of Figuig, the Liassic sea was littoral or lagoonal. deposits are now azoic unfossiliferous dolomites int calated with gypsum and marls; eastward the transiti the open-sea environment of the ammonite-bearing nestones of the Ksour mountains is made through lcareous shales and oölitic and biostromal limestones. nilar changes presumably occur in the characteristics the strata assigned to the Dogger and early Malm, lecting deepening of the sea eastward, but only the meridgian and Portlandian crop out from one end the chain to the other so that the changes can be served. In the Aurès mountains of eastern Algeria e Kimeridgian consists of a lower thousand foot unit limestones and marls with characteristic ammonites erlain by fifteen hundred feet of soft limestones and irls that contain no diagnostic fauna. The Portlandian, ich rests unconformably on these soft limestones, is presented by the light-colored, sublithographic calci-ites with *Calpionella* so typical of the Late Tithonian d Berriasian. On the south side of the trough open- deposits are replaced by dolomitized reefal lime-ones. Westward, at the longitude of Laghouat, the iestones of Kimeridgian and Portlandian age contain ly oysters, and still farther west give way to the idstones of a clastic Jurassic section, of which the per portion is clearly littoral.

Tunisia. East of Constantine the Saharan and Tellian las troughs are not separated by an intermediate ssif. Consequently, in Tunisia the boundary between e two is obscure, and its location a matter of inter-etation. Bonnefous (1967, pls. 4-7) indicates that prior the Kimeridgian the Tellian trough was the more oidly subsident, although local uplift in it is shown the presence of limestone conglomerates of Bathon- age. Although Bonnefous' elucidation of facies rela-nships is somewhat different from that in the paper Bismuth *et al.* (1967), the two are compatible.

The Jurassic section exposed in the tract that can rly be assigned to the Saharan Atlas trough consists nost entirely of carbonates, but in a discrete struc-ral province, the "Chotts-Sahel depression", which cupied the eastern coast near Sfax and Gabes and e Chotts region, the late Bathonian, Callovian, and cfordian are represented by nearly five thousand feet greenish gray to dark gray shale (Bonnefous, 1967, 115). The clays and silts of these neritic shales were obably derived from an emergent tract to the east, no large source of terrigenous clastics appears to ve existed to the north and west, and to the south goonal and lacustrine strata were being deposited at is time. However, a small emergent area, the "Mat-ita high", was present immediately south of the otts (Bonnefous, 1967).

Local uplift at the end of Pliensbachian and early Domerian times in the Tunisian portion of the two oughs is shown by the complete absence of strata of e Domerian-Toarcian age in some localities, and of merian-Aalenian beds in others. Condensation in e several units of the Dogger has been found in a mber of outcrops along a north-south trending high, ggesting prolonged stability of a shallow sea floor that tract. Similar stability late in Oxfordian times suggested by the occurrence of *ammonitico rosso,* hough late Oxfordian uplift is reported at Jabal ghouan. During the Kimeridgian-Portlandian, tintin-nid-bearing fine-grained limestones were deposited in both troughs, but shallows existed locally (Bismuth *et al.,* 1967, p. 165) on which clastic limestones were laid down.

The Chotts-Sahel depression (see fig. 14) was sub-sident more or less as a unit throughout Dogger and early Malm times. The Chotts region continued to sink throughout the remainder of the Malm, although the thick shales and sandstones of the Dogger-early Oxford-ian are replaced gradually there by littoral limestones of Portlandian age with some clastic and evaporite interbeds, whereas the Sahel was less actively subsident.

On the relatively stable shelf of the Saharan foreland south of the Chotts, evaporites like those of the late Triassic, intercalated with thin dolomites and varicolor-ed clay shales, were deposited without interruption until Bathonian times. The salt content in an essentially anhydritic sequence increases to the southwest, as dol-omites disappear. (Busson, 1967, p. 40).

This anhydrite is capped in outcrop and in wells by limestones containing a meager Bathonian fauna, in turn succeeded by some three hundred fifty feet of sandstones and clay shales with intercalations of shell beds and limestones.

The clastic sequence is assigned to the Bathonian by Busson. On it rest abundantly fossiliferous littoral limestones, locally dolomitized, containing Callovian am-monites along with foraminifera, pelecypods, echino-derms, and corals. This unit represents the Callovian and Oxfordian. The lower hundred feet of this seven hundred foot limestone body may be Bathonian, how-ever, according to the microfauna.

In stratigraphic continuity on these partially dol-omitized limestones are more than four hundred feet of Purbeckian-Wealden sandstones and shales. Dolomites are intercalated in quantity in the lower portion of this predominantly clastic unit. This sequence termi-nates in a green clay shale, above which are Barremian conglomerates and sandstones lying uncomformably on the older shales.

Libya. In northwestern Libya the Jurassic section in the escarpment paralleling the coast is cut out east-ward by overlapping and unconformable Barremian sandstones, so that at Garian these truncating clastics, known as the Chicla Formation, rest on Triassic. West-ward along the scarp, however, evaporites of Lias age appear above those assigned the late Triassic, followed successively by Bathonian limestones and shales, Callov-ian sandstones, Callovian-Oxfordian dolomites, and Purbeckian-Wealden sandstones and capping shales, all thinner than the corresponding formations in southern Tunisia, but sufficiently well dated to permit their correlation.

Marine Jurassic has also been encountered in deep wells near the Libyan coast in eastern Cyrenaica where shales and limestones make up most of the section. Details of the stratigraphy of these wells have not been published, but the total thickness of Jurassic strata must be several thousands of feet. As shown on figure 6, the shoreline of the Tithonian sea was presum-ably in northern Cyrenaica and was deflected northward by the seaward extension of the Sirte arch.

Algerian and Libyan Sahara. The changes in facies of Jurassic strata from the subsident troughs of the Saharan Atlas and the Chotts-Sahel depression to the Saharan platform were mentioned briefly in the preceding discussion of Tunisia. Similar changes undoubtedly exist in the Algerian Sahara where Jurassic rocks have been found in many wells but do not crop out.

The meager information available shows that in the Algerian Sahara, as in Tunisia, the evaporites of the late Triassic form part of a single depositional cycle that continued into the Lias, with salt being replaced late in Lias and early in Dogger times by anhydrite interbedded with limestone and dolomite. The change in facies occurs earlier in the south. The evaporites grade upward to clay shales, and the carbonates themselves become argillaceous in younger strata. Late in Bathonian times deposition of sandstones began in the Algerian Sahara as a result of uplift of the Precambrian shields to the south and of Morocco to the west. During the Callovian-Kimeridgian, clastics from the south and west reached the western portion of the Hauts Plateaux, but did not attain the Tellian trough nor extend east of the longitude of Algiers (Carantini, 1968). Clastic deposition ceased during the Portlandian on the western Hauts Plateaux, but the marine deposits of this stage are intercalated in sandstones in the Algerian Sahara, and are not found south of the Hassi Messaoud and Hassi R'mel fields (Ortynski *et al.,* 1959, p. 712).

Withdrawal of the sea from the Sahara probably began at about the same time as in Morocco (medial Bathonian, Choubert and Faure-Muret, 1952), and clastics of late Bathonian age in southern Tunisia suggest that the uplift responsible for the regression and the deposition of sandstones and shales affected most of the Saharan foreland. However, the supply of clastics to southern Tunisia and northwestern Libya was interrupted in Callovian and Oxfordian times, whereas in the Algerian Sahara it was constant throughout the late Jurassic and early Cretaceous.

In southern Libya and in adjacent portions of Algeria strata assigned to the Jurassic are entirely continental. They include all of one red clay shale formation and a portion of an underlying variegated shale unit. The two are separated by a minor unconformity. The overlying "bone beds", also unconformable, are referred to the Wealden (Burollet, *et al.,* 1960, p. 18).

As shown on figure 6, most of Libya and contiguous portions of Algeria and western Egypt were above sea level throughout the Jurassic period, for the Sirte arch, formed during Hercynian orogeny, was still in existence. However, in Egypt the Jurassic sea advanced over an area that had been emergent since late Paleozoic times.

Egypt. According to some reports the Jurassic sea in the Western Desert was progressively transgressive southward, attaining its maximum extent during the Malm when it reached 29° N. latitude, before Portlandian regional regression and intensive block faulting began. All of the information concerning the Jurassic of the Western Desert comes from wells drilled in the search for oil. According to Said (1962, fig. 40), a predominantly calcareous facies is present only north Cairo, the proportions of clay and sand increas toward the shoreline. However, the thickness of strata of Jurassic age ranges widely, probably beca of variations in the rate of subsidence of the sea fl as well as in the supply of detritus to the several ar with positive and negative tendencies. A general sou ward thinning is obvious, however. More than thousand feet of Jurassic shales and fine-grade lir stones were found in the subsurface southwest Alexandria.

In the Gulf of Suez region, marine strata of Juras age crop out in several localities. The formation r ognized most widely is an epineritic limestone of Bath ian age, on which lie "Nubian" Sandstones (fluv deltaic?) commonly assigned to the Cretaceous in t region, but lying in apparent conformity on the Bath ian. The greater part of the strata of Jurassic age the Gulf of Suez region in both well and outcrop clastic, mainly sandstones and shales with coal sea demonstrating the presence of an ancestral gulf, clos to the south.

On the Sinai peninsula outcrops of rocks of Juras age are confined to the northern portion where, in axes of several breached anticlines, a sixty-five hund foot Pliensbachian-Oxfordian predominantly carbon section is known. Most of the rock units exposed neritic or littoral limestones, but thin sandstones a shales occur throughout the sequence, and one ei hundred foot deltaic and lagoonal formation occ in the section. The existence in a calcareous successi of this clastic element made up of cross-bedded sa stones, siltstones, and shales with coal seams sho shallowing and rejuvenation of sources of clastics the Sinai peninsula during Bathonian times, for litto limestones with a fauna of Bathonian lamellibran and ammonites rest conformably on it.

No evidence exists in Egypt to indicate the prese of a land mass anywhere offshore during Jurassic tim This is in marked contrast to the sandstones of l Jurassic age along the Algerian coast of which only feasible provenance was an emergent tract to north. To the contrary, the distribution of Juras facies in Egypt suggests that all terrigenous mate came from the south.

A part of the Nubian Sandstone which crops widely in southern Egypt and the Sudan may be Jurassic age, but most commonly it is referred to Lower Cretaceous. Probably in local depressions the Paleozoic erosion surface Permian, Triassic, a Jurassic continental formations were deposited bef the sheet sands consequent on regional, late Cimmer uplift covered the entire area.

Jordan. As on the Sinai peninsula the Rhaetian a Lias of southern Jordan consist of light colored qua sandstones with intercalations of slate blue shales a near the top, limestones that are commonly oöl (Wetzel and Morton, 1959, p. 127). These clastics ripple-marked and cross-bedded indicating their dep tion in a fluvial, deltaic, littoral, or lagoonal envir ment. The Dogger is represented by neritic limesto that indicate an advance of the sea to the southe so that during medial and late Jurassic times the shoreline was only slightly west of that of the Muschelkalk, and roughly parallel to the existing coast (*idem,* p. 172). Strata representing the Dogger are light colored, littoral and epineritic limestones and sandy calcareous shales with an abundant fauna of lamellibranchs and echinoids, and a few gastropods, brachiopods, and ammonites. These strata thicken sharply northward, but are cut off by an unconformity southward, so that the Malm is not recognized in most of the country.

The Levant. In Palestine marine strata of Jurassic age crop out only in the axes of eroded anticlines, but have been encountered in many wells drilled for oil. According to Bentor *et al.,* (1960) in one outcrop in the south, basal Jurassic clastics rest unconformably on the Triassic, indicating some local movement before Liassic transgression.

The thickness of the predominantly carbonate Jurassic section ranges up to nearly ninety-five hundred feet in the southern coastal plain of Palestine, but decreases to less than five thousand feet eastward and southward where a number of sandstones are intercalated in a limestone and shale section which is progressively completely replaced by continental clastics in Jordan.

In general, the Lias consists of sandstones in central and southern Palestine, and the Dogger of shallow-water limestones and calcareous shales with some sandstone intercalations. Reefs are present locally. In the south the Malm is represented only through the Oxfordian by gray and brown calcareous shales and limestones succeeded unconformably by Neocomian sandstones. These rest on progressively older Jurassic rocks southward into Jordan (Coates, *et al.,* 1963, p. 33).

In northwestern Jordan, marine Jurassic like that in central Palestine extends eastward a short distance (Bender, 1968, table 16). The section consists of shales and limestones of late Lias through Oxfordian age with characteristic Dogger faunas reported in several localities. These marine strata undoubtedly gave way eastward to continental deposits, but as they are capped unconformably by sandstones assigned to the lower Cretaceous (Bender, 1968, p. 129), preservation of pre-Cretaceous beds of this type is unlikely except in local depressions.

In southern Lebanon and western Syria outcrops of Jurassic age occupy large portions of the Lebanon and Anti-Lebanon ranges. In Mt. Hermon west of Damascus the complete Jurassic section is over sixty-two hundred feet thick. Of this total, only the five hundred feet of lower Lias (?) is of terrigenous origin — green shales with lacustrine gastropods intercalated in sandstones and lignites. The upper Lias is represented by two thousand feet of dark-colored limestones and dolomites, and the Bajocian by five hundred feet of brown, partly dolomitized oölitic and sandy limestones. The Bathonian-Callovian is also calcareous, some twenty-three hundred feet of slightly arenaceous neritic limestones, but the Malm includes some ammonite-bearing shales in a seven hundred and fifty foot carbonate unit capped unconformably by Neocomian sandstones (Bender, 1968, p. 47). The Portlandian is not represented in this section which thins markedly southward to less than two hundred feet of oölitic, partly dolomitized limestones (Renouard 1955, p. 2132).

In northern Lebanon the Lias is not exposed. Dogger limestones and dolomites are less sandy than in the south, and marine Portlandian is present, resting on seven hundred feet of Kimeridgian volcanics made up of basalts and cinerites intercalated with conglomerates, marls, and beds of lignitic clay. The Portlandian consists of epineritic and littoral limestones — biostromes and oölites — intercalated with conglomerates, marls, and clays as well as fine-grained limestones with chert nodules, and is succeeded unconformably by Wealden sandstones.

Iraq. In Iraq, Jurassic strata crop out in the Zagros and Taurus ranges. Resting conformably on Rhaetian shales are dolomites and limestones, in part showing the effects of leaching of evaporites in typical cargneules. These are intercalated with oölitic and detrital limestones and occasional shales, the whole becoming darker in color upward. This poorly fossiliferous Lias is succeeded by black bituminous limestones and papery shales with ammonites, brachiopods, and lamellibranchs that demonstrate a late Lias-Bathonian age and are capped by a highly condensed section of black shales and bituminous limestones, and dark gray or blue limestones representing either much of the lower Malm or an incomplete section in which several interruptions in sedimentation have not been detected. The overlying lower Kimeridgian consists of laminated locally cherty limestones and dolomites, with some autoclastic breccias caused by leaching of evaporites. The terminal unit of the Jurassic section is made up of thin-bedded limestones and shales with abundant ammonites, radiolaria, and tintinnids demonstrating a medial Tithonian-Berriasian age. About half of this unit is assigned to the Jurassic. Unconformity has not been recognized in the type section but is reported in other outcrops between Jurassic and Cretaceous strata.

In this Jurassic section exposed in Kurdistan no evidence of unconformity between units assigned to the Dogger and Malm has been adduced, although Wetzel (in Dunnington, *et al.,* 1952) suggests that it may exist. In any event, in all of central Iraq evidence of Bathonian-(?) Callovian erosion has been encountered in wells.

In central Iraq the Lias found in wells consists of oölitic and detrital limestones resting on late Triassic shales, and succeeded by two anhydrites separated by a pseudo-oölitic "normal" marine limestone intercalation. As in outcrop, the late Lias-Bathonian is represented by black papery shales and limestones on which rests unconformably another sequence of partially dolomitized oölitic and pseudo-oölitic limestones capped by Kimeridgian anhydrites. These are overlain by the thin-bedded, ammonite-bearing limestones of the medial and late Tithonian, similar lithologically and faunally to those in Kurdistan. An important pre-Berriasian unconformity exists in all wells, and formations of Aptian age rest directly on Dogger shales northwest of Mosul.

These same unconformable relationships occur in northeastern Syria on the Jauf-Mardin high (Weber, 1963, p. 252-253). Several wells along the axis of the high found lower Cretaceous limestones on Dogger shales, and one near the Turkish frontier encountered lower Cretaceous on Triassic (?) - Jurassic evaporitic dolomites. In the Toueman wildcat at the south end of this series of wells (36° N., 41° E.), Upper Cretaceous limestones and dolomites rest on thin Jurassic dolomites. A similar relationship exists in outcrop in Wadi Hauran (33° 30'N., 41° 15'E.) in western Iraq where Bathonian limestones are capped by Cenomanian sandstones. Little doubt can exist that at the end of Jurassic times the Jauf-Mardin high, as well as a broad tract to the west of it, was above sea level. The report of marine late (?) Jurassic in the Palmyra region (Dubertret and Vautrin, 1937) suggests that some kind of seaway existed at least that far east, but in general during the late Malm the sea was shallow in the Levant, and restricted to a narrow belt along the Mediterranean coast.

Turkey. Outcrops of Jurassic age are few and scattered in Turkey, except in the southeast near the Iraq border, and it is not possible to determine whether the absence of deposits is due to non-deposition or to denudation. Mesozoic transgression on the Hercynian surface began with the Triassic along the north coast, while in the central region it began with the Lias. Very probably, as shown on figure 6, the principal Paleozoic massifs remained emergent throughout the period and strata of Jurassic age were deposited only in troughs between them. In northern Turkey the Jurassic fauna shows a close affinity to that of the Mediterranean and Alpine regions, but in southeastern Turkey and Iraq Lias and Dogger fossils are like those of the Caucasus and northern Iran.

One of the most complete sections of the Jurassic in Turkey is at Bayburt in the mountains south of Trabzon on the eastern Black Sea coast. Here the seventy-five hundred foot section consists of the Lias, made up of sandstones, shales, and limestones with some intercalated lavas and tuffs, and a late Malm comprised mainly of limestones, with a sandstone and conglomerate at the upper limit. The Dogger seems not to be present. Ammonites are abundant, and permit recognition of many of the standard European zones of the Lias (Arkell, 1956, p. 350). Other less complete outcrops occur farther west, one group being just north and west of Ankara, where, however, Callovian has been recorded along with the fragmentary Lias. Callovian is also known on the coast of the Sea of Marmara, along with Oxfordian and Tithonian. The latter is present as sublithographic limestones with *Calpionella,* resting in some places on Permo-Carboniferous shales and Permian limestones with *Fusulina.*

In southeastern Turkey near the Iraq border Jurassic strata are included in a comprehensive limestone sequence that ranges from Permian through Cretaceous, but has not been studied in as much detail as in Iraq (Schmidt, 1964). Apparently no interruption in sedimentation took place in this Harbol region during the Jurassic. Similar comprehensive limestone sequences are exposed far to the west in the vicinity of Antalya on the south coast of Turkey. Both occurences are in the Taurus trough, now the Taurus ranges. Some radiolarian cherts and ophiolites also near Antalya may be of late Jurassic age (Blumenthal, 1963, p. 660; Orombelli, *et al.,* 1967, p. 832).

Cyprus. The island of Cyprus is considered to be in a western extension of the Taurus trough which then passes through Crete (Aubouin and Dercourt 1966) and swings northward into Greece and Yugoslavia as the Dinaric trough (see fig. 14). Strata of Jurassic age crop out in the Kyrenia ranges along the north coast of Cyprus in a long, narrow, eastwest trending belt. The formation, known as the Hilarion Limestone is about one thousand feet thick. It consists of slightly metamorphosed limestones and dolomites, some of which are black or dark blue, others gray and white Fossils are rare, but a medial and late Jurassic age is confirmed. Underlying strata are oölitic limestones or gray calcareous shales that presumably represent the Lias. An occurrence of similar limestone in southwestern Cyprus is intercalated with volcanic agglomerates and tuffs. Henson *et al.* (1950, p. 13) are doubtful that this unit is Jurassic.

Balkan Peninsula. The continuation of the Taurus geosynclinal trough in Greece has been studied in detail by Aubouin (1965a) who, with his students and associates, has defined a number of discrete structural elements in the Hellenic portion of the Dinaric trough and demonstrated westward migration of tectonic pulses in a eugeosynclinal-miogeosynclinal couple (*eu*-igneous activity; *mio*-no igneous activity). During the early and medial Lias the Pindus trough remained slowly subsident (see discussion of Triassic of Greece: The earliest Mesozoic linear depression, situated on the flank of the persistently positive Pelagonian massif). In the shallow portion of this trough neritic limestones were laid down, while in the center siliceous, fine-grained limestones were deposited. Late in Lias times (Pliensbachian?) the "Ionian" trough formed west of the Pindus trough, separated from it by the "Gavrovo" ridge. In the deeper portion of this new linear depression shales with *Posidonomya* and siliceous limestones and cherts were formed, while on its flanks limestones and marls were deposited, part of which are in an *ammonitico rosso* facies. Through the remainder of Jurassic times both the Pindus and Ionian troughs were persistently subsident. Minor flows of spilite may have been extruded on the sea floor along the inner margin of the Pindus trough ("sub-Pelagonian" zone) during Lias times, and after the Kimeridgian much more extensive and thicker submarine flows of ophiolite were laid down there. These igneous rocks are covered by radiolarites of late Tithonian age. On the stable tracts on either side of both troughs neritic limestones represent the entire Jurassic, whereas in the Ionian trough black shales with *Posidonomya* are common in the Dogger and in the Pindus trough radiolarites are associated with thin-bedded limestones throughout the section representing the medial and late Jurassic. The western limit of the Ionian trough was the Adriatic foreland (see fig. 4).

In Albania the troughs and ridges described in Greece extend the length of the country, the move

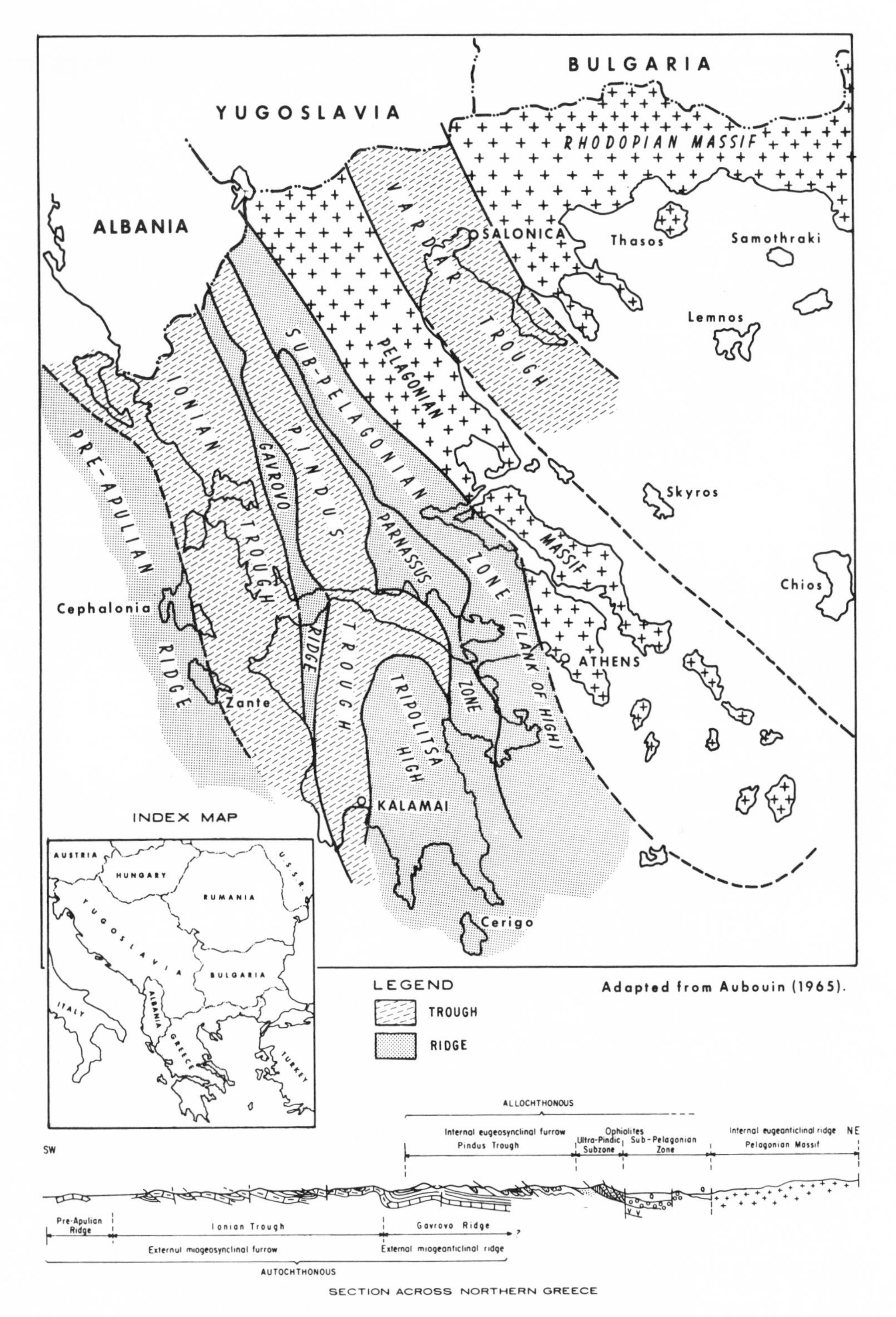

Fig. 4. Greece. Mesozoic structural elements.

ments on the Scutari-Pec line having no effect on the continuation of the Pindus and Ionian troughs and the Apulian foreland (Aubouin and Nodjaj, 1965). The ophiolites of the sub-Pelagonian zone (west flank of Pelagonian massif) are prominently developed.

In Yugoslavia, the Dinaric Alps are divided by Ciric (1963, p. 566) into two major units, "Inner" and "Outer" which correspond to Aubouin's division of these ranges in Greece into eugeosyncline (Pindus trough) and miogeosyncline (Ionian trough). The distinction is based on the presence or absence of igneous flows and intrusions. In both the Inner and Outer Dinaric Alps limestones of Lias age rest conformably on Triassic reefs and biostromes. Ammonites and the problematic *Lithiotis* are common in many localities, their presence suggesting open-sea conditions, and *ammonitico rosso* is well developed, indicating sea-floor stability.

In the Outer Dinaric Alps the Dogger and Malm are represented by partially dolomitized shallow-water limestones and biostromes, together attaining a thickness of nearly three thousand feet. Local emergence late in Jurassic times is indicated by bauxites intercalated in the littoral and neritic carbonates. In some localities in the Outer Dinaric Alps dark colored bituminous limestones of Dogger and Malm age show that sea-floor circulation may have been restricted.

The Inner Dinaric Alps were involved in disturbances during the Jurassic which may have begun late in Lias times, but were much more important in the Kimeridgian and later Malm. These involved uplift of the more labile areas, and were accompanied by deposition of tuffs and submarine flows. Also, intrusions of ultrabasic magma have been found in calcareous strata of Lias, Dogger, and early Malm age. The flows in some localities cover *ammonitico rosso* and hard-grounds and are intercalated with varicolored radiolarian cherts (Aubouin, *et al.,* 1965). The time of eruption was certainly post-Kimeridgian, for *Calpionella*-bearing limestones interbedded with some flows represent the late Tithonian.

Only meager information is available concerning the extension of the Carpathians through Rumania as the Transylvanian Alps, and their juncture with the Pontic mountains of Turkey through the Balkan mountains of Bulgaria. In this great arc of mountains which bounds the Moesian platform on three sides (see fig. 8), the Lias is predominantly in a near-shore facies — sandstones, shales, micaceous, sandy limestones, and occasional conglomerates. Coal seams are mined in some localities. Ammonites of Pliensbachian and Toarcian age are found in the upper part of this section. The Bajocian is not well represented, but where present includes shales with *Posidonomya* and some sandstones. The Bathonian is a very thin limestone, obviously condensed, containing ferruginous oölites and an abundant ammonite fauna. Callovian and Oxfordian also are represented by thin and probably incomplete sections, or are absent. The Kimeridgian is more widespread, as red or green, in part arenaceous, nodular limestones, and the Tithonian, here transgressive, consists of more than eleven hundred feet of massive white limestones with rare corals and *Diceras,* but no ammonites. Southwest of Sofia the reef facies changes into nodular calcareous and argillaceous sandstones with intercalations of sandy shales, marls, sandy limestones and conglomerates. The lower Tithonian there includes some diabase and radiolarian cherts.

The lithologic succession of the Jurassic in the Pontic and Balkan troughs is similar, and supports the inference that they formed a single tectonic entity. The transgressive nature of Tithonian beds, the condensation or absence of strata of Dogger age, and the widespread occurrence of clastics in the Lias are common to both.

The Moesian platform is enclosed by the Balkan arc. Following deposition thereon of evaporites and intrusions and flows of diabase and melaphyres during the Keuper, it was submerged during the Dogger, the sea first laying down sandstones and organic limestones on a gently tilted surface. Calcareous shales of late Dogger age are present in the western part of the platform, but absent in the east. The Malm sequence, transgressive and unconformable on tilted Dogger or Lias beds, is predominantly carbonates up to twenty-five hundred feet in thickness in the eastern and southern parts of the platform, thinning to less than two hundred feet in the center and north, where it consists of nodular limestones with ammonites (Patrut *et al.,* 1963).

To summarize, a central Paleozoic and Precambrian massif on the Balkan peninsula was flanked during the Jurassic by nascent subsident troughs, the Dinaric trough to the southwest, and the Balkan trough to the northeast. The Dinaric trough was already in existence in Triassic times, but the Balkan trough may have been defined only during the Jurassic.

At its northern extremity the Dinaric trough joined the Alpine trough, certainly a major structural element throughout the early Mesozoic. The eastward extension of the Alpine trough known as the Carpathian trough swung sharply southward to join the Balkan trough.

The Alps. In the Alpine trough the facies of Jurassic rocks range through a broad spectrum reflecting a great diversity of depositional environments. This variety is a result of the appearance of elongated depressions and ridges, presumably horsts and grabens (Aubouin, 1965, p. 216), much more numerous than those developed during the Triassic, and formed at a more rapid pace. In spite of the mobility of its floor, the Alpine sea was on the whole deeper than in Ladinian and Carnian times. Igneous activity was manifest only in the deepest trenches, but the prevalence of cherty and siliceous rocks suggests that silica was supplied to the sea water in large quantities. Sequences with numerous condensations or diastems are widespread; currents presumably kept the deposition of sediments to a minimum in these stable tracts.

In the southern Alps, Jurassic facies are like those in the "Outer" zone of the Dinaric Alps. The Jurassic section is exposed near Lake Como and crops out with minor interruptions eastward through Lombardy and Venetia. Dip is southward. The lower Lias is very

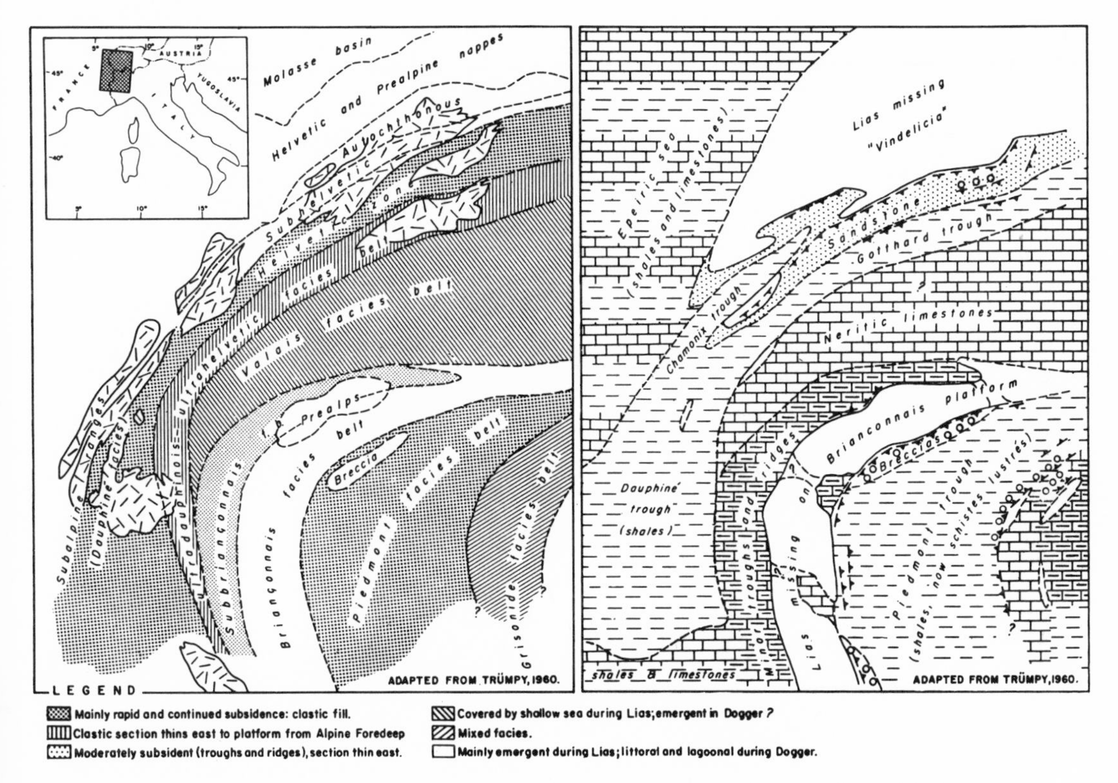

Fig. 5. The Western Alps. (a) Jurassic facies belts. (b) Liassic paleogeography and facies.

thick (up to two thousand feet) and consists of gray, siliceous, bedded limestones and dolomites easily distinguishable from the more massive limestones of the Upper Triassic with which their contact is commonly conformable. Late Lias, Dogger, and early Malm are included in a thin (condensed) and oölitic or siliceous sequence. Toarcian, Aalenian, and Bajocian are in some localities represented by *ammonitico rosso,* and in others by varicolored crystalline limestones that may total only twenty feet in thickness. Bathonian through Kimeridgian limestones are also thin, but apparently less condensed. They are commonly interbedded with cherts, and the Kimeridgian may be represented by *rosso ammonitico veneto,* a brick red and violet, cherty, nodular limestone. Throughout the southern Alps the late Tithonian is marked by the appearance of white sublithographic tintinnid-bearing limestones about two hundred feet thick, that represent also the basal Neocomian. The complete Jurassic sequence in the southern Alps averages about twenty-five hundred feet in thickness, of which more than two thousand feet is Lias.

In the northern Alps, proximity to the Bohemian massif and its westward extension is shown by the deposition during the Lias of coarse arkose and sandy shales with a few coal seams, in a series like that of the Balkan trough. Farther to the south these terrigenous clastics are replaced by light colored limestones with brachiopods and crinoids, red nodular limestones, gray siliceous limestones, or dark gray calcareous shales and marly limestones, depending on the depth of the sea floor. Strata of Dogger age are similar but facies are displaced northward because of transgression of the sea in that direction. The sea was deepest during the Malm, and radiolarian cherts reflect this deepening and the increased incidence of igneous activity. The light colored, sublithographic limestones of the late Tithonian are fringed to the north and south by reefs. In general, the Jurassic sequence in the northern Alps, except at the northern margin of the Alpine trough, is similar to that in the southern Alps, although condensation is less marked. However, over the central Alpine basement ridge the Dogger section is incomplete. The Jurassic in the northern Alps, like that in the south, is about twenty-five hundred feet thick, but only half is of Lias age.

The western Alps have been studied intensively by French geologists for many years. The Jurassic section there differs from that of the predominantly calcareous southern and northern Alps in that shales and thin sandstones are common throughout. Facies belts are developed in arcs concave eastward, although not precisely in the direction of existing mountain chains. Trümpy (1960) gives an excellent discussion of the area. He defines five structural and facies entities: 1) the Helvetic and Dauphinois provinces, making up an outer miogeosyncline 2) the Valais facies, laid down in an outer eugeosyncline 3) the Briançonnais belt, divided into outer Sub-Briançonnais, moderately uplifted zone separated from the Dauphinois by a basement high, and thinning eastward to the Briançonnais, a broader basement platform which gives way to 4) the Piedmont or *schistes lustrés* province, an inner eugeosyncline. This grades through 5) the Grisonide zone, to the south Alpine province described briefly above. (See fig. 5a).

The Liassic of the Helvetic and Dauphinois provinces consists of up to two thousand feet of shales, sandstones, and arenaceous limestones. In the Dauphinois region the thick argillaceous limestones are capped by dark shales. In the Helvetic province the medial Jurassic is made up of black pyritic shales followed by sandstones with ferruginous oölites and crinoidal limestones, the whole thinning and condensed eastward, whereas in the Dauphinois black and gray shales in a very thick sequence represent the Bathonian, Callovian, and lower Oxfordian. In both provinces the remainder of the Malm is dark colored limestones overlain by reefoid limestones in the north and light colored, fine-grained limestones in the south.

In the Valais outer eugeosyncline the Lias is made up of thin limestones, and the Dogger and Malm are commonly absent. Malm limestones resting on granite boulders are present only in the southeastern part of the Valais. This region was presumably a starved linear depression during most of Jurassic times (see fig. 5b).

The Sub-Briançonnais is characterized by the existence of nearly two thousand feet of cherty, silty, crinoidal limestones and calcareous shales of Lias age that are missing in the Briançonnais proper. The Dogger consists of thick shales and silty limestones, in part oölitic and coral-bearing, and the Malm is predominantly light colored, relatively thin, neritic limestones.

The Briançonnais, a sixty-mile wide persistently positive platform in much of late Triassic times, and again emergent in the Lias (bauxites, possibly terrestrial breccias) was covered during the Bathonian by a shallow sea in which oölitic limestones and calcareous shales containing *Mytilus* and corals were laid down. In paludal tracts coals were formed at this time. Late Jurassic transgression is shown by basal red muds derived from weathered Dogger and Triassic surfaces, now red and green nodular limestones. The remainder of the Malm section consists of light colored neritic limestones with some red radiolarites near the base. The upper portion of this unit contains tintinnids and ammonites. The entire Jurassic on this high is less than five hundred feet thick, as compared to nearly five thousand feet in the Dauphinois.

In the Piedmont or *schistes lustrés* province, less than three thousand (Barbier, *et al.,* 1963, p. 336) or perhaps as much as seven thousand (Trümpy, 1960, p. 855) feet of calcareous sericitic schists represent probably the Lias and Bajocian. These were originally thin-bedded, argillaceous, silty, and rarely sandy limestones, interbedded with calcareous shales, laid down in a subsident trough. The fauna found to date is confined to one level near the base and consists of sparse radiolaria, dwarf molluscs, bryozoa, crinoid plates, and arenaceous foraminifera. Ophiolites associated with the *schistes lustrés* are now considered as late Jurassic and early Cretaceous in age (Barbier, *et al.,* 1963, p. 368). These were laid down not only as extrusives on the sea-floor, but also as intrusives. They include peridotites, gabbros, porphyrites, spilites, and their metamorphic derivatives.

The remainder of the Dogger, the Malm, and probably the lower Cretaceous of the Piedmont are represented by thin radiolarian cherts and fine-grained limestones.

Trümpy and some other authors consider the thin Dogger and Malm sections made up of siliceous limestones and radiolarites to have been deposited in great depths of water, for the paucity of fossils other than radiolaria and *aptychi* (ammonite opercula) they ascribe to the effects of pressure solution of calcium carbonate.

Italian Peninsula. The thickness and lithology of the strata of Jurassic age laid down in the northern part of the Italian peninsula are similar to those of the southern Alps. In Tuscany the lowest unit of the Jurassic, assigned to the Hettangian-Sinemurian, is massive dolomite or a white and gray waxy, crystalline limestone. In some localities this unit apparently ranges higher, but commonly it is succeeded by a Hettangian *"rosso ammonitico toscano"* which in turn is capped by bedded white, gray, or pink, cherty, siliceous limestones. In Umbria and the Marches a Toarcian *ammonitico rosso* is present in many outcrops of this predominantly siliceous limestone sequence.

Near the upper limit of these siliceous limestones is a very persistent red or green shale containing *Posidonomya alpina* Gras, here referred to the Bathonian. Above these shales but below the overlying radiolarian cherts is a thin unit of siliceous limestones presumably representing the Callovian-Oxfordian. The radiolarian cherts range up to five hundred feet in thickness, and include a few megafossils suggesting a Kimeridgian-Berriasian age. In some localities the chert gives way to white sublithographic limestones with tintinnids of Tithonian-Berriasian age. In Liguria ophiolites are associated with the radiolarites.

In the northern and central Apennines the thickness of the Jurassic section ranges between thirteen hundred and thirty-five hundred feet of which strata assigned the Lias make up nearly four-fifths.

South of Rome and of a line extending north-northeast of that city across the Italian peninsula (see fig. 6), middle and upper Jurassic strata reflect a marked change in environment from open sea, relatively deep water to epineritic and littoral conditions. The transition occurs in a belt from fifteen to twenty-five miles wide. The lower Lias, least affected, includes some corals, and the siliceous limestones and *ammonitico rosso* of the upper Lias are replaced by oölitic ınd clastic limestones, with abundant benthonic fossils. The radiolarian cherts of the Malm give way to yellow iliceous limestones, but the calpionellid-bearing subithographic limestones of the late Tithonian extend unhanged across the transition zone and far to the south.

In many localities in central and southern Italy the hallow-water limestones of Jurassic age are partially or completely dolomitized, but an original oölitic or letrital content can be detected in many localities, articularly in the southern Apennines where the secion attains a thickness of twenty-three to thirty-three undred feet. Ophiolites of late Jurassic age are found ocally (Merla, *et al.,* 1964, p. IV-21).

In the Lagonegro-Potenza region on the coast south of Naples, Jurassic strata are included in an allochhonous, highly contorted unit of siliceous nodular limetones and shales on which rest siliceous bedded limetones intercalated with radiolarites. The autochthonous urassic in this area is made up of littoral and epineriic limestones. These are widely exposed farther south n the Calabrian massif where Jurassic strata are thin nd characterized by interruptions in the continuity of he sequence as well as intercalations of terrigenous lastics, whereas in adjacent regions the section is much hicker and made up almost exclusively of carbonates Quitzow, 1935). Late Tithonian limestones are found earest the central portion of the massif, resting on a ew feet of sandstone or conglomerate above basement.

Sicily. In the subsurface of the Iblean plateau in outhern Sicily the lower Lias is represented by black hales and limestones that constitute the upper portion of the shales with *Estheria* mentioned in the discussion of the Triassic, but the Triassic-Jurassic limit has not een defined precisely. The overlying unit is a grayreen, compact, well-bedded limestone with intercalaions of gray-green and gray-brown marly limestones, rading northward to reefal limestones, the whole rangng in thickness from seven hundred to thirteen hunred feet. Fossils include ammonites and algae demontrating a Liassic age.

The succeeding formation, of Dogger and early Malm ge, is again mostly gray-green, compact, well-bedded imestone, but includes intercalations of red and grayreen calcareous shale and black and red radiolarian hert. The thickness of this Dogger-lower Malm unit, which farther north includes biostromes, is between six undred and twenty-nine hundred feet. A tuff-volcanic reccia interbed in it ranges from a few feet to nearly hirteen hundred feet. This igneous stratum is found in he Ragusa field.

According to Schmidt di Friedberg (1962, p. 644), he late Malm-early Cretaceous on the Iblean plateau onsists of three hundred to twelve hundred feet of tylolitic, white, dense, sublithographic limestone with enses of chert and interbeds of green and white marl. This unit rests unconformably on the older Jurassic ut the duration of the hiatus must have been short, nd any angular discordance minor. Megafossils are umerous.

Thus, the Jurassic section in southern Sicily, accordng to published information, is from two thousand to our thousand feet thick, and is comprised almost entirely of carbonates with some siliceous intercalations.

In northeastern Sicily, the existence of the Calabro-Sicilian massif as a positive element during Jurassic times is shown in the succession of the strata of this age which thin and contain lagoonal elements north of Taormina (Schmidt de Friedberg, 1962, p. 644). The presence of oölites and gray-green clay shales in a Lias section made up predominantly of siliceous limestones indicates shallow water and proximity to a source of clastics. Similarly oölitic limestones and red and green clay shales of Dogger and Malm age, although intercalated with radiolarites, indicate a relatively shallow environment. As in southern Sicily, Malm white and green limestones are transgressive, but here include a basal limestone conglomerate. The thickness of the Jurassic near Taormina is about two thousand feet.

The Jurassic exposed in the ranges of northern Sicily consists predominantly of biostromal and reef limestones, detrital limestones, breccias and conglomerates, about twelve hundred feet thick, obviously formed in shallow water in which waves and currents were active. Farther to the south, but still in the mountains of northern Sicily, the Jurassic is represented mainly by some thousand feet of clay shales with chert and detrital limestone interbeds, with occasional tuff and basalt intercalations in the portion assigned to the Dogger. Still farther south, a deeper water section like that on the Iblean plateau crops out, with evidence of an unconformity between the Tithonian and older Malm, as on the Iblean plateau and near Taormina.

Ammonites are common in the Jurassic outcrops, and serve to date the strata in which they occur. The minor vulcanism in Dogger times was accompanied by uplift that caused local interruption in sedimentation. At least a portion of these igneous manifestations are Bajocian, for Fabiani and Trevisan (1937) report stratified tuffs with Bajocian fossils at Roccapalumba and at Vicari. A basal limestone conglomerate with basalt pebbles exists above these tuffs and is capped by Bathonian limestones.

The Jurassic in Sicily, with minor unconformities in both Dogger and Malm accompanied by Bajocian vulcanism, differs from that of peninsular Italy where, although the Dogger may be thin, no evidence of igneous activity is reported. Bajocian disturbances and tuffs and flows are, however, prominent in the Caucasus. Note that this vulcanism is older than the uplift on the African foreland (Bathonian).

Sardinia. Formations assigned to the Jurassic crop out extensively in northern Sardinia. In the Nurra region on the northwest coast thin blue and gray fine-grained Lias limestones lie conformably on the Keuper-Rhaetian and are succeeded by two thousand feet of Dogger oölitic and pisolitic limestones with a littoral fauna. This clastic unit grades up to light colored fine-grained, compact, dense limestones representing the Malm. On the east coast the Malm is transgressive on Hercynian granite and schist. A thin red basal sandstone is succeeded by some thousand feet of gray dolomite and about twelve hundred feet of white fine-grained limestone. On the southwest coast a similar limestone is less than fifty feet thick. In southern

Sardinia terrigenous sandstones with well preserved plant fossils of Bathonian age rest on basement or on Lias limestones. These transgressions and the existence of terrigenous clastics demonstrate that Bajocian tectonism affected Sardinia as well as Sicily, lifting some tracts and causing others to sink.

Corsica. The marine Jurassic in Corsica is represented only by small outcrops west and southwest of Bastia. In these, the Lias rests on Rhaetian limestones with *Avicula contorta* and consists of dark-colored partially dolomitized azoic limestones grading up to siliceous limestones with chert nodules containing molluscs and brachiopods of Sinemurian age. Emergence followed, and continued until late in the period (Delcey, *et al.,* 1965, p. 324).

During the Kimeridgian, or immediately thereafter, submarine albitic basalts (spilites) were extruded. On these lie thin beds of red and green radiolarian cherts and gray and pink more or less siliceous limestones with *Calpionella alpina.* Resting on basement in other localities are pseudo-oölitic and oölitic limestones with benthonic foraminifera and algae indicating a Kimeridgian or Portlandian age. Thereafter, uplift on a regional scale must have occurred, for the succeeding beds in one locality are clastics made up of limestone pebbles and cobbles with some Paleozoic elements, and in other outcrops breccias and conglomerates (made up of fragments of Triassic and Jurassic carbonates, granite, rhyolite, schists, and black cherts, the whole intercalated with clay shales and dark gray limestones) rest on shallow-water limestones of the late Jurassic. Farther south, marine sedimentation may not have begun until early in Cretaceous times (*ibid,* p. 327), when, however, no disturbances occurred.

The age of the *schistes lustrés* of eastern Corsica has been disputed for many years. Delcey *et al.* suggest that the ophiolites intercalated in it may be late Jurassic, and consequently that the *schistes lustrés* are in part of Malm and Neocomian age. Lapadu-Hargues and Maisonneuve (1964) consider that the granites intruding the *schistes lustrés* are of Paleozoic age, and that consequently the intruded strata must be older. They cite other evidence, including the presence of *schistes lustrés* pebbles in strata of Mesozoic age to support their argument. Bloch (1964) denies the metamorphism of the *schistes lustrés* by granite intrusions, refuting, however, only one such occurrence, and refers them to the Eocene.

In addition, the amount of allochthony and the direction of the forces that formed the nappes of the *schistes lustrés* are in dispute.

Southeastern France. During the Jurassic, Provence and the tract farther north between the *Massif central* and the arc of Hercynian granite massifs bounding the Alpine geosyncline was for the most part under a shallow sea, in which deposits were mainly carbonates. Corroy (1963) and Aubouin and Menessier (1963) discuss the deposits laid down in this sea in Provence.

Hettangian white or ash-gray, crystalline well-bedded limestones intercalated with marls and clays rest conformably on Rhaetian limestones with *Avicula contorta.* Dolomitization has occurred in some localities and cargneules are not uncommon. Sinemurian and Lotharingian are not represented in southern Provence but are sporadically developed near Aix. The retreat of the sea responsible for the absence of strata assigned to these stages ended during the Pliensbachian. The Domerian and later Lias are represented by red ferruginous limestones with chert intercalations. The Toarcian is in some localities distinguished by black, papery shales.

The sea deepened during Dogger times, and all of the Bajocian and lower Bathonian are included in a sequence of white or off-white marly limestones and calcareous shales. Sharp uplift early in Bathonian times in southern and central Provence is shown by the development of yellow or red organic, commonly coarsely oölitic limestones, and by the existence of Bathonian lignitic clays north of Antibes. In northwestern Provence this uplift did not occur, and the light colored marly limestones represent the whole of the Dogger.

The Malm sequence is made up entirely of carbonates which consist of light colored and somewhat marly limestones near the base, and bedded and sublithographic limestones in strata considered to be lower Kimeridgian. Gray, massive, well-bedded dolomites, occasionally sandy, and with cargneules and red clay intercalations represent the late Kimeridgian. As it did during the Dogger, the sea shallowed progressively during the Malm, through early Portlandian times. The late Portlandian is either white sublithographic limestones (Aubouin and Menessier, 1963, p. 51), or white organic limestones, which include the first rudistic reefs, and, east of Marseilles, Purbeckian lagoonal limestones with gastropods and *Chara.* Emergent tracts are characterized by the presence of molluscan burrows below the exposed surface (Corroy, 1963, p. 23-24).

North of Gap and Digne in the area occupied by the Dauphinois miogeosyncline the sea was deep throughout Jurassic times both west and east of the Belledone and Pelvoux massifs. Between this deep and the *Massif central* subsidence on a lesser scale began during the Tithonian in the *Vocontian* trough (see fig. 10) an east-west elongated depression at the latitude of the Mercantour massif (Remane, 1967; Tempier 1957).

Still farther north the facies and sequence of Jurassic strata are like those in the southern Jura mountains where the section assigned to the Lias is made up of dark-colored limestones and calcareous shales, with ferruginous oölites in strata referred to the Toarcian and Aalenian. The Bajocian consists of thick, shallow water limestones containing crinoids and corals, succeeded by oölitic limestones. The Bathonian-Oxfordian is represented by calcareous shales and marls, and the Kimeridgian-Portlandian by similar strata intercalated with reefal limestones that are increasingly numerous upward.

Thus, in the Rhodanian corridor shallowing southward is indicated by changes in the facies and thickness of Jurassic strata. Local emergence occurred west of Nice late in Malm times, and it is possible that low-lying land area existed in what is now the offshore as an extension of the Maures-Esterel massif, itself mostly above sea level in Portlandian times (Corroy, 1963, p. 24).

Along the southern edge of the *Massif central* Jurassic limestones crop out extensively in the Causses. The Lias is typically developed, with black ammonite-bearing calcareous shales representing the Pliensbachian-Toarcian. The Dogger and Malm consist of limestones, some of which are reefal and biostromal.

Southwestern France. In the Aquitaine basin of southwestern France the Jurassic section attains a maximum thickness of about five thousand feet. From west to east in the subsurface it changes in facies from a sequence consisting predominantly of calcareous shales and marls with minor amounts of dolomitic limestones and dolomites to less argillaceous clean limestones and dolomites. Near the upper limit of the section sandstones occur. These are referred to the Purbeckian (Cuvillier, *et al.*, 1962).

Early in Lias times sandstones were laid down in some localities near the *Massif central,* and Hettangian eruptions in the Pyrenees caused the deposition of some two hundred feet of tuffs near the mountain front. Most of the Lias is represented by littoral and epineritic limestones, but Toarcian and Aalenian black shales with *Posidonomya* were laid down in the vicinity of the Poitou and Minervois straits, along with ferruginous oölites. The Lias is about two thousand feet thick near Bordeaux.

Dogger strata in the Aquitaine basin are essentially limestones and marls in the west with increasing amounts of oölites intercalated in the east. North of Toulouse lignitic beds with a brackish water fauna indicate an approach to a shoreline. Near the Pyrenees some thousand feet of dark-colored fetid dolomites suggest local subsidence. Elsewhere the Dogger sea in the Aquitaine basin was shallow.

At the end of Bathonian times the strait between the Tethys and the Aquitaine basin was cut off (Bonnard, 1958, p. 1102). However, communication was maintained with the Paris basin through the Poitou strait. Immediately south of the strait shales and argillaceous limestones represent most of the late Jurassic, the uppermost marine strata being dated early Portlandian. In the central portion of the basin about thirty-five hundred feet of limestones and dolomites are assigned to the Malm, the lower portion made up of sublithographic limestones (Callovian-Oxfordian) and the Kimeridgian-Portlandian by dolomitic limestones and dolomites.

Along the *Massif central* shoreline, reefs are common in the lower Malm, and the Kimeridgian is made up of argillaceous limestones. The Portlandian begins with a basal conglomerate, which lies in slight unconformity on older strata. Succeeding beds are fine-grained dense limestones with some ammonites which give way to green calcareous shales and limestones with brackish water fossils. Continental sandstones assigned to the Purbeckian and Wealden terminate this sequence.

Uplift on the south side of the western Pyrenees during the Oxfordian-Kimeridgian, mentioned in the discussion of the Iberian peninsula with which this review of Jurassic stratigraphy began, is matched in the Basque country of southern France. Marls and shales represent the Callovian and Oxfordian there, and coarsely sandy and pebbly limestones with oysters and shell beds characterize the Kimeridgian, with which the marine cycle ends. This would suggest that marine conditions persisted slightly longer on the north side of the Pyrenees than on the south flank where Oxfordian sandstones are the uppermost marine Jurassic.

Salt tectonics were continually effective along certain trends during Jurassic times in the Aquitaine basin, their effect being to produce structures on which no Jurassic is present, without any change in the facies of the strata immediately adjacent to the structure (Cuvillier, *et al.*, 1962, p. 370).

It appears that most or all of the Aquitaine basin rose above sea level at the end of the Jurassic period.

JURASSIC DIASTROPHISM, PALEOGEOGRAPHY, AND SEDIMENTARY ENVIRONMENTS

Diversity in the facies and sequence of Jurassic strata reflects the presence of new structural entities in the Mediterranean region and the continued manifestations of mobility in those already in existence. However, because a discussion of the origin and development of tectonic elements during Jurassic times is an integral part of the review of Mesozoic structural evolution that concludes this essay, the following paragraphs are concerned only with the succession of diastrophic events, their influence on paleogeography, and the location and magnitude of igneous activity related to these events.

The Rhaetian-early Lias marked a return to a "normal" marine environment throughout those portions of western Europe and Morocco in which late Triassic seas were hypersaline. As the invasion of waters of oceanic salinity began with the Rhaetian, French geologists include that stage in the Jurassic, stressing its lithologic similarity to the early Lias and the fact that an unconformity separates Keuper from Rhaetian in parts of the French Alps (Debelmas and Lemoine, 1957, p. 150). German geologists assign the Rhaetian to the Triassic, for in the German North Sea basin continental beds intervene between marine Rhaetian and Hettangian. In many localities in the Alpine trough no interruption occurred in the deposition of marine strata assigned to the late Triassic-Lias, but the prevalence of terrigenous clastics of Lias age on the north flank of the Alpine trough, and in Rumania, Bulgaria, and northern Turkey shows that uplift of adjacent lands must have preceded the deposition of these sandstones. The only known occurrence in the Mediterranean region of strong tectonism that might have taken place between Triassic and Jurassic times, in other words, Suess' "early Cimmerian" orogeny, is in northern Dobrogea on the Black Sea coast of Rumania on the foreland of the Balkan mountains (see fig. 8). There the Triassic section is sharply folded and overturned. Late Triassic sandstones and conglomerates are overlain by relatively undisturbed Callovian (?) or Oxfordian sandstones which are succeeded by Malm limestones with an abundant ammonite fauna (Arkell, 1956, p. 189).

Liassic Tectonism. Manifestations of diastrophism are rare and restricted areally in rocks of early Jurassic age, with the exception of the great linear, fault-bounded depressions of the Moyen and Haut Atlas and Tellian and Saharan Atlas which were well established before the end of the Sinemurian. Late Hettangian uplift occurred in southern Provence where Sinemurian and Lotharingian strata are not present, but the sea again covered this region during the Pliensbachian. In northern Corsica, possibly part of the same structural entity, emergence took place after the Sinemurian, but marine deposition there was not renewed until Kimeridgian times.

Uplift began in northern Majorca during the Pliensbachian, first causing terrigenous clastics to succeed carbonates in the section, and then producing a brief emergence. At the same time Minorca was raised above sea level and remained exposed to erosion until Cretaceous times. In northern Tunisia, Domerian movements resulted in the local absence of strata of late Domerian and Toarcian age in otherwise generally subsident tracts. In Greece, the Ionian trough began to form at about this time, its genesis perhaps being accompanied or followed by ophiolite flows in the Pindus trough. In Yugoslavia tectonism was presumably coeval with that in Greece. In the western Alps the structural depressions making up the " miogeosynclines " and " eugeosynclines " were more actively subsident during the Lias than in the late Triassic, and the two submarine ridges in the southern Alps on which Domerian and Toarcian *ammonitico rosso* were formed began their development in Pliensbachian times (Aubouin, 1964, p. 745). Toarcian *ammonitico rosso* also exists in the Dinaric Alps, the Betic mountains, the Tellian Atlas, and on Majorca, suggesting the development of similar ridges in these localities at that time.

Lias tectonism did not cause large tracts to rise above sea level except in Morocco and western Algeria where Toarcian red beds covered much of the Meseta, the western Hauts Plateaux, and the flanks of the Haut and Moyen Atlas troughs. Regression of the sea was brief, for Aalenian littoral carbonates cover the red beds, and a tongue of marine Toarcian is intercalated in the continental sequence in some localities.

Dogger Tectonism. Dogger uplift in western North Africa was stronger and more persistent than that of the Lias. Minor Bajocian folding in the Midelt region of the Haut Atlas was followed in medial Bathonian times by regional regression. The sea left the Haut Atlas trough, the Eastern Meseta, and most of the Moyen Atlas trough to return only briefly late in the Cretaceous period. At the same moment, the Precambrian massifs south of the Sahara and the Hercynian massifs of Morocco were subjected to erosion as evidenced by the thick dark gray sandstones of Bathonian age laid down in the western half of the Saharan Atlas and in the Algerian Sahara. Terrigenous clastics of medial and late Bathonian age crop out also in northwestern Libya, southern Tunisia, and on the Sinai peninsula in sections otherwise predominantly carbonates, their presence suggesting that all of North Africa was affected to some degree by Bathonian epeirogeny.

On the Hauts Plateaux of western Algeria Bathonian regional uplift caused only minor deposition of sandstones in a carbonate section, but continuation of the movement in Callovian through Oxfordian times resulted in the distribution of deltaic sandstones and shales over most of the Hauts Plateaux west of the longitude of Algiers. Well data indicate that similar deposits occupy much of the western Sahara.

In the Middle East, evidence of Dogger epeirogeny is widespread. In the Zagros ranges of Iraq, a hiatus between middle Bathonian and Callovian is suspected owing to the absence of fossils of late Bathonian age in a condensed, dark colored, argillaceous section. In the bore holes of the central Tigris region of Iraq, upper Jurassic limestones and evaporites rest unconformably on eroded Bathonian, demonstrating a period of emergence during late Bathonian-Callovian (?) times, and in the area west and northwest of Mosul, Albian or older Cretaceous rests on eroded Bathonian over a broad tract which extends westward into northeastern Syria.

Thus, Bathonian movements affected much of the Tethyan depression between the Turkish and Arabian massifs. However, in the Levant marine sedimentation was continuous throughout the Dogger, although coarser textures in carbonate rocks suggest shallowing in Bajocian and Bathonian times.

In Europe, Bathonian uplift involved smaller areas than in Africa and the Middle East and its effects appear to have been less lasting. In central and southern Provence, Bathonian oölites and lignitic clays intercalated in a predominantly finer-grade carbonate section indicate shallowing. In parts of the Alps and in the Tuscan-Umbrian basin of northern Italy dark colored calcareous shales of Bathonian age overlie Bajocian limestones, some of which are oölitic or detrital. The terrigenous components of these shales must have been derived from an area subject to erosion, albeit low and distant. One such area may have been the Briançonnais, and another perhaps the Piedmont.

Callovian uplift affected northern Spain and a part of southern France in the vicinity of the Montagne Noire where the direct connection between the Tethys and the Aquitaine basin was cut. Larger tracts were caught up in this epeirogeny in Malm times, so that the Pyrenean and Iberian ranges formed the low-lying shores of an "Aragon Gulf". After the Kimeridgian all of northern and central Spain rose above sea level.

Callovian epeirogeny was also active in the northern Lusitanian and western Algarve basins of Portugal. In the Lusitanian region, an emergent block in the west, now mainly offshore, provided clastics to the strait between it and the mainland. In the Algarve basin emergence was less prominent, so that few clastics occur in the Malm section there, although unconformity marks the Callovian-Kimeridgian contact.

Malm Tectonism. Except in northern Spain little diastrophism occurred in Oxfordian times, but Kimeridgian epeirogeny and minor orogeny, more or less coincident with Nevadan orogeny in western North America, was effective in many places throughout the Mediterranean region. On the Iberian peninsula, in addition to the continued uplift of the Pyrenees and western Iberian ranges during the Kimeridgian, the emergent tract offshore in the Lusitanian basin was rejuvenated and the northern part of the basin itself rose above sea level.

In North Africa a land mass came into existence of what is now the Algerian littoral during the Kimeridgian. On the Hauts Plateaux deposition of terrigenous clastics ceased with the Kimeridgian, showing that the source area was no longer topographically higher than this region. In the Aurès mountains of southeastern Algeria, Portlandian calcilutites rest unconformably on Kimeridgian calcareous shales and limestones. In the Western Desert of Egypt and on the Sinai peninsula marine deposition ceased after early Kimeridgian times. Block uplift was common. The thickness of the strata on the several blocks is a measure of their uplift and the erosion to which they were subjected early in the Cretaceous.

In the Levant, the effects of regional uplift and erosion are shown by the progressively older Jurassic exposed southward from Palestine into Jordan. Kimeridgian epeirogeny may have raised most of the area above sea level, for marine Portlandian is present only in northern Lebanon where it rests on Kimeridgian volcanic ejecta. However, the Portlandian in southern Lebanon may have been removed by erosion.

In Iraq and southern Turkey the thin and incomplete black shales and carbonates of Malm age found in some localities in the Taurus trough suggest interruption in sedimentation. Elsewhere in the trough the Malm is thicker and in a neritic and littoral carbonate facies. Malm uplift is probably in part responsible for the absence of post-Dogger Jurassic in northwest Iraq and northeast Syria, for Bathonian epeirogeny in central Iraq was followed by resumption of marine deposition during the Oxfordian-Portlandian.

Near Bayburt in the Pontic trough of northern Turkey sandstones and conglomerates in a predominantly carbonate section indicate pre-Tithonian, probably Kimeridgian movements, for the overlying strata are calpionellid-bearing sublithographic limestones which crop out extensively south of the Sea of Marmara. These limestones grade laterally to pseudo-oölites and coral-bearing limestones obviously laid down in shallow water.

In many localities in western Greece Dogger shales are capped unconformably by Tithonian limestones (Institut de Géologie, 1966, pp. 40, 294). This relationship suggests post-Bathonian uplift that accentuated the development of horsts which originated during the Lias. Some of these horsts may have been islands for extensive periods. Similar tectonism occurred in the "Inner" Dinaric Alps, while in the "Outer" Dinaric Alps neritic, littoral, and reefal limestones were laid down with little evidence of local movements (Ciric, 1963, p. 569).

In the French Alps widespread block faulting was coeval with that in the Dinaric Alps. The sequence of Jurassic strata on the horst blocks shows evidence of differential movement in all except Portlandian times when a calpionellid-bearing limestone transgressed on older strata. The principal tectonic activity was in the Lias and Dogger, however.

In Italy and Sicily evidence of unconformity or prolonged interruption in sedimentation in the Jurassic section is rare in spite of Dogger igneous activity. However, in Sicily Tithonian sublithographic limestones rest in angular discordance on Kimeridgian limestones, sometimes with a basal conglomerate. Kimeridgian-Portlandian transgression is also reported on the Calabrian massif and in Corsica.

In Provence, strata of Malm age are thin and incomplete in many localities, but the lacunae observed are presumably the result of current erosion in a shallow sea (Corroy 1963, p. 23). In the Aquitaine basin slight unconformity exists between Kimeridgian argillaceous limestones and Portlandian sublithographic limestones that give way upward to brackish water calcareous shales and limestones. According to Coustau, *et.al.* (1969) unconformity separates these strata from sandstones of early Cretaceous age. The unconformity is between progressively older and younger strata eastward along the north front of the Pyrenees, the Sinemurian being overlain by Senonian at St. Marcet, where however, thrusting may be involved.

This second unconformity is basically Purbeckian/Berriasian in the Aquitaine basin, and is clearly a result of uplift at the end of Jurassic times. This movement concides with Suess' "Late Cimmerian" orogeny which is well expressed in Morocco for locally conglomeratic basal Cretaceous rests on moderately folded Jurassic in the Haut and Moyen Atlas. Nevadan movement was also effective there, because thick continental clastics crop out in the Haut Atlas under Cretaceous geosynclines (Choubert and Faure-Muret 1962, p. 511). In the Tellian trough of Algeria shallowing at the end of Jurassic times is shown by the development of *Chara*-bearing lacustrine limestones succeeding Tithonian sublithographic calpionellid-bearing limestones

ne places. In the Saharan Atlas trough
iilar shallowing is indicated by the occur-
ce of littoral sandstones. In southern
nisia and northwestern Libya Purbeckian-
ealden shallow-water shales and sand-
nes in an otherwise predominantly car-
nate section demonstrate prolonged uplift
a source of clastics to the south. In Egypt,
: withdrawal of the sea begun during the
meridgian continued in Portlandian times,
d blockfaulting was even more prevalent
e in the period. In the grabens thus form-
, marine sedimentation persisted in north-
Egypt into early Cretaceous times, but
ewhere "Nubian" fluviatile and littoral
ndstones were laid down as a result of
lift to the south.

Consequently, it appears that Cimmerian
lift was fairly general throughout North
rica, following Bathonian regional epeiro-
ny and widespread Kimeridgian move-
nts.

In much of the Levant and Iraq the
ects of Cimmerian movement are masked
immediately preceding Kimeridgian
tonism, but in the subsurface of central
aq unconformity is reported only in strata
early Berriasian age.

According to Kehtin (1962, p. 100) evi-
nce of late Jurassic folding and uplift are
nspicuous along the Black Sea coast of
urkey in the Pontic ranges, but not farther
uth. As indicated above, this diastrophism
esumably took place in Kimeridgian times,
it may have continued locally into the
ortlandian.

In Europe, Tithonian sublithographic
ntinnid-bearing limestones appear generally
be transgressive, and in many of the
oughs rest disconformably or in minor
nconformity on older strata, commonly of
imeridgian age, but locally, particularly in
e Alpine and Dinaric regions, on forma-
ons assigned to the Dogger or even the
ias. In most of Europe and in parts of the
ear East these fine-grained, porcellaneous
nestones represent not only the Tithonian
it a portion of the Neocomian, so that
ortlandian-Berriasian movement was prob-
bly not appreciable in those regions.

Jurassic Vulcanism. The minor tectonic events of the Lias were accompanied only infrequently by igneous manifestations. The earliest of these is on the north flank of the Pyrenees where basalts and tuffs are intercalated in Hettangian dolomites. Sinemurian lavas and tuffs are reported near Bayburt in northern Turkey where they are intercalated in sandstones and shales. Others probably exist in connection with Pliensbachian and Toarcian movements, but information concerning them is not available. The occurrence of ophiolites of Liassic age in Greece is questionable, for these flows may be much younger than the strata on which they lie.

Bajocian igneous activity in the form of tuffs and basalt flows is well documented in Sicily, and as mentioned previously, in the Caucasus and the Crimea. Dogger vulcanism may have occurred elsewhere, but is not well identified.

The principal manifestations of Jurassic igneous activity in the Mediterranean region occurred during and immediately after Kimeridgian times. Extrusive submarine ophiolites of this age are common in the Alpine, Dinaric, Balkan, and Pontic troughs. Intrusions of basalt occurred then and later in the Haut Atlas of Morocco, and late Jurassic diorites are reported in the Pontic trough. Kimeridgian tuffs and flows crop out in northern Lebanon, probably exist in southern Turkey, and possibly are present in Cyprus.

The eugeosynclinal portions of the troughs of southern Europe in which ophiolites with typical pillow structure were deposited on the sea floor are also the sites of very extensive deposits of radiolarian cherts. The association between radiolarites and volcanic rocks lends credence to the hypothesis that submarine vulcanism, perhaps supplemented by volcanic ash, enriched the silica content of sea water, thus providing an environment favorable to the propagation of radiolaria. In most of the Mediterranean region extrusion of ophiolites ceased during the Portlandian, so the fact that cherts of late Portlandian age are less common than those assigned the Kimeridgian may support the hypothesis.

Jurassic Paleogeography. The return to normal marine conditions in the epeiric seas of western Europe and Morocco that began in Rhaetian times was completed early in the Lias. In the Algerian Sahara, however, deposition of anhydrite continued until late in the Bathonian, so that in this region hypersaline conditions persisted throughout the Lias and early Dogger. In the Western Desert of Egypt the seas of the early Jurassic advanced southward over an essentially peneplaned Paleozoic surface, whereas in the Levant transgression occurred on generally undisturbed Triassic. Similar conformable relationships between Triassic and Jurassic are widespread in Syria and Iraq but the late Triassic is absent on portions of the Jauf-Mardin high.

In the Alpine and Dinaric troughs Carnian and Norian seas were essentially normal in salinity, so that no significant change in environment occurred in these tracts with the onset of Jurassic times. Some evidence suggesting Lias transgression on Vindelicia is reported, however.

The minor tectonism during the early Lias in Provence and Corsica, and the medial Lias in Majorca apparently had no effect on the general sea level, but the sharp retreat of the sea from Morocco during the Toarcian may have been a result of lowering of sea level.

As strata of Dogger age are incomplete or absent in many localities in the Alpine, Dinaric, Balkan, Pontic, Taurus-Zagros, and Tellian troughs, late Lias and Bajocian play on fault blocks in the mobile belts probably occurred, but is particularly evident in the Balkan and Pontic regions. A general retreat of the sea may have accompanied these movements, as was certainly the case in Morocco and the western Algerian Sahara. On the other hand, in portions of the western Alpine trough, the Dogger sea covered blocks previously emergent. Furthermore, the thin and generally condensed Dogger of the Tuscan-Umbrian trough is referred to as a deep-water deposit in contrast to the neritic limestones representing the Middle Jurassic farther to the south on the Itali peninsula.

A possible explanation of some of the apparently contradictory implications concer ing sea level during Dogger times may exi in Aubouin's recognition of the existence a stage in the development of troughs wh a linear depression exists that is not immec ately inundated with clastic fill. This is h "période de vacuité" in geosynclinal evol tion. During this time of low relief in adj cent lands, only sediments derived from t sea itself are available for deposition in t trough. These physico-chemically or orga cally derived sediments need not bulk larg so that a long period of time can represented by those deposits, which throu current scour or lateral transport may some areas either not be laid down, wholly or partially removed.

The difficulty with this explanation is th it does not account for the relatively thi Lias limestones which appear to have be formed in an environment nearly identic to that of the Dogger. Great depth of wat is commonly proposed as a concomitant thin deposits of Dogger age because of t absence of certain foraminifera and t shells of ammonites, which are represent in some of these deposits only by the opercula. The absence of these fossils explained as the result of the solution calcium carbonate in abyssal depths. Oth reasons have been invoked and may equally factitious.

Strata of Malm age are also thin in ma of the localities where the Dogger is poor represented. The predominance of radiolari cherts in these rocks is taken by many indicate their deposition in abyssal depth but this conjecture is certainly not val where these cherts are interbedded wi oölites (*e.g.* northeastern Sicily). The sa may be said of the overlying Tithoni sublithographic limestones with calpionelli which are interbedded with sandstones northern Algeria and grade laterally biostromal limestones in northern Turkey.

Figure 6 is an attempt to depict t extent of the Tithonian sea, of whi calpionellid-bearing sublithographic lim

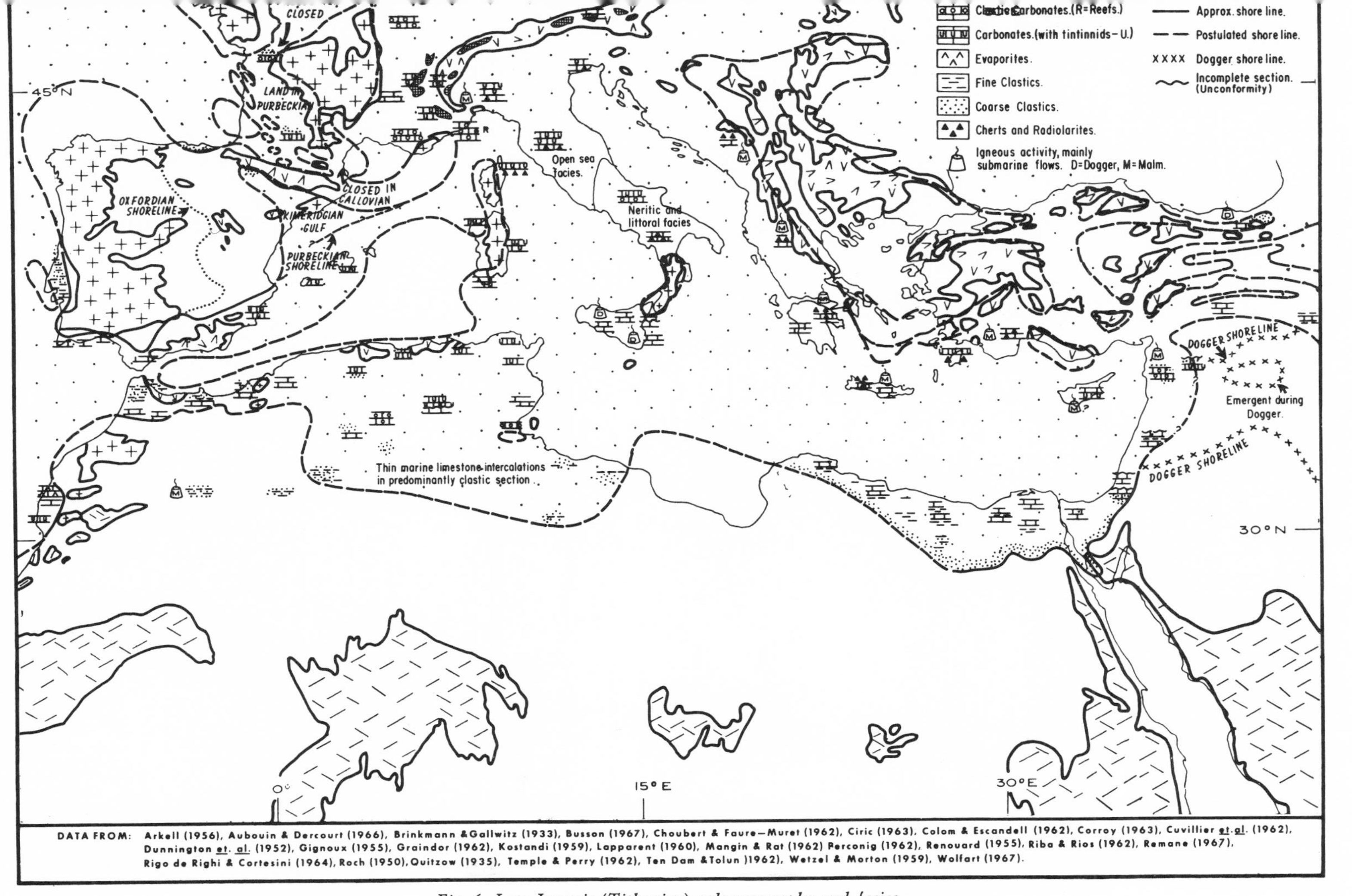

Fig. 6. Late Jurassic (Tithonian) paleogeography and facies.

stones were characteristic. Data concerning Tithonian paleogeography were taken from many publications, but much uncertainty remains, particularly concerning the shorelines in the Algerian Sahara where only well data can provide accurate information, and on the Balkan massifs where metamorphosed marine Jurassic is reported in several localities, presumably in areas lying within the "land" shown on the figure.

In general, Malm seas were less extensive than those of the Keuper in Spain, western North Africa, and the Levant, where the Jauf-Mardin high was probably largely above sea level. In Egypt, on the other hand, early Malm transgression reached tracts which had been emergent since Paleozoic times.

CRETACEOUS FACIES

In early Cretaceous times most of the Aquitaine basin, the Iberian peninsula, Morocco, and the western Sahara were above sea level, as well as Libya, nearly all of Egypt, and the Levant. In Morocco, late Jurassic-early Cretaceous uplift and minor folding was particularly effective (Choubert and Faure-Muret, 1962, p. 473). In northern Africa only the Rif, Tellian, and eastern Saharan Atlas troughs were the site of continuous marine deposition from Tithonian into Neocomian times.

In southern Europe, deposition of the calpionellid-bearing limestones of the Malm sea continued without interruption into the Berriasian. Late Cimmerian movement is reflected in the increased argillaceous content of predominantly calcareous strata only on the flanks of mobile belts. In marine tracts near emergent areas the early Cretaceous is represented mainly by terrigenous clastics, of which a large proportion is sandstone, but in the open sea facies carbonates make up most of the Neocomian section. Local movements controlled the thickness and type of strata deposited, particularly in the relatively shallow waters of southeastern France and adjacent portions of Switzerland where the type sections of all of the stages of the Lower Cretaceous were selected.

The most important single diastrophic event during Cretaceous times was Albian-Cenomanian tectonism (Austrian oroge that began the cycle of mountain-build which was to culminate in the Tertiary w the development of the Alpine ranges in of the mobile belts. In North Africa t tectonism was expressed by the collapse the Sirte arch. Somewhat earlier block fa ing on a large scale had begun in Algerian Sahara and similar moveme occurred in northern Egypt. The trends these faults demonstrate reactivation of v ancient lineations of fractures on which p continued into Cenozoic times.

In Europe, Albian tectonism consis mainly of block movements in the Alp trough. It was preceded locally in the east Alps by less prominent disturbances Hauterivian age. However, with the Alb began the compression responsible for development of cordilleras and island cha in the Alpine and Dinaric troughs, on wh erosion was rapid, with the consequ formation of thick masses of *flysch*. Th are synorogenetic deposits consisting for most part of regularly alternating beds clay shale and sandstone. At least a part the sandstones are graded, but some are n and the coarse-textured beds may inclu cobbles up to ten centimeters in diamet The sandstones are commonly micaceous a feldspathic, with poor rounding and sorti Their cement is generally calcareous, l may be argillaceous. Basement rock pebb are numerous in the coarse fraction. Fos are extremely rare and consist mainly burrows and tracks, along with the sh of benthonic animals, mostly large fora nifera and calcareous algae. Planktonic fo minifera and radiolaria are also found. So in graded beds are obviously redeposited

The thickness of flysch deposits ran from a few hundred to five or six thousa feet. They were laid down with no appar interruption in a relatively short period time. Individual beds are rather thin, l continuous over large areas.

Used in this restricted sense for syno genetic clastics deposited rapidly the te flysch has a definite connotation, but ma authors have used it so loosely as to ma it a catch-all for any thin-bedded alternat

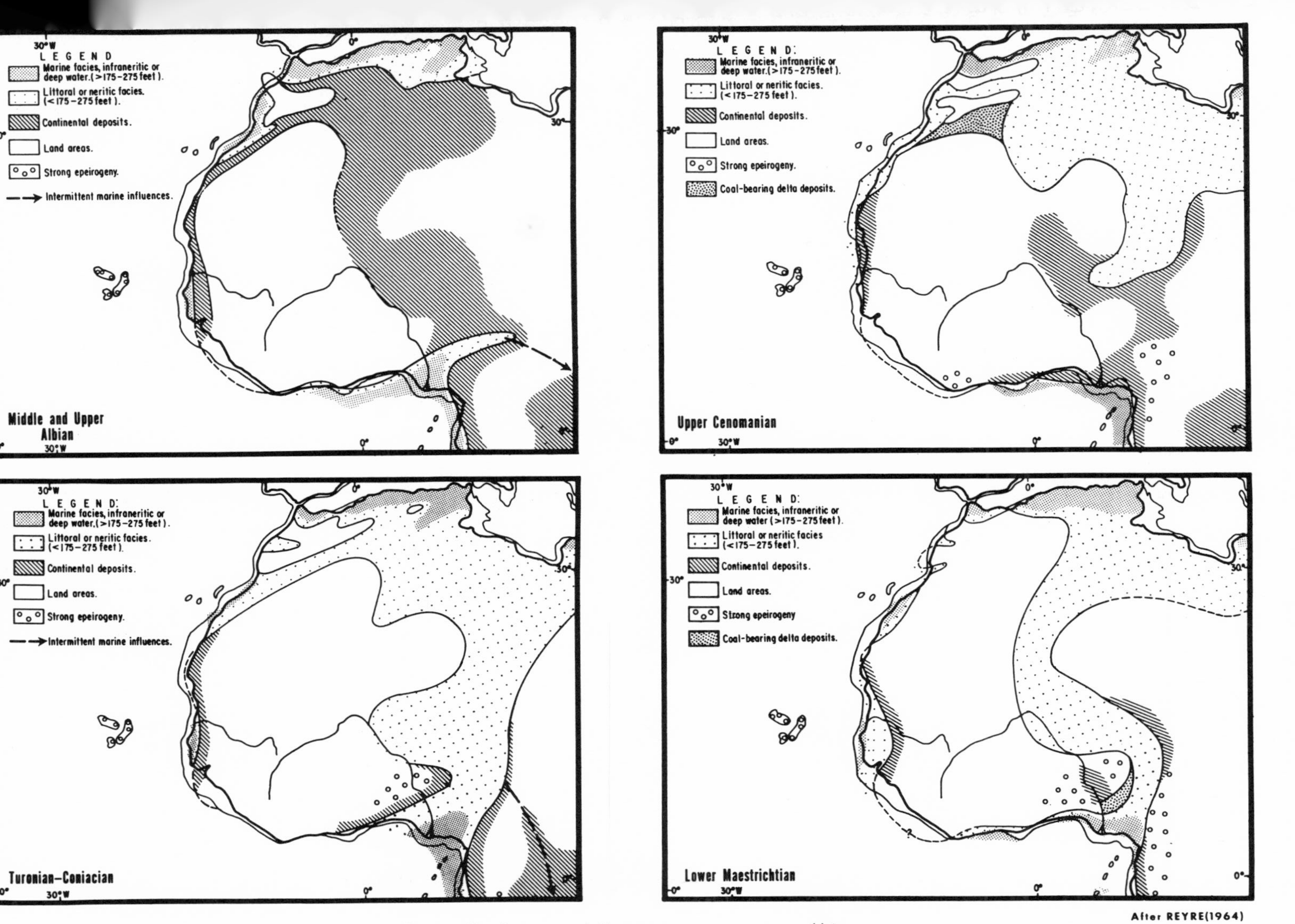

Fig. 7. *Development of Cretaceous seaway across Africa.*

sequence of sandstone and shale, shale and limestone, or limestone and marl or chert. Consequently the word flysch is in bad odor with many geologists, some of whom have urged abandoning its use. In this discussion, flysch means a synorogenetic rapidly deposited clastic sequence with the characteristics given above, laid down only in mobile belts.

At the same time as the first manifestations of true orogeny in the mobile belts of Europe, the sea began a great transgression in Africa. This had been preceded by a much shorter and areally smaller incursion during late Aptian times in the Algerian and Tunisian Sahara. Cenomanian seas covered very large portions of North Africa, reaching the Precambrian massifs of the Tibesti and the Hoggar and extending southward between them to link with a seaway that formed a channel northward from Nigeria. Thus a seaway connected the south Atlantic and the Mediterranean during late Cretaceous times. Deposits of this sea were for the most part littoral and neritic limestones, but in subsident tracts like the Sirte embayment and the Tellian trough basal sandstones are succeeded by calcareous shales. Carbonates predominate only on the relatively stable platforms flanking these depressions, and on uplifted fault blocks within the generally subsident troughs.

Transgression was also general in Europe during the Cenomanian, but had been preceded by early Neocomian and Aptian incursions in many localities. As in Africa the sea was largest during the Cenomanian, but flooding attained a second apogee during the Campanian-Maestrichtian, when some areas were submerged that had not previously been covered by Cretaceous seas. In the shallow portions of this sea in southern Europe organic limestones were laid down in quantity; in deeper water tracts calcareous shales in considerable thickness were deposited, most of which are characterized by the presence of an evolutionary sequence of the planktonic foraminifer *Globotruncana* and allied genera. In coastal Portugal the sea withdrew during the late Turonian. Near the Pyrenees and in Provence regression began in Maestrichtian times leading to the deposition of continental-lacustrine strata early in the Tertiary. Elsewhere in Europe biostromal and reefal limestones of Maestrichtian age are common, and they also exist in portions of North Africa.

In the Middle East and the Levant a predominantly carbonate section is characteristic of the Cretaceous away from the shield areas which were subjected to erosion in Barremian and Albian-Cenomanian times and consequently are fringed by aureoles of sandstones of these ages. In the eastern Taurus and Zagros troughs the Cretaceous is represented mainly by fine-grained limestones and shales with planktonic faunas which grade laterally to the south and west to neritic limestones, with some oölitic interbeds. Aptian transgression in tracts emergent since Jurassic times in northwestern Iraq is matched by a limestone tongue in the sandstones of Barremian-Albian age in southern Syria. Local block movement is indicated by unconformity between Albian and Cenomanian in some localities, and more widespread movement by unconformity in much of Iraq and Syria between Cenomanian or Turonian and Santonian or Campanian.

In the Middle East, Aptian transgression and Turonian-Campanian regional unconformity make a tripartite division of the Cretaceous system natural, but in Europe the Cenomanian transgression, coincident with the first compressional phases of diastrophism makes a twofold division more appropriate. However, the two are easily reconciled.

CRETACEOUS STRATIGRAPHY

Like the discussions of Triassic and Jurassic stratigraphy, this review of the Cretaceous around the Mediterranean begins with the Iberian peninsula. Owing to the development of new structural entities during Cretaceous times, facies changes are more numerous than those which occurred earlier in the Mesozoic. Consequently, the stratigraphic summaries are even less accurately representative of the diversity of Cretaceous sedimentary environments than those which concerned the Triassic and Jurassic. Nevertheless, relationships between the development of the larger tectonic elements and the facies characterizing them are still recognizable.

Iberian Peninsula. The ancestral Pyrenees, emergent during the late Jurassic, along with most of the peninsula, remained above sea level throughout the early Cretaceous, the axial portion presumably partially covered only by Cenomanian and more extensively by Senonian seas (but see Rat, 1964, p. 214). In the trough flanking the Paleozoic massifs of the central Pyrenees to the south early Cretaceous deposits are mainly Wealden continental sandstones in the west, and may be represented by sandy unfossiliferous dolomites in the east. However, in coastal Catalonia a massif already present at the end of Jurassic times presumably remained emergent throughout the Cretaceous, so that thinning and wedge-out are common in its vicinity.

Transgression began during the Aptian, the sea encroaching in Cantabria from a small embayment east of Bilbao over Wealden sandstones and conglomerates. Thick reefal limestones of Aptian-Albian age were formed in the Santander and Navarre regions, with sandy calcareous shales laid down between them. South of the shoreline of this sea the shales are replaced by sandstones almost indistinguishable from those of the underlying Wealden but known as the "Utrillas" beds. On the south flank of the Pyrenees between Pamplona and Boltana a broad tract remained emergent. This "Aragon sill" was covered by the sea only during Campanian-Maestrichtian times. Elsewhere, the Aptian-Turonian sea laid down sandstones, followed in shallow-water tracts by sandy limestones with oysters, rudistids and orbitolinas, and in deeper-water areas by limestones and calcareous shales with ammonites and planktonic foraminifera. Transgression occurred in Cenomanian times in the axial portion of the Pyrenees, for conglomerates and sandstones of Cenomanian age have been found in this region.

Although the Iberian massif was apparently not involved, northwest-southeast directed fault troughs and trends of facies developed in Cantabria during the Cenomanian. In one of these troughs extending southeast from Santander great thicknesses of sandstones, calcareous shales and limestones were deposited (reported to exceed thirty thousand feet; Rios, 1956, p. 88). Marine fossils are rare in this great pile except in the limestones, and lignitic beds and lacustrine molluscs occur in some of the shales. Mangin and Rat (1962) report that lavas, tuffs, and scorias of late Cenomanian-Turonian age are intercalated in the lower part of the sequence.

During Maestrichtian times the sea began to withdraw from the Cantabrian-Pyrenean region. In most of Cantabria this regression is demonstrated by the deposition of varicolored clays and dolomitic limestones. "Normal" marine Maestrichtian is confined to the coast where it is represented by sandstones and sandy limestones with a characteristic littoral fauna of larger foraminifera, as well as some plankton, including *Globotruncana.*

In the Pyrenees proper Maestrichtian conglomerates and lignitic sandstones demonstrate uplift, but over the Aragon sill, the sandy "calcaires nankin" of Maestrichtian age show no evidence of regression. Nevertheless, by the end of the Maestrichtian the sea had withdrawn from all of Spain except a portion of the western Pyrenees in Aragon and Navarre where the "Aturian Gulf" of the Aquitaine basin persisted from Maestrichtian into Danian times.

At the beginning of the Neocomian all of the area now occupied by the Iberian ranges was emergent. Consequently, strata of early Cretaceous age there are continental-Wealden fresh water limestones and clastics. These beds were deposited in structural depressions, presumably associated with faulting, primarily in the Sierra de la Demanda and Teruel basins, although thinner deposits exist elsewhere. Near the Mediterranean coast they are intercalated with marine sandstones and limestones that form an increasingly important part of the section southward, and probably eastward. To the northeast, communication with the sea was cut off by the Catalan massif (Rios, *et al.,* 1944).

During Aptian times the sea advanced inland, occupying approximately the Wealden Teruel basin along the Mediterranean, and lapping against the northwest side of the Sierra de la Demanda from the Cantabrian region. These seas laid down limestones relatively free of clastic detritus, but during Albian times renewed uplift caused deposition of the coarsely clastic "Utrillas" sandstones, although the area covered by sea increased in size, as shown by the intercalation of marine and continental beds far inland, resting indiscriminately on Trias, Lias, or Dogger. To the northeast these Albian strata wedge out near Beceite.

Upper Cretaceous seas occupied roughly the same basins but extended even farther inland, laying down littoral and epineritic limestones of which the maximum thickness is about two thousand feet. All basins were joined during the maximum transgression of the sea in Santonian-Campanian times, and marine deposition ceased following Maestrichtian regression (Riba and Rios, 1962, p. 281).

Late Albian uplift on the flank of the Catalan massif caused northern and central Valencia to rise above sea level, and to remain so throughout the remainder of Cretaceous times. Elsewhere along the coast the Cenomanian-Maestrichtian is represented by limestones and calcareous shales, the latter making up most of the section to the southeast where the transition to the wholly marine Cretaceous of the Betic trough begins.

Most of the area occupied by the Betic mountains was under water during the transition from Jurassic to Cretaceous, for the uplift that brought all of central and northern Spain above sea level had little or no effect in the central portion of the trough. All of the Neocomian is represented there by gray fine-grained limestones with a fauna of ammonites and tintinnids. This facies is developed from Cadiz to Alicante, a distance of more than three hundred miles.

North and south of this central trough facies change rapidly, and the Neocomian sea laid down shallow water limestones intercalated with and lying on sandstones at least part of which are continental. Aptian shallow-water limestones with *Orbitolina* cover these sandstones, and extend farther south to lie on Neocomian marls with ammonites. In the central portion of the trough the Aptian fine-grained limestones contain *Nannoconus* and radiolaria, but Albian green sands suggest shallowing and Alastrué (1956, p. 320) says that Aptian and Albian are absent in all of the central portion of the trough. This seems unlikely (see Dürr, *et al.,* 1962, p. 223, 226), for the late Cretaceous in a transgressive clastic facies exists only on the flanks of the "Rondaids" north of Malaga where sandstones with Paleozoic components are overlain by the red marly shales with *Globotruncana* characteristic of the Senonian elsewhere in the Betic trough. Some movement of the Betic-Rif massif presumably took place during Senonian times for terrigenous clastics are interbedded in the calcareous shales of the late Cretaceous in the Campo de Gibraltar region.

The total thickness of the Cretaceous in what was the deeper portion of the Betic trough ranges from fifteen hundred to twenty-five hundred feet, but increases to four thousand to fifty-five hundred feet on its flanks where thick limestones of biostromal and reefal type represent the Senonian. During Maestrichtian times the sea withdrew from the Betic trough, as it did from nearly all of Spain.

In the Lusitanian basin of Portugal marine sedimentation continued unbroken into the Cretaceous only in the vicinity of Cintra where Malm dark-colored limestones are followed by similar limestones of Neocomian age that contain shallow-water molluscs and echinoids. Everywhere north of Cintra, uplift during the late Jurassic brought the basin above sea level, and erosion cut progressively deeper northward into the Jurassic clastics and limestones on which Lower Cretaceous strata rest in slight unconformity. The Neocomian sea did not deposit limestones north of Torres Vedras, for although near Lisbon the Albian "Almargem" Sandstones intercalated with thin marine limestones rest on Aptian biostromal carbonates, they lie on Jurassic sandstones north of that locality. Probably the lower limit of the clastics is considerably older near Torres Vedras than farther south, for Wealden fossils have been found there in the continental sandstones and shales resting on Malm sandstones.

Cenomanian incursion took place in central Portugal, and the limestones and calcareous shales laid down in the shallow waters of the transgressive sea contain an abundant fauna that includes ammonites, oysters, and rudistids. Transgression continued into Turonian, but then ceased, for nearly all of the Lusitanian basin rose above sea level at that time, earlier than most of Spain. Only south of Aveiro do Coniacian and younger marine littoral deposits exist in a small coastal tract. The upper portion of these strata is not securely dated, but may include some Maestrichtian.

In the Algarve basin sedimentation was continuous from Malm into Neocomian times only in the eastern sector - where Valanginian-Hauterivian littoral a[illegible] epineritic limestones rest on Malm shallow-water ca[illegible] bonates with corals, *Diceras,* and *Nerinea.* In the we[illegible] the lower Cretaceous, resting on estuarine Mal[illegible] consists of littoral sandstones of Barremian-Aptian ag[illegible] followed by similar sandstones intercalated in a lim[illegible] stone unit with Aptian-Albian fossils. These san[illegible] stones are also represented in the eastern Algarve whe[illegible] they rest on Hauterivian argillaceous limeston[illegible] Cenomanian limestones cover the entire basin, in o[illegible] vious transgression.

Clearly, the western Algarve was above sea lev[illegible] during the late Malm and early Cretaceous, while [illegible] the east sedimentation was continuous during this tim[illegible] Tuffs in Albian limestones indicate moderate igneo[illegible] activity. Presumably the entire tract rose above s[illegible] level after or during the Turonian.

Balearic Islands. During the Valanginian-Barremi[illegible] nearly all of Majorca and Ibiza were covered by a s[illegible] in which deposits were similar to those of the la[illegible] Malm: very fine-grained limestones with radiolari[illegible] *Nannoconus,* tintinnids, and open sea ammonites. The[illegible] limestones resemble closely the majolica and bianco[illegible] of Italy. Only in western Ibiza and in eastern Major[illegible] occur biostromal limestones with rudistids and ech[illegible] noids of early Cretaceous age. During the Barremi[illegible] the sea invaded Minorca, first laying down san[illegible] shales on which rest Aptian biostromal limeston[illegible] with rudistids. Elsewhere in the Balearics Aptian stra[illegible] consist mainly of fine-grained limestones with globige[illegible] inas and *Pithonella.* At the end of Aptian tim[illegible] shallowing occurred throughout the Balearics so th[illegible] the Albian is represented mainly by calcareous shal[illegible] with pyritized ammonites, *Inoceramus,* and brachiopod[illegible] Cenomanian-Turonian deposits are fine-grained lim[illegible] stone with *Globotruncana,* and no later Cretaceous [illegible] known. The entire Cretaceous section is less than fi[illegible] hundred feet thick (Colom and Escandell, 1962, p. 13[illegible]

Morocco. South and west of the Paleozoic massi[illegible] of the Rif lies the *Dorsale calcaire* in which the enti[illegible] essentially allochthonous Cretaceous consists of lim[illegible] stones and calcareous shales. In the broad "Zone marn[illegible] schisteuse" west and south of the *Dorsale calcaire* t[illegible] Cretaceous is again allochthonous, according to Durar[illegible] Delga *et al.* (1962, p. 407), and is included in seve[illegible] nappes derived from the northeast. Early Cretaceo[illegible] strata are made up of an alternating sequence [illegible] greenish colored fine-grained micaceous and arkos[illegible] sandstones and dark colored calcareous shales, wi[illegible] some thin interbeds of light colored limestones wi[illegible] calpionellids and *Nannoconus.* Near the top some lin[illegible] sandstones with *Orbitolina* represent the Aptian. D[illegible] rand Delga refers to this unit as "flysch", and it inde[illegible] has many of the characteristics of the strata call[illegible] flysch in the Alps. If so, however, orogenic mov[illegible] ments would necessarily have begun between Afri[illegible] and Europe late in Jurassic times, and this seem[illegible] unlikely, for no evidence of such disturbance exis[illegible] in Spain.

In another nappe the late Cretaceous is represen[illegible] ed. Formations of Cenomanian through Maestrichti[illegible] e consist of marly, sandy limestones, alternating with ricolored calcareous shales. Some cherty limestones ar the base are dated as Cenomanian, and bituminous erts are referred to the Turonian. The Senonian cludes calcareous sandstones and limestone micro-eccias on which rest gray Paleocene limestones. The tal thickness of the Cretaceous in the nappes is on e order of three thousand feet.

The "Outer Rif" in which the strata are autochthon-s or parautochthonous lies south of the "Zone arno-schisteuse". The Neocomian there consists of cally sandy, variegated marly limestones with pyritiz- ammonites, and the Cenomanian and later Cretaceous so are represented by limestones and marls. Albian d younger beds include planktonic and benthonic icrofaunas, and most of the Senonian is characterized *Inoceramus.*

Late in Jurassic times the sea had withdrawn from l of Morocco except the Rif trough, a strip along the editerranean coast east of it, and the Haba basin the Atlantic coast north of Agadir. The diastrophism at accompanied this withdrawal involved minor fold-g and vulcanism. Throughout the remainder of esozoic and early Tertiary times the sea entered orocco mainly from the west, the Mediterranean sea cupying only the Rif trough and a narrow strip along e coast east of it.

Transgression from the Atlantic early in Cretaceous nes occupied several newly formed structural depres-ons along the coast in areas that had been above sea vel during the Triassic and Jurassic. These were: the Preafrican trough, developed along the north ont of the Anti-Atlas ranges, 2) the Haouz basin, eastward extension of the Haha basin in the south-n part of the formerly emergent Moroccan Meseta, which the central portion remained as the Jebilet-ehamna high, and 3) the Settat corridor, a less arply subsident east-west trending tract across the orthern portion of the Moroccan Meseta (see fig. 11). he Settat corridor and the Rif trough were separated a persistently positive tract of considerable dimen-ons that Choubert and Faure-Muret call the "Terre es Idrissides".

In areas not covered by the transgressions from the tlantic, red continental sandstones up to a thousand et thick accumulated throughout early Cretaceous mes in structural and topographic depressions pre-ominantly elongated east-west. The Cenomanian sea lvanced eastward across central Morocco and west-ard from the Saharan Atlas trough so that all of e central portion of the country was submerged. An m of this sea extended southward across the eastern nd of the Anti-Atlas as far as Zegdou. From the tlantic another incursion entered the northern Tin-ouf basin south of the Anti-Atlas.

Cenomanian-Senonian deposits in the seaway across lorocco attain a maximum thickness of less than two ousand feet. They consist of calcareous shales and mestones which thin and become arenaceous near e shores of the strait. Near the Atlantic coast the ction is thicker and more complete than in the terior farther east.

The Senonian is separated from the Cenomanian-Turonian by an important disconformity representing the Coniacian. In the strait the Santonian consists predominantly of light colored limestones with many sea urchins, but includes a number of lagoonal inter-beds which become more numerous eastward. After a number of oscillations in sea level a second regression occurred at the end of Campanian times. The advanc-ing Maestrichtian-Danian sea occupied well-defined troughs in western Morocco only, its limestone and phosphatic sandstone deposits locally in slight un-conformity on older beds. During the Maestrichtian the sea was not as extensive as in earliest Cenozoic times.

The Haut Atlas was persistently positive throughout the Cretaceous, and the area involved in its uplift gradually became greater, so that the Haha basin west of it was progressively reduced in dimensions. Concom-itantly, the interior troughs in which continental and lagoonal strata accumulated continued their subsidence, for near Ksar es Souk gypsum and salt deposits of Senonian age are nearly three thousand feet thick.

Algeria. Although in Morocco the Haut Atlas and Moyen Atlas troughs were no longer subsident after Bathonian times, the continuation of the Haut Atlas trough in Algeria — the Saharan Atlas trough — was, throughout late Jurassic and Cretaceous times, the site of deposition of sedimentary strata thicker than those of the same age encountered on the Hauts Plateaux and Ain M'lila môle to the north and in the Algerian Sahara to the south.

The Tellian Atlas trough, the eastward continuation of the Rif trough of Morocco, also maintained its identity as a subsident mobile belt throughout the Cretaceous, with the *Dorsale calcaire* a persistently positive element on which calcareous strata in a thin and incomplete sequence were deposited, whereas farther north great thicknesses of shale were laid down in what was presumably a deeper portion of the trough. Sand grains are increasingly sparse northward, thus indicating that their source was to the south. Near the southern edge of the trough some two thousand feet of black cherty limestones and siliceous shales represent the period of transition from Jurassic to Cretaceous times. A Valanginian marly limestone caps them in this region, but in the central portion of the trough the entire pre-Barremian Cretaceous is argilla-ceous, with a few calcareous intercalations near the base of a thirty-five hundred foot section.

A similar change from limestones near the southern edge of the trough to shales in the deeper portions occurs in strata of Aptian age, whereas the Barremian is argillaceous wherever it is encountered, the Barrem-ian-Aptian together attaining a thickness of more than ten thousand feet. Aptian strata are not present on the *Dorsale calcaire.*

In Albian times deposition of clastics was essentially ubiquitous in the Tellian trough. These dark-colored platy shales, in some places interbedded with thin quartz sandstones, range up to more than five thou-sand feet in thickness and locally rest unconformably on older strata as a result of intra-Aptian uplift and erosion (*e.g. Grand Pic du Ouarsenis* where Albian rests on Neocomian, Oxfordian, or Dogger).

Near the close of Albian times sandstone and dark shale deposition ceased in the Tellian trough. In all of northern Algeria Albian limestones containing diagnostic ammonites and foraminifera overlie this older sequence and form the lower portion of a transgressive cycle that continued into the Upper Cretaceous. Late Albian-Cenomanian limestones and marls attain a thickness of nearly three thousand feet in the Tellian trough. The Lower Cenomanian is characterized by calcareous shales and marls, the Upper Cenomanian by massive limestones.

The Turonian section begins with some fifty feet of black, platy bituminous shales that grade up to more than three thousand feet of gray calcareous shales of the same age. The overlying Senonian also consists of calcareous shales with thin limestones interbeds, distinguishable only faunally from the Turonian. Oysters are common in the lower part of the five thousand foot section. No unconformity exists at the Mesozoic-Cenozoic contact, and the Danian is very similar lithologically to the Senonian.

The thickness of the Cretaceous section in the Tellian trough must exceed fifteen thousand feet, whereas on the Hauts Plateaux it is not greater than two thousand feet of which Lower Cretaceous continental sandstones constitute about one-third. The transgressive Cenomanian there is represented by less than three hundred feet of oyster-bearing calcareous shales. Apparently, strata of Turonian age are not present on the Hauts Plateaux for the Senonian is transgressive, the basal conglomerates of the seven-hundred foot section of clay shales and biostromal limestones resting unconformably on Lias limestones at Jebel Nador (Lucas, 1952) and under Chott Chergui. Gypsum interbeds near the upper limit of the Senonian sequence suggest a gradual withdrawal of the sea, and the Hauts Plateaux have been above sea level since the end of Cretaceous times.

Like the Hauts Plateaux, the Ain M'lila môle was a positive element throughout the Cretaceous, separating the eastern portion of the Saharan Atlas and Tellian troughs. The Neocomian in the Hodna mountains at its western end consists of biostromal and epineritic limestones ranging up to twenty-five hundred feet in thickness (reefal development?). Farther south sandstones are intercalated in the sequence, clastics appearing south of Jabal Teioualt (36°03'N, 6°23'E), and making up forty per cent of the Neocomian section near the Aurès mountains.

The Barremian-Aptian too is made up of carbonates on the Ain M'lila môle. Dolomites up to twenty-five hundred feet in thickness represent these stages in the west, but thin eastward to Ain M'lila where equivalent strata are less than one thousand feet of dolomitized biostromes. Local movements on fault blocks caused sharp changes in thickness and facies, however. Sandstones appear in the limestone sequence southward, while north of the môle the reefal limestones are replaced abruptly by dark-colored calcareous sha with pyritized ammonites on the south flank of t Tellian trough.

A similar relationship obtained during the Albia Turonian, for some fifteen hundred feet of reef lim stones represent these stages on the Ain M'lila mô whereas to the south thick Albian sandstones a Cenomanian-Turonian calcareous shales occur in t Saharan Atlas trough, and to the north similar sha and marls exist in the Tellian trough. On the môl beds of Turonian age have in some localities be removed by erosion as a result of post-Turoni (Coniacian-Santonian?) tilting and uplift, which w particularly effective in the Hodna mountains. In t western ranges the Senonian sequence is nearly fi thousand feet of calcareous shales with a plankton microfauna intercalated with oyster beds and thin lim stones, whereas to the east conglomerates demonstra local emergence. There the entire Senonian section gypsiferous calcareous shales and cherty limestones less than seven hundred feet thick, and part of the ar remained emergent until Maestrichtian times. Eoce strata are commonly transgressive in this region, f the Danian is recognizable as black, platy calcareo shales present only locally. Elsewhere the Ypresi rests on Cretaceous.

In the western Saharan Atlas, strata assigned to t Lower Cretaceous differ in lithology from those in t eastern ranges of these mountains, for regional upl in late Bathonian-Callovian and Portlandian times Morocco and on the North African massifs caused t continuous deposition of thick sandstones in the Sahar trough of western Algeria, whereas farther east in th subsident tract the Cretaceous section includes greater proportion of carbonates.

Portlandian-Albian sandstones and shales with on a few limestone and dolomite intercalations reach thickness of some ten thousand feet in the weste Saharan Atlas. The basal Cretaceous consists of abo two thousand feet of continental sandstones on whi lie seventeen hundred feet of marly limestones, poo dated but probably Valanginian. The Hauterivi includes a thin limestone at the top of a five hundre foot continental sandstones section, and the Barremia Albian is represented by two thousand to five thousa feet of continental sandstones (Cornet, 1952, p. 1

In the eastern ranges of the Saharan Atlas Low Cretaceous formations include a greater proportion limestones than they do west of the Hodna bas where sandstones appear in the marine portions the section and increase in abundance until th predominate, as described above. The thickness of t Lower Cretaceous in the eastern Saharan trough rang widely. The entire Valanginian-Albian measures mo than thirteen thousand feet if maximum thickness are totalled, but probably averages about eight tho sand feet. The basal Cretaceous, the Valanginian-Hau erivian, is represented by marine limestones and d omites in the Aurès mountains, but becomes littor and lagoonal westward, while the Barremian and A bian there consist of cross-bedded continental sandston with pebble beds which are separated in the east b marine limestone of Aptian age containing *Orbitolina*. his limestone wedges out westward in the vicinity Laghouat.

Cenomanian-Turonian marine strata are transgressive all of the Saharan Atlas. They consist predominantly limestones and dolomites in the east, but grade to ypsiferous calcareous shales westward. The thickness the sequence increases rapidly from west to east, om three hundred and fifty feet at Figuig to more an seventy-six hundred feet at Jabal Fernane. East this region it thins to about one thousand feet.

In the eastern Saharan Atlas the Senonian is presented entirely by shallow-water marine limestones nging up to seventy-five hundred feet in thickness. Vest of Laghouat the section is much thinner and cludes limestone conglomerates, sandstones, and ypsum, all indicating an approach to a shoreline.

The facies, stratigraphic succession, and thickness Cretaceous rocks in Algeria reflect clearly the ontinued existence throughout the period of the Tellian nd Saharan Atlas troughs and the stable blocks of ne Hauts Plateaux and Ain M'lila separating them.

Tunisia. In Tunisia the Tellian and Saharan Atlas oughs are also distinguishable throughout the Creta-eous although there they are not separated by a stable lock, for the Tellian trough was the site of deposi-ion of open sea and presumably deep water sediments, early recognizable in the rather small and infrequent utcrops, whereas the Saharan Atlas trough was much ss actively subsident, and in fact portions of it howed positive tendencies (Central Tunisia High). he Chott-Sahel depression south of the Saharan Atlas ough was the site of deposition of thick clastics early n Cretaceous times whereas the Saharan platform was nore stable.

In the Tellian trough the basal Cretaceous is epresented predominantly by light colored, sublitho-raphically limestones with tintinnids. This basal unit, epresenting the upper portion of the Tithonian-Berriasian calpionellid-bearing fine-grade limestone videspread in the Tethys, is overlain by about four undred feet of gray-blue and olive-colored marls and imestones of Valanginian age succeeded by some one housand feet of dark brown calcareous shales with hin sandstones and limestone intercalations and twelve undred feet of gray-blue calcareous shales, the two nits being considered as Hauterivian-Barremian, based n ammonite faunas. The Aptian, with *Orbitolina enticularis* and other characteristic fossils, consists of leven hundred feet of dark green and blue calcareous hales on which lies two thousand feet of brown and ellow calcareous shales with some sandstone interbeds, nore numerous upward, capped by seventeen hundred eet of blue nodular calcareous shales and limestones. The sandstones presumably are not present in the axial ortion of the trough.

Strata referred to the Albian include a basal three undred-fifty foot unit, of gray-black limestone follow-d by a similar thickness of somewhat shaly and lauconitic limestone which grades up to calcareous hales. The twelve hundred foot platy limestones overly-ng these shales are probably of late Albian-early Cenomanian age and certainly represent the first stage of late Cretaceous transgression. Dated Cenomanian consists of gray calcareous shales overlain by yellow massive limestones, the whole about twenty-three hun-dred feet thick.

Strata of Turonian age are predominantly calcareous shales in the Tellian trough, commonly gray or ash blue, with thin interbeds of gray limestones with *Inoceramus*. They range up to two thousand feet in thickness. The Coniacian-Santonian is also represented by calcareous shales and marls with some limestone interbeds, the whole nearly theree thousand feet thick. The Campanian-Maestrichtian consists of about a thousand feet of light colored calcareous shales and limestones, the limestones somewhat chalky and com-monly massive. Overlying Danian brown and black calcareous shales are conformable.

The total thickness of the Cretaceous section in the Tunisian portion of the Tellian trough is at least thirteen thousand feet, and may range up to sixteen thousand feet. This thickness is about the same as that found in eastern Algeria. No evidence of a bathyal environment exists in the lithology or fauna of these strata, although obviously no sources of terrigenous clastics existed nearby.

In Tunisia a portion of the Saharan Atlas trough was subject to local uplift during Cretaceous times on the "Central Tunisia High", south of which the Chott-Sahel depression continued to sink during the Neocomian. As a result of these differential move-ments facies are more varied in central Tunisia than in the north. In general, strata representing the Neocom-ian in the Saharan Atlas trough are fine-grade sand-stones intercalated in calcareous shales and dolomitized carbonates, the whole thinning markedly over the high. The sandstones include many interbeds of lignit-ic detritus. Northward in the trough the sand content decreases, and a gradual transition to the calcilutites of the Tellian province occurs, while southward sand-stones predominate. Hauterivian-Barremian regression is shown by the widespread occurrence of cross-bedded, poorly sorted and weakly cemented sandstones. These are succeeded by Aptian marine limestones, which occur in biostromes and reefs immediately south of the Tellian trough, and as bioclastic calcarenites, calci-lutites, and dolomites in the remainder of central Tu-nisia. These limestones are transgressive on Wealden continental sandstones in some localities. The lower Albian is predominantly sandstones with some shale and limestone interbeds, and locally includes evaporites, whereas the upper Albian is principally limestone and dolomite.

According to Burollet (1967, p. 55) Cenomanian strata in central Tunisia rest unconformably on eroded Albian sandstones and limestones, the amount of ero-sion being greatest on the Central Tunisian High. The Cenomanian-Turonian is represented by limestones, shales, and evaporites which thin on the high. Strata younger than Turonian are absent in the central por-tion of this positive feature. Campanian-Maestrichtian massive limestones on its flanks rest directly on Turon-ian or older strata.

South of the Central Tunisian High in the Chott Fedjadj region, called the "Chott depression", Neocomian strata are very thick. More than seven thousand feet of Purbeckian-Wealden sandstones, gypsum, and red clay shales exist in the depression, evidence of continued slow subsidence in a desiccating environment, whereas on the south flank of the Central Tunisian High the basal Cretaceous is represented by only thirty-five hundred feet of littoral and epineritic limestones, sandstones and dolomites, and on the Saharan platform by two hundred-fifty feet of sandstones.

The overlying Aptian-Albian in the Chott region is represented by about three hundred and fifty feet of dolomites and sandy limestones with *Orbitolina,* interbedded with gypsiferous calcareous shales on which rest two hundred feet of conglomeratic sandstones referred to the early Albian and six hundred feet of shallow-water limestones with ammonites of late Albian age.

Cenomanian deposits begin with a massive dolomitic limestone which grades up to green calcareous shales and gypsum capped by light-colored, sublithographic limestones, the whole some sixteen hundred feet thick. The Turonian is represented by eight hundred feet of light colored, coarsely clastic, partially dolomitized limestone with rudistids and corals, and the Senonian by six hundred to two thousand feet of calcareous shales with some gypsum and occasional limestone interbeds capped by Campanian-Maestrichtian calcarenites with *Omphalocyclus* and other orbitoids. The Danian is conformable, but is represented by less than seventy-five feet of calcareous shales and limestones.

On the Saharan platform, sandstones and green clay shales of Purbeckian-Wealden age are widespread, with some intercalated dolomites near the base of the section. Fossil wood is relatively common, and the remains of theropod dinosaurs have been found, as well as those of crocodiles. The thickness of the unit in southern Tunisia ranges up to two hundred and fifty feet, but it thickens southwestward into the Algerian Sahara to more than one thousand feet.

The Barremian has a limited distribution but consists of a black basal conglomerate conformable with the underlying green shales of the Wealden, on which lie cross-bedded sandstones and red shales, obviously laid down by rivers. The abrupt coarsening of texture at the base of this sequence suggests uplift in a source area, for the presence of large exotic blocks of clay shale, and the generally coarse texture and irregularity of the deposits suggests rapid torrential deposition separated by intervals of erosion. The thickness of this sequence (never deposited on the southern Tunisian and western Libyan coast) attains a maximum of somewhat more than one thousand feet in the Algerian Sahara, but of less than three hundred fifty feet in southern Tunisia and western Libya.

The early Aptian may be represented in the coarse clastic unit described above, for late Aptian-Cenomanian age is assigned the overlying sequence which begins with limestones containing *Orbitolina* and in a large tract in southern Tunisia grades up to shales and gypsum capped by limestones with a late Cenomanian fauna. The Albian sandstones common in Alge and in the remainder of Tunisia are not present this "Southern Jeffara Dome" over which the must have been very shallow during the Apti Cenomanian. The site of this dome had previou been actively subsident, and its appearance in the l Aptian certainly had some connection with the sub quent Cenomanian transgression that covered much North Africa, for tracts in southern Tunisia that h previously been relatively stable became subsident this same time.

The Turonian is represented by two to three h dred feet of crystalline dolomites which occupy extensi tracts in southern Tunisia, western Libya and the Alg ian Sahara. Fossils are rare but Busson states (19 p. 134) that the Garian Dolomite of northweste Libya is Turonian, and its attribution to the Cenom ian is erroneous.

The Senonian is represented by locally import deposits of gypsiferous shales in southern Tunisia.

Algerian Sahara. Published data concerning distribution of lower Cretaceous strata in the Sah are sparse, the most significant information concerni them being unpublished well logs and company repo most of which are not to hand. The early Neocomi is represented by continental sandstones and sha that are stratigraphically continuous with those of l Jurassic age. The deposition of these clastics continu without interruption until Barremian times when, indicated by the appearance of coarse clastics in section on the Tunisian portion of the Saharan platfo and in the Saharan Atlas of eastern Algeria, uplift source areas resulted in a renewal of the supply coarse terrigenous detritus to tracts previously d cient in them. This uplift was accompanied by mo ment on ancient faults in the Sahara, the most pro nent of which trend north-south. The resulting hor and grabens are covered unconformably by sandston in which is intercalated a thin marine limestone Aptian age. Folding accompanying this faulting w minor, but was sufficiently strong to prevent the depo tion of this Aptian limestone over the highs whi must have been emergent at that time (Claret a Tempère, 1967, p. 6). The faulting may have occurr early in the Aptian, for the marine intercalation thought to be of late Aptian age. By Albian tim movement had ceased. Albian sandstones cover t accidented surface in a sheet of essentially unifo thickness. Cenomanian transgression is indicated the deposition of clay shales and evaporites, succeed by limestones and dolomites, a part of which presu ably represent the Turonian. Senonian deposits a mainly anhydrite, gypsum and limestones.

In the southern Sahara between Fort Flatters a the Libyan frontier gypsiferous red and green sha lie on green sandy calcareous bone beds with *Onch pristis, Ceratodus* and *Lepidosteus,* as well as rept teeth (*Megalosaurus*) and plant remains (*Weichseli* These fossils suggest a Wealden age for these stra At the upper limit of the bone bed is a shelly lim stone with *Exogyra, Cuneolus,* and *Eocallista,* a similar mollusc-bearing levels exist in the overlying r ıd green gypsiferous shales. The shales are in part ırly Cenomanian in age for they are capped by ıassive Cenomanian limestones with characteristic ssils that form an escarpment which can be followed the Algerian Sahara over great distances. The overly- g Turonian is also represented by a basal gypsiferous ıale and capping hard cliff-forming limestones. Locally hite chalky limestones mark the initial stage of Turon- ın transgression, and presumably the maximum depth the Cretaceous sea in this region. Together the enomanian-Turonian measure about eight hundred feet. he Senonian section repeats the alternation of red and een gypsiferous shales and limestones of the Albian- ırly Cenomanian, but some sandstone intercalations cur locally in dated Maestrichtian. The thickness of ıe Senonian is about five hundred feet.

The importance of Barremian-early Aptian disturb- ıce is clear from the above discussion, and is confirm- by Lelubre (1949, p. 258) who considers the Tibesti assif to have been uplifted early in Cretaceous times, ıd the non-marine Mesozoic in the Fezzan to have een folded gently prior to Cenomanian transgression. The "Southern Jeffara Dome" which presumably ıme into existence late in Aptian times indicates ontinuation of the tectonism begun in Barremian times, ıd farther north on the unstable foreland. Albian ovement is shown by the unconformable relationship etween Cenomanian and Albian on the Central Tuni- a High. Local epeirogenic disturbances during the late retaceous are reported in the Central Sahara by usson.

Libya. Barremian-early Aptian tilting took place in orthwestern Libya, for Purbeckian-Wealden sand- ones thin and wedge out and may be to some degree uncated by the overlying sandstones of the Chicla ormation which farther east lie unconformably on ırassic and Triassic marine strata. The poorly sorted ındstones of the Chicla are intercalated with red clay ıales and contain silicified wood. They vary in thick- ess markedly from place to place but probably average bout three hundred feet. Their age is controversial. usson (1967) considers them Barremian-Aptian; Burol- t and Manderscheid (1965) ascribe them to the Aptian- lbian. Yellow dolomitic limestones conformably over- ing the Chicla contain a fauna of late Cenomanian ge near their upper limit. Where this unit is thick, in northern Tunisia, Aptian fossils occur near its ase (Busson, 1967, p. 177). Busson considers these mestones to represent Aptian through Cenomanian ven where they are thin, as in Libya where they easure about seven hundred feet.

Above these Cenomanian limestones is a conformable equence of greenish yellow shales and yellow argilla- eous dolomites with some interbedded gypsum, consid- ered transitional between Cenomanian and Turonian. his is capped by massive gray, microcrystalline cherty olomites of Turonian age (Garian Dolomite) which over extensive areas. The remainder of the Cretaceous represented by alternating dolomitic limestones, ypsiferous shales, and dolomites, the whole capped y detrital limestones with Maestrichtian orbitoids hich grade upward to green shales and gypsiferous limestones. The total thickness of the Cretaceous section in northwestern Libya is about twenty-seven hundred feet.

The Sirte embayment began its existence as a discrete entity in medial Cretaceous times, probably during the Cenomanian, although the tilting of Triassic and Jurassic strata in northwestern Libya had taken place somewhat earlier, perhaps during the Barremian or early Aptian. If so, this tilting was coincident with faulting in the Algerian Sahara. In any event the sandstones laid down on Paleozoic or Precambrian clastics or basement by the transgressive sea contain *Thomasinella punica,* an arenaceous foraminifer presumably a marker of the Cenomanian (Barr and Gohrbandt, 1967).

As explained by Klitzsch (1968, p. 76), the embayment, some two hundred and eighty miles wide, was formed by block faulting where three old uplifts converge. The most prominent direction of faulting is north northwest-south southeast. The major fault systems that delimit the depression to the east and west have this trend. Other trends exist, nearly at right angles to these primary fractures, but vertical movement along them is commonly less pronounced, although strike-slip displacements on a considerable scale have been reported. The relative vertical displacement of horsts and grabens developed by this faulting increased throughout late Cretaceous and early Tertiary times. In general, the floor of the embayment deepened eastward toward the great fault system bounding the embayment to the east and separating it from the Cyrenaican platform. In the grabens thick calcareous shales were deposited throughout much of the late Cretaceous while on the adjacent horsts coeval strata are thinner and include bioclastic and detrital limestones, now in part dolomitized. Generally, limestones are more common in strata of Campanian-Maestrichtian age. During this time reefs were developed locally on the horsts. In some localities their upper limit is marked by Maestrichtian-Danian unconformity. In other tracts chalk was laid down at about this same time, and in many places Maestrichtian fine-grade limestones are succeeded without interruption by Danian limestones of the same lithologic type. Similar transitions occur in sequences consisting predominantly of shale. Facies relationships are complex, particularly on the flanks of the basin. The thickness of the marine Cretaceous in the Sirte embayment ranges up to twelve thousand feet or more in the grabens, and is probably at least two to four thousand feet on the horsts, except where the section is reduced owing to late transgression on tectonically produced highs on the sea floor.

Strata of late Cretaceous age crop out in northern Cyrenaica in Jabal al Akhdar. According to Kleinsmiede and van den Berg (1968) the oldest beds exposed are late Cenomanian and consist of chalky or dolomitic limestones and dark green calcareous shales with some yellow marl. Pietersz (1968) confirms this age attribution, but Barr (1968, p. 140) reports one occurrence of early Cenomanian gray shale on the coast. The thickness of the Cenomanian-Senonian section ranges up to thirty-three hundred feet in incomplete outcrop and to more than five thousand feet in wells. Facies are obviously neritic and few deep-sea influences are apparent in the restricted outcrops, most of which can be reliably dated by planktonic foraminifera (Barr, 1968). Barr (p. 131) states that Cenomanian-Coniacian consists of dark shales and light colored limestones and dolomites whereas the uppermost Cretaceous (Santonian-Maestrichtian) is represented primarily by white, chalky limestones with *Inoceramus*.

The lower Cretaceous found in wells in northern Cyrenaica consists of thick limestones intercalated with lesser amounts of shale and sandstone. Some of the Aptian and Albian strata include characteristic *Orbitolina* and *Choffatella* which confirm the shallow nature of the transgressive sea which in early Cretaceous times laid down at least two thousand feet of section near the existing coast. Thus in the Cyrenaican portion of the Libyan-Egyptian unstable platform (see fig. 14), Cretaceous strata are some twenty-five hundred feet thick.

Farther south, on the Cyrenaican platform, the marine Lower Cretaceous grades within a few miles to continental sandstones, whereas the Cenomanian-Turonian seas advanced much farther southward, laying down carbonates and subordinate shales and evaporites over very extensive tracts. Approximately at the beginning of Santonian times deposition of shales ceased and white chalky limestones represent the remainder of the Cretaceous. Locally these have been completely dolomitized. Uplift at the end of Maestrichtian times gently tilted northern Cyrenaica, and brought some areas above sea level. Thus, the overlying early Tertiary siliceous limestones are locally unconformable in the Cyrenaican mobile belt, but on the Cyrenaican platform no evidence of inter-era disturbance has been found in the uninterrupted carbonate section.

In Cyrenaica, microfaunas indicate that the Turonian sea was deeper than that of the Cenomanian. This agrees with the paleogeographic sketches by Reyre (1964, figs. 7, 8, 9) who shows the connection with the Atlantic across central Africa as having been established only during Turonian-Coniacian times east of the Hoggar massif. The embayment west of the Hoggar massif was, according to Reyre, opened as the main channel of communication across Africa only during the Maestrichtian when the eastern strait was closed (see fig. 11).

Egypt. As in Cyrenaica the Neocomian of the Western Desert of Egypt is represented predominantly by terrigenous clastics. These were laid down on a relatively undisturbed subsident platform broken near the coast by northwest-southeast trending faults which developed at least one large graben in which some six thousand feet of Neocomian-Aptian shales were deposited. A high extended eastward from Libya into Egypt just north of the 30° N. parallel, a rejuvenation of a Paleozoic feature. South of it the lower Cretaceous is represented only by continental and deltaic sandstones; north of it late Aptian transgression is shown by the presence of shallow-water detrital carbonates with *Orbitolina* along the northwest coast.

Aptian incursion was succeeded by a renewal widespread deposition of terrigenous clastics duri the Albian. This sheet sand deposition was presumat preceded and accompanied by block faulting that ca ed marked variations in the amount of sandsto deposited on the blocks. The maximum thickness these Albian sandstones is more than fifteen hundr feet.

Cenomanian-Turonian transgression is shown by t deposition of a basal sandstone followed by gray sha and calcareous sandstones and green shales and lig colored calcarenites. Together these reach a maximu thickness of four thousand feet. All of these stra were laid down in a shallow marine environme Deepening early in Turonian times is indicated by increase in the number of plankton, and uplift imme iately thereafter by local unconformity. Late Turoni deposits are predominantly shallow-water carbonat

Folding in an east-northeast direction occurred the eastern sector of the Western Desert during Coni ian times, and may have started somewhat earli These folds are bounded on the south side by hi angle faults, downthrown to the south, and are *en echelon* trends.

As in Libya, deposition of chalky limestones beg during the Senonian. The chalks attain a maximu thickness of nearly three thousand feet in rapidly su sident tracts, but are missing over the faulted folds which they were probably deposited in an onl relationship, but were removed by erosion late Maestrichtian times when the high parts of the structures rose above sea level and remained emerge throughout the Paleocene.

Owing to tectonism, the thickness of strata of Cre ceous age in the Western Desert ranges widely, but a maximum is certainly more than twelve thousa feet.

In the Gulf of Suez region the basal Cretace is represented by continental sandstones referred the "Nubian" Formation. They are cross-bedded a include intercalations of carbonaceous shales. Thic nesses are of the order of a thousand feet. T overlying Cenomanian is conformable. As is common North Africa, the early Cenomanian is represented shales, commonly yellow, gray, or green, and the la portion by limestones and marls. Together these un have a thickness of about five hundred feet. Mo ment on fault blocks during the deposition of t Cenomanian caused sharp local variations in thickn and facies. The overlying Turonian-Santonian is ge erally conformable with the Cenomanian, but loca shows evidence of discordance. The lower part of t Turonian consists predominantly of limestones, wh grade upward to yellowish brown shales and sandston The remainder of the Cretaceous section consists chalky limestones of Campanian-Maestrichtian age ra ing in thickness from some six hundred to nearly o thousand feet. On some horsts Campanian-Maestri tian beds were never deposited, because of Coniaci Santonian uplift.

On the western shore of the Red Sea the Cretacec section is thin, with Campanian shales resting

Nubian" Sandstones and capped by Campanian-aestrichtian limestones with thin phosphate-bearing tercalations. East of the Red Sea in central Sinai the ubian Sandstones are succeeded by similar calcareous ales with occasional sandstone intercalations but in is area littoral fossils demonstrate a Cenomanian age r the unit which thickens northward from about venty-five to more than twelve hundred feet. The nformable Turonian-Santonian is represented by five undred feet of limestones, also formed in very shallow ater, and the three hundred foot Campanian-Santon-n chalk completes the section. Uplift during early oniacian times is demonstrated by the presence of ndstones of this age in a predominantly carbonate ction.

In northern Sinai, the Syrian arc folds are typically veloped in a broad synclinal belt. Each fold is an ymmetric anticline with its major axis trending north ortheast. The north flanks dip at five to twenty grees to the northwest, and the south flanks at rty-five to ninety degrees to the southeast. Faulting d thrusting occur on the steep flank. Folding only rely effects beds younger than Cretaceous, and most the growth of the structures appears to have occur-d late in Maestrichtian times.

In the northern portion of the Sinai peninsula ma-ne strata of Barremian through Aptian age crop out, contrast to the central portion of the peninsula here marine Cenomanian rests on Wealden continen-l sandstones. The relationship between these marine rly Cretaceous strata and underlying Jurassic marine ocks is not observable, but presumably "Nubian" andstones of Valanginian age separate the two. The mestones of the Barremian-Aptian marine section are bout seven hundred and fifty feet thick. Conformably n them lie fifteen hundred feet of white compact enomanian limestones with an eighty-foot basal calcar-ous shale. Elsewhere Cenomanian rests in apparent onformity on continental "Nubian" Sandstones.

The Turonian-Santonian of the northern Sinai pen-sula consists of thin chalky limestones and shales. evelopment of compressional folds during this time indicated by the absence of strata of Santonian ge on some structures. About five hundred feet of ampanian-Maestrichtian chalks lie unconformably on uronian on these structures, but in structural ows are conformable with the Santonian.

Thus, during Cretaceous times tectonism on a consid-rable scale occurred in Egypt. Early movements were ssentially all associated with normal faulting, and ere on a considerable scale in the mobile belt along e existing coast. Compressional folding, associated ith faulting, presumably became important only dur-g and after Turonian times. The principle compres-onal pulses appear to have taken place early in Coniac-n and late in Maestrichtian times. Cenomanian ansgression was ubiquitous in Egypt. The Aptian arine limestones with *Orbitolina* characteristic of arlier transgression in much of the Algerian Sahara ere confined to tracts near the existing coast in ibya and Algeria.

Jordan. In southern Jordan the Lower Cretaceous is represented by continental red and variegated sand-stones with red, green, and gray shale intercalations. Plant scraps are common in the upper portion of this sequence which includes some marine intercalations north of Al Zarqa. The thickness of this clastic section ranges from less than six hundred to more than thirteen hundred feet in its outcrops on either flank of the Dead Sea graben. Igneous sills exist in the section near its base in some localities (Shaw, 1947, p. 22).

These Lower Cretaceous sandstones rest unconform-ably on Callovian limestones, and in southern and eastern Jordan represent the Valanginian-early Ceno-manian. On the Palestinian side of the Dead Sea graben in Wadi Farah, Albian marly limestones with *Knemniceras* appear, and farther north marine Aptian crops out as well. However, in most of Jordan late Cenomanian-Santonian limestones and marls mark a sharp transgression, earlier indications of a marine environment being confined to ripple-marked sand-stones. These Upper Cretaceous calcareous strata thin and are intercalated with sandstones to the south and east. Their maximum thickness is about sixteen hun-dred feet. The overlying Campanian-Maestrichtian con-sists of siliceous chalky limestones which are phospha-tic at several levels. Local unconformity of these strata over south southwest-north northeast to southwest-northeast trending structures that began their devel-opment late in Turonian times is known in Jordan, but is particularly evident in Palestine. The occurrence of phosphates suggests regression during Campanian-Maestrichtian times, and in fact the shoreline was displaced westward during the late Senonian from its farthest eastward position (see fig. 11). However, the Maestrichtian sea was deeper and presumably more extensive than that in existence during the Campanian. The transgressive nature of the Maestrichtian is clearly demonstrated in northernmost Saudi Arabia near Jauf where Maestrichtian limestones rest indiscriminately on Cretaceous sandstones and Paleozoic limestones, shales, and sandstones.

The Levant. In Palestine uplift at the end of Juras-sic times was followed by the deposition of continental sandstones ranging in thickness from three hundred to fourteen hundred feet. These extend northward into Lebanon, but in the coastal region marine sedimenta-tion was continuous throughout Lower Cretaceous time, a transition being observable from continental strata in the south and east to a progressively thicker and more marine sequence in the west and north. The marine Lower Cretaceous consists of fossiliferous detri-tal and sandy limestones and dark colored shales rang-ing up to four thousand feet in thickness. A marine transgression from the northwest expanded the area covered by sea during the Aptian, and in Cenomanian times covered all of the country, so that all Upper Cretaceous rocks are marine. The sea shallowed during the Turonian, however, accompanied by the devel-opment of low anticlinal structures between which gypsum was deposited in the south where open sea communications were interrupted. The folding contin-ued throughout the remainder of Cretaceous times, but its effects on facies and the development of unconformities on the anticlines was strongest in the south where the sea was shallow. Strata of Coniacian age are absent except in one basin, so that in general Senonian beds rest unconformably on Turonian limestones (Bentor, 1960, p. 8).

During the Santonian conditions in southern Palestine favored the deposition of chalky limestones, but in late Campanian times cherts and phosphates were developed between levels made up predominantly of chalk. Similar deposits of phosphate occur also in Syria, Jordan, and Egypt. Farther north, both cherts and phosphates are replaced by chalky limestones which represent both Campanian and Maestrichtian.

The thickness of the Cretaceous in Palestine ranges from about fifteen hundred feet at a minimum in the south to more than eight thousand feet along the coast where a thick marine section of Lower Cretaceous age is present replacing the continental sandstones found elsewhere in Palestine, Jordan, and southern Lebanon.

In coastal Lebanon the basal Cretaceous consists of some one hundred to seven hundred feet of these red and white continental sandstones. In both northernmost Lebanon and adjacent portions of Syria these clastics are either absent or much reduced in thickness, thus suggesting that their source was to the south. They are also missing in northwestern Iraq and over part of the Jauf-Mardin high in northeastern Syria. In these localities marine Aptian rests directly on Jurassic. Aptian transgression is thus clearly established in all of the Levant. Aptian strata consist mainly of limestones with occasional anhydrite and chert nodules. Characteristic fossils are *Orbitolina discoidea* and *Cyclammina whitei.* In coastal Lebanon sandstones are intercalated in the Aptian sequence, but elsewhere the entire Aptian-Albian consists of carbonates in which rudistids and ammonites have been found to confirm the age attribution. These marine strata are intercalated with anhydrite west of the emergent tract near Deir es Zor (see fig. 11) and east of the Lebanon coastal basin. The thickness of the Aptian-Albian ranges from less than three hundred feet to more than one thousand feet in this relatively stable tract, whereas in Lebanon it reaches at least twenty-five hundred feet.

The Cenomanian-Turonian sea extended farther toward the Arabian shield than that of the Aptian-Albian both west and east of the Jauf-Mardin high. Near the tracts that remained emergent, the otherwise predominantly carbonate deposits are replaced by sandstones in both Albian and Cenomanian time-rock units. In the Taurus-Zagros trough of southern Turkey and northern Iraq fine-grained limestones and calcareous shales were laid down almost continuously throughout Cretaceous times until near the end of the period, although Cenomanian and Turonian and Coniacian strata may be separated by disconformity. South of the trough on the unstable platform of northern and central Iraq and southeastern Turkey, Campanian and Maestrichtian limestones and calcareous shales rest unconformably on Turonian limestones which in turn lie with slight angular discordance on Albian limestones (Dunnington, *et al.,* pl. III). The uplift and tiltir that caused the absence of Cenomanian and low Senonian strata in parts of Iraq occurred also in Syri and Lebanon. In Lebanon, however, the princip movement appears to have been Turonian, for Turo ian strata do not exist over structural highs develope during Cenomanian times. (Arambourg, Dubertret, *al.,* 1959, p. 214), whereas the Cenomanian-Turonia "Judea Limestone" is an unbroken sequence of fin grained carbonates in structural lows, although th fauna of oysters and rudistids suggests a shallow marir environment.

The Campanian-Maestrichtian chert-bearing chalk limestones of Lebanon give way eastward in east-we trending linear belts north and south of Deir es Z to gray calcareous shales with a characteristic Tethya microfauna. In western Syria the Maestrichtian is n represented in these shales. Over extensive tracts th shales are replaced by massive biostromal limeston which predominate where the Maestrichtian transgressive as it is in the Ga'ara depression on th east flank of the Jauf-Mardin high, and over the cre of this high north of Jauf and Sakaka. The tot thickness of Upper Cretaceous (Cenomanian-Maestric tian) strata in the northern shale belt is nearly fi thousand feet. The limestones of coastal Lebanon ne Beirut also approach this thickness but elsewhere th Upper Cretaceous does not exceed two thousand fee

Uplift occurred during the late Senonian immediat ly northeast of the northeastern border of Iraq, shown by accumulations of clastics made up of Tria sic to Turonian carbonate pebbles, green rock fragmen and radiolarian cherts in a linear tract along the fronti with Turkey and Iran. These clastics, the "Tanje Formation", are progressively younger southwestwar They rest conformably on Campanian-Maestrichti calcareous shales and are intertongued with conter poraneous reefal limestones.

Turkey. Evidence of diastrophism more or le coeval with Laramide orogeny also exists farther we on the south flank of the Taurus trough in southeaste Turkey where clastics of Mesozoic age are intermixe with great volumes of ophiolites, the whole in tecton contact with autochthonous clastics of late Cretaceo age (Temple and Perry, 1962, p. 1604-05). Rigo Righi and Cortesini (1964, p. 1922) describe the chao assemblage of Mesozoic strata in a manner similar Temple and Perry but in considerably more deta They suggest that igneous activity occurred at interva from Jurassic through late Cretaceous time in th Taurus trough ("eugeosyncline") in which a gre variety of sedimentary rocks was deposited. During th Campanian-early Maestrichtian uplift occurred in th northern part of the trough that triggered southwa gravity sliding of the strata in it. The gliding plar appears to have been turbidites of Campanian ag and several olistostromes and "gravity nappes" a distinguished. Similar slides affected strata of Tertia age. However, an unconformity marks the top of th Cretaceous in much of southeastern Turkey and nort ern Iraq.

Dubertret (1955) describes the ophiolites present in the Hatay region of southern Turkey near the Mediterranean coast and in contiguous areas of Syria north and west of Aleppo where they crop out in a north northeast-south southwest elongated belt. These igneous strata are predominantly peridotites, gabbros, and dolerites, along with some pillow-lavas. They are associated with radiolarites and locally are capped by Maestrichtian limestones. Undoubtedly they were poured out on the sea floor contemporaneously with the incipient orogeny in the Taurus trough. Their thickness ranges from thirty-five hundred to ten thousand feet.

Still farther west, on the Datça peninsula (36°40'N., 27°40'E.) the Cretaceous section consists predominantly of limestones laid down in the open sea. No interruption in sedimentation occurred between Jurassic and Cretaceous times, the fine-grade limestones of this age being characterized by tintinnids and radiolaria. Gray cherts are intercalated at several levels in the basal Cretaceous. Clastic limestones of Aptian age occur some six hundred feet above the base of the section, but in general the entire seventeen hundred foot unit is micritic. Microfaunas confirm a Maestrichtian age for the upper levels. Although no unconformity separates late Maestrichtian marly limestones and calcarenites from these older strata, the upper beds are easily distinguishable lithologically, for their coarser texture obviously is a result of intra-Maestrichtian disturbance, presumably centered farther to the north.

From the above it is clear that the Taurus trough was involved throughout its length from the west coast of Turkey eastward through northernmost Iraq in orogeny of Maestrichtian age. This preliminary tectonic phase (known to Turkish geologists as *Van orogeny*) caused the development of gravity slides and olistostromes and the deposition of great masses of coarse clastics on the south flank of the trough. The allochthonous material was principally from Mesozoic strata laid down in the trough. Ophiolite submarine flows accompanied this movement which had little effect except in and near the ancestral geosyncline. South of the trough the old Mardin massif, the northernmost portion of the Jauf-Mardin High, was covered by the seas of Lower Cretaceous age, Neocomian deposits resting in some localities on Precambrian and Cambrian.

The Pontic trough along the north coast of Turkey now includes two mountain chains, the Pontic and Anatolid ranges in which the Cretaceous section is not identical. Late Jurassic uplift apparently caused portions of the Pontic region to rise above sea level, whereas in the Anatolids to the south sandy neritic limestones of early Cretaceous age were laid down. The Cenomanian sea covered nearly all of the emergent tracts of the Pontic region as well as most of the Paleozoic massifs of the central highlands, and Senonian transgression was even more extensive.

In the Pontic region the advance of the seas was accompanied by orogenic movements and in the Anatolid portion of the trough by extensive intrusions and extrusions of ophiolites, which are found in association with radiolarites and siliceous shales. On the flanks of the ancient massifs and in the slowly sinking Pontic portion of the trough conglomerates, sandstones, and calcareous shales lie on the ophiolites. Most of these clastics appear to be of Cenomanian-Turonian age and are locally capped by Turonian reef limestones. Campanian - Maestrichtian sandstones, shales, and conglomerates demonstrate the recurrence of uplift during the late Senonian which was accompanied by additional submarine flows of andesites and basalts intercalated in Campanian-Maestrichtian limestones.

According to some reports, granitic intrusion of the crystalline massifs of central Turkey occurred late in Cretaceous times, and conglomerates composed in part of ophiolite fragments separate Cretaceous and Eocene sedimentary sequences (Kehtin, 1963, p. 63). Presumably coeval syenites and gabbros have intruded limestones and calcareous shales of late Cretaceous age. In general, these transgressive strata are thinner in the massifs of central Turkey than in the troughs flanking them, and in many localities rest directly on Permian or older rocks. Commonly, the trangressive sequence includes conglomerates, limestones, radiolarites and some basic eruptive rocks. In some localities it has been subjected to both dynamic and thermal metamorphism.

Cretaceous developments in Turkey are difficult to summarize because of their complexity. It appears probable, however, that orogeny in the Pontic trough north of the central massifs began in Albian times (Austrian phase) and was renewed during the Maestrichtian (Laramide or "Van" phase). In the Taurus trough the first disturbances accompanied by strong uplift occurred in Maestrichtian times, although Austrian, Pre-Gosau (post-Turonian), and Subhercynian (Coniacian-Santonian) movements may have occurred, accompanied by ophiolite intrusion.

Cyprus. The island of Cyprus has few or no sedimentary rocks of Lower Cretaceous age, for reports of their occurrence are conflicting and based on the reported discovery of *Orbitolina* in exotic blocks. Presumably uplift at the end of Jurassic times brought most of the island above sea level. Senonian transgression was widespread with calcareous shales and neritic limestones of Maestrichtian age abundant. The transgression was accompanied by great outpourings of pillow lavas on the sea floor. These are presumably Senonian, possibly mainly Maestrichtian, in age. They may attain a thickness of several thousands of feet.

Crete. On the island of Crete rocks of Cretaceous age crop out in several facies, mainly in highly faulted southward directed nappes. The thrusting occurred in late Eocene-early Miocene times and caused the superimposition of one allochthonous sheet on another. The most extensive outcrops of Cretaceous strata are made up mainly of carbonates in the upper portion of which Senonian faunas occur, including rudistids and characteristic foraminifera. Another less common facies of the Cretaceous in Crete consists of deeper water, open-sea siliceous limestones and radiolarites succeeded by Cenomanian-Senonian sandstones made up of grains of ophiolites and metamorphic micaschists. Locally, finely clastic and mud-grade limestones with marl intercalations contain Campanian-Maestrichtian microfaunas, and are followed by sandstones made up of ophiolite and schist grains. A third sequence of Cretaceous strata is made up mainly of ophiolites, including serpentine, gabbro, diorite, dolerite, basalt and spilite. These have been thrust, according to Aubouin and Dercout (1966, p. 796), southward over the other facies; their original position was presumably near the massif in the Aegean sea (now foundered), which was a prolongation of the Rhodopian massif. According to Aubouin's scheme the several troughs and ridges comprised in the evolution of the geosyncline developed successively on the periphery of this basement element. In any event the occurrence in Crete of marine strata of Cretaceous age in an apparently continuous sequence suggests that the island is within the persistent linear depression making up the Dinaric-Taurus trough, particularly in that intra-Maestrichtian disturbance is indicated by the presence of clastics of this age resting on fine-grained carbonates.

Balkan Peninsula. The Hellenic portion of the Dinaric trough continued to evolve during Cretaceous times. West of the Rhodopian massif in Thrace Aubouin *et al.* (1963) define nine "zones" consisting of three troughs and four ridges of which the flanks are named as zones when characterized by a distinctive lithology. (see fig. 4). The Pelagonian massif mentioned in the discussion of the Jurassic of the Balkan Peninsula is in fact separated from the Rhodopian massif to the east by a trough, the "Vardar zone", which was in existence at least as early as the Neocomian and is distinguished by intrusions of ophiolites, granodiorites, and dacites. The Pelagonian massif, on which metamorphosed Paleozoic is capped by crystalline limestones of Triassic and Jurassic age was the site of deposition of neritic limestones early in Cretaceous times while radiolarites were being laid down in the Pindus trough west of it. On the Gavrovo ridge which separated the Pindus from the Ionian trough neritic limestones were deposited during the Neocomian while at the same time siliceous limestones were formed in the Ionian trough.

Essentially vertical uplift began on the Pelagonian massif during Barremian times (*cf.* coeval faulting and uplift in the Algerian Sahara and in the Saharan Atlas trough). It caused sandstones and shales to be deposited in the adjacent Pindus trough. The axis of this uplift migrated westward so that in Aptian-Albian times the late Jurassic radiolarites on the west flank of the Pelagonian massif were in turn subjected to erosion.

This Aptian-Albian movement was followed by Albian-Cenomanian transgression during which the Pelagonian massif was covered by a shallow sea, with the exception of a western recently uplifted portion which continued until Senonian times to provide radiolarian chert debris to the tracts on either side of the land mass. In the Ionian trough deposition of neritic limestones continued undisturbed, and neither the Gavrovo nor Apulian ridges were involved in this early Cretaceous movement. During the Cenomanian transgression thin-bedded platy limestones and cherts were deposited in the Pindus and Ionian troughs and rudistid limestones formed on the Pelagonian, Gavrovo, and Ap lian ridges.

In Coniacian-Santonian times these three ridges we subject to strong vertical uplift and erosion so th rudistid debris was deposited in quantity on the flan of the Pindus trough, and appeared for the first tin in the Ionian trough where it then formed a distincti component of the sediments deposited.

During the Maestrichtian the Pelagonian massif w again brought above sea level, perhaps through compre sion, and the products of erosion from the emerge tract accumulated in the Pindus trough. Most of t strata of Maestrichtian age laid down in the trou are littoral limestones and calcareous sandstones for ing the lower two hundred feet of a five-thousand fo section, the upper part of which is of Eocene age.

The thickness of Cretaceous rocks in the Heller portion of the Dinaric Alps ranges widely dependi on the location of the exposure measured. It is estin ted at about two thousand feet in the troughs and to nearly seven thousand feet on the ridges where re limestones were prominently developed. Note that ree limestones can develop in great thicknesses only tracts where subsidence is continuous. Therefore, t distinction between troughs and ridges made Aubouin is to some degree a distinction betwe *rates* of subsidence. Aubouin considers that both t Pindus and Ionian troughs were in a "période vacuité" during the Jurassic and Cretaceous, wi minimal sedimentation persisting through medial Eoce times in the Ionian trough.

In any event, it is clear that the structural elemer close to the Rhodopian massif were involved in oroge before those nearer the Adriatic, the eugeosynclir Pindus trough having been filled by late Cretaceo Eocene clastics from the Pelagonian massif where the miogeosynclinal Ionian trough was filled by ter genous clastics only in late Eocene-Oligocene times.

In Albania, the sequence of events during t Cretaceous was similar to that in Greece and the tectc ic elements present are like those defined in Gree (Aubouin and Ndojaj 1965, p. 602; Patzelt, 196 The Cretaceous transgression in the inner zones m have begun earlier than in Greece, however.

In Yugoslavia, Ciric's "Inner" and "Outer" Dina ranges can be related to Aubouin's tectonic elemen albeit with difficulty because the Pelagonian mas disappears north of the 42nd parallel. According Aubouin (1965, p. 79) the now allochthonous c bonates along the Adriatic coast correspond to t Gavrovo ridge in Greece, but some of the coas region, and in particular the Istrian peninsula wh strata are autochthonous appears more nearly to c respond to the Pre-Apulian zone.

The Cretaceous in this sixty-mile wide exposure carbonates consists of some ten to fifteen thousand fe of biostromal and reefal carbonates with rudistids a *Orbitolina*. This great thickness of limestone and d omite appears to have been deposited without sign icant interruption until Laramide movements, includi gentle folding, brought much of coastal Yugosla above sea level. Numerous bauxite deposits bear witn

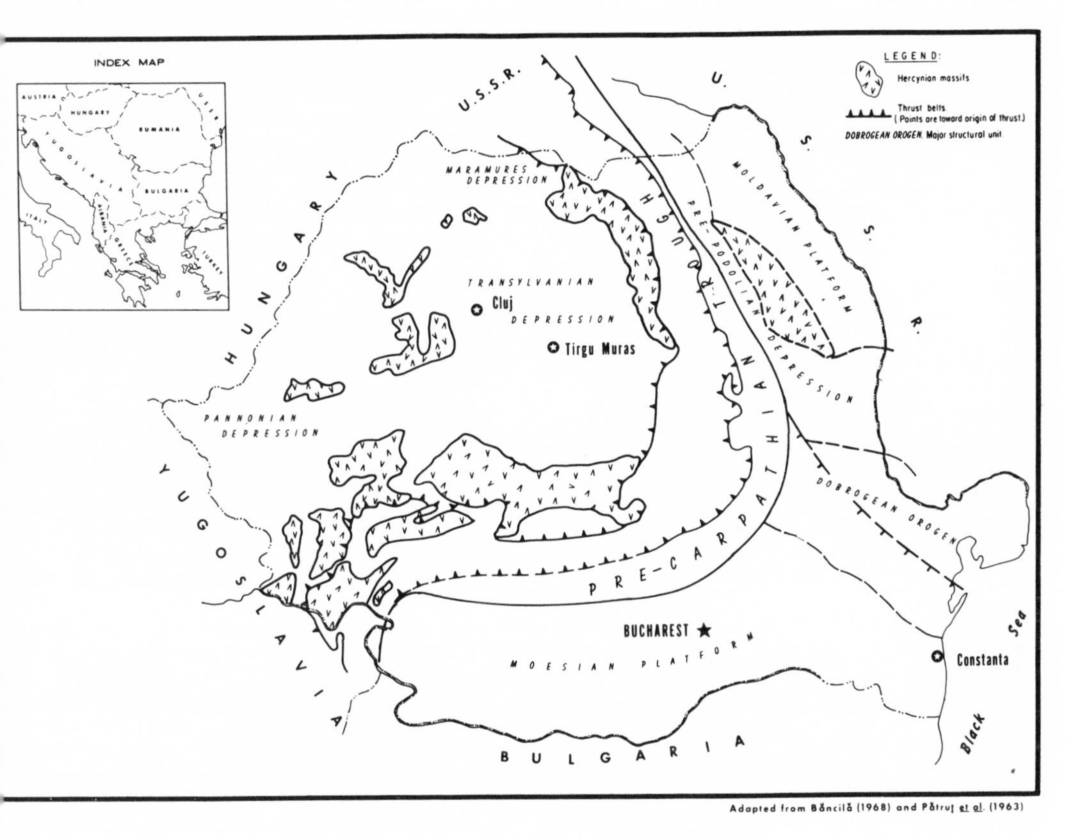

Fig. 8. Structural elements of Rumania.

o this emergence. Away from the coast, Maestrichtian rogeny caused the deposition of the sands, shales, and onglomerates typical of flysch on the flanks of emergent folds which rose farther above sea level in the nterior.

In the "Inner" Dinaric ranges early Cretaceous strata re irregularly developed and represented in many lifferent facies (Ciric 1963, p. 570). However, sandtones, shales, and calcareous breccias, with occasional ntercalations of conglomerate, occur in a typical lysch sequence from Belgrade south through Skopje in belt some twenty miles wide paralleling the Serbo-Macedonian igneous massif. This unit is more than hirty-five hundred feet thick (Ciric, 1966, p. 503), nd is more argillaceous than the overlying Maestrichtian-Danian, also a flysch sequence, but occupying a nuch greater area.

The Maestrichtian-Danian flysch, which attains a hickness of at least five thousand feet, rests unconformbly on Jurassic or Cretaceous limestones in those ocalities where it is not associated with similar Albian-Cenomanian deposits. It includes not only the alternaions of sandstones and shales characteristic of flysch, out also very large exotic blocks and conglomerates. Locally, rhyolitic tuffs and flows and radiolarian cherts are found in this sequence (Ciric, 1966, p. 500). Obviously, in contrast to the restricted area in which Austrian orogeny occurred, Laramide movements affected nearly all of Yugoslavia. After the emergence of the area late in Jurassic times some tracts in the "Inner" Dinaric region remained above sea level until the Cenomanian sea advanced eastward from the ancestral Adriatic. The deposits of this sea are predominantly limestones which lie unconformably on Malm radiolarites, ophiolites, and serpentines. Maximum transgression occurred during the Campanian-Maestrichtian when rudistid limestones were laid down over most of the still emergent portions of the Dinaric trough and contiguous massifs. Laramide orogeny and concomitant development of sharp folds brought some cordilleras above sea level, as demonstrated by the widespread occurrence of late Maestrichtian-Danian flysch.

During the Cretaceous, the development of the Carpathian-Balkan trough in the eastern Balkan peninsula differed in many respects from that of the Dinaric trough in the west, thus emphasizing the independence of the two structural elements already demonstrated during the Jurassic. As previously mentioned, the

Carpathian mountains, following the trace of the ancestral trough, enter the Balkan peninsula across a sliver of the Soviet Union that separates Czechoslovakia and Rumania. In Rumania these mountains, the southern portion of which is sometimes called the Transylvanian Alps, are bounded to the east by the Moldavian and Moesian platforms, and to the west by the Maramures, Transylvanian, and Pannonian depressions. They form the upper half of a huge reversed "S", the remainder of which is in Yugoslav and Bulgarian territory (see fig. 8). In Bulgaria the continuation of the Carpathians (see Grubic, 1968), the Balkan mountains or *Stara Planina anticlinorium,* swings sharply east along the north side of the Rhodopian and Istranca massifs, from their essentially north-south trend in the central portion of the reversed "S".

The three structural depressions west of the Carpathians in Rumania were, until Cenomanian times, a single stable basement massif covered only intermittently during the early Mesozoic by shallow seas. The advent of Austrian orogeny caused the platform to break up and to sink through normal faulting during late Cretaceous and Tertiary times, so that some twenty-five thousand feet of sedimentary strata have accumulated in the more rapidly subsident portions of these depressions. Of this total, only a small part is of Cretaceous age.

The Moesian platform which occupies much of the lower loop of the reversed "S" was tilted eastward late in Jurassic times so that the Cretaceous is transgressive on Malm limestones in the west, but is conformable with them in the east. The Neocomian consists of organic, oölitic and detrital limestones interbedded with finer-grade carbonates, and ranges from two hundred to nearly two thousand feet in thickness. Apparently uplift occurred after Aptian times for the Albian-Senonian is discordant on older strata. The Albian-early Cenomanian ranges from eight hundred to nearly two thousand feet in thickness with facies changing from glauconitic sandstones in the east to calcareous shales and marls in the west, with an intermediate transition zone of alternating limestones and sandstones. The late Cretaceous is represented predominantly by chalky limestones deposition of which ceased when epeirogenic uplift brought the entire Moesian platform above sea level at the end of Cretaceous times.

The Moldavian platform, which bounds the northern, north-south trending portion of the Rumanian Carpathians to the east, was already a stable massif in late Paleozoic times. Sedimentary cover on this block consists of gently folded Ordovician and Silurian sandstones and shales on which rest nearly undisturbed Cenomanian-Senonian limestones and sandstones, the whole tilted gently southwest toward the pre-Carpathian depression and covered by clastics during the early Neogene.

Between the Moldavian platform and the Carpathians is a tract in which the Mesozoic is represented by Triassic continental red beds, Dogger calcareous shales with *Posidonomya,* and Malm limestones, sandstones and evaporites. Late Cretaceous-Eocene transgression is evidenced by thin limestones, but in general this trac newly designated the "Pre-Prodolian depression" (Bar cila, 1968, p. 118), was moderately subsident durin the early Mesozoic, and stable throughout Neocomia and most of late Cretaceous times. It presumabl represents the downwarped edge of the Moldavia platform. The northern portion of this fifty-mile wid entity is known as the "Llow depression", and th southern portion as the "Predobrogean depression' They are separated in outcrop by the "Central Moldav ian massif". South of the Predobrogean depression is th "North Dobrogean massif" (orogen) now a part of th foreland. It consists predominantly of crystalline schist and old eruptive rocks plunging west northwest, on whic lie marine basal Triassic conglomerates and sandstone capped by medial and late Triassic limestones and dol omites and by the sandstones and conglomerates of th latest Trias-Lias, evidence of early Cimmerian orogeny In most of this region the late Cretaceous rests un conformably on the strongly folded Triassic, but a mentioned previously, Arkell reports Callovian or Ox fordian sandstones and Malm limestones in this region with their trends of folding at angle to those whic affected the Triassic.

The Carpathians themselves include three Paleozoi massifs made up of crystalline schists and granites rendered monolithic during Hercynian orogeny (se fig. 8). These are probably joined at depth (see prio discussion). The Carpathian-Transylvanian-Balkan revers "S" trough developed during the Triassic between thi massif and the Moldavian and Moesian platforms, an continued in existence throughout the Jurassic, wit transgressive relationships developed between the expos ed crystalline basement and the marine strata lai down in the trough, deposition of which was interrupt ed during the Dogger.

In most of the central and southern Carpathia trough the Tithonian-Neocomian is represented by light colored, fine-grained limestones with ammonites, cocco liths, and radiolaria. These limestones are transgressiv on the crystalline and metamorphic basement, an locally the Cretaceous portion includes thin sandston and conglomerate beds. Reefs and biostromes occur ir many localities. Local uplift in the northern portio of the Rumanian Carpathians at the end of Maln times is shown by the occurrence of olistoliths o Triassic and Jurassic sedimentary rocks associated wit diabase and serpentine, as well as radiolarian chert an red clays, but movements of this kind apparently di not occur in the central and southern parts of th trough. However, late Jurassic movements took plac in the southern portion of the western Carpathians, a isolated unit separating the Transylvanian and Pannon ian basins. In the northern portion of these range emergence at the end of Jurassic times is shown b the existence of Valanginian-Hauterivian bauxites an lacustrine limestones, on which lie Barremian reef lime stones.

Strong uplift apparently began in the Rumania Carpathians early in Aptian times and was renewe during the Albian when folded cordilleras presumabl rose above the shallow sea. In any event, conglomerate ıade up of fragments of Tithonian and Neocomian mestones, crystalline schists, diabase, granite, etc. (in-ercalated locally with Aptian-Albian fossiliferous lime-tones) represent the first widely distributed coarse lastics of Cretaceous age in these ranges. Tectonism t this time is also reported in the Northern Limestone lps west of Vienna, and in the outer Flysch zone of ıe Czechoslovakian Carpathians (Andrusov, 1967, . 521). However, the first thrusting and gliding in e northern Carpathians associated with Alpine orogeny now considered to have occurred in the "Pre-Gosau" rogenic phase (between Turonian and Coniacian), hereas Bancila (1968, p. 110) considers it medial retaceous in the Rumanian portion of these ranges.

Patrut *et al.,* (1963, p. 144) suggest a short emer-ence after Aptian uplift near the Paleozoic massif the Rumanian Carpathians, followed by renewed eposition of marine strata — Cenomanian sandstones nd conglomerates — after Austrian orogeny. Uninter-ıpted Turonian-Senonian deposition of limestones, ındstones, conglomerates, biostromal limestones and oal-bearing non-marine deposits is indicated in most of ıe Rumanian Carpathians by Patrut *et al.,* (idem).

East of the Paleozoic massifs the Carpathian portion f the Balkan trough sank almost continuously through-ut the Cretaceous. During this period, the axis of aximum sedimentation migrated away from the Paleo-oic massifs, and, in general, the sedimentary rocks ecome finer-grained in the same direction.

Near the Paleozoic massifs the early Neocomian (and possibly the uppermost Jurassic) in this trough consists of limestones with radiolarian cherts, or red and green argillaceous shales associated with diabase, all resting on the older Mesozoic and on basement. These rocks obviously represent Aubouin's *période de vacuité* in the development of a synclinal trough. They are followed by about thirty-five hundred feet of Hauterivian-Aptian limestones, calcareous shales and sandstones, locally with intercalations of reefal limestones. Disturbances of the source areas during the Barremian-early Aptian is suggested by the prevalence of sandstones of this age, and early Albian uplift is shown by Albian polygenic conglomerates which grade eastward into alternating sandstones and shales. Short emergence then occurred, and sandstones with calcareous shale intercalations of late Albian-Cenomanian age rest unconformably on the older Cretaceous. The Turonian consists of limestones capped by red and green sandstones, and the Senonian of red and gray calcareous shales with rare sandstone intercalations.

Farther east in the trough the Neocomian is represented by clay shales with siderite concretions and rather rare intercalations of glauconitic sandstones and cherts, all black in color, and rarely fossiliferous. Westward, this sequence is replaced by sandstones of Barremian age, but presumably it includes Aptian and Albian, and may in some localities be as old as Valaginian.

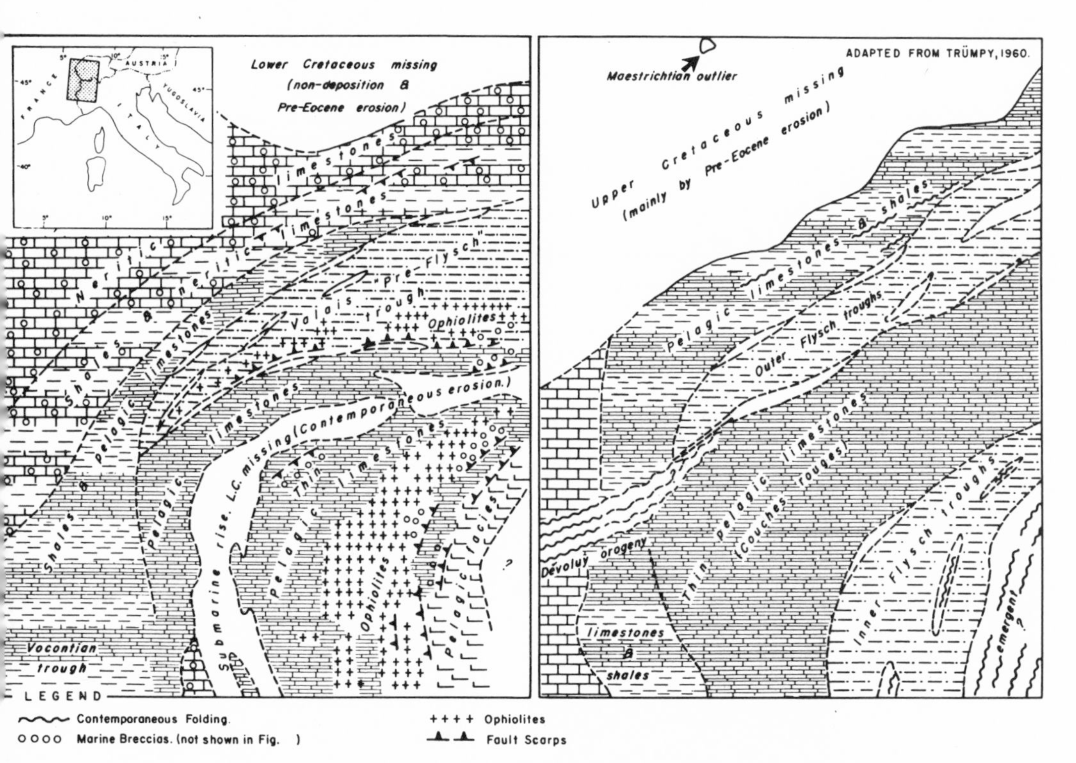

ig. 9. Western Alps. (*a*) Early Cretaceous paleogeography and facies. (*b*) Late Cretaceous paleogeography and facies.

Ashy clay shales represent the late Albian-Cenomanian in the central portion of the trough. Toward the outer edge the Cenomanian includes platy limestones and calcareous shales, and near its periphery the medial Cretaceous becomes siliceous, consisting of alternating red, gray and green cherty limestones. Overlying these relatively fine-grained deposits are sandstones and shales with fucoids and *Inoceramus* which continue without interruption into the Eocene. Subsidence in the trough continued well into the Tertiary.

Little is available concerning the extension of the Carpathians in Bulgaria, but it appears according to Beloussov (1962, p. 713-717) that north of the Rhodopian massif, itself a positive area after Hercynian orogeny, a shallow sea existed early in Triassic times in which evaporites and red sandstones were deposited followed by continental sandstones and coal beds. During the Jurassic the nascent trough was the site of deposition of sandstones, limestones, and shales in considerable thickness. An interruption in sedimentation is indicative of late Jurassic uplift. During the early Cretaceous renewed transgression is shown by the deposition of carbonates intercalated with sandstones and shales. Beloussov, (after Mouratov) indicates that the axis of the trough was brought above sea level during the late Cretaceous so that two subsident depressions were then present: one on the north flank of the Rhodopian massif, and the other between the uplifted ridge and the southern edge of the Moesian platform. Upper Cretaceous deposits in the two troughs are sandstones, shales, marls, and detrital limestones, demonstrating continuous uplift on the ridge (Stara Planina uplift). Similar conditions persisted into the Eocene. (*cf.* the Pelagonian massif in the Hellenic portion of the Dinaric Alps where emergence began after Neocomian times. The Vardar trough separates it from the Rhodopian massif and the Pindus trough bounds it to the west).

In general, the Rumanian Carpathians and the Dinaric Alps appear to be bilaterally symmetrical tectonic elements, one north and east of the Rhodopian, Fagarras and Marmaros massifs, and the other south and west of these and contiguous massifs. However, the sequence of tectonic events appears to be somewhat different in the two tracts, although the information available is too scanty for certainty.

The Alps. In the outer zones of the Swiss Alps, north and west of the more or less autochthonous Paleozoic massifs (see fig. 1) the lower Cretaceous is represented by three hundred to five thousand feet of neritic to littoral, oölitic, calcarenitic and pellet limestones with thin glauconite interbeds. Farther southeast (see fig. 9a) on and near the massifs these shallow-water strata give way to what are considered to be open-sea limestones and calcareous shales, and to the south, in the Vocontian trough, to calcareous shales. In the Valais facies belt representing the main eugeosynclinal trough up to ten thousand feet of "Bündnerschiefer" (calcareous shales) were deposited during the Neocomian. The lower part of this section is intruded by ophiolites. The upper portion of the Bündnerschiefer includes sandstones and shales in alternating sequences that gradually replace the calcareous shales and approa typical flysch in character. Aptian limestones are int calated locally in this formation, lying unconformab on older strata, and green quartzites and black shal referred to the "Gault" also occur.

The Briançonnais high south of the Valais euge syncline was largely emergent during early Cretaceo times, but where the Neocomian is represented t strata are mainly siliceous, fine-grained limestones, a in the Piedmont, apparently in a *période de vacui* thin radiolarian cherts and fine-grained limeston associated with basic and ultrabasic ophiolites a some shales and breccias were laid down during t time.

Farther east, along the north front of the Austri Alps, the Lower Cretaceous consists mainly of th shales and limestones, commonly black and red. Int Albian unconformity is known locally, and the Cen manian, transgressive, consists of calcareous shal sandstones, and dolomite breccia. The Pre-Gosau o genic phase (between Turonian and Coniacian) caus the development of the first thrust sheets, and t subsequent transgressive late Coniacian-Campanian s laid down the Gosau Beds, nearly six thousand fe of conglomerates, sandstones, rudistid limestones, calc eous shales, and coals and other non-marine deposi The Maestrichtian sea continued to occupy this regio laying down some fifteen hundred feet of litto limestones.

In the Alps the beginning of synorogenetic depo tion of clastics — flysch — was coincident with t development of anticlinal island cordilleras in the th existing basins. These seem to have appeared first Albian times, and to have become larger and mo numerous later when they provided great masses detritus to the troughs remaining on either side whi were rapidly filled by flysch. This synorogenetic filli began in the Piedmont and Valais troughs, but mov north and west with time, "finally encroaching on t Briançonnais platform, the Helvetian miogeosyncline, a even the nongeosynclinal areas beyond" (Trümpy, 196 p. 898).

The cordilleras became involved in thrusting a gravity sliding. The earliest nappes in the eastern Al are Late Cretaceous (but see Tollmann, 1964, w considers them Pre-Cenomanian); in the western Al Late Eocene. Major thrusting movements put an end the deposition of flysch, but flysch is locally trangressi on folds and minor thrusts. It is of interest th flows of ophiolite ceased when compressive stress grew strong.

Late Cretaceous deposits in the western Swiss A and the Dauphiné are commonly open-sea limesto and shales, while in the Valais eugeosyncline th consist of one to three thousand feet of flysch, and the Briançonnais of calcareous shales and marly lin stones with planktonic foraminifera. In the Piedmo the tract adjacent to the Briançonnais includes sha with plankton, whereas flysch is present in the ax region and in the Grisonides belt to the southea (see fig. 9b).

In contrast to Tollmann's views concerning the pre-Cenomanian development of thrusting in the northern Austrian Alps, Andrusov (1963, p. 521) states that in the Carpathians the early phase of Austrian orogeny (Albian) although it produced angular unconformity between the Neocomian and Cenomanian, was commonly followed by Albian transgression. Andrusov suggests that the main folding movements took place principally between the Turonian and Coniacian and at the beginning of the Miocene. The older movement was confined to the inner Carpathians; the other was effective in the foredeep. Other authors consider the principal movement to have been Tertiary (Ksiazkiewicz, 1963, p. 544).

Aubouin (1964) points out in the southern Alps of Italy and Austria the existence in Jurassic and later Mesozoic times of troughs and ridges like those in Greece, consisting from west to east of the Lombard trough, the Tridentine ridge, the Belluno trough, the Friulan ridge, and the Julian trough. These occupy the tract lying between the metamorphosed Paleozoic outcrops near Stresa in northern Italy and those northeast of Trieste in Yugoslavia, a span of about five hundred miles. These paleostructures were nearly at right angles to the trend of the existing Tertiary structural elements in the northern Alps, that is, their axes were directed approximately north-south.

The Lombard trough was already present in Dogger and Malm times when red radiolarites were deposited in it succeeded by Tithonian white fine-grained calpionellid-bearing limestones with black chert interbeds. Similar limestones continued to be formed through the Barremian, the whole five hundred foot section comprising the "Majolica" or "Biancone" of Italian authors. The conformably overlying black and gray calcareous shales include Aptian benthonic microfaunas near the base and Albian benthos near the top of the five hundred foot section.

The ash-gray marly occasionally argillaceous and micaceous limestones which follow, the "sass della luna", range up to fifteen hundred feet in thickness and contain abundant microfaunas indicating an Albian-Turonian age. The Cenomanian is locally distinguishable by thin black shale intercalations with *Mantelliceras mantelli* as well as plant scraps, and the Turonian includes gray clay shales, alternating cyclically with the "sass della luna".

With the exception of the Maestrichtian, the Senonian is represented predominantly by sandstones. The Coniacian consists of eighteen hundred feet of gray and blue-green somewhat argillaceous sandstone. The base of the fifteen hundred-foot Santonian sandstone is marked by a hundred-foot conglomerate containing Majolica, Triassic dolomite, and red-brown Permian conglomerate pebbles, thus demonstrating a strong phase (Subhercynian) in the succession of orogenic events in the southern Alps of which the pre-Gosau was the first strong enough to cause deposition of clastics.

The basal Campanian consists of a basal one hundred-seventy feet of marly limestones with chert nodules on which lie two hundred fifty feet of highly micaceous sandstone and more than two thousand feet of interbedded sandstone, clay shales, and thin limestones in a cyclic sequence. The Maestrichtian is represented by four hundred feet of gray and red well-bedded marly limestone, breaking into triangular chips on weathering, the so-called "Scaglia".

Aubouin considers the *période de vacuité* in the Lombard trough to have ended with the deposition of the black calcareous shales of the Aptian, which indeed are clastics, although of a pelitic grade.

On the Tridentine ridge to the east the entire Cretaceous is represented by less than two hundred feet of limestones, mainly sublithographic, containing planktonic microfauna. Some ten feet of basaltic tuff occurs at the upper limit of the Cretaceous section. This ridge must have been in deep water throughout the Cretaceous, in spite of its designation as an uplifted tract, for otherwise reef growth should have occurred on it.

In the Belluno trough the entire Cretaceous section is again limestone, here about twelve hundred feet thick, and consisting predominantly of calcilutites intercalated with white and black cherts. In these fine-grained rocks the typical plankton of the Neocomian and younger Cretaceous strata occur. The upper unit, assigned to the Senonian consists of red, bedded, marly limestone with *Globotruncana.*

On the Friulan ridge Lower Cretaceous biostromal limestones are at least fifteen hundred feet thick and Upper Cretaceous rudistid limestones attain four thousand feet. Late Cretaceous emergence is shown by the presence of "hard-grounds".

In the Julian trough the early and medial Cretaceous section is made up of bedded limestones and clastic limestones. The Senonian is represented by sandstones and calcareous shales with intercalations of clastic limestones, the latter mainly derived from erosion of the rudistid limestones on the Friulan ridge to the west.

Aubouin provides a mass of supporting data for his establishment of the ridge-trough sequence which he compares with tracts in Greece and peninsular Italy. Whether or not his interpretation is entirely correct, the demonstration that terrigenous clastics occur in the Cretaceous section of the southern Alps only in the extreme west and extreme east suggests that some barrier existed to the north that stopped the flysch derived from the rising cordilleras in the French, Swiss, and Austrian Alps from reaching the Adriatic region. Presumably this barrier was the east-west elongated ranges of metamorphosed Paleozoic strata, at that time a relatively stable and only slightly positive tectonic element in the central Alps.

In short, the southern Alps remained undisturbed during the development in medial and late Cretaceous times of cordilleras in the Alpine eugeosyncline farther to the north and west. The incipient Senonian orogeny responsible for the first northward-directed nappes in the northern portion of the Austrian Alps had no influence on the open-sea deposits laid down north of the existing Adriatic.

Italian Peninsula. The tectonic calm that prevailed in most of the southern Italian Alps during the Cretaceous existed also on the Italian peninsula. However, according to Parea (1968), a tract in the northern Tyrrhenian sea was brought above sea level in Albian times. This tract included Corsica and northern Sardinia and its emergence and subsequent exposure to erosion caused the deposition in the subsident area (Parea calls it a eugeosyncline) between the newly formed land and what is now the Italian peninsula of arenaceous and calcareous turbidites which were raised and subsequently moved eastward during Oligocene and Miocene times. These allochthonous commonly calcareous sandstones now crop out on the Italian peninsula as discrete large blocks "pietraforte" in masses of younger similar strata present in parts of Tuscany and Emilia and in the Florence basin.

The occurrence of Cretaceous marine neritic limestones in both northwestern and eastern Sardinia militate against this region's having been emergent during the late Mesozoic. In Corsica, uplift during the early Cretaceous is indicated by the existence of coarse polygenic conglomerates in shales and siliceous limestones. The conglomerate is covered, however, by cherty limestones and by platy limestones containing planktonic foraminifera of Albian through Maestrichtian and occasionally of Eocene age (Delcey, *et. al.*, 1965, p. 327). Eocene uplift and erosion is demonstrated by conglomerates and sandstones the upper portion of which contain nummulites and discocylinas of late Lutetian age in a calcareous matrix. Major orogeny in Corsica occurred during the Oligocene. It seems probable, therefore, that if indeed the tract between Corsica and the Italian mainland was supplied with terrigenous clastics during the Cretaceous they were not derived from the area now forming Sardinia, but northern Corsica could have been a source.

In northern Italy the Cretaceous section consists for the most part of calcareous rocks. The basal unit is the fine-grained, calpionellid-bearing limestone typical of the open-sea Tithonian-Neocomian of the Mediterranean, and includes Aptian microfossils at its upper limit. The Albian-Cenomanian is represented by the "scisti a fucoidi", light gray, red or green thin-bedded, marly, cherty limestones, commonly with one or more intercalations near the top of black bituminous shale containing fish remains. The remainder of the Cretaceous is represented by "scisti policromi", a more vividly colored variant of "Scaglia" but containing an identical microfauna. The thickness of the Cretaceous in northern Italy ranges from about four hundred to more than two thousand feet.

An obvious change in facies from open-sea to littoral comparable to that seen in rocks of Jurassic age on the Italian peninsula occurs in a belt from twelve to twenty miles wide running north-northeast across the peninsula from the Tyrrhenian coast south of Rome. This transition zone occupies roughly the same tract as that in Jurassic rocks. South of it almost the entire section consists of littoral, biostromal, and reefal limestones, mainly calcarenitic and oölitic, occasionally dolomitized. Locally the Cenomanian includes black shales or yellow and gray marls. The average thickness of the Cretaceous in all of southern Italy is more than three thousand feet, and in some tracts in the south reaches more than six thousand feet. Rudistids are common, along with corals, larger foraminifera, and *Nerinea*. In Calabria the upper Cretaceous limestones are transgressive on basement.

In the Lagonegro region the allochthonous Mesozoic sequence includes Neocomian siliceous limestones and radiolarites at the base of the Cretaceous on which rest three hundred to one thousand feet of alternating sandstones and dark gray limestones of Cenomanian-Maestrichtian age, the whole described by Grandjacquet (1962, p. 613) as "flysch noir crétacé". Owing to the abundance of limestone in the section and the probability that true synorogenetic conditions were not responsible for the deposition of this unit, the term flysch is probably misapplied. However, if the Lagonegro sequence is allochthonous, it was presumably derived from the southwest in an area now below the Tyrrhenian sea, but raised above sea level and shoved eastward early in Miocene times.

The prevalence of limestones throughout its length suggest that the Italian peninsula was relatively stable and undisturbed during the Cretaceous, sinking only sufficiently in the south to insure continuity of deposition of thick littoral carbonates, while in the north limestones were not laid down so rapidly, probably because reef-building organisms were not present owing to the greater depth of water in the region.

Sicily. On the Iblean platform in southern Sicily the Cretaceous in boreholes consists of a lower sublithographic light-colored limestone with rare chert lenses of which the lower portion is referred to the Malm. This calcilutite grades upward to grayish green calcareous shales ranging from two hundred to one thousand feet in thickness, mainly of Aptian-Albian age, on which rest conformably white siliceous limestones with brown chert nodules intercalated with gray, red, or green tuffaceous shales, the whole including an evolutionary sequence of *Globotrunca* demonstrating a Cenomanian-Maestrichtian age for the unit which has a maximum thickness of slightly over one thousand feet. Apparently uplift took place at the end of Cretaceous times in this region for no strata of early or medial Eocene age are known there.

In northeastern Sicily on the flank of the Calabro-Sicilian massif in the Peloritan mountains the basal unit of the Cretaceous section is about three hundred feet of white or greenish fine-grained limestone resting on a limestone and basement pebble conglomerate, the whole assigned to the Tithonian-Neocomian. The remainder of the Cretaceous is represented by about one thousand feet of siliceous limestones and gray calcareous shales, but tectonic complications make it difficult to determine their sequence and thickness accurately. Uplift in this region at the end of Cretaceous times seems likely, for the overlying unit is of Oligocene age.

In the northern mountains of Sicily the Jurassic and Cretaceous are in some localities (western Madonie mountains south of Palermo) represented by variegated clay shales intercalated with cherts and detrital limestones. Strata of Upper Cretaceous age are absent, and late Eocene limestones rest unconformably on the medial Cretaceous (Cenomanian?). Elsewhere (eastern Madonie mountains southeast of Termini Imerese) the shales are replaced by reefal and biostromal limestones ranging up to fifteen hundred feet in thickness, overlain by late Eocene limestones and shales. Farther west, in the Castellammare mountains the Cretaceous sequence is in an open-sea facies like that on the Iblean plateau but much thinner. In this region marine sedimentation continued unbroken into the early Tertiary.

Although the outcrops of Cretaceous age in Sicily cover a relatively small area, they can be interpreted to suggest that the Tithonian transgression responsible for the calpionellid-bearing limestones so widespread in the Mediterranean established an open-sea régime over much of Sicily which continued unbroken through Neocomian times. Along parts of the present north coast, however, the sea was shallower for biostromes developed, and in some localities the presence of shales suggests a source of terrigenous clastics, probably to the north.

The Cretaceous/Paleocene regional uplift that brought most of Sicily above sea level was accompanied by vulcanism locally in the northern part of the island, mainly as basalt flows, but with some ash beds and tuff in shales.

Sardinia. Outcrops of strata of Cretaceous age cover reasonably large areas on the east coast of Sardinia near the Gulf of Orosei where in some localities they rest on basement and in others are conformable with the underlying Jurassic. Much smaller outcrops occur on the northwest coast, west of Alghero on Cape Caccia, and on St. Antioco island off the southwest coast where pseudo-oölitic limestones with recrystallized rudistids have been described.

On the northwest coast on the Murra peninsula the Lower Cretaceous consists of some two hundred feet of white compact limestone including bryozoa, pelecypods, and small foraminifera, resting conformably on Jurassic limestones. This unit is followed by another two hundred feet of light colored clastic limestone with some brown calcareous shales intercalations. The fauna of larger foraminifera indicates a Turonian-Senonian age for this upper unit, as well as indicating that it was formed in relatively shallow water.

On the east coast late Jurassic-early Cretaceous strata form a single unit of gray, marly limestone where Jurassic is present, on which lie white cherty limestones, now referred to the Upper Cretaceous.

The existence in Sardinia of marine strata of Cretaceous age made up almost entirely of carbonates precludes the possibility of any major part of the island

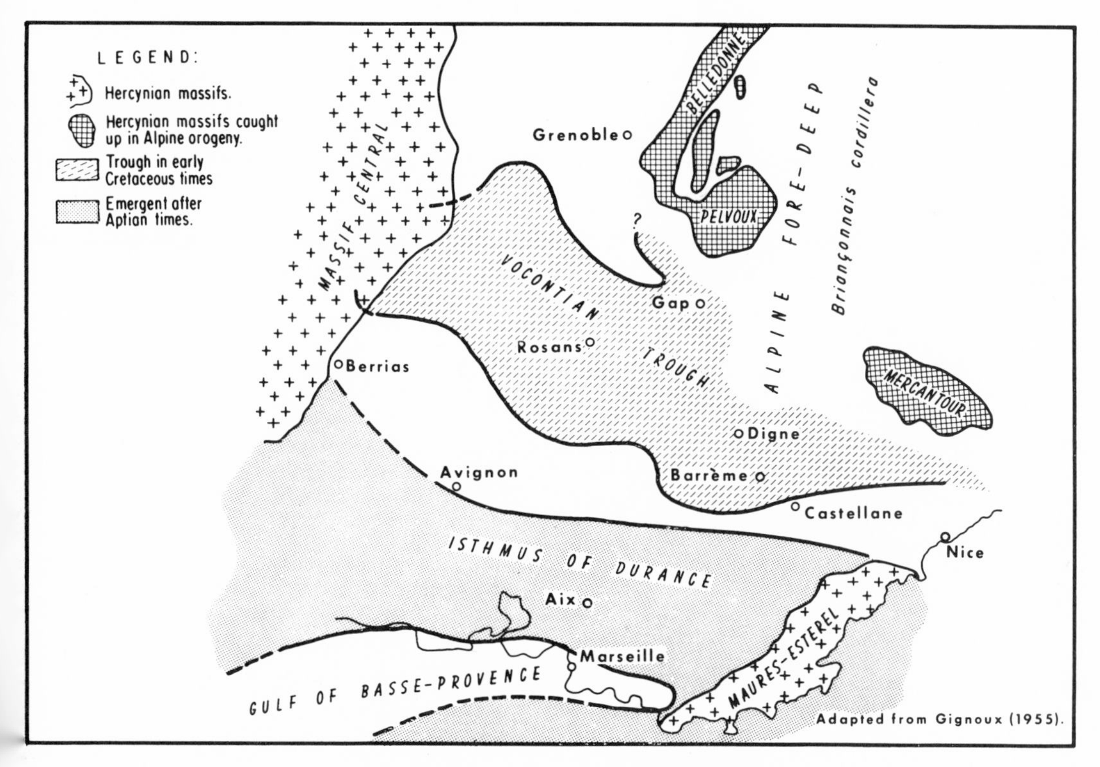

Fig. 10. Cretaceous structure in southeast France.

having served as the provenance of terrigenous clastics to surrounding areas during that period.

Corsica. After the emergence of northern Corsica in Liassic times the sea returned to northern Corsica and spilites were laid down on the sea floor during the Malm accompanied by radiolarites and capped by gray and pink limestones with *Calpionella* in some localities and clastic limestones in others. On these Tithonian strata is a thin-bedded sequence of black cherts alternating with clay shales and dark gray to black siliceous limestones. Some sandy limestones are intercalated as lenses, along with limestone-cemented breccias made up of fragments of granite, rhyolite, schists, Triassic dolomites, limestones of late Jurassic age and spilites. These breccias are coeval with the black chert sequence for they are contained in it. This chert-limestone-shale-breccia unit lies conformably on the limestones of the late Jurassic and includes fossils that demonstrate its Neocomian-Albian age. The disturbance responsible for the development of the breccia is thus at least in part pre-Cenomanian and can be compared with early Austrian tectonism in the Alps. Another flow of submarine pillow basalts followed.

The siliceous limestone-breccia sequence is found in several localities, and Delcey *et al.* (1965, p. 326) consider one occurrence as coeval with the upper levels of the *"schistes lustrées"*, a portion of which would then be late Jurassic-early Cretaceous.

Farther south, the coarse clastics of the early Cretaceous are not present, replaced by gray, finely crystalline limestones and gray black argillaceous limestones. The upper levels of this unit are intercalated with sandstones. Its microfauna of planktonic foraminifera demonstrates its age as Albian through Eocene. The sandstones are very probably of Tertiary age, for elsewhere the basal Eocene is represented by conglomerates and arkose capped by alternating sandstones and black shales.

The rather strong tectonism in northern Corsica during Cretaceous times suggests that the region may have been near an area in which orogeny was effective. The proximity of the Alpine geosyncline suggests a relationship between the two regions.

Southeastern France. The stratigraphy of southeastern France has been studied in detail for many years. Here only an incomplete résumé of the Cretaceous sequence is feasible. During early Cretaceous times the southern part of this region was still a part of the foreland of the Alpine geosyncline, although farther north the Vocontian trough remained in existence between the *Massif central* and the Mercantour massif as a subsident tract athwart the Rhodanian corridor. In the central portion of this trough the open-sea, rather deep water strata of the Neocomian consist predominantly of calcareous shales and marls with interbeds of slightly argillaceous limestone that become more numerous upward. However, dark colored calcareous shales and marls predominate in strata assigned to the upper Aptian, and Albian beds are characterized by lenses of sandstone. The megafauna of these Neocomian (from Neuchâtel in Switzerland) strata is predominantly ammonites which are numerous enough to be used for precise zoning. Many of the designations of the stages of the Lower Cretaceous are taken from geographic locations in or on the south flank of the Vocontian trough, in the central portion of which the total thickness of the Neocomian does not exceed fifteen hundred feet. The trough, which began to shallow in Albian times, had retreated eastward from the *Massif central* and narrowed sharply by the end of the Turonian. Thereafter it is not recognizable.

North of the Vocontian trough and west of the Paleozoic massifs of the French Alps, the calcareous shales and marls characteristic of the open-sea facies of the trough are intercalated with organic limestones that first appeared late in Valanginian times. The Hauterivian basal sandstones are glauconitic, and on them lie marly limestones that include some sandstones near their upper limit. A thousand feet of Barremian-Aptian biostromal and reefal limestones rest on these softer beds that total some seven hundred feet in thickness. Emergence late in Aptian times is suggested by the absence in some localities of strata of this age, and by the transgressive nature of the thin glauconitic crystalline limestones of the Albian.

The Cenomanian in the region west of the Paleozoic massifs is represented by a complete section only as far north as Villar de Lans where it consists of white calcareous sandstones with belemnites. Farther north, Maestrichtian limestones, more than five hundred feet thick rest on Albian. Both Turonian and lower Senonian are presumed to be absent. This Maestrichtian transgression is contemporaneous with the retreat of the sea in Provence at this time.

In the Jura mountains to the north the Lower Cretaceous is entirely littoral, the first marine beds overlying the Purbeckian representing the base of the Valanginian, of which the type section is near Neuchâtel. As in the ranges west of the Alpine massifs, the Albian, consisting of fossiliferous glauconitic sandstones, is transgressive, the Aptian being reduced in thickness or absent. Only very small remnants of the original cover of Upper Cretaceous remain in the Jura.

In Provence, the sequence and facies of Cretaceous strata are to a large extent controlled by the Isthmus of Durance which formed a land bridge between the *Massif central* and the Maurès-Esterel massif, and thus separated the sea in lower Provence from that in the remainder of the Rhodanian corridor during Albian through Santonian times (see fig. 10). The isthmus was flanked by marine littoral and epineritic deposits The sea on its south side transgressed northward, for the axis of the isthmus moved northward throughout its existence as a discrete entity. During the Campanian-Maestrichtian all of southern Provence rose above sea level and only lacustrine beds were then deposited on the former isthmus.

In the western part of Provence the Valanginian (from the village of Valangin) is represented by a basal unit of fine-grained limestones with thin greenish gray shale intercalations on which rest light-colored, well-bedded limestones. The Hauterivian, above, consists of a basal gray calcareous shale with limestone inter-

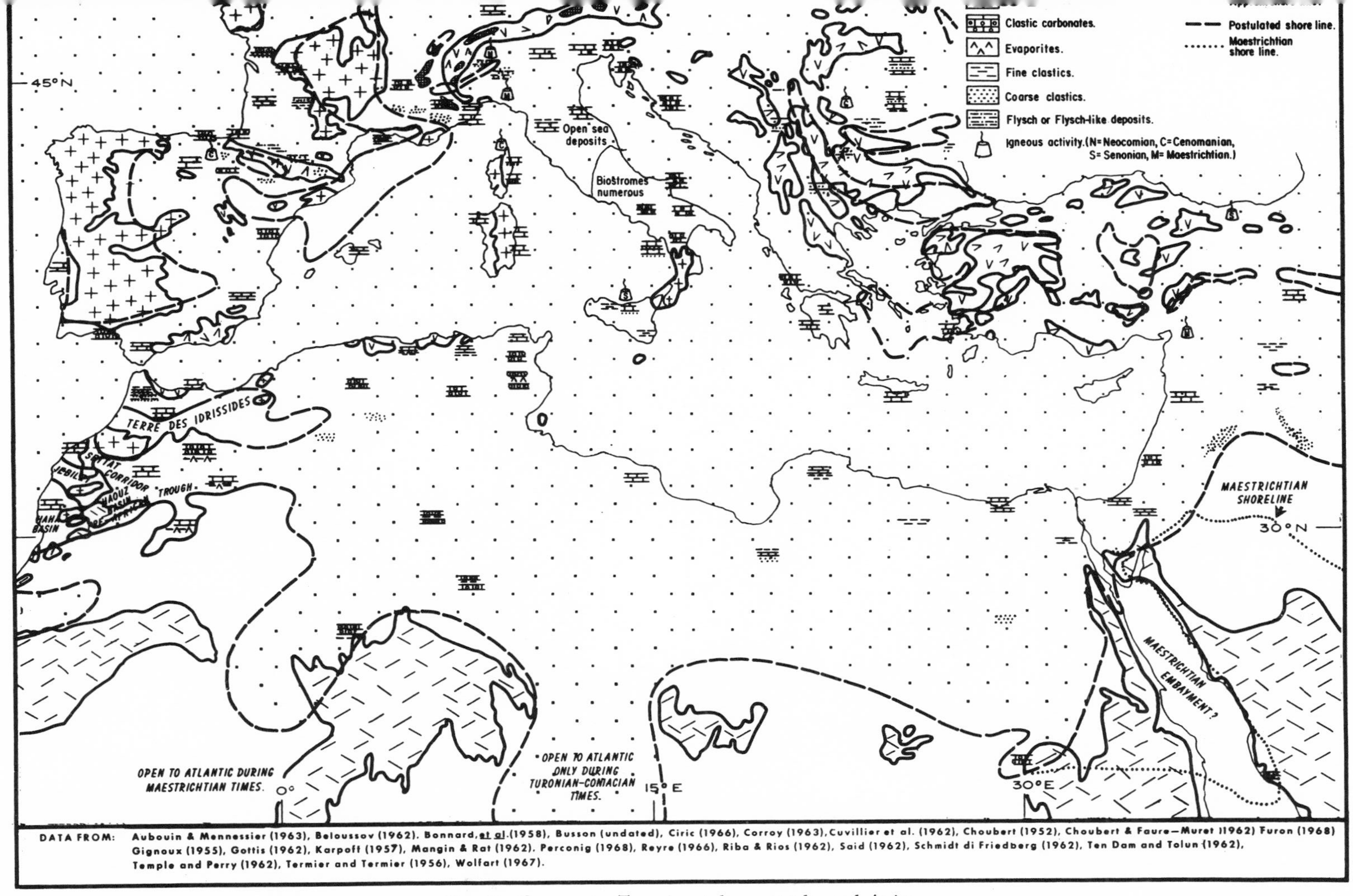

Fig. 11. Cenomanian-Turonian paleogeography and facies.

beds capped by gray siliceous limestones. Farther east in Provence the entire Valanginian-Hauterivian is made up of bedded limestones, some of which are biostromal.

Barremian and lower Aptian strata in Provence are commonly white and yellow massive or poorly bedded limestones, for the most part fragmented invertebrate skeletal and test debris. These are referred to the "Urgonian facies", although in southern Provence they begin at a lower level in the Barremian than in outcrop at Orgon on the south bank of the Durance east of Tarascon where the type Urgonian (no longer accepted as a stage) was originally selected. These organic clastic deposits extend eastward past Marseilles and the Rhone at Aix. In other localities the lower Aptian ("Bedoulian", from La Bédoule, east of Marseilles) consists of bluish marly limestones with large chert nodules on which lie the upper Aptian dark colored calcareous shales and the dark colored shaly limestones called "Gargasian" (from Gargas, near Apt) of which the type section lies just to the north of the Isthmus of Durance.

The Albian is not represented in much of southern Provence owing to widespread emergence at that time. In the west, however, deposits of this age are present as dark colored calcareous shales and calcareous glauconitic sandstones. They are also present in the east near Nice and Grasse as thin glauconitic sandstones extending northward on the Isthmus of Durance as terrigenous sands and clays. Marine Albian in a neritic facies of black calcareous shales exists only in the central portion of the much diminished Vocontian trough north of the isthmus.

The emergent isthmus was subjected to weathering and erosion so that the bauxite deposits formed on the surface were concentrated in topographic depressions by the advancing Cenomanian-early Senonian seas, and late Senonian lakes. According to Aubouin and Mennessier (1963, p. 52) the bauxite (from the town of Les Baux, hence bozite) lies on carbonates ranging in age from Dogger through Barremian, the differences in age of the depositional surface being a result of erosion of varying intensity.

The Cenomanian sea advanced from the southwest to the northeast, transgressing on the Isthmus of Durance from the "Gulf of Basse Provence" formed at the same time as the isthmus by uplift of the Maurès Esterel massif and its extension to the southwest. Perhaps, as shown on figure 11, the emergent tract extended to the Catalan massif. Deposits in the gulf are in any event like those in the eastern Aquitaine basin north of the Pyrenees, and, as mentioned, became lacustrine late in Cretaceous times.

On the north flank of the Isthmus of Durance the sea was shallow throughout the late Cretaceous. Cenomanian deposits are glauconitic sandstones, and the Senonian is represented by rudistid limestones. In some localities most of the Late Cretaceous deposits are sandstones with lignitic intercalations, along with a few rudistid limestone interbeds. The Campanian-Maestrichtian are generally not represented, but where strata of these stages are present they consist of sandstones or clays which grade upward into a continent Eocene.

East of Nice the littoral facies of Cenomanian-Senonian deposits are replaced by argillaceous limestone and calcareous shales with an open-sea fauna, but t the northeast on the Mercantour massif Upper Cretaceous deposits consist of rudistid limestones. On th remainder of the Paleozoic massifs the Cretaceous i absent. However, between the Pelvoux and Mercantou massifs the outer thrust sheet of the Dauphinois in cludes a Lower Cretaceous section made up of dark colored calcareous shales with belemnites, and an Uppe Cretaceous of calcareous shales with *Globotruncan*

Differentiation in Cretaceous times between th unstable platform of the Rhodanian gulf and the ope sea of the Alpine geosyncline is clearly shown by th lithology and fauna of the strata laid down in th two areas. Furthermore, the Rhodanian region ros above sea level at the end of the Mesozoic era an was never again fully covered by the sea, wherea much of the Alpine geosyncline was under marin waters throughout Lutetian and Priabonian times, th marine régime ending only during the Oligocene whe orogeny caused a northward and westward migratio of the sea to the Molasse basin of Switzerland an Bavaria and to the outer bastions of the French Alp west of the Paleozoic massifs.

Southwestern France. Much of the Aquitaine basi was above sea level at the end of Jurassic times fo Purbeckian continental sandstones, lignites, and lacus trine limestones have been found in many places Continental Wealden demonstrates that the emergenc persisted into the Cretaceous period. In long-lastin depressions gypsiferous shales and salt indicate th presence of a late Jurassic regressive sea, however.

Marine Neocomian, presumably post-Valanginian rests conformably on Wealden sandstones both alon the Atlantic coast near Parentis, and in the Pyrenea foothills in the Lacq region. In both of these presum ably discrete basins the Neocomian is represented b light colored marine limestones but in the foothills th thousand-foot Hauterivian-Barremian carbonate sectio includes some lagoonal deposits. In the Pyrenean regio this early Cretaceous section is cut out estward b Aptian transgressive overlap, the result of the develop ment of a long trough from fifteen to thirty mile wide along the north front of the ancestral Pyrenee (Bonnard *et al.*, 1958, p. 1104). In this trough, ir some places with a basal conglomerate, Aptian-Albiar strata are present in thicknesses ranging up to twelve thousand feet, the section including both dark-colorec shales with ammonites and rudistid biostromal anc reefal limestones that in some localities comprise the entire section. West of Dax the Albian is represented by sandstones, as it is in Nevarre in Spain.

In the Parentis region lower Aptian limestones, conformable with those of the Neocomian, are capped by upper Aptian calcareous shales, that although ranging up to two thousand feet in thickness, have been removed by unconformity in the west that cuts deep into the Neocomian limestones. Albian calcareous shales lie on the unconformable surface.

During the Cenomanian-Senonian, the sea, previously estricted to the vicinity of Bayonne, the Adour river nd the area north of the axial massifs of the Pyrenees, overed the whole of the Aquitaine basin. The Cenomanian incursion began near Isle Madame north of the Gironde estuary and laid down sandstones and lignitic ays over a large tract including the flanks of the *Massif central.* Rudistid limestones were deposited by ne seas of late Cenomanian age, and the Turonian ea, more extensive and deeper, reached the Paris basin hrough the Strait of Poitou. Senonian seas occupied ne entire Aquitaine basin, reaching their apogee during ne Campanian.

In most of Aquitaine Senonian-Turonian deposits onsist predominantly of more or less arenaceous organ- limestones, some levels of which are siliceous and herty. Rudistids are common throughout. In the subsurface the average thickness of these strata is eighteen hundred to two thousand feet, but in some fault roughs they thicken to five thousand feet. In these hicker sections calcareous shales exist, particularly in oniacian and younger beds.

The Senonian becomes increasingly arenaceous eastward, and late Maestrichtian regression was almost universal, for no marine Paleocene is known except in ne Aturian gulf that extended from Bordeaux to t. Marcet. In this gulf, marine deposits of Maestrichtian age extend eastward as far as the Corbières where andstones with littoral fossils were deposited. The Danian, marine in the west near Biarritz and Dax, s represented by lacustrine limestones south of Tououse in the Petites Pyrenees and eastward in the Hérault. Similar lacustrine limestones exist in Basse Provence, so that it is clear that marine connections were roken between the Mediterranean and the Gulf of Basse Provence in late Senonian times. The continental "Garumnian" deposits of this region include these Danian lacustrine limestones, along with continental andstones and clays.

The axial Pyrenees remained emergent throughout the Neocomian, for transgressive Aptian limestones occur along the entire north slope, and south of them ear the crest, sandstones and conglomerates with *Orbitolina* of Albian-Cenomanian age demonstrate transgression. Maximum submergence of the axial region ppears to have taken place during the Campanian, nd regression began immediately thereafter, so that by the end of Maestrichtian times the entire chain was mergent except in the extreme west where a branch f the shrinking Aturian gulf extended into Navarre.

The southern part of the Aptian-Albian basin north of the Pyrenees was involved in tectonism at the same time that the Cenomanian transgression began. A long narrow trough nearly two hundred miles long by welve miles wide developed in which ten to fifteen thousand feet of calcareous sandstones, calcareous shales, nd clastic limestones were deposited in Cenomanian-Senonian (locally Cenomanian-Eocene in the west) times. The sandstones are best developed near St. Marcet. Elsewhere they are largely replaced by argillaceous limestones. The terrigenous clastic portion of this sequence may have been derived in part from the erosion of tracts newly submerged by the advancing sea, but Gignoux (1955, p. 427) indicates that long narrow cordilleras were developed in the trough, along which breccias comparable to Alpine "Wildflysch" (conglomerates and exotic blocks in sizes up to tens of feet from collapse of submarine cliffs, Rech-Frollo, 1960) were developed, particularly near Hendaye and Bidart, and also in the St. Marcet gas field.

In some localities the transition zone between the subsident trough and the shelf facies to the north is less than one mile wide (Bonnard *et. al.*, 1958, p. 1107), and it is possible to assume that the trough was formed by faulting. If so, the "cordilleras" could have been horst blocks. In any event, Termier and Termier (1957, p. 642 and 652) indicate pre-Aptian, intra-Aptian, and pre-Cenomanian tectonism in the Pyrenees, all on a minor scale. Metasomatism in a small tract was associated with the pre-Cenomanian phase, and some folding occurred at that time, according to Casteras (quoted in Termier and Termier). Termier and Termier are certain, however, that the site of the Pyrenees was not a geosyncline after Carboniferous times, and all evidence available confirms this opinion.

CRETACEOUS DIASTROPHISM, PALEOGEOGRAPHY, AND SEDIMENTARY ENVIRONMENTS

Both the facies and stratigraphic relationships of strata of Cretaceous age in the Mediterranean region show that the tempo and strength of tectonic events increased as the Mesozoic Era approached its end. The effects of diastrophism on shelves and platforms are clear, and are even more obvious in the mobile belts. Some tectonic events in the developing geosynclines were contemporaneous with weaker movements on shelves. On the other hand, tracts contiguous to geosynclines were not necessarily affected appreciably by marked tectonism in the labile areas (*e.g.* Alpine trough and peninsular Italy).

Although Kimeridgian regional uplift was widespread in southern Europe, North Africa, and the Middle East, Tithonian transgression was almost equally ubiquitous, and the fine-grade, tintinnid-rich limestones laid down by Portlandian seas suggest the absence of sources of terrigenous clastics in their vicinity. These conditions lasted into early Cretaceous times, and in some localities persisted through the Barremian (western Italian Alps), or Aptian (southern Sicily).

Barremian Diastrophism. In much of North Africa movement on pre-existing faults began in Barremian-Aptian times. Epeirogenic uplift in northwestern Libya caused unconformity between Barremian and younger strata; in southern Tunisia and adjacent portions of the Algerian Sahara similar disturbances resulted in the deposition of conglomerates locally and of sandstones over a much larger area. Hauterivian-Barremian clastics occur also in central Tunisia on the "Central Tunisia High". In the western Algerian Sahara horsts were formed by Barremian reactivation of old north-south trending faults and remained above sea level during Aptian times. Farther east, the Tibesti massif was rejuvenated. In the Tellian trough some portions of the *Dorsale calcaire* were emergent in Barremian times.

Barremian uplift was also effective in many parts of Europe, and particularly in the mobile belts. The Pelagonian massif in the Dinaric trough was raised above sea level, and sandstones deposited in the adjacent Pindus trough. Movements of about the same intensity reportedly occurred in the Rumanian portion of the Carpathian trough where Barremian sandstones occur in a carbonate section. Transgressive Aptian limestones in the French Alps are unconformable on the older shales of the Bündnerschiefer.

Barremian uplift of the Arabian shield is demonstrated by the aureole of coarse clastics of this age present north and east of it. This uplift may have been general in the basement massifs of northern Africa, for, as mentioned, Lelubre reports movement of the Tibesti at about this time.

Aptian Diastrophism. Intra-Aptian tectonism in the Mediterranean region affected fewer areas than that of the Barremian. However, uplift in the *Dorsale calcaire* of the Tellian trough occurred during the Aptian, and the "Jeffara Dome" of southern Tunisia came into existence late in Aptian times. In Europe some disturbance dated as Aptian occurred in the Pyrenees, in the Aquitaine basin along the front of the Pyrenees, and perhaps in Corsica. Near the Paleozoic massifs east of the Rhodanian corridor uplift and emergence occurred lat in Aptian times, and may have started the on the Isthmus of Durance, emergent durin the Albian. In the Rumanian and Czechoslo vakian Carpathians tectonic movement appear to have begun during the Aptiar for older papers report thrusting at tha time. Most recent works state that nappe did not come into existence before th Senonian (Andrusov, 1963, 1967; Ksiakie wicz, 1963, 1966; but see Tollmann, 1964 who reports Hauterivian thrusting in Aus tria).

Albian-Early Cenomanian Diastrophism. "Austrian orogeny", named for the area i which it was recognized, is commonly consid ered to cover only Albian tectonism. Tw pulses are reported, one immediately before and one during the stage. Some authors however, broaden the limits to include th Barremian, so that nearly all significant pre Cenomanian diastrophism would thus b embraced in this phase (Termier and Ter mier, 1957, p. 685).

Here, the Austrian phase is restricted t tectonic events in Albian and early Ceno manian times. Thus defined, it can be consid ered as the initial phase of Alpine orogeny that is, the first of the series of compres sional pulses that formed the Alps. In an event, diastrophism during Albian times wa exceedingly widespread in the Mediterra nean region not only in mobile belts, bu also on shelves and platforms.

In the Rumanian Carpathians, tectonisn called, "Cretacé moyen" by Bancila (1968, p 110) appears to have been most effective du ing the Albian when uplift, perhaps accom panied by some folding took place in th Carpathian trough, and on the shelves bound ing it to the east and south Patrut *et al* 1963, p. 147-149). However, according t Ksiazkiewicz (1966, p. 453) the "cordille ras" that provided coarse clastics and Wild flysch to the northern Carpathian trough a about this time were probably horst block Similar blocks may have been present i Rumania. In any case, intra-Albian un conformity is reported in the Czecho slovakian Carpathians, and at the eastern e the country thrusting may have occurred at this time (Andrusov, 1967, p. 1058).

Intra-Albian unconformity is known also in the Austrian Alps. In the French and Swiss Alps the Briançonnais horst, already largely emergent, probably was further uplifted during the Albian for the thick Bündnerschiefer of the Valais eugeosyncline includes sandstones in its upper levels, above Aptian limestones themselves unconformable on older strata. Trümpy's 1960 paper can be read to indicate that most flysch is Cenomanian or younger. Nevertheless, as Trümpy indicates (p. 898), the island cordilleras from which flysch deposits came presumably began their development earlier in some localities.

In the Dinaric Alps, Albian-Cenomanian flysch is present in a belt twenty miles wide extending from Belgrade to Skopje. This flysch presumably was derived from the debris of island cordilleras, although its argillaceous content is high and its proximity to the Serbo-Macedonian massif suggests that this old basement high may have contributed to the deposits.

In the Hellenic portion of the Dinaric Alps uplift on the Pelagonian massif continued during the Albian, the emergent tract migrating westward with time. In the Turkish prolongation of these mountains evidence for Albian movement is not available. The continuous sedimentation throughout Cretaceous times until the Maestrichtian reported there may represent conditions only in an outer portion of the trough, unaffected by uplift and igneous activity in the inner ranges, for in the Pontic mountains of northern Turkey Albian-Cenomanian tectonism was strong, and accompanied by ophiolite intrusions and flows.

On the shelves and platforms in the Mediterranean region Aptian transgressive limestones are in many large areas succeeded by Albian terrigenous clastics, sure evidence of the rejuvenation of old source areas or the development of new ones. The entire Iberian peninsula, including the Betic trough, was raised during Albian times, and sandstones were deposited over very large areas. Shallowing is reported in the Balearic islands. Albian strata in the Tellian and eastern Saharan Atlas troughs reflect renewal of the supply of coarse detritus, ended by late Albian transgression.

In southern Tunisia and the Algerian Sahara Albian sandstones form a blanket over older strata. In central Tunisia the Cenomanian rests unconformably on Albian sandstones on the "Central Tunisia High". Albian sheet sandstones are known in the Western Desert of Egypt, where their deposition was probably accompanied by block faulting. On the Sinai peninsula part of the "Nubian" sandstones are presumably of Albian age as they are in Jordan.

Late Albian uplift in northern and central Iraq and southeastern Turkey is shown by the unconformity between Turonian and Albian limestones seen in wells. On the flanks of the Arabian shield Albian and Barremian sandstones are separated by Aptian limestones in subsurface, but are in contact, presumably unconformably, near the outcrop where in some localities Cenomanian sandstones rest on Barremian sandstones.

On the Balkan peninsula the Pannonian and Transylvanian massifs began to founder during the course of Albian tectonism, and early Albian movements tilted the Moesian platform.

In contrast, few effects of Albian diastrophism can be found on the Italian peninsula or in Sicily, in spite of the postulated emergence of a tract including Corsica in the northern Tyrrhenian sea at this time. However, the Maurès-Esterel massif in southeastern France, northwest of the postulated land mass, did emerge, along with the Isthmus of Durance. The Catalan massif in northeastern Spain rose farther above sea level at this time and may have been linked with the Maurès-Esterel massif by a land bridge which cut off direct connections between the Ebro and Aquitaine basins and the Mediterranean. In the Parentis region of the Aquitaine basin, along the Atlantic coast south of Bordeaux, early Albian uplift was followed by deep erosion, so that Albian shales rest on Neocomian limestones. In the western French Pyrenees Albian sandstones reflect movement in the region, in the same way as similar sandstones in northern Spain.

Early in Cenomanian times a trough was formed in the southern Aquitaine basin paralleling the north front of the Pyrenees, and another in the Cantabrian basin southeast of Santander. These long, narrow subsident belts in which great thicknesses of rocks of late Cretaceous age accumulated, may have been controlled by faulting, as was the Sirte embayment of Libya which is more or less contemporaneous. All presumably were formed when compressional stresses in the three areas were relieved.

Cenomanian-Turonian Diastrophism. Block-faulting, prominent in the Sirte region of Libya early in Cenomanian times presumably was more or less continual there throughout the late Cretaceous and early Tertiary. Similar faulting occurred in coastal Egypt where Said (1962, p. 250) reports also some compressional folding. Ophiolite flows of Cenomanian age (possibly Albian?) are mentioned by Gattinger (1962) in his study of the Trabzon portion of the Pontic trough.

However, the next orogenic phase that involved the entire Mediterranean region occurred at the end of Turonian times. This is Pre-Gosau orogeny, named in the Austrian Alps where the molasse-like Gosau Beds overlie the first nappes. Similar conditions obtain in the Carpathians of Czechoslovakia. Although nappes were formed in the French Alps only late in Eocene times, the cordilleras from which flysch deposits were derived were numerous, and locally, as on the Dévoluy massif between Grenoble and Gap, Senonian beds rest unconformably on Turonian and older strata.

The Hellenic portion of the Dinaric Alps was subjected to strong vertical uplift during the Coniacian, and presumably the Yugoslav portion was also involved, along with the Taurus where depositional hiatus is reported between Turonian and Campanian strata even in the deeper portion of the trough.

The effects of Pre-Gosau movement on shelves and platforms have been found throughout the Mediterranean. Post-Turonian uplift raised the Portuguese Atlantic coastal basin and the Balearic islands above sea level. In the central Moroccan seaway Turonian and Senonian strata are separated by unconformity involving the Coniacian. On the Hauts Plâteaux Senonian shales are unconformable on Cenomanian limestones, and it is known that the western portion of the adjacent Saharan Atlas trough emerged at the end of Turonian times. At the same time the eastern portion of the Hodna ranges and a tract between Ain M'lila and Constantine rose above sea level. Strata younger than Turonian are absent over the apex of the Central Tunisia High. In the northeastern sector of the Western Desert of Egypt folding and faulting were accentuated late in Turonian-Coniacian times, and were ever more prominent on the Sina peninsula in the Syrian Arc structures. Similar movements occurred in Jordan, coasta Palestine, and in Lebanon. Campanian anc Maestrichtian limestones rest unconformabl on Turonian limestones over large portion of the unstable platform of central anc northern Iraq, northern Lebanon and southern Turkey, and similar relationships exis in the western Persian Gulf. In souther Lebanon, Cenomanian-Maestrichtian strat lie unconformably on Cenomanian lime stones.

Uplift in portions of the Italian Alps i shown by the sharp change in sedimentar facies from Turonian argillaceous limestone to Coniacian sandstones in the Lombar trough and in the Julian trough near th Dinaric Alps. However, the Italian penir sula, Sicily, and southern France do not see to have been much affected by Pre-Gosa tectonism.

Senonian Diastrophism. Two oroger phases are commonly recognized in the la Cretaceous of Europe - *Subhercynian* ar *Laramide.* Unfortunately, the name Subhe cynian has been assigned not only to mov ments between the Coniacian and Santonia but also to disturbances between the Sant nian and Campanian. In many localiti here Campanian or Maestrichtian rests on uronian, the effects of the Subhercynian nase cannot be ascertained.

In the Carpathians, nappe-formation took ace in small areas at the end of Santonian mes. This thrusting and gliding was accomnied by intrusions and flows of both acid d basic lavas and deposition of tuffs. lovements on a less imposing scale but ill involving strong unconformity occurred the Tellian trough (Durand Delga, 1967, 62).

On the unstable platforms of North Afri- disturbances at the end of the Santonian ve been reported in several places. In the odna ranges on the Ain M'lila môle at bal Bou Taleb, Santonian conglomerates ove renewed uplift at that time. Some ovement also occurred in the Tunisian tlas. Structures on the Sinai peninsula and Palestine rose above sea level during the ntonian, for Santonian-lower Campanian rata are absent on the crest of some antiines but present in adjacent synclines.

In the Mediterranean region the last ogenic phase of the Mesozoic Era, rather aptly called Laramide orogeny, took place uring and at the end of Maestrichtian times. the mobile belts the expression of this ogeny is very obvious. Deposits of flysch which the basal part is of Maestrichtian e are known in two localities in the Swiss lps as part of a sequence of flysch deposits at become progressively younger northard. Consequently, island cordilleras must ve formed at this time. In the "Inner" inaric Alps of Yugoslavia Maestrichtian ysch is very common, and includes the cotics characteristic of olistostromes and Wildflysch". The flysch is associated with yolitic flows and tuffs. In the "Outer" inaric Alps late Cretaceous gentle folding d uplift raised much of Dalmatia and the trian peninsula above sea level. In the ellenic portion of these mountains the elagonian massif emerged once again during e Maestrichtian this time possibly as a sult of moderate compression rather than eirogenesis. In the Taurus extension of e Dinaric trough Maestrichtian disturbces caused very extensive gravity slides and olistostromes associated with outpourings of ophiolites both in and near the trough. In more stable tracts intra-Maestrichtian unconformity is evident, or at least coarsening of clastic textures occurred during this time.

In the Pontic ranges the late Senonian is represented by sandstones, shales, and conglomerates accompanied by widespread flows of andesites and basalts. This disturbance reportedly began before the Maestrichtian however and thus could represent both Subhercynian and Laramide ("Van") orogeny. On the central massifs of Turkey conglomerates separate Cretaceous and Eocene sedimentary sequences, and syenites and gabbros have intruded limestones and shales of late Cretaceous age, but not younger strata.

In the Rumanian Carpathians Maestrichtian uplift is indicated by the deposition of sandstones and shales on limestones in the outer portion of the Balkan trough. Information is not available concerning individual phases of late Cretaceous orogeny near the Paleozoic massifs, but apparently Senonian movements are not identifiable individually.

In the Betic trough, in the Campo de Gibraltar region, terrigenous clastics in Senonian calcareous shales indicate uplift on the basement massif. The sea withdrew from the Betic area at the end of Cretaceous times. A similar withdrawal took place in the Pyrenees where the sea remained only at the western end of the tract.

In the Rif trough, although the Senonian includes sandstones and clastic limestones, emergence did not occur at the end of the Mesozoic, nor did it in the Tellian trough of Algeria and Tunisia. Obviously, Senonian disturbances had only local effects in this long mobile belt, and in western Sicily. Along the coast of northern Sicily, however, emergence at the end of Cretaceous times is suggested by unconformity between Cenomanian or younger strata and Eocene limestones and shales. Uplift there was accompanied by vulcanism including basalt flows and ash beds.

On the unstable platforms and shelves of the Mediterranean region evidence of late

Maestrichtian disturbance is nearly ubiquitous. All of Spain and most of southern France were above sea level at the end of the Mesozoic; all of Morocco was land except the Rif trough and small gulfs and bays along the Atlantic coast in which phosphates were deposited. The Hauts Plateaux and western Saharan Atlas were part of the Moroccan promontory at the beginning of the Cenozoic, whereas the eastern Saharan Atlas and portions of the Ain M'lila môle along with most of central and northern Tunisia remained under the sea. Local epeirogenic disturbances are reported in the Algerian Sahara during the late Cretaceous. Unconformity separates Cretaceous and Eocene strata in most of western Libya. In the northern portion of the Sirte embayment and in some grabens and stable tracts farther south and east in this persistently sinking region inter-era deposition was continuous in both shale and carbonate sections, while on horsts unconformity commonly marks the Cretaceous-Tertiary boundary. Parts of northern Cyrenaica were brought above sea level by gentle tilting and folding (?) at the end of Maestrichtian times, but farther south on the stable platform the carbonate-depositing regime was not disturbed.

On the stable shelf of Egypt transgression that began during the Campanian continued unbroken into Tertiary times, the shales and limestones laid down in the shallow sea containing the microfossils denoting a complete sequence. However, some structures on the stable shelf rose above sea level at the end of the Mesozoic era. Similar local movements occurred in the Gulf of Suez, near Cairo, and on the Sinai peninsula, and in all of these localities anticlinal structures became at least partly emergent. In the subsurface of the northern portion of the Western Desert movement on faults caused great differences in relief between horst and graben blocks, but not all of this movement was Laramide.

Most of Palestine appears to have risen above sea level at the close of Maestrichtian times, and northern Iraq, southeastern Turkey, and northern Syria also emerged. Parts of Jordan remained submerged, as we as the Damascus basin and the Palmy region of Syria.

The northern part of the Italian peninsu was little disturbed by Laramide movement but in the south emergence was commc over extensive areas at the end of tl Cretaceous, the Eocene transgression havir been preceded by the development bauxites locally. The Calabro-Sicilian mass rose above sea level at this time, along wit most of Sicily, where uplift was accompanie by vulcanism along the north coast. In Sard nia, relatively strong northwest-southea trending folds involve late Cretaceous lim stones, and the compressional forces a assumed to be Laramide, for Eocene lim stones lie almost horizontally on them.

In southeastern France Maestrichtia emergence was preceded by shallowing du ing the Campanian. Maestrichtian rocks a locally conglomeratic, the clastics consistir of Jurassic and Cretaceous limestone pebble derived from local structural highs produce by weak folding. At the same time the regic was raised epeirogenically, so that Proven was above the sea throughout the Eocen

The Aquitaine basin was tilted westwa during Maestrichtian times so that at tl close of the Mesozoic the sea had retreate to a gulf of which the shoreline ran sout east between the Gironde esturary and S Marcet. Certain islands remained in th sea, in particular one now comprising tl area occupied by the Laq field.

Paleogeography and Facies. Land and se relationships were discussed briefly in tl introduction to the review of Cretaceo stratigraphy so that the following is to som degree a recapitulation. At the start Cretaceous times light-colored, sublith graphic limestones were being laid down most of the mobile belts of the Mediterr nean region. The lower part of these stra was deposited by the late Tithonian se and they are transgressive on formation involved in Kimeridgian diastrophism. How ever, a number of the stable platforms we emergent, as well as the Paleozoic massi between the mobile belts. Among other e emergent tracts included most of the erian peninsula and southwestern France, orocco, Algeria, the Sirte arch, nearly all the Western Desert of Egypt, and the atform between the Arabian shield and e massifs of central Turkey. Smaller emernt areas were present on the Balkan ninsula and in southeastern France.

The Valanginian-Hauterivian seas in hich these distinctive fine-grade limestones ere deposited were obviously not supplied ith terrigenous clastics from the emergent eas, which, however, did provide Wealden ndstones in large quantities to nearby asins. Barremian movements disturbed nd-sea relationships only slightly, but Apan transgression characterized by the wideread occurrence of rudistid limestones was rominent throughout North Africa and the liddle East as well as in the mobile belts. ustrian orogeny provided clastics in great nounts to the sea, which retreated from me areas, presumably as a result of local id regional uplift.

After the second pulse of Austrian rogeny in mid-Albian times the great Cenomanian" transgression began, aided y relaxation of compressional tensions as iown by the development of graben stems in northern Spain, in France north the Pyrenees, and in Libya. At this time, o, began the sinking of the Pannonianransylvanian massif or "Zwischengebirge". he Late Cretaceous sea (see figs. 7, 10) vered much of North Africa, and during uronian-Coniacian and Maestrichtian times as linked through channels between the loggar and Tibesti and Hoggar and Eglabetti massifs with the Atlantic off Nigeria.

Pre-Gosau disturbances caused a temorary retreat of the sea in parts of North frica, the Middle East, and Europe, and e accentuation of compressional folding in gypt and the Levant where some structures mained partially emergent in the advancing enonian seas which reached their apogee Campanian-Maestrichtian times. Large eas, land since Paleozoic times, were vered by shallow waters in which both halky and biostromal limestones were deposited. The Paleozoic massifs between the mobile belts already partly submerged in the Cenomanian-Turonian transgression were invaded once again.

At the close of the Mesozoic Era, rapid regression took place and early Tertiary seas retreated in many places to the mobile belts. The Iberian peninsula, including the Betic trough, rose above sea level, leaving only a small tract at the west end of the Pyrenees under the waters of the Aturian Gulf. Nearly all of Morocco and western and central Algeria south of the Tellian trough were land.

Much of Tunisia, the Sirte embayment, Cyrenaica, and the Western Desert remained submerged, whereas brief emergence was the rule in the Levant which was affected by the strong Maestrichtian tectonism in the Taurus trough.

Characteristic of transgressive Cretaceous strata laid down in shallow water are the rudistids, a thick-shelled pelecypod group with one valve much enlarged and assuming a roughly conical shape. These are rare in early Neocomian beds, but in Barremian times and thereafter made up an important part of the fauna of the advancing seas. True rudistids appear only in the Cenomanian and persist throughout the remainder of the Mesozoic; the earlier pachydonts are related but belong to discrete genera such as *Requiena* and *Toucasia.*

The deeper water formations commonly include planktonic foraminifera in large quantities, and from Cenomanian times onward, the foraminifer *Globotruncana* and its relatives are numerous, particularly in the calcareous shales and chalks so characteristic of the Senonian. Other distinctive foraminifera, those found in shallow-water deposits, are *Orbitolina* prominent in Barremian-Cenomanian beds, *Orbitoides,* confined to the *Campanian-Maestrichtian,* and *Omphalocyclus,* a marker of the Maestrichtian.

All of these and many others help to identify the several stages of the Cretaceous represented in the extensive outcrops of the Mediterranean region. These owe their existence to the vast areas covered by the late Cretaceous seas, in which the chalks that gave their name to the period were so commonly laid down.

Cretaceous Vulcanism. Ophiolites of Neocomian age occur in the western Alps intercalated in the Bündnerschiefer of the Valais trough and associated with the radiolarian cherts and fine-grained limestones of the Piedmont. Similar ophiolites exist in the eastern Alps, and in the Rumanian Carpathians in the "diabases" of the eugeosyncline.

Some of the ophiolites of the Taurus and the Dinaric Alps of northern Greece and Yugoslavia are of early Cretaceous age. In Morocco, gabbros, diorites, and syenites were intruded in the Haut Atlas region during the Valanginian. However, throughout the Mediterranean region, early Cretaceous vulcanism was not strong in comparison to that of the late Jurassic.

Cenomanian flows and intrusions are reported in several localities, among them the Pyrenees and the great fault trough of Cantabria, but Austrian movements were accompanied by intrusions and flows of igneous rocks on a large scale only in the Anatolid portions of the Pontic trough of northern Turkey, although in the Carpathians several levels of tuffs and intrusions of dacite exist in Cenomanian and younger strata (Ksiazkiewicz, 1963, p. 560).

The largest and strongest manifestation of Cretaceous vulcanism in the Mediterranean region were connected with Laramide orogeny. Flows and intrusions of ophiolite of great thickness are known in and near the Taurus trough. Campanian-Maestrichtian limestones and clastics in the Pontic trough are intercalated with submarine flows of andesite and basalt. The ancient massifs of central Turkey were intruded by syenites and gabbros, and, according to some authors by granite. The Maestrichtian-Danian flysch of Yugoslavia is interbedded with rhyolite flows and tuffs. In the southern Italian Alps basaltic tuffs exist in the Maestrichtian-Danian section.

STRUCTURAL EVOLUTION OF THE MEDITERRANEAN REGION IN MESOZOIC TIMES

INTRODUCTION

The preceding review of Mesozoic se imentary environments provides more less factual information that can be used develop a hypothesis concerning the evo tion of the structural framework of t Mediterranean region. Conjecture plays important role, nevertheless, for the proc of the validity of some key assumptions c be found only in the stratigraphic successi of tracts now beneath the sea. Furthermo evaluation of the significance of "facts" a the assumptions made from them are su jective, and hence suspect.

This caveat given, premises fundamen to the development of the thesis can be st ed as follows. First, during the whole of t Mesozoic Era the basement massifs of Nor Africa, Asia Minor, and southern Euro did not change appreciably their positio relative to each other, and these positio were not far removed from those they no occupy. In other words displaceme between massifs has been minor, and th between the African and European co tinents not more than two hundred mil Most of the movement that has occurr took place in Tertiary times in connecti with crustal shortening during the princip phases of Alpine orogeny. In support of th view, bathymetric and deep crust seism velocity and thickness measurements in t western Mediterranean are not compatib with seafloor spreading. If such spreadi has begun in the essentially discrete bas between Sicily and the Levant, the relati position of the shores has not yet been chan ed appreciably. Furthermore, although strik slip faulting exists in North Africa, it do not involve displacements of great ma nitude.

Second, the proto-Atlantic was already existence late in Triassic times off both Spa and Morocco, the presence of evaporites this age suggesting restricted circulation waters off Portugal, and consequently th the seaway may have been narrow. By Jurassic times free circulation had been established, as shown by the existence of carbonates with open-sea faunas in both Portugal and Morocco, and the seaway had presumably broadened.

Third, published interpretations of both biofacies and lithofacies, in relation to sedimentary environments are on the whole correct. As a corollary, the *periode de vacuité* in geosynclinal development, recognized and named by Aubouin, is a valid concept.

Fourth, sedimentary processes in the past involved the same agents and took place at approximately the same rate as those active today. Consequently, to make inferences concerning tectonic activity from the stratigraphic succession is both feasible and reasonable.

The European practice of assigning standard stage designations to formational and larger units and referring to them by the stage name or names is not in accordance with the rulings of the Committee on Stratigraphic Nomenclature of the American Association of Petroleum Geologists. Although this practice is scientifically unsound, it leads to serious error only rarely. Difficulty arises most commonly when a formation changes in age from place to place. European practice has been adopted in this paper in order to use the voluminous literature, for otherwise this work could not have been undertaken.

ELEMENTS OF REGIONAL STRUCTURE

For the purposes of this study the major structural elements present in the Mediterranean region during the Mesozoic can be classified as:

1) *Precambrian and Paleozoic massifs.* These consist of metamorphosed sedimentary rocks and intrusive igneous rocks, mainly granite, along with minor amounts of extrusives. They are for the most part monolithic and some can be classed as *Zwischengebirge* or median massifs.

2) *Mobile Belts.* These linear belts, which were subsident troughs in Mesozoic times, include great thicknesses of sedimentary strata. Rapid vertical and lateral changes in the facies and thickness of these strata demonstrate the labile nature of their floors, on which ridges and furrows were common. Nearly all of these mobile belts are now the sites of mountain ranges and most can be called geosynclines. Both intrusive and extrusive igneous phenomena characterize portions of the belts during a part of their existence, and justify their differentiation into eugeosyncline and miogeosyncline. Throughout this discussion mobile belts are referred to as troughs, their actual state during the Mesozoic.

3) *Large fault-troughs.* In these major grabens and graben systems sedimentary rocks are commonly thick. All known grabens were actively subsident during only a portion of Mesozoic times. Some of them have since been raised into mountains, mainly by reversal of movement on bounding faults. Minor low angle thrusting has accompanied the main high angle reverse faulting, all of which is post-Mesozoic.

4) *Platforms and shelves.* Platforms are large, more or less linear areas, only slightly disturbed by orogeny, in which sedimentation was relatively uniform over extensive tracts. Epeirogeny and faulting may have recurred at intervals, however. Platforms are classed as either stable or unstable. Stable platforms are those on which the sedimentary section is relatively thin and incomplete because subsidence was slow and intermittent. Unstable platforms are those in which subsidence was relatively persistent, and the section is thick and more or less complete, although movement on faults and minor orogeny may have caused local hiatus. The term shelf, although inappropriate in the sense of continental shelf, is applied to the eastern Mediterranean sea-floor of which the structure is unknown.

5) *Basins.* As used here the term basin describes an irregularly shaped more or less persistent depression commonly bounded on

one or more sides by a massif or by a mobile belt. Sedimentation in most basins is interrupted by regression at times of regional epeirogeny. An exception is the Tuscan-Umbrian basin, caught up in Alpine orogeny, in which the section resembles that of the southern Alps, with which it should perhaps be included, as shown on figure 14. However, the Tuscan basin was not labile during Mesozoic times.

6) *Persistent Highs.* In addition to the Paleozoic massifs over which the sea advanced only at intervals during Mesozoic times, seldom covering them completely, the persistently emergent Sirte arch and Jauf-Mardin horst system are worthy of remark. Only Cenomanian and later seas occupied the area of the Sirte arch, which then was replaced by a complex graben system, and only those of the Maestrichtian covered the whole of the Jauf arch, although the Mardin basement high was invaded during the Neocomian.

7) *Faults.* Major fault systems and their trends are particularly significant in Africa where they reflect ancient fracture systems in the basement. The fractures played an important role in determining the location not only of horsts and grabens, but also of major features such as shorelines and igneous flows.

INDIVIDUALIZATION OF REGIONAL STRUCTURES

As a result of Hercynian orogeny the several massifs and the foreland of the African shield were already in existence when the Mesozoic Era began. The mobile belts that were to evolve on the peripheries of some of the massifs were foreshadowed by Carboniferous and Permian subsident troughs south of the Alpine massif and west of the Pannonian and other massifs in Yugoslavia. The development of these and similar but younger troughs is the topic of the following paragraphs.

During the Triassic, the Alpine and Dinaric troughs persisted as more or less linear sinking tracts on the edge of a shallow sea between the Bohemian massif to the north (the Alpine massif was submerged by the Anisian sea), the emergent Italian peninsula to the west, and the Pannonian and Rhodopian massifs to the east. This seaway was joined to that of southeastern Europe and the Levant through the Tethys which had remained in existence in the eastern Mediterranean region from Permian times.

The western Mediterranean was land during the late Paleozoic, but after the Werfenian the sea presumably began to cover it, for a strait very probably linked the Tethys with an area now occupied by the *Chaine calcaire* of the Rif mountains and the southern Betic mountains. Evidence for the existence of this seaway is tenuous, but reasonably convincing. The Betic-Rif area includes up to four thousand feet of Triassic carbonates with some faunal elements in common with those of the Alpine region, whereas in adjacent tracts the section is thinner and includes terrigenous clastics and evaporites. The presence of dolomites of late Triassic age in northern Tunisia at Jabal Ichkeul is in sharp contrast to evaporites of the same age farther south, and therefore Jabal Ichkeul may have been on the south flank of the connecting strait which must have been fairly broad and deep in order that a normal marine environment would persist when on both sides evaporites were being laid down. No evidence for the existence of this strait can be adduced in the Algerian portion of the Tellian ranges. On the other hand, if Aubouin's proposals concerning the outward migration of geosynclinal troughs and ridges are valid, the subsident belt may have been some distance north of the existing coast of Algeria.

On the west coast of Italy the allochthonous cherty, siliceous limestones of Triassic age at Lagonegro demonstrate the existence of open-sea conditions west of the peninsula while it was still emergent, for the base of the allochthonous section contains Ladinian fossils, whereas the autochthonous reefal limestones are Carnian and younger and in Calabria rest on basement. In Sardinia, the Triassic section is thin and in a Germanic facies. Did a subsea depression xist as early as Ladinian times between ardinia and the Italian mainland?

Thickening in a predominantly carbonate ection in northern Iraq and southern 'urkey, flanked by the Paleozoic massifs to he north and by thinner and somewhat ncomplete and locally evaporitic Triassic ections in central Iraq and northern Syria, uggests that the eastern portion of the 'aurus trough was a subsident tract late in 'riassic times.

In short, we have some basis for suppos- ng that the Betic-Rif region and the Alpine-)inaric area were persistently subsident dur- ng the medial and late Triassic while urrounding areas were either emergent (the nassifs) or relatively stable sea-covered racts (platforms and shelves). Tenuous vidence exists for the presence of other sub- ident tracts, one west of Italy, and another n the Taurus trough.

Note that no indication of persistent sub- idence is shown by the facies of Triassic trata in the Pyrenees and Atlas moun- ains, although terrigenous clastics collected n considerable thicknesses in basins and rabens in these areas.

The most striking single event in early urassic times was the appearance of the Haut and Moyen Atlas fault-troughs in Mo- occo and the Saharan Atlas fault trough in Algeria, while nearby in the Algerian Sahara leposition of evaporites continued without nterruption. The Moroccan troughs were ctively subsident only through the medial Bathonian, but in the relatively short period of time between the Sinemurian and Batho- nian more than twelve thousand feet of carbonates with subordinate clastics accu- nulated in the deeper portion of these inear, fault-bounded depressions. The Saha- an Atlas trough remained in existence hroughout the remainder of the Mesozoic, nd the eastern portion was below sea level ntil late in Eocene times. In it were leposited about thirty thousand feet of redominantly marine strata.

Of interest is the fact that marine deposi- ion in the Moroccan troughs ceased in Bathonian times, when coarse clastics deriv- ed from uplifted Paleozoic and Hercynian massifs in Morocco and Algeria were first laid down in the western portion of the Saharan Atlas trough.

The mobile belt of the Rif and its eastward prolongation in the Tellian Atlas trough, the Alpine trough, the Dinaric trough, the Taurus trough, and the Betic trough were all clearly defined by characteristic sedimentary environments in Jurassic times, the strata of Dogger and Malm age in particular exhibiting facies restricted to subsident tracts. Presumably the Carpathian-Balkan mobile belt and its extension in the Pontic trough were also in existence at this time, if not earlier, but data are meager.

Very common in some portions of the troughs listed is a thin and sometimes incomplete Dogger section (lower Malm in the eastern Taurus) associated in some localities with *ammonitico rosso.* According to Aubouin, the strata of Dogger age are thin because of their deposition during a *période de vacuité,* and are incomplete because currents removed some of the sediments laid down, particularly on submarine highs.

Kimeridgian vulcanism is prominent in the "inner" eugeosynclinal portions of many of the troughs, and was obviously accompanied by uplift which was not confined to the mobile belts but also affected many of the platforms and basins.

During Cretaceous times the mobile belts continued to evolve but no new ones came into existence. Widespread progressive broadening of the labile regions at the expense of the foreland started in Barremian times. Albian disturbance was considerably stronger and more nearly ubiquitous. In the Alpine and Dinaric troughs the first compressional island cordilleras presumably developed then, and the first true flysch deposits were laid down. This Austrian orogenic phase affected the entire Mediterranean region, for when compressional forces were relaxed the Sirte horst and graben system, the Cantabrian graben, and the north Pyrenaean graben came into existence. At this same time the central portion of the Pannonian-Transylvanian massif began to sink, leaving stable remnants relatively undisturbed north, east, and south of the subsident tract.

PERIOD		CHRONOLOGY AND STAGES	EQUIVALENCIES	EFFECTS OF DIASTROPHISM ON FACIES AND PALEOGEOGRAPHY	DIASTROPHISM AND VULCANISM	
		65				
CRETACEOUS	UPPER	MAESTRICHTIAN	SENONIAN	Sea withdraws from much of Europe, Africa, Middle East in late Maestrichtian. Sirte trough, Cyrenaica, North Italy, and Rif and Tellian troughs remain submerged. Widespread chalky limestones deposited by transgressive Campanian-Maestrichtian sea. Senonian biostromes common. Vulcanism mainly Maestrichtian on platforms, late Senonian in troughs.	LARAMIDE Ophiolite intrusions and flows in Turkey, Syria. Olistostromes in Taurus. Metamorphism and vulcanism on massifs.	ALPINE
		70				
		CAMPANIAN				
		76				
		SANTONIAN		Santonian uplift in Tunisia and on Ain M'lila môle, folding in N.E. Egypt. Reefal limestones common in North Africa. Locally shales, gypsum.	SUBHERCYNIAN Thrusting in Carpathians. Faulting and igneous intrusions in southern Carpathians. First nappes in Austrian Alps.	
		82				
		CONIACIAN		Coniacian is absent in many localities because of uplift, minor folding during and after Turonian. Senonian transgression, accompanied by basal clastics locally, attained maximum in Maestrichtian when sea covered some regions not reached by very extensive Cenomanian-Turonian flooding which laid down limestones and shales. In Tellian Atlas transgression began in late Albian. European massifs generally submerged. Fault troughs formed early in Cenomanian times in Libya, northern Spain, Aquitaine basin. Faulting in northern Egypt, Morocco.		
		88				
	MIDDLE	TURONIAN			PRE-GOSAU Unconformity on shelves Africa and Middle East. Post-Turonian folding in Egypt and Levant. Disturbance in Carpathians. Unconformity in Alps. Ophiolites and flysch in Alps, Carpathians	
		94				
		CENOMANIAN	GARDONIAN			
		100	VRACONIAN			
		ALBIAN		Albian uplift and disturbances nearly ubiquitous in Mediterranean region caused deposition of terrigenous clastics widely. Isthmus of Durance emerges, linked with Catalan massif (?) West Portugal rises.	AUSTRIAN	
		106				
	LOWER	APTIAN	GARGASIAN / BEDULIAN	Aptian transgression very common throughout Mediterranean region, mainly littoral limestones. Intra-Aptian disturbance in the Pyrenees, Tunisia.	INTRA-APTIAN Tectonism in Rumanian Carpathians. Local uplift in North Africa	
		112				
		BARREMIAN	NEOCOMIAN	Barremian disturbance in Dinaric and Rumanian Carpathian troughs shown by sandstones in carbonate section. Faulting and uplift in northwest Libya, Algeria, southern Tunisia shown by widespread cgls. and sandstones.	BARREMIAN Uplift and faulting in Algeria and Tunisia. Pelagonian massif brought above sea level. Ophiolite intrusions and tilting in western Alps and Carpathians. Uplift of African shield.	MESOZOIC
		118				
		HAUTERIVIAN	WEALDEN			
		124				
		VALANGINIAN		Wealden sandstones common in areas raised by late Cimmerian orogeny (Spain, western Algeria). Deposition of tintinnid-bearing calcilutites continued in undisturbed tracts. Calcareous shales in part of Alpine trough. Catalan massif remained emergent.		
		130				
		BERRIASIAN				
		136				
JURASSIC	UPPER	PURBECKIAN	MALM / TITHONIAN	Most of Iberian peninsula and Aquitaine basin rise above sea level. General withdrawal of sea from much of North Africa accompanied by deposition of clastics. Jauf-Mardin high mainly emergent. Transgression of Tithonian calpionellid-bearing limestones in Europe.	LATE CIMMERIAN Uplift in Aquitaine basin and Iberian peninsula. Intrusions of diorite and gentle folding locally in Morocco.	
		141				
		PORTLANDIAN				
		146				
		KIMERIDGIAN		Retreat of sea starts in Egypt, Levant. Widespread radiolarites in major European troughs. Epeirogenic uplift and minor folding in many localities, including Morocco.	NEVADAN Ophiolite flows and intrusions widespread in European mobile belts. Diorite intrusions in Morocco. Lavas in Lebanon	
		151				
		OXFORDIAN	RAURACIAN ARGOVIAN DIVESIAN	Continued uplift in northern and central Spain. Minor transgression in northeastern Morocco.		
		157				
	MIDDLE	CALLOVIAN	DOGGER	Beginning of uplift in northern Spain and coastal basins of Portugal. Transgression in western Morocco.	CALLOVIAN	
		162				
		BATHONIAN		Retreat of the sea from interior of Morocco. Sandstones widespread in North Africa. Dogger missing or much reduced in European and Tellian troughs, and in Pontic trough. Interruptions in sedimentation and condensation in Iraq.	INTRA-DOGGER Uplift of Paleozoic and Pre-Cambrian massif in the Sahara. Vulcanism in Sicily. Local unconformity.	
		167				
		BAJOCIAN				
		172				
	LOWER	AALENIAN	LIASSIC	Development of Pindus trough in Greece. Toarcian <u>ammonitico rosso</u> widespread in all mobile belts. Short withdrawal of sea from Morocco.		
		TOARCIAN			TOARCIAN Block movements in major European troughs.	
		178				
		DOMERIAN	CHARMOUTH.	Block movement in Tunisia (Domerian). Sandstones in Majorca. Emergence of Minorca.		
		PLIENSBACHIAN		Return of sea to Provence. Limestone deposition in western Europe	PLIENSBACHIAN Uplift north of Majorca.	
		183				
		LOTHARINGIAN				
		SINEMURIAN		Corsica emerges and remains above sea level until Kimeridgian. Lias transgression in Egypt. Normal salinity returns to the hypersaline seas of western Europe and Morocco. Limestones of Lias age widespread. Provence emergent. Basal Lias clastics on flanks of mobile belts.	Development of Tellian and Atlas troughs in North Africa.	
		188				
		HETTANGIAN			Uplift in Provence.	
		195				
TRIASSIC	UPPER	RHAETIAN	KEUPER	Fine terrigenous clastics (Rhaetian) in predominantly carbonate section locally in Alpine and Kinaric troughs. Evaporites common in Carnian strata of Greece, Italy, Sicily, eastern Europe, succeeded by carbonates. Salt and anhydrite represent most of the Keuper in Algeria, northern Spain. Carnian transgression in Algeria and Tunisia.	EARLY CIMMERIAN Folding in Dobrogea. Ophiolite flows and intrusions widespread. Basalt flows in Morocco.	HERCYNIAN
		NORIAN				
		CARNIAN				
		205				
	MIDDLE	LADINIAN	MUSCHELKALK	Almost universal regression at end of Ladinian times. Shallow water marine limestones characteristic of southwestern Europe. Deeper water limestones in the east European and Turkish troughs. Widespread transgression due to change in sea level. Mainly limestones deposited.	Minor orogeny and folding in Dobrogea. Tuffs and flows in Ladinian of Italy; European troughs.	
		ANISIAN (VIRGLORIAN)				
		215				
	LOWER	WERFENIAN (SCYTHIAN)	BUNTSANDSTEIN	Terrigenous clastics ubiquitous in western Europe, laid down under continental and fluviatile-littoral conditions. Slight transgression in eastern Europe where most strata are carbonates. Sedimentation continuous from Permian in Alpine troughs. Conglomerates and sandstones on flanks of Alpine trough where Triassic rests on Paleozoic unconformably.	Local uplift in Yugoslavia. Numerous flows of basalt and spilite in Spain, Morocco, southern France. Other flows in Greece, Turkey, the Levant. PALATINE	
		225				

Fig. 12. Mesozoic diastrophism in the Mediterranean region.

The effects of Pre-Gosau tectonism are s obvious in the mobile belts than on the ıtforms and in the basins where general lift and minor folding occurred during and the end of Turonian times. However, the ;t nappes then developed in the eastern ps, and short-lived emergence was com- ›n in all mobile belts. Vulcanism on a ge scale began in the Pontic ranges.

In Coniacian-Santonian times strong lift took place on the ridges of the Hellen- portion of the Dinaric trough in connec- n with Subhercynian tectonism, but in ›st troughs the effects of this phase are ısked by those of the much stronger Lara- de orogeny when vulcanism was very ›minent in the Pontic and Taurus troughs, d flysch deposits were common in most the mobile belts of southern and eastern ırope. Extensive areas rose above sea level the end of the Mesozoic, including many the troughs.

Faults are an important tectonic element the North African foreland where the sement and its cover of nearly flat-lying liments have been fractured in a complex ttern. Four major trends of faulting have en found. None is confined to one geograph- area, but certain trends are more com- ›n in one locality than in another. The ıut Atlas of Morocco, which began to rise ring late Cretaceous times is bounded by ılts trending east-northeast. The Gargaf ticline of Libya has about the same trend, do the Tinrhert and Messak scarps. This ection, or one even more easterly (N. 85°), is that of many faults on the Hauts ateaux and in other localities in Algeria, rticularly on the flanks of the Tellian ›ugh. The major axis of the Eglab-Yetti ıssif also has this trend.

The Moyen Atlas in Algeria is bounded faults directed northeast-southwest or rth northeast-south southwest. Similar ections are known in faults on the Hauts ateaux (N. 55°E.), on the east flank of the ırzuk basin, and on the Tibesti massif, as ll as in the Ennedi-Auenat uplift east of Kufra basin. They are also found in the te embayment, and in the Gulf of Aqaba- ad Sea graben system.

The bounding faults of the Sirte embay- ment are directed north northwest--south southeast. They are more or less parallel to those delimiting the Gulf of Suez, and the western edge of the Messak.

A generally north-south fault trend is reported in many structures in the Algerian Sahara. The best known of these is the Tihemboka-Edjele axis, on wich some individual structures are directed north northwest. Many north-south trending faults are known in subsurface, where they help to delimit oil-producing structures. Faults bounding the Jauf-Mardin high also prob- ably trend north-south, or slightly east of north.

Generally speaking, the north-south and east of north fault directions appear to be more numerous in Algeria and Morocco than they do in Libya and Egypt, where direc- tions west of north predominate, with southwest-northeast trends less conspicuous, although present. Strike-slip displacement along faults with this trend have been reported in Libya.

Some of the movement on the faults exhibiting the several trends can be tied to Mesozoic tectonism. Tertiary diastrophism has certainly caused additional play on many faults and probably formed new ones.

PERSISTENCE AND STABILITY IN STRUCTURAL ELEMENTS

The introductory remarks of this paper stress the fact that Precambrian and Hercy- nian massifs by their presence alone restrict areas of potential crustal mobility. The valid- ity of this concept is confirmed for most of the Mediterranean region in the review of its stratigraphic and diastrophic history. This summary shows that the mobile belts (Mesozoic troughs) commonly found on the flanks of massifs became wider only at the expense of the foreland, never of the massif. However, not all massifs are bounded by mobile belts (*e.g. Massif central*), nor were all massifs completely stable. Some small massifs in the Alps have been caught up in Alpine orogeny; others, much larger, have sunk to considerable depths. One of these

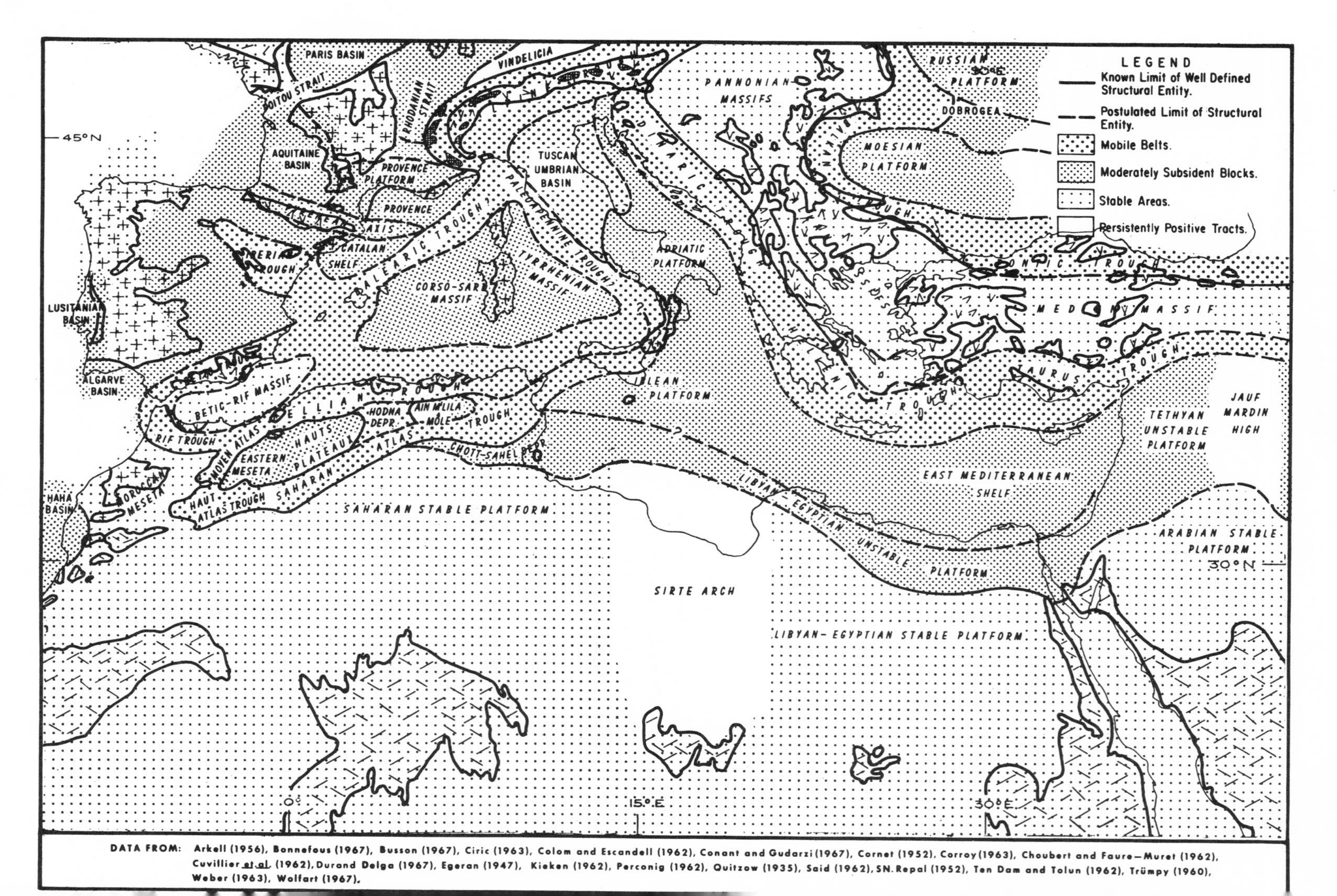
LEGEND
Known Limit of Well Defined Structural Entity.
Postulated Limit of Structural Entity.
Mobile Belts.
Moderately Subsident Blocks.
Stable Areas.
Persistently Positive Tracts.
PARIS BASIN
POITOU STRAIT
AQUITAINE BASIN
RHODANIAN STRAIT
PROVENCE PLATFORM
PROVENCE
VINDELICIA
TUSCAN UMBRIAN BASIN
PALEOAPENNINE TROUGH?
TYRRHENIAN MASSIF
CORSO-SARD MASSIF
BALEARIC TROUGH
CATALAN SHELF
LUSITANIAN BASIN
ALGARVE BASIN
BETIC-RIF MASSIF
RIF TROUGH
MOYEN ATLAS
EASTERN MESETA
HAUTS PLATEAUX
HODNA DEPR.
AIN M'LILA MÔLE
TROUGH
ATLAS
SAHARAN
HAUT ATLAS TROUGH
MOROCCAN MESETA
HAHA BASIN
CHOTT-SAHEL
SAHARAN STABLE PLATFORM
ADRIATIC PLATFORM
PLATFORM
PANNONIAN MASSIFS
RUSSIAN PLATFORM
DOBROGEA
MOESIAN PLATFORM
BALKAN
MEDITERRANEAN MASSIF
TAURUS
TROUGH
JAUF MARDIN HIGH
TETHYAN UNSTABLE PLATFORM
EAST MEDITERRANEAN SHELF
ARABIAN STABLE PLATFORM
LIBYAN-EGYPTIAN UNSTABLE PLATFORM
SIRTE ARCH
LIBYAN-EGYPTIAN STABLE PLATFORM
45°N
30°N
0°
15°E
30°E
DATA FROM: Arkell (1956), Bonnefous (1967), Busson (1967), Ciric (1963), Colom and Escandell (1962), Conant and Gudarzi (1967), Cornet (1952), Corroy (1963), Choubert and Faure-Muret (1962), Cuvillier et al. (1962), Durand Delga (1967), Egeran (1947), Kieken (1962), Perconig (1962), Quitzow (1935), Said (1962), SN. Repal (1952), Ten Dam and Tolun (1962), Trümpy (1960), Weber (1963), Wolfart (1967).

s now the Pannonian-Transylvanian basin. The thin sedimentary cover of small massifs on the periphery of this basin indicates that Mesozoic seas covered them only at intervals until Cenomanian times when the block between them began to sink. This subsident tract is now covered by thirty thousand feet of late Mesozoic and Tertiary sedimentary rocks.

Another sunken tract, which probably began to subside only late in Tertiary times, is the northern Aegean sea in which granite islands suggest the existence of a flooded portion of the Rhodopian massif.

In contrast to these down-dropped blocks are the Haut Atlas and Pyrenaean regions, both linear, and both sites of Hercynian orogeny which to some degree cratonized both tracts. The two are now mountain ranges because of the effects of strong reverse faulting during the Tertiary. This uplift began at the same time in the Pyrenees and in the Haut Atlas (post Lutetian). The history of the two ranges differs in that the eastern Haut Atlas was actively subsident during the early and medial Jurassic, whereas the Pyrenees sank no more rapidly than adjacent basins and rose above them late in Jurassic and late in Cretaceous times.

The Saharan Atlas differs from both the Haut Atlas and the Pyrenees for it was a rapidly subsident, eastward-tilted graben during Jurassic, Cretaceous, and early Tertiary times, but is like them in that it exists as a mountainous region because of Tertiary reverse faulting.

The compressional forces presumably responsible for the reverse faulting in the graben systems probably also caused acceleration of the rate at which folds and overthrusts were formed in the mobile belts. Some of these island cordilleras and thrust-slide plane sheets were already in existence late in Cretaceous times, and others developed in the early Tertiary of the European mobile belts which rose above sea level during the Oligocene and Miocene.

The platforms of North Africa, which are the foreland of the African shield, have maintained their identity and relative stability since the end of Paleozoic times, with faulting and epeirogeny responsible for most of their deformation. Other platform-like areas have been less calm tectonically. For example, a portion of southeastern France began to be downwarped late in Malm times, and continued to sink throughout the early Cretaceous. The subsidence of this Vocontian trough was countered during Albian-Senonian times by the uplifting of the Isthmus of Durance. Similar strong differential movements occurred in the Aquitaine basin, and others on the Moesian platform and in Dobrogea.

TECTONISM AND TRANSGRESSION

Figure 12, "Mesozoic Diastrophism in the Mediterranean Region", illustrates graphically the relative importance of regional diastrophic events. The length of the recumbent V's on the right shows the relative strength of the tectonic pulses, and the width of the "V's" the approximate duration of each phase. Slanting lines in the apex indicate igneous activity. The commentary describes, albeit with inaccurate generalization and the omission of some important points, the location of the phenomena and their effects on facies and paleogeography.

Alternation of diastrophic disturbance and transgression during late Jurassic and Cretaceous times is immediately apparent from figure 12. Transgression followed each tectonic pulse, but the advance of the sea was not necessarily greatest in those tracts which had been most strongly involved in tectonism. Nevadan (Kimeridgian) movement (characterized by ophiolite flows and intrusions in many mobile belts and widespread epeirogenic uplift) was immediately succeeded by Tithonian transgression. The tintinnid-bearing calcilutites laid down by the late Malm-early Neocomian seas are widespread in the Mediterranean. Weak late Cimmerian movement had little effect in the tracts covered by these seas, although regional uplift was widespread at the end of the Malm. Barremian epeirogenic movement in much of North Africa and eastern Europe was followed by Aptian transgression. Much stronger and more nearly ubiquitous Aus-

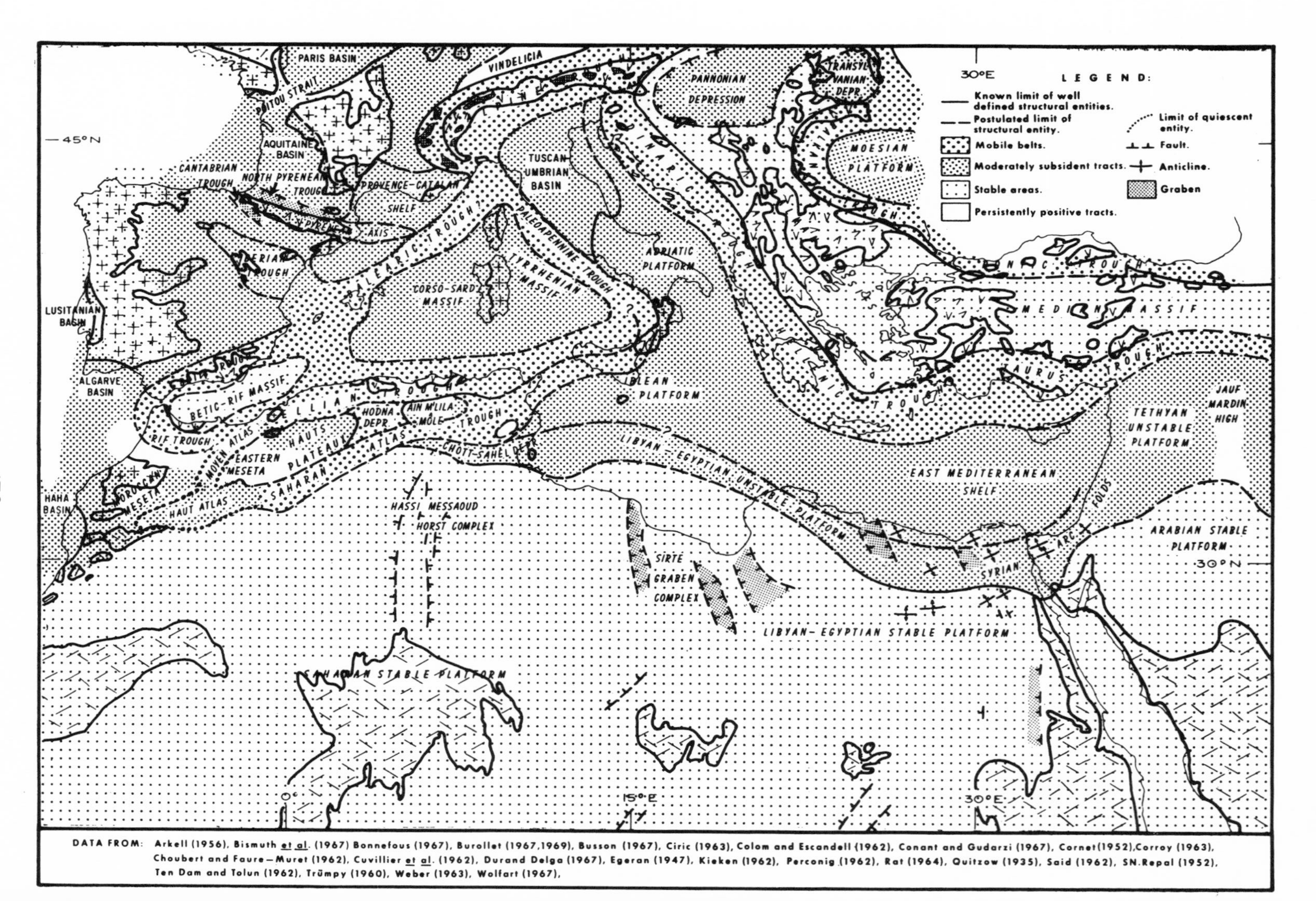

DATA FROM: Arkell (1956), Bismuth et al. (1967) Bonnefous (1967), Burollet (1967,1969), Busson (1967), Ciric (1963), Colom and Escandell (1962), Conant and Gudarzi (1967), Cornet(1952),Corroy (1963), Choubert and Faure—Muret (1962), Cuvillier et al. (1962), Durand Delga (1967), Egeran (1947), Kieken (1962), Perconig (1962), Rat (1964), Quitzow (1935), Said (1962), SN.Repal (1952), Ten Dam and Tolun (1962), Trümpy (1960), Weber (1963), Wolfart (1967).

·ian tectonism, here considered the first hase of the Alpine cycle, gave way to the :eat late Albian-Cenomanian transgression. .t this time the area of the Mesozoic sea ·as at its apogee. Turonian-Coniacian (Pre-·osau) diastrophism caused disturbances of ıe sea floor that resulted in the emergence ot only of broad areas of platform, but also f newly formed anticlinal crests. Weaker ıovements during the Santonian were ıcceeded by Campanian-Maestrichtian in-ırsions. Maestrichtian seas were second in rea only to those of the Turonian, and, ıdeed, many tracts were inundated during ıe late Senonian that had been land since)evonian or Carboniferous times. Other ·acts brought above sea level during the 'uronian remained emergent throughout the :mainder of the Cretaceous, however.

In short, the increasing strength of iastrophism in late Jurassic-Cretaceous ·mes was matched by the progressively lar-er areas covered by the trangressive seas in ıe intervals of time between major tectonic hases. At the end of the Mesozoic Era, early all of the platforms rose above sea :vel, but some of the mobile belts were :ill occupied by marine waters.

YPOTHESIS OF STRUCTURAL EVOLUTION N THE MEDITERRANEAN

The western and eastern Mediterranean :a differ in many respects and must be onsidered as discrete structural entities. Iany lines of evidence substantiate this state-ıent, including paleogeography, bathymetry, nd seismic data from various sources. The ·estern Mediterranean is by far the more omplex, for it is bounded on two sides by ıobile belts now uplifted into folded moun-ains. During the Cretaceous another mobile ·elt extended northeastward from the Betic rough through the Balearic islands and ·erhaps linked with the Alpine geosyncline see fig. 14). The eastern Mediterranean is ·ounded only to the north by a mobile hrust front — the Hellenic and Taurus anges. To the south is the unstable Egyp-ian shelf and to the east lie the north-south rending fault-block mountains of the Le-·anon.

Bathymetry shows the floor of the western Mediterranean to be composed of a number of flat basins separated by intermediate higher areas. The largest basin is the Balearic Plain which extends from the northern margin of the Ligurian Sea southward between Sardinia and the Balearic Islands to the continental rise of North Africa. This plain lies almost eleven thousand feet below sea level. A second smaller plain in the center of the Tyrrhenian Sea is over fourteen thousand feet below sea level. The flat floors of these basins result from recent sedimentation, and individual layers can be correlated for twenty miles or more (Hersey, 1965). Basement lies at a considerably greater depth, and the intervening strata are presumably sedimentary rocks. Between Corsica and the coast of France the deeper sedimentary layers include seismic reflection anomalies thought to represent diapiric salt (Leenhardt, 1969) which is considered to be of Triassic age by Glangeaud, *et al.* (1967, p. 932). In other localities in the western Mediterranean the sea floor is underlain by tectonically disturbed sedimentary strata in which diapirism is not involved.

According to Payo (1967, p. 171) the deep crust of the western Mediterranean is almost oceanic in character, for it is thin, and deep seismic wave velocities in it are high.

In the eastern Mediterranean the thrust front of the Hellenic and Taurus ranges is paralleled to the south by the "Hellenic Trough", an area of negative gravity anomalies. This trough is in turn bounded to the south by the "Mediterranean Ridge", an elevated linear feature nearly one hundred miles wide that extends from southern Italy in an arcuate curve paralleling the ranges of Greece and southern Turkey. The Mediterranean Ridge, almost six thousand feet below sea level, rises nearly two thousand feet above the narrow abyssal plain to the south. The ridge is broadly equidistant from Africa and Europe. From the abyssal plain, an almost uniform slope rises to the continental shelf of Egypt. Deltaic deposits from the Nile are thought to underlie nearly all of this slope (Emery, *et al.,* 1966). Al-

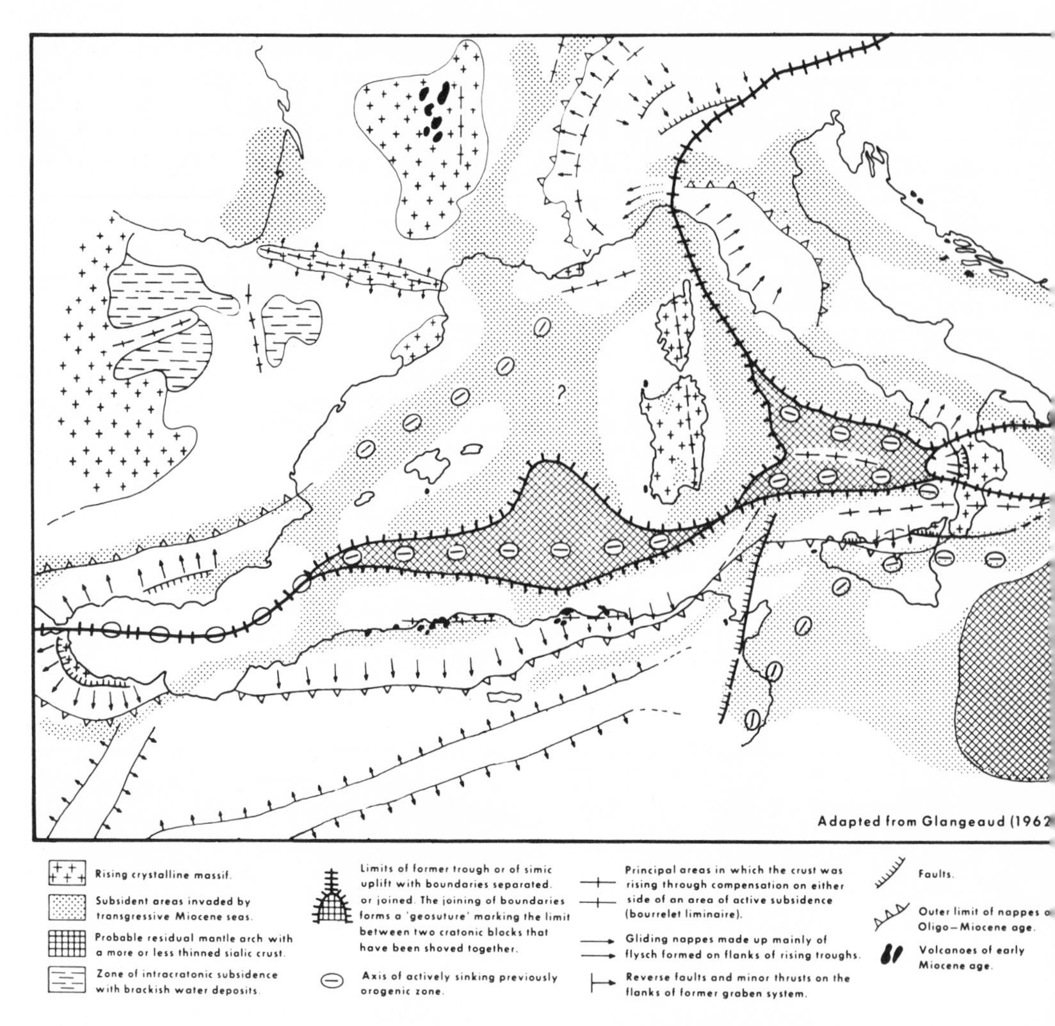

Fig. 15. Miocene structure in the Western Mediterranean.

though coastal Egypt is known to be broken into a number of horsts and grabens active during the Mesozoic and early Tertiary, no evidence of their existence can be seen in the sea floor profiles owing to the masking effect of the deltaic deposits.

The tectonic nature of the Mediterranean Ridge is unknown. Some opinion has it that it may represent the first stage of sea-floor spreading. Others hold that it is the result of underthrusting from the south, with consequent piling up of strata in the area of thrust. One recent suggestion is that t trough fronting the mountains is like t trenches in the Pacific where downwa movement of crustal material from sea-flo spreading occurs. In this event the rid represents a piling up of sediments scrap from the conveyor belt of descending cru An older view is that the ridge is an ea ward prolongation of the Apennines.

In any event, the floors of the weste and eastern Mediterranean are different aspect. Payo (1967, p. 171) says that t ıst is thicker in the eastern Mediterranean, d has lower velocities, especially in the dimentary strata.

These meager data are not in disagree- ent with a hypothesis of structural evolu- n based on criteria completely discrete m those mentioned above. The hypothesis ı be summarized as follows:

The existing structural configuration of e Mediterranean basin is a result of the ects of alternations in the predominance tensional and compressional forces on tonic elements brought into existence by ercynian orogeny. An increase in the ength and tempo of the compressional ases during the Cretaceous and early Ter- ry caused mobile belts on the peripheries cratons to narrow, so that their sedimen- y fill buckled and ultimately was partly ected by thrusting and gravity sliding. mpression of sediment-filled ancient fault oughs and of other linear fault-bounded cts that had been sites of Hercynian oro- ny caused similar phenomena. These how- er, were not predominantly unidirectional, e those in the mobile belts, but rather ted equally on both flanks. In these oughs and sites of ancient orogeny the dimentary fill was raised through reversal the direction of movement on old faults. rusting and gliding are secondary effects, led in many localities by the existence of iassic salt or anhydrite which served as a ane of thrusting and as a lubricant for avity sliding.

During the times that tensional forces edominated, fractures induced or enlarged earlier compression permitted play on ult blocks. These movements began early the Lias and continued intermittently roughout the Mesozoic and Tertiary. The lative movement on fault blocks was par- ularly intense after the completion of the ajor phases of Alpine orogeny late in Mio- ne times.

A relationship appears to exist between rong subsidence and tracts bounded on ree or more sides by mobile belts. Two these tracts, the Pannonian-Transylvanian sin and the central portion of the western editerranean sank very rapidly, the first starting its descent during the late Creta- ceous, and the second probably late in Mio- cene times. Is it possible that deep convec- tion currents associated with the develop- ment of mobile belts eventually absorbed or stoped away the continental crust underlying these tracts? The oceanic-type crust under the western Mediterranean suggests that something of this sort occurred. In any event, subsidence in both areas began after the termination of a compressional phase of considerable strength.

Transcurrent faulting on the foreland of the African shield, as opposed to normal faulting, presumably was of importance mainly in Tertiary times when most of the movement between the African and Europe- an continents appears to have occurred. A discussion of this matter may be found in Dubourdieu (1962).

In the presentation of this hypothesis, the fact that portions of the geosynclines bounding the Corso-Sard and Tyrrhenian massifs (see fig. 14) sank with these massifs was not mentioned. That these massifs exist- ed is indisputable for allochthonous terri- genous clastics of Jurassic through Oligocene age that can only have come from an emer- gent tract to the north are found in the onshore (miogeosynclinal) portion of the Rif-Tellian-North Sicilian orogen, and Dog- ger sandstones in the essentially open-sea Jurassic sequence of Elba were certainly derived from the west (Bodechtel, 1964, p. 28). Similarly, allochthonous late Mesozoic and Tertiary terrigenous clastics found on- shore farther north on the Italian coast must have come from the west. The existence of a eugeosyncline in the Tyrrhenian and Ligu- rian Seas is suggested by the occurrence of Malm diabase and spilite on Elba, as well as the allochthonous open-sea Mesozoic sec- tion at Lagonegro.

No similar evidence is available to dem- onstrate the existence of the Balearic trough shown on figures 13 and 14. However, Colom and Escandell (1962, p. 133) pos- tulate a seaway in the general area during the Jurassic and again in the Miocene. As the Miocene thrusting on Majorca is predom- inantly to the northwest, a stable massif to the southeast (Corso-Sard massif) and a foreland to the northwest (Catalan massif) can be postulated.

Only a small part of the hypothesis is original, for it combines and attempts to reconcile the opinions of many authors concerning various facets of the evolution of the Mediterranean basin and the structural elements thereof. Almela, Aubouin, Beloussov, Boucart, Burollet, Caire, Castany, Colom, Dubertret, Durand Delga, Glangeaud, Grandjacquet, Kieken, Llopis-Llado, Merla, Van Bemmelen, Trümpy, and a number of others have studied the several aspects of this problem. Older workers whose opinions are to some extent no longer considere valid include Argand, Cueto, Kober, Russ Staub, Stille, and Termier.

ACKNOWLEDGEMENTS

In conclusion, the author must expre his thanks to the Amoco International C Company, and in particular to Dr. Willia E. Humphrey, Vice President, Exploratio for permission not only to publish this pa er, but also to use the stenographic, bibli graphic and drafting facilities of that orga ization in its preparation.

BIBLIOGRAPHY

ALASTRUÉ, E., 1956, *Las Cordilleras Béticas*: *In* Memorias de Instituto Geológico y Minero de Espana, Tomo LVII; El Cretáceo en Espana, p. 303-322, 1 table.

ALMELA, A., 1969, *Las corrientes de convección en el manto y la tectónica pirenaica*: *In* libro homenaje al Prof. D. Obdulio Fernandez con motivo del cinquantenario de su ingreso en la Real Academia de ciencias exactas, fisicas y naturales, p. 475-480.

ANDRUSOV, D., 1963, *Les principaux plissements alpins dans le domaine des Carpathes occidentales*: *In* Livre à la mémoire du Prof. Paul Fallot, t. 11, p. 519-528, (Mém. hors série de Soc. Géol. Fr.).

——, 1967, *Aperçu général sur la géologie des Carpathes occidentales*: Bull. Soc. Géol. Fr., ser. 7, t. 7, p. 1029-1062, figs. 1-12 (1965).

ARAMBOURG, C., DUBERTRET, L., SIGNEUX, J. and J. SORNAY, 1959, *Contributions à la stratigraphie et à la paléontologie du Crétacé et du Nummulitique de la marge N.W. de la péninsule arabique*: Notes et Mémoires sur le Moyen Orient, Museum National d'Histoire naturelle, t. VII, part. I. Stratigraphie (L. Dubertret), p. 193-220.

ARKELL, W.J., 1956, *Jurassic Geology of the World*: p. i-xv, 1-806, figs. 1-101, tables 1-27, pls. 1-46, Oliver and Boyd, Ltd. Edinburgh.

AUBOUIN, J., 1959, *Sur la géologie de la zone littorale du Monténégro: les bouches de Kotor (Yougoslavie). Comparaison avec les séries héllèniques correspondante*: Bull. Soc. Géol. Fr., p. 833-840, figs. 1-2.

——, 1960, *Essai sur l'ensemble italo-dinarique et s rapports avec l'arc alpin*: Bull. Soc. Géol. F ser. 7, t. II, p. 487-526, figs. 1-3, table 1.

——, 1961, *Propos sur les géosynclinaux*: Bull. S Géol. Fr., ser. 7, t. III, p. 629-711, figs. 1-18.

——, 1963, *La tectonique de la Méditerranée moyen et les séismes*: Bull. Soc. Géol. Fr., ser. 7, t. p. 1124-1129.

——, 1964, *Essai sur la paléogéographie post-triasiq et l'évolution sécondaire et tertiaire du versa sud des Alpes orientales (Alpes méridionale Lombardie et Vénétie, Italie, Slovénie occide tale, Yougoslavie)*: Bull. Soc. Géol. Fr., ser. t. V, p. 730-766, fig. 1, foldout 1, (1963).

——, 1965, *Réflexions sur le faciès "ammonitico r so"*: Bull. Soc. Geol. Fr., ser. 7, t. VI, p. 47 501, figs. 1-10, (1964).

——, 1965 a, *Geosynclines*: p. i-xv, 1-335, figs. 1-6 Elsevier Publishing Company, Amsterdam.

AUBOUIN, J., BRUNN, J.H., CELET, P., DERCOURT, GODFRIAUX, I. and J. MERCIER, 1963, *Esquis de la géologie de la Grèce*: *In* Livre à la m moire du Prof. Paul Fallot. t. 11, p. 538-61 figs. 1-5 (Mém. hors série de la Soc. Géol. Fr

AUBOUIN, J., CADET, J.P., RAMPNOUX, J.P., DUBAR, and P. MARIE, 1965, *A propos de l'âge de série ophiolitique dans les Dinarides Yougoslave la coupe de Mihajlovici aux confins de la Serb et du Monténégro (région de Plevlja, Yougos vie)*: Bull. Soc. Géol. Fr., ser. 7, t. VI, p. 10 112, text figs. 1-2, (1964).

AUBOUIN, J. and DERCOURT, J., 1966, *Sur la géologie de l'Egée: regard sur la Crète (Grèce)*: Bull. Soc. Géol. Fr., ser. 7, t. VII, p. 787-821, figs. 1-10, (1965).

AUBOUIN, J. and MENNESSIER, G., 1963, *Essai sur la structure de la Provence*: *In* Livre à la mémoire du Professeur Paul Fallot, t. 11, p. 45-98, figs. 1-4, pl. I.

AUBOUIN, J. and NDOJAJ, I., 1965, *Regard sur la géologie de l'Albanie et sa place dans la géologie des Dinarides*: Bull. Soc. Géol. Fr., ser. 7, t. VI, p. 593-625, figs. 1-8.

AWAD, G. and SAID, R., 1966, *Lexique stratigraphique international,* Vol. IV, Afrique, Fasc. 4b, Egypt. p. 1-73, table I.

BANCILA, I.N., 1968, *Aperçu sur la tectonique de la R.S. Roumanie*: XXIII International Geological Congress, Proc. Sect. 3, Orogenic Belts, p. 107-120, figs. 1-2.

BARBIER, R., BLOCH, J.P., DEBELMAS, J., ELLENBERGER, F., FABRE, J., GIDON, M., GOGUEL, J., GUBLER, Y., LANTEAUME, M., LATREILLE, J. and M. LEMOINE, 1963, *Problèmes paléogéographiques et structuraux dans les zone internes des alpes occidentales entre Savoie et Méditerranée*: *In* Livre à la Mémoire du Professeur Paul Fallot, t. 11, p. 331-377, figs. 1-7, pls. I-II.

BÄR, C.B. and KLITSCH, E., 1964, *Introduction to the Geology of Egypt*: In Guidebook to the Geology and Archaeology of Egypt, Petroleum Exploration Society of Libya, Sixth Annual Field Conference, p. 71-98, figs. 1-6, pls. I-IV.

BARR, F.T., 1968, *Upper Cretaceous Stratigraphy of Jabal al Akhdar, Northern Cyrenaica*: *In* Geology and Archaeology of Northern Cyrenaica, Libya, Petroleum Exploration Society of Libya, Tenth Annual Field Conference, p. 131-142, figs. 1-9, pls. 1-3.

BARR, F.T. and GOHRBANDT, K.H.A., 1967, *Thomasinella punica, a Tethyan Foraminifer from the Cenomanian of Central Tunisia*: *In* Guidebook to the Geology and History of Tunisia, Ninth Annual Field Conference, Petroleum Exploration Society of Libya, p. 153-158, figs. 1-4.

BEHRMANN, R.B., 1938, *Appunti sulla geologia della Sicilia centro-meridionale*: Vacuum Oil Co. S.A. 1. p. 1-60, figs. 1-2, pls. 1-5, including colored geologic map at 1:100,000.

BELOUSSOV, V.V., 1962, *Basic Problems in Geotectonics*: p. I-XVI, 1-816, figs. 1-318, English Edition, McGraw-Hill Book Co., Inc., New York.

BENDER, F., 1968, *Geologie von Jordanien*: p. i-vii, 1-230, figs. 1-156, pls. I-V, tables 1-14; in pocket, geologic map, table 16, figs. 117-a-b-c. Band 7, Beiträge zur Regionalen Geologie der Erde, Borntraeger, Berlin-Stuttgart.

BENTOR, Y.K., GRADER, P., PARNES, A., REISS, Z., SHIFTAN, Z. and A. VROMAN, 1960, *Lexique stratigraphique international,* Vol. III Asie, Fasc. 10c2, p. 1-150, 2 maps, 1 table in pocket.

BEYDOUN, Z.R., 1965, *A Review of the Oil Prospects of Lebanon*: Fifth Arab Petroleum Congress, Proceedings, ser. No. 29 (B-3), p. 1-18, pls. 1-2.

BIROT, P. and J. DRESCH, 1953, *La Méditerranée et le Moyen-Orient*: t. 1, La Méditerranée occidentale, Presses Universitaires de France, p. i-viii, 1-552, figs. 1-56, pls. I-XII.

BISMUTH, H., BONNEFOUS, J. and DUFAURE, P., 1967, *Mesozoic Microfacies of Tunisia*: *In* Guidebook to the Geology and History of Tunisia, Petroleum Exploration Society of Libya, Ninth Annual Field Conference, p. 159-214, pls. I-XXVII.

BLOCH, J.P., 1964, *Rélations géomètriques et chronologiques entre les séries sédimentaires, les schistes lustrés et le granite dans le Sud de la Corse alpine*: Bull. Soc. Géol. Fr., ser. 7, t. V, p. 363-369, figs. 1-3.

BLUMENTHAL, M.M., 1963, *Le système structural du Taurus sud-anatolien*: *In* Livre à la mémoire du Professeur Paul Fallot, t. II, p. 611-662, figs. 1-19 (Soc. Géol. Fr.).

BODECHTEL, J., 1964, *Stratigraphie und Tektonik der Schuppenzone Elbas*: Geologische Rundschau, Bd. 53, h. 1, p. 25-41, figs. 1-6, pls. 1-2.

BOLZE, J., 1954, *Age des séries schisteuses et dolomitiques de l'Ichkeul et de l'Harrech (Tunisie septentrionale)*: Comptes rendus de l'Academie des Sciences, Paris, t. 238, p. 2008-2010.

BONNARD, E., DEBOURLE, A., HLAUSCHEK, H., MICHEL, P., PEREBASKINE, V., SCHOEFFLER, J., SERONIE-VIVIEN, R. and M. VIGNEAUX, 1958, *The Aquitaine Basin, Southwest France*: *In* Habitat of Oil. Amer. Assoc. Pet. Geol. Sp. Pub., p. 1091-1122, figs. 1-9.

BOUSQUET, J.C., 1961, *Position des diabases-porphyrites dans la région de Cetraro - Intavolata et de Sangineto (Calabre, Italie méridionale)*: Bull. Soc. Géol. Fr., ser. 7, t. III, p. 603-610, figs. 1-4, one table.

BRINKMANN, R., 1931, *Betikum und Keltiberikum in Südostspanien*: *In* Beiträge zur Geol. der Westmediterrangebiete, Bd. 6. Abh. Gesell. Wissen. Göttingen, Math-naturw, Cl. n.s., vol. 3, p. 1-108.

——, 1960, *Geologic Evolution of Europe*: p. i-vi, 1-160, 18 unnumbered tables, Hafner Publishing Co., New York.

——, 1962, *Aperçu sur les chaines ibériques du nord de l'Espagne*: *In* Livre à la mémoire du Professeur Paul Fallot, t. I, p. 291-299, figs. 1-5.

BRINKMANN, R. and LÖGTERS, H., 1968, *Diapirs in Western Pyrenees and Foreland*: *In* Symposium Diapirism and Diapirs, Amer. Assoc. Pet. Geol. Memoir, No. 8, p. 275-292, figs. 1-20.

BROQUET, P., CAIRE, C. and MASCLE, G., 1966, *Structure et évolution de la Sicile occidentale (Madonies et Sicani)*: Bull. Soc. Géol. Fr., ser. 7, t. VIII, p. 994-1013, figs. 1-9.

BROUWER, H.A., 1957, *Contributions à la géologie de la Corse*: Geologie en Mijnbouw, (NS) Jaarg. 19, No. 8, p. 317-328, figs. 1-10.

BUROLLET, P.F., *et al.*, 1960, *Lexique stratigraphique international,* Fasc. IV a Libye, p. 1-62, 2 maps, Centre national de la Recherche scientifique, Paris.

BUROLLET, P.F., 1967, *General Geology of Tunisia*: *In* Guidebook to the Geology and History of Tunisia, Petroleum Exploration Society of Libya, Ninth Annual Field Conference, p. 51-58, figs. 1-6.

BUROLLET, P.F. and MANDERESCHIED, G., 1965, *Le Crétacé inférieur en Tunisie et en Libye*: *In* Colloque sur le Crétacé inférieur, Mem. Bur. Recherches Géol. et Min., No. 34, p. 785-794, 1 pl. in pocket.

BUROLLET, P.F., 1969, *Petroleum Geology of the Western Mediterranean Basin*: Preprint, Joint Conference Amer. Assoc. Pet. Geol. and Institute of Petroleum, Brighton, 30 June - 2 July 1969, p. 1-12, figs. 1-12.

BUSSON, G., 1967, *Mesozoic of Southern Tunisia*: *In* Guidebook to the Geology and History of Tunisia, p. 131-151, figs. 1-13, Petroleum Exploration Society of Libya, Ninth Annual Field Conference.

——, 1967a, *Le Mésozoique saharien. 1re partie: l'Extrème-Sud tunisien*: Centre National de la Recherche Scientifique. Centre de Recherches sur les Zones arides. Serie: Géologie, No. 8. p. 1-194, figs. 1-18, pls. I-IV, 3 black and white plates and two colored plates in pocket.

CAIRE, A., 1961, *Remarques sur l'évolution tectonique de la Sicile*: Bull. Soc. Géol. Fr., ser. 7, vol. III, p. 545-558, figs. 1-6.

CARATINI, C., 1968, *Evolution paléogéographique et structurale de la région de Chellala - Reibell (Départements de Médéa et Tiaret, Algérie)*: Bull. Soc. Géol. Fr., ser. 7, t. ix, p. 850-858, figs. 1-2 (1967).

CASTANY, G., 1951, *Etude géologique de l'Atlas tunisienne oriental*: Annales des Mines et de la Géologie, No. 8, p. 1-632, figs. 1-243, psl, I-XXX, photographs I-XXVII separate folder.

——, 1956, *Essai de synthese du territoire Tunisie-Sicile*: Annales des mines et de la geologie, No. 16, p. 1-101, figs. 1-37, tables 1-6, pls. I-XII.

CHARLES, FL. A., 1954, *Les actions tectoniques dans la région charbonniere du nord de l'Anatolie et leur influence sur la sédimentation crétacée*: XIX Intern. Geol. Congress, Proceedings, Fasc. XIV. p. 307-336, pls. I-II.

CHOUBERT, G., 1952, *Géologie du Maroc*: Fasc. II. Histoire géologique du domaine de l'Anti-Atlas, Notes et mémoires du Service géologique du Maroc, No. 100, p. 77-195, figs. 1-15.

CHOUBERT, G. and A. FAURE-MURET, 1962, *Evolution de domaine atlasique marocain depuis le temps paléozoiques*: *In* Livre à la mémoire du Professeur Paul Fallot, t. I, p. 447-527, pls. I-VII.

CHOUBERT, G. and H. SALVAN, 1950, *Essai sur la paléogéographie du Sénonien du Maroc*: Notes du Service géologique du Maroc, t. II, p. 13-50, figs. 1-6, unnumbered tables.

CHOUBERT, G. and J. MARÇAIS, 1952, *Géologie du Maroc*: Fasc. I. Aperçu structural, Notes et mémoires du Service géologique du Maroc, No. 100, p. 1-73, 2 maps.

CIRIC, B., 1963, *Le développement des dinarides yougoslaves pendant de cycle alpin*: *In* Livre à la mémoire du Professeur Paul Fallot, t. II, p. 565-582, 1 pl.

——, 1966, *Sur les flyschs et les molasses du cycle alpin dans les Dinarides yougoslaves*: Bull. Soc. Géol. Fr., ser. 7, t. VII, p. 499-510, figs. 1-3 (1965).

CIRY, R., 1940, *Etude géologique d'une partie des des provinces de Burgos, Palencia, Léon et Santander*: Bull. Soc. Hist. Nat. Toulouse, vol lxxiv, p. 5-511, figs.

CLÉMENT, B., 1968, *Observations sur le Trias de Pat seras et du Parnis en Attique (Grèce)*: Comptes rendus sommaires Soc. Géol. Fr., Fasc. 9, p. 332 334, 2 figs.

COATES, J., GOTTESMAN, E., JACOBS, M. and E. Ro SENBERG, 1963, *Gas Discoveries in the Western Dead Sea Region*: Proceedings, 6th World Pe troleum Congress, vol. 1, p. 21-36, figs. 1-7.

COLCHEN, M., 1966, *Sur la tectonique du massif pa léozoique de la Sierra de la Demanda (Espagne et de sa couverture mésozoique et cénozoique* Bull. Soc. Géol. Fr., ser. 7, t. VIII, p. 87-9 figs. 1-6.

COLLEY, B.B., 1963, *Libya: Petroleum Geology an Development: Sixth World Petroleum Congres* sect I, Proceedings. p. 1-10, figs. 1-4.

COLOM, G., 1950, *Más allá de la prehistoria; una ge logía elemental de las Baleares*: p. 1-285, Conse Superior Inves. Cient., Inst. San José de Cal sanz, Madrid.

——, 1955, *Jurassic-Cretaceous pelagic sediments the western Mediterranean zone and the Atlant area*: Micropaleontology. vol. No. 2, p. 109-12 text figs. 1-4, pls. 1-5.

COLOM, G. and B. ESCANDELL, 1962, *L'évolution géosynclinal baléare*: *In* Livre à la mémoire Professeur Paul Fallot, t. I, p. 127-136, figs. 15 (Soc. Géol. Fr.).

NANT, L.C. and G.H. GOUDARZI, 1967, *Stratigraphic and Tectonic Framework of Libya*: Bull. Amer. Assoc. Pet. Geol., vol. 51, No. 5, p. 719-730, figs. 1-5.

RNET, A., 1952, *L'Atlas saharien suboranais*: XIX International Geological Congress, Monographie régionale No. 12. p. 1-48, figs. 1-9, sections I-IV.

RROY, G., 1963, *L'évolution paléogeographique post-hercynienne de la Provence*: *In* Livre à la mémoire du Professeur Paul Fallot, t. II, p. 19-43.

USTAU, H., GAUTIER, J., KULBICKI, G. and WINNOCK, E., 1969, *Hydrocarbon Distribution in the Aquitaine Basin of SW France*: Preprint, Conference, Institute of Petroleum and Amer. Assoc. Petroleum Geologists, Brighton, England, p. 1-12, figs. 1-12.

VILLIER, J., FOURMENTRAUX, J., HENRY, J., JENNER, J., PONTALIER, Y. and J. SCHOEFFLER, 1962, *Etat actuel des connaissances géologiques sur le bassin d'Aquitaine au sud de l'Adour*: *In* Livre à la mémoire du Professeur Paul Fallot, t. I, p. 367-382, figs. 1-19 (Mémoire hors série de la Société Géologique de France).

BELMAS, J., 1964, *Essai sur le déroulement du paroxysme alpin dans les alpes franco-italiennes*: Geologische Rundschau, Bd. 53, h. 1, p. 133-153, figs. 1-2 (1963).

—, 1966, *Progrès recents et perspectives nouvelles de la géologie des Alpes occidentales franco-italiennes*: Ann. Soc. Geol. Belg. t. 89, Bull. 9, p. 423-446, figs. 1-5.

BELMAS, J. and LEMOINE, M., 1957, *Discordance angulaire du Rhétien sur le Trias dans le massif de Peyre-Haute au sud de Briançon. Importance de l'érosion antérhétienne dans la zone briançonnaise*: Comptes rendus, Soc. Géol. Fr., No. 9, p. 150.

BELMAS, J., LEMOINE, M. and M. MATTAUER, 1966, *Quelques remarques sur le concept de "Geosynclinal"*: Revue de géographie physique et de géologie dynamique, (2) vol. 8, fasc. 2, p. 133-150.

ELCEY, R., LIMASSET, J.C. and ROUTHIER, P., 1965, *Les bassins sédimentaires du nord de la Corse; essai de synthèse stratigraphique et aperçu tectonique*: Bull. Soc. Géol. Fr., ser. 7, t. VI, p. 322-333, fig. 1, table 1.

ELEON, G., GOJKOVIC, S. and M. VUKASOVIC, 1961, *The Age Determination of a Certain Number of Granitic Rocks in Yougoslavia*: II Kongr. Geol. Jugo., Proc. p. 443-447.

EMAISON, G.J., 1965, *The Triassic Salt in the Algerian Sahara*: *In* Salt Basins Around Africa, Institute of Petroleum, p. 91-100, figs. 1-3.

ESIO, A. and ROSSI RONCHETTI, C., 1960, *Sul Giurassico medio di Garet el Bellaa (Tripolitania) e sulla posizione stratigrafica della formazione di Tacbal*: Rivista Italiana di Paleont. Stratigr. t. 66, No. 2, p. 173-196, 3 pls.

DESIO, A., ROSSI RONCHETTI, C. and P.L. VIGANO, 1960, *Sulla stratigrafia del Trias in Tripolitania e nel Sud-Tunisino*: Riv. Italiana di Paleont. Stratigr. t. 66, No. 3, p. 273-322, 2 pls.

DESTOMBES, J., 1962, *Stratigraphie et paléogéographie de l'Anti-Atlas (Maroc): un essai de synthèse*: Bull. Soc. Géol. Fr., ser. 7, vol. IV, p. 453-463, 1 fig., 1 table (unnumbered).

DIENER, C., 1916, *Die marinen Reiche der Triasperiode*: Denkschriften der Kaiserlichen Akademie der Wissenschaften, Mathematisch-Naturwissenschaftliche Klasse, Bd. 92, p. 407-549, 1 map.

DIMITROV, S., 1959, *Kurze Ubersicht der Metamorphen Komplexe in Bulgarien*: Freiberger Forschungshefte (Die Bergakademie, Beihefte, Ausgabe C, Angewandte Naturwissenschaften), C 57, p. 62-72, sketch map.

DUBAR, G., 1938, *Sur la formation des rides à l'Aalenian et au Bajocen dans le Haut-Atlas de Midelt*: C.R. Acad. Sciences, Paris, vol. 206, p. 525.

DUBAR, G. and PEYRE, Y., 1960, *Observations nouvelles sur le Jurassique inférieur et moyen dans les cordillères betiques sur la transversale de Malaga (Andalousie, Espagne)*: Bull. Soc. Géol. Fr., ser. 7, t. III, p. 330-339, 1 fig., 1 table.

DUBERTRET, L. and VAUTRIN, A., 1937, *Sur la présence du Jurassique marin dans la région plissée palmyréenne*: Comptes rendus Soc. Géol. Fr., No. 10, p. 135-136.

DUBERTRET, L., 1947, *Problèmes de géologie du Levant*: Bull. Soc. Géol. Fr., ser. 5, vol. XVII, 3-32, figs. 1-10.

——, 1955, *Géologie des roches vertes du nord-ouest de la Syrie et du Hatay (Turquie)*: CNRS, Notes et Memoires sur le Moyen-Orient, tome VI, p. 1-179, pls. I-XXI, pls. A-B geologic maps plus Antioch sheet.

DUBOURDIEU, G., 1962, *Dynamique wegenérienne de l'Afrique du Nord*: *In* Livre à la Mémoire du Professeur Paul Fallot, t. 1, p. 627-644, figs. 1-5.

DUNNINGTON, H.V., WETZEL, R. and MORTON, D.M., 1959, *Iraq*: Fasc. 10a, Lexique stratigraphique international, p. 1-333, Centre national de la recherche scientifique, Paris.

DURAND DELGA, M., 1967, *Structure and Geology of the Northeast Atlas Mountains*: *In* Guidebook to the Geology and History of Tunisia, Petroleum Exploration Society of Libya, Ninth Annual Field Conference, p. 59-83, figs. 1-10.

DÜRR, S., HOEPPENER, R., HOPPE, P. and F. KOCKEL, 1962, *Géologie des montagnes entre le Rio Guadalhorce et le Campo de Gibraltar (Espagne méridionale)*: *In* Livre à la mémoire du Prof. Paul Fallot, t. I, p. 209-227, figs. 1-4, pl. 1.

EGERAN, N.E., 1947, *Tectonique de la Turquie et relations entre les unités tectoniques et les gîtes metallifères*: Thesis, p. i-ix, 1-97, text figs. 1-16, 18 photographs (thin-sections), 3 maps in back, Georges Thomas, Nancy.

EICHER, D.B., 1946, *"Conodonts from the Triassic of Sinai, Egypt"*: Bull. Amer. Assoc. Pet. Geol., vol. 30, p. 613-616.

EMERY, K.O., HEEZEN, B.C., ALLEN, T.D., 1966, *Bathymetry of the eastern Mediterranean Sea*: Deep Sea Research, vol. 13, p. 173-192, figs. 1-12.

ERENTÖZ, C., 1956, *A general review of the geology of Turkey*: Bull. Mineral Research and Exploration Inst. Turkey, Foreign Ed., No. 48, p. 40-54, 1 fig.

FABIANI, R. and TREVISAN, L., 1937, *Di alcune novità geologiche nel territorio del foglio Termini Imerese*: Boll. della Società di Scienze Naturali ed Economiche di Palermo, N.S. vol. 19, p. 1-6, pls. I-II.

FALLOT, P., 1931, *Essais sur la répartition des terrains secondaires et tertiaires dans le domaine des Alpides espagnoles. I. Le Trias*: *In* Géologie de la Méditerranée occidentale, vol. IV, No. 1, part II, p. 10-27, plate II.

——, 1934, *Essais sur la répartition des terrains secondaires et tertiaires dans le domaine des Alpides espagnoles. IV. Le Jurassique supérieur*: *In* Géologie de la Méditerranée occidentale, vol. IV, No. 1 (part II), p. 75-115, figs. 1-12, pls. IV, V, table.

FALLOT, P. and MARIN, A., 1952, *Maroc Septentrional (Chaine du Rif)*: Livret Guide des Excursions A31 et C31, Partie C. XIX Congres Géologique International, p. 1-34, pls. I-IV, figs. 1-15.

FOUCAULT, A., 1967, *Le diapirisme des terrains triasiques au secondaire et au tertiaire dans le subbétique du NE de la Province de Grenade (Espagne. méridionale)*: Bull. Soc. Géol. Fr., ser. 7, t. VIII, p. 527-536, figs. 1-9 (1966).

FURON, R., 1953, *Introduction à la géologie et a l'hydrogéologie de la Turquie*: Mem. Museum National d'Histoire naturelle, ser. C., t. III, Fasc. 1, p. 1-128, pls. I-VII.

——, 1959, *La Paléogéographie. Essai sur l'évolution des continents et des océans*: Payot, Paris, p. 1-405, text figs. 1-76, pls. I-XII (pocket).

——, 1960, *Géologie de l'Afrique*: Payot, Paris, p. 1-400, figs. 1-32.

GAERTNER, H.R., 1931, *Geologie der Zentralkarnischen Alpen*: Denkschriften der Akad. der Wissenschaften in Wien. Math.-Naturwiss. Kl., Bd. 102, p. 113-199, text figs. 1-16, pls. 1-5.

GATTINGER, T.E., 1962, *Explanatory Text of the Geological Map of Turkey at 1:500,000. Trabzon sheet*: p. I-IV, 1-75, 1 fig. pls. I-III, Geological map 1:500,000 in pocket.

GEMMELLARO, R., 1921, *Sul Trias della regione occide[n]tale della Sicilia*: Mem. Accad. Lincei, ser. II vol. 12, p. 451-473, pls. I-V.

GÈZE, B., 1963, *Caractères structuraux de l'arc de N[i]ce*: *In* Livre à la Mémoire du Prof. Paul Fallo[t] t. II, p. 289-300, figs. 1-7.

GIGNOUX, M., 1955, *Stratigraphic Geology*: Translatio[n] of Fourth French Edition, 1950, p. I-XVI, 1-68[2] figs. 1-165, Freeman and Company San Francisc[o].

GILL, W.D., 1965, *The Mediterranean Basin*: *In* Sa[lt] Basins Around Africa, Institute of Petroleum p. 101-111, figs. 1-5.

GLANGEAUD, L., 1951, *Interprétation tectono-physiqu[e] des caractères structuraux et paléogéographiqu[es] de la Méditerranée occidentale*: Bull. Soc. Géo[l.] Fr., ser. 6, t. 1, fasc. 8, p. 735-762, text figs. 1-.

——, 1957, *Correlation chronologique des phenomèn[es] géodynamiques dans les Alpes, l'Apennin, [et] l'Atlas Nord-Africain*: Bull. Soc. Géol. Fr., ser. [6] Vol. VI, p. 867-891, figs. 1-5.

——, 1957 (a), *Essai de classification géodynamiqu[e] des chaînes et des phénomènes orogéniques*: Re[v.] de Géographie physique et de Géologie dyn[a]mique, ser. 2, Vol. 1, p. 201-221.

——, 1962, *Paléogéographie dynamique de la Médite[r]ranée et de ses bordures. Le rôle des phas[es] Ponto-Plio-Quaternaire*: *In* Colloques Nationau[x] du CNRS, "Océanographie géologique et gé[o]physique de la Méditerranée occidentale", p. 12[5]-165, figs. 1-11.

GLANGEAUD, L., ALINAT, J., POLVÈCHE, J., GUILLA[U]ME, A. and O. LEENHARDT, 1967, *Grandes stru[c]tures de la mer Ligure, leur évolution et leu[rs] rélations avec les châines continentales*: Bu[ll.] Soc. Géol. Fr., ser. 7, t. VIII, p. 921-937, figs. [1-]7, 1966.

GLANGEAUD, L., SCHLICH, R., PAUTOT, G., BELLARCH[E], G., PATRIAT, P. and M. RONFARD, 1965, *Morph[o]logie, tectonophysique et évolution géodynamiq[ue] de la bordure sous-marine de Maurès et de l'[E]sterel. Relations avec les régions voisines*: Bu[ll.] Soc. Géol. Fr., ser. 7, t. VII, p. 998-1009, figs. 1-

GOTTIS, C., 1962, *Stratigraphie, structure et évolutio[n] structurale de la Kroumirie et de ses bordure[s]*: *In* Livre à la Mémoire du Professeur Paul Fa[l]lot, t. I, p. 645-656, 1 fig.

GOTTIS, C., 1962a, *Architecture tertiaire en Bas-La[n]guedoc*: *In* Livre à la mémoire du Professe[ur] Paul Fallot, t. 1, p. 383-395, figs. 1-4, pl. 1.

GRACIANSKY, P.C., DE, *Données stratigraphiques et te[c]toniques nouvelles sur la montagne de Tausch*: Bull. Soc. Géol. Fr., ser. 7, t. IV, p. 509-52[.] figs. 1-10.

GRANDJACQUET, C., 1962, *Aperçu morphotectonique [et] paléogéographique du domaine calabro-lucani[e] (Italie méridionale)*: Bull. Soc. Géol. Fr., ser. [7] t. III, p. 610-618, figs. 1-3 (1961).

Grandjacquet, C. and L. Glangeaud, 1963, *Structures mégamétriques et évolution de la mer Tyrrhénienne et des zones perityrrhéniennes*: Bull. Soc. Géol. Fr., ser. 7, t. IV, p. 760-773, figs. 1-5 (1962).

Grubic, A., 1968, *Sur le problème de la limite entre les Karpates méridionales et les Balkanides dans la Serbie orientale (Yougoslavie)*: XXIII International Geological Congress, Proc. Sect. 3, p. 129-138, fig. 1.

Hassan, A.A., 1963, *The Distribution of Triassic and Jurassic Formations in the Middle East*: Fourth Arab Petroleum Congress, Paper 25 (B-3), p. 359-368, figs. 1-2.

Hecht, F., Fürst, M. and E. Klitsch, 1964, *Zur Geologie von Libyen*: Geologische Rundschau, Bd. 53, p. 413-470, figs. 1-2, pls. 1-5.

Henson, F.R.S., Browne, R.V. and McGinty, J., 1950, *A Synopsis of the Stratigraphy and Geological History of Cyprus*: Quart. Journ. Geol. Soc., vol. CV, p. 1-41, pls. I-II (1949).

Hersey, J.B., 1965, *Sedimentary Basins of the Mediterranean Sea*: Woods Hole Oceanographic Institute Contribution No. 1628, p. 75-91, figs. 32-35.

Hollister, J.S., 1934, *Die Stellung der Balearen im Variscischen und Alpinen Orogen*: Abhandlungen der Gesellschaft der Wissenschaften zu Göttingen, Math-Phys. Kl. Folge III, Heft 10, p. 117-154.

Huene, F. von, 1942, *Die Tetrapodenfährten im toscanischen Verrucano und lhre Bedeutung*: Neues Jahrbuch für Mineralogie, Géologie und Paläontologie, Beilage Bd. 86, Heft 1, p. 1-34 (1941).

Institut de Géologie et Recherches du Sous-Sol — Athènes and l'Institut français du Petrole — Mission Grèce, 1966, *Etude Géologique de l'Epire*: p. 1-306, figs. 1-101, separate Atlas, Editions TECHNIP, 7 Rue Nelaton, Paris XVI.

Ippolito, F., 1949, *Contributo alle Conoscenze geologiche sulla Calabria*: Mem. e Note dell'Istituto di Geologia Applicata dell'Università di Napoli, vol. 2, p. 17-35, one sketch map at 1:2,000,000.

Jones, W.D.V., 1968, *Results of Recent Geological Surveys in Western Greece*: Proceedings Geol. Soc. London, No. 1645, p. 306-310, May, 1968.

Karpoff, R., 1965, *Les grands époques de fracture et de bombement du Sahara central*: Bull. Soc. Géol. Fr., ser. 7, t. VII, p. 469-473, 1 fig.

——, 1967, *Sur l'existence du Maestrichtian au N. de Djeddah*: Comptes rendus Acad. Sci., t. 245, p. 1322-1324.

Kafka, F.T. and Kirkbride, R.K., 1959, *The Ragusa Oil Field, Sicily*: Proceedings Fifth World Petroleum Congress, Sect. I, Geology and Geophysics, p. 233-257, figs. 1-11.

Kehtin, I., 1962, *Explanatory Text of the Geological Map of Turkey, Sinop Sheet*: p. 1-111, pls. I-III, geol. map at 1:500,000 in pocket.

——, 1963, *Explanatory Text of the Geological Map of Turkey at* 1:500,000; *Kayeri Sheet*: p. 1-82, five unnumbered pls. photographs, pls. I-III, Maden Tetkik ve Arama Enstitüsü Vayinlarindan, Ankara.

Khain, V.E. and Milanovsky, E.E., 1963, *Structure tectonique du Caucase d'àprès les données modernes*: *In* Livre à la mémoire du Professeur Paul Fallot, t. II, p. 663-703, figs. 1-18, pls. I-II.

Kieken, M., 1962, *Les traits essentiels de la géologie algérienne*: *In* Livre à la mémoire du Professeur Paul Fallot, t. I, p. 545-614, text figs. 1-10, pls. I-VI.

Kleinsmiede, W.F.J. and Berg. N.J. van den, 1968, *Surface Geology of the Jabal al Akhdar, Northern Cyrenaica, Libya*: *In* Geology and Archaeology of Northern Cyrenaica, Libya, Petroleum Exploration Society of Libya, Tenth Annual Field Conference, p. 115-123, figs. 1-5.

Klemme, H.O., 1958, *Regional Geology of the Circum-Mediterranean Region*: Bull. Amer. Assoc. Pet. Geol., vol. 42, No. 3, p. 477-512, figs. 1-9.

Klitzsch, E., 1968, *Outline of the Geology of Libya*: *In* Geology and Archaeology of Northern Cyrenaica, Libya, Tenth Annual Field Conference, Petroleum Exploration Society of Libya, p. 71-78, figs. 1-3.

Kostandi, A.B., 1959, *Facies maps for the study of the Paleozoic and Mesozoic sedimentary basins of the Egyptian region*: First Arab Petroleum Congress, Cairo, vol. 2, p. 54-62.

Ksiazkiewicz, M., 1956, *Geology of the Northern Carpathians*: Geologische Rundschau, Bd. 45, heft 2, p. 369-411, figs. 1-12.

——, 1963, *Evolution structurale des Carpathes polonaises*: *In* Livre à la mémoire du Professeur Paul Fallot, t. II, p. 529-562, figs. 1-20, (Soc. Géol. Fr.).

——, 1966, *Les cordillères dans les mers crétacées et paléogènes des Carpathes du nord*: Bull. Soc. Géol. Fr., ser. 7, t. VIII, p. 443-455, figs. 1-4, tables 1-2 (1955).

Laemmlen, M., 1958, *Lexique stratigraphique internationale*, vol. 1 Europe. Fasc. 5 Allegmagne 5 (d2) Keuper, P. 1-235, 2 colored maps.

Lapadu-Hargues, P. and Maisonneuve, J., 1964, *L'âge des schistes lustrés de la Corse*: Bull. Soc. Géol. Fr., ser. 7, t. V, p. 1012-1028, 1 table, text plates I-II (1963).

Lapparent, C. de, 1960, *Esquisse paléogéologique du Sahara*: 4.ième Congrès national du Pétrole. Assoc. Franc. des Techniciens du pétrole, t. II, p. 53-60, pls. V-X.

LEENHARDT, O., 1969, *Le problème des domes de la Méditerranée occidentale: étude géophysique d'une colline abyssale, la structure A*: Bull. Soc. Géol. Fr., ser. 7, t. X, p. 497-509, figs. 1-10, (1968).

LEFRANC, J.P., 1958, *Stratigraphie des séries continentales intercalaires au Fezzan nord occidentale (Libye)*: C.R. Acad. Sci., t. 247, No. 17, p. 1360-1363.

LELUBRE, M., 1949, *Géologie du Fezzan orientale*: Bull. Soc. Géol. Fr., ser. 5, t. XIX, p. 251-262, fig. 1, pl. VIII.

LEONARDI, P., 1963, *Les recifs coralliens triasique des Dolomites*: *In* Livre a la mémoire du Professeur Paul Fallot, t. II, p. 237-244, figs. 1-3, pls. I-II.

LÉVY, R.G. and TILLOY, R., 1952, *Maroc Septentrionale (Chaîne du Rif)*: XIX Congres Géologique International, Série Maroc No. 8. Livret-Guide des Excursions A 31 and C 31, Partie B, p. 1-65, pls. 1-8.

LITTLE, O.H., 1945, *Handbook on Cyrenaica. Part I. Geology of Cyrenaica*: p. I-VI, 1-104, Printing and Stationery Services. M.E.F.

LLOPIS LLADO, N., 1954, *Types des chaines alpidiques du littoral méditerranéen franco-espagnol et leurs rapports avec les Alpes françaises*: 19th International Geological Congress, Comptes rendus, Fasc. XIV, p. 271-279, figs. 1-6.

——, 1958, *Lexique Stratigraphique International*, vol. I, Fasc. 10a, Espagne, u. 1-92, 1 map in text.

LORENZO, G. DE, 1894, *Le montagne mesozoiche di Lagonegro. Atti dell'Accademia delle Scienze Fisiche e Matematiche di Napoli*: vol. 6, ser. II, No. 15, p. 1-124, text figs. 1-84, 2 pls. geologic map at 1:500.000, cross-sections.

LOWELL, J.D., 1962, *Tectonic Framework of Yugoslavia*: *In* Petroleum Exploration Society of Libya, Fourth Annual Field Conference; Excursion to Yugoslavia, p. 27-33, figs. 3-5, table I.

LUCAS, G., 1942, *Déscription géologique et pétrographique des Monts du Ghar Rouban et du Sidi bel Abed*: Bull. Serv. Carte géol. Algérie, ser. 2, No. 16, 2 vols., text and atlas, vol. I, p. 1-539, figs. 1-131, vol. II, pls. I-XXIV, photographs, maps, sections.

——, 1952, *Bordure nord des hautes plaines dans l'Algèrie occidentale*: Pub. XIX International Geological Congress, Monographie régionale No. 21, p. 1-139, figs. 1-59.

MACOVEI, G., 1956, *Lexique stratigraphique international*, vol. 1, Europe, Fasc. 132, Roumanie, p. 1-119, 1 geol. map 1:2,500,000.

MAGNIER, P., 1963, *Etude stratigraphique dans le Gebel Nefousa et le Gebel Garian (Tripolitaine, Libye)*: Bull. Soc. Géol. Fr., ser. 7, vol. 5, p. 89-94, figs. 1-2.

MALARODA, R., 1964, *Les faciès à composante détritique dans le Crétacé autochtone des Alpes-Maritimes italiennes*: Geologische Rundschau, Bd. 53, h. 1, p. 41-57, figs. 1-5, pls. 1-6 (1963).

MANGIN, J.P. and P. RAT, 1962, *L'évolution post-hercynienne entre Asturies et Aragon*: *In* Livre à la mémoire du Professeur Paul Fallot, t. I, p. 333-349, figs. 1-10.

MASCLE, G., 1968, *Remarques stratigraphiques et structurales sur la région de Palazzo-Adriano, monts Sicani (Sicile)*: Bull. Soc. Géol. Fr., ser. 7, t. IX, p. 104-110, figs. 1-3 (1967).

MATHIEU, G., 1949, *Contribution à l'étude des Monts troglodytes dans l'extrème Sud-Tunisien*: Ann. Mines et Géologie, Ser. 1, No. 4, Régence de Tunis, Dir. des Travaux publiques, p. 1-74, text figs. 1-11, A-H, pls. I-III.

MATTAUER, M., 1964, *Le style tectonique des chaînes tellienne et rifaine*: Geologische Rundschau, Bd. 53, h. 1, p. 296-313, figs. 1-6, (1963).

MENNING, J.J., VITTEMBERGA, P., and P. LEHMANN, 1963, *Etude sédimentologique et pétrographique de la formation Ras Hamia (Trias moyen) du NW de la Libye*: Revue de l'Institut française du Pétrole, t. 18, No. 11, p. 1504-1519, figs. 1-5.

MERLA, C., MAXWELL, J., SCARSELLA, F., TREVISAN, L. and R. SELLI, 1964, *Guidebook, International Field Institute*: Italy, 1964, 190 unnumbered pages, many maps. charts, figs.

MORRE, N. and THIEBAULT, J., 1963, *Les roches volcaniques du Trias inférieur du versant nord des Pyrénées*: Bull. Soc. Géol. Fr., ser. 7, t. IV, p. 539-545 (1962).

MOURATOV, M.V., 1960, *Tectonic Structures of the Alpine Geosynclinal Area in Eastern Europe and Asia Minor and the History of Their Development*: XXIst International Geological Congress, Part XVIII, p. 137-148, 1 fig.

——, 1962, *Histoire de l'évolution tectonique de la zone plissée alpine de l'Europe orientale et de l'Asie mineure*: Bull. Soc. Géol. Fr., ser. 7, t. IV, p. 182-200, text figs. 1-3.

NORTON, P., 1965, (editor), *Guide to the Geology and Culture of Greece*: Seventh Annual Field Conference, Petroleum Exploration Society of Libya, p. 1-146, many unnumbered figures and photographs; three sections in back cover.

OGNIBEN, L., 1963, *Stratigraphie tectono-sedimentaire de la Sicile*: *In* Livre à la mémoire du Professeur Paul Fallot, t. II, p. 203-216, figs. 1-2.

OROMBELLI, G., LOZEJ, G.P., and L.A. ROSSI 1967, *Preliminary notes on the Geology of the Datça peninsula (SW Turkey)*: Atti Accad. Naz. dei Lincei, Cl. Sc. Fis., Mat., Naturali, vol. XLII, Fasc. 6, p. 830-841, figs. 1-2.

ORTYNSKI, I., PERRODON, A. and C. DE LAPPARENT, 1959, *Esquisse paleogéographique et structurale des bassins du Sahara septentrionale*: Proc. Fifth World Pet. Cong., sect. I, p. 705-727, figs. 1-9, (plus 1-3).

OSWALD, F., 1912, *Armenien*: Handbuch der regionalen Geologie, Vol. 3, Heft 10, p. 1-40, pls. I-IV, geol. map.

PAMIR, H.N., 1960, *Lexique stratigraphique international,* Vol. III, Asie, Fasc. 9c, Turquie, p. 1-95, four unnumbered maps.

PAREA, G.C., 1968, *The Possibility of Reconstruction of the Pre-Tectonic Current Trend in Allochthonous Turbidite Formations*: XXIII International Geological Congress, Proc. Sect. 3, p. 225-234, figs. 1-6.

PATRUT, I., POPESCU, M., TEODORESCU, C., PETRISOR, I., and G. ANTON, 1963, *Potential Mesozoic Oil Deposits in the Rumanian People's Republic*: Sixth World Petroleum Congress, Proceedings, sect. 1, p. 141-153, figs. 1-4.

PATZELT, G., 1968, *Zur Geosynklinalentwicklung und Tektonik der äusseren Helleniden: die Ionische Zone in Albanien*: XXIII International Geological Congress, Proceedings Sect. III, Orogenic Belts, p. 139-152, figs. 1-2, 1 table.

PAYO, G., 1967, *Crustal Structure of the Mediterranean Sea by Surface Waves. Part. 1: Group Velocity*: Bull. Seismological Society of America, vol. 57, No. 2, p. 151-171, figs. 1-10.

PERCONIG, E., 1962, *Sur la constitution géologique de l'Andalousie occidentale en particulier du bassin de Guadalquivir*: *In* Livre à la mémoire du du Professeur Paul Fallot, t. I, p. 228-256, figs. 1-6.

——, 1968, *Microfacies of the Triassic and Jurassic of Spain*: International Sedimentary Petrographical Series, vol. X, p. 1-63, figs. 1-11, pls. 1-CXXIII.

PICARD, L., 1959, *Geology and Oil Exploration of* Proceedings Fifth World Petroleum Congress. Sect. I, Geology and Geophysics, p. 311-336, figs. 1-10.

PINHAR, N., and E. LAHN, 1954, *La position tectonique de l'Anatolie dans le système orogénique méditerranéen*: Comptes rendus, XIX Intern. Geol. Congress, Fasc. XVII, p. 171-180, fig. 1.

POWERS, R.W., RAMIREZ, L.F., REDMOND, C.D. and E. L. ELBERG, 1966, *Geology of the Arabian Peninsula, Sedimentary Geology of Saudi Arabia*: US Geological Survey Professional Paper 560-D, i-vi, D1-D147, figs. 1-14, sections 1-58.

QUENELL, A.M., 1951, *The Geology and Mineral Resources of (Former) Trans-Jordan*: Colonial Geology and Mineral Resources. Vol. 2, No. 2, p. 85-115, pls. I-VIII, photographs, unnumbered fig. 1.

QUITZOW, H.W., 1935, *Der Deckenbau des Calabrischen Massif und seiner Randgebiet*: Abh. Gesell. Wissen. Göttingen. Math-Phys. Kl. Folge III, Heft 13, p. 63-179, text figs. 21-56, pls. III-V.

RAT, P., 1964, *Problèmes du Crétacé inférieur dans le Pyrénées et le nord de l'Espagne*: Geologische Rundschau, Bd. 53, p. 205-220, figs. 1-2, pls. 18-19.

RECH-FROLLO, M., 1960, *Le Cénomanian a blocs exotiques nord-pyrénéen et le Wildflysch des Préalpes suisses*: Bull. Soc. Géol. Fr., ser. 7, t. I., p. 809-816. (1959).

REMANE, J., 1967, *Note préliminaire sur la paléogéographie du Tithonique des chaînes subalpines*: Bull. Soc. Géol. Fr., ser. 7, t. VIII, p. 448-453, figs. 1-2, (1966).

RENOUARD, C., 1951, *Sur la découverte du Jurassique inférieure (?) et du Jurassique moyen au Liban*: Comptes rendus, Acad. Sciences, t. 232, p. 992-994, Paris.

RENOUARD, G., 1955, *Oil Prospects of Lebanon*: Bull. Amer. Assoc. Pet. Geol., vol. 39, No. 11, p. 2125-2169, figs. 1-20.

RENZ, C., 1955, *Die Vorneogene Stratigraphie der Normal Sedimentaren Formationen Griechenlands*: p. I-XVI, 1-637, pls. I-IV, 17 maps and sections in folder, Institute for Geology and Subsurface Research, Athens.

REYRE, D., 1966, *Particularités géologiques des bassins cotiers de l'ouest africain*: *In* Bassins sédimentaires du littoral africain, symposium, 1.re partie, Littoral atlantique, Union Int. Sci. Géol. Assoc. Serv. Géol. Africaine, p. 253-301, figs. 1-13, one unnumbered map (XXII International Geological Congress, New Delhi).

RIBA, O. and RIOS, J.M., 1962, *Observations sur la structure du secteur sud-ouest de la chaîne ibérique*: *In* Livre à la mémoire du Prof. Paul Fallot, t. 1, p. 275-290, figs. 1-8 (Mémoire hors-série de la Soc. Géol. Fr.).

RICOUR, J., 1956, *Lexique stratigraphique internationale,* vol. 1, Europe, Fasc. 4a, France, Belgique, Pays Bas, Luxembourg, Fasc. 4am Trias. p. 1-54, 5 unnumbered illustrations-maps, tables, charts.

——, 1959, *Stratigraphie du Trias du Bassin de Paris*: Bull. Soc. Géol. Fr., ser. 7, t. 1, p. 1-12, figs. 1-3.

——, 1962, *Contribution à une révision du Trias français*: Mémoires pour servir a l'éxplication de la carte détaillée de la France, Ministère de l'industrie, p. 1-471, pls. I-XXIV, pls. 1-8 in pocket, figs. 1-122, Imprimerie Nationale, Paris.

——, 1963, *Particularités paléogéographiques des Alpes occidentales françaises aux temps triasiques*: *In* Livre à la Mémoire du Professeur Paul Fallot, t. II, p. 395-405, figs. 1-9.

RIGO DE RIGHI, M. and BARBIERI, F., 1959, *Stratigrafia pratica applicata in Sicilia*: Bol. Serv. Geol. d'Italia, vol. 80 (1958), Nos. 2-3, p. 351-442, figs. 1-10, pls. I-XIV.

RIGO DE RIGHI, M. and CORTESINI, A., 1964, *Gravity Tectonics in Foothill Structure Belt of Southeast Turkey*: Bull. Amer. Assoc. Pet. Geol., vol. 48, p. 1911-1937, figs. 1-10.

RIOS, J.M., GARRIDO, J. and ALMELA, A., 1944, *Reconocimiento Geológico de una parte de las provincias de Cuenca y Guadalajara; primera parte, La Région de Cuenca-Priego-Cifuentes, segunda parte, Paleogeografía e historia geológica del sistema ibérico segun Richter y Teichmüller*: Soc. Espanola Hist. Nat. ser. B, t. 42, nos. 1-4, p. 107-128, 263-286.

RIOS, J.M., 1956, *El sistema Cretáceo en los Pirineos de Espana*: *In* Memorias del Instituto Geológico y Minero de Espana, tomo LVII, El Cretáceo en Espana, p. 7-128, figs. 1-36.

ROCH, EDOUARD, 1950, *Histoire stratigraphique du Maroc*: Service géologique du Maroc, Notes et Mémoires No. 80, p. 1-433, figs. 1-77, pls. I-XXII.

RUSSO, P., 1927, *Sur les rapports stratigraphiques et tectoniques entre le Rif et les pays situés plus a l'est (Maroc oriental)*: Bull. Soc. Géol. Fr., ser. 4, t. XXVI, p. 43-47.

RUSSO, P. and L.R. RUSSO, 1927, *Recherches sur la géologie du Rif central*: Bull. Soc. Géol. Fr., ser. 4, t. XXVII, p. 419-435, 1 fig.

SAID, R., 1962, *The Geology of Egypt*: 377 p. 71 figs., Elsevier Publishing Company, Amsterdam - New York.

——, 1964, *Trip to Gulf of Suez*: *In* Guidebook to the Geology and Archaeology of Egypt, Petroleum Exploration Society of Libya, Sixth Annual Field Conference, p. 141-148, 14 photographs.

SANDER, N.J., 1968, *The Premesozoic Structural Evolution of the Mediterranean Region*: *In* Petroleum Exploration Society of Libya, Tenth Annual Field Conference, p. 47-70, figs. 1-3, tables 1-3.

SCHMIDT, G.C., 1964, *A review of Permian and Mesozoic formations exposed near the Turkey-Iraq border at Harbol*: Bull. Mineral Research and Exploration Institute of Turkey (Foreing edition), No. 62, p. 103-119, 2 unnumbered sections, 1 map.

SCHMIDT DI FRIEDBERG. P., 1962, *Introduction à la Géologie Pétrolière de la Sicile*: Revue de l'Institut Français du Pétrole, vol. XVII, No. 5, p. 635-668, pls. I-XV, May 1962.

SCHMIDT DI FRIEDBERG, P. and TROVO, A., 1963, *Contribution à l'étude structurale du Groupe de Monte Judica (Sicile orientale)*: Bull. Soc. Géol. Fr., ser. 7, t. IV, No. 5, p. 754-759, 1 fig. (1962).

SCHMIDT-THOMÉ, P., 1963, *Le bassin de la molasse d'Allemagne du Sud, avec des considerations particulières sur la Molasse plissée de Bavière*: *In* Livre à la Mémoire du Prof. Paul Fallot, t. II, p. 431-452, figs. 1-7.

SHAW, S.H., 1947, *Southern Palestine*: Geological map on a scale of 1:250,000 with explanatory notes, p. 1-41, Government printer, Jerusalem.

SIKOSEK, B. and MEDWENITSCH, W., 1965, *Neue Daten zur Facies und Tektonik der Dinariden*: Zeitschrift der Deutschen Geologischen Gesellschaft, Band 116, 2 Teil, p. 342-358, text figs. 1-7, pls. 1.

SITTER, L.V. DE, 1956, A *Cross-Section through the Central Pyrenees*: Geologische Rundschau, Bd. 45, Heft 2, p. 214-233, figs. 1-16.

SOLIMAN, S.M. and FARIS, M.I., 1963, *General Geologic Setting of the Nile Delta Province, and its Evaluation for Petroleum Prospecting*: Fourth Arab Petroleum Congress, Paper 23 (B-3), 327-337, pls. I-IV.

S.N. REPAL, 1952, *Régions sud-telliennes et Atlas saharienne*: XIX International Geological Congress, Monographie régionale No. 20, p. 1-39, pls. I-VI.

TEICHMÜLLER, R., 1931, *Zur Geologie des Tyrrhenisgebietes. I. Alte und junge Krustenbewegungen in südlichen Sardinien*: Abh. Ges. Wissen. Gött., Math-Phys. Kl. Folge III, Heft 3, No. 7, p. 857-950, text figs. 1-36, 1 geol. map 1:100,000 (pl. I), pls. II-III.

TEMPIER, C., 1967, *Les faciès du Jurassique terminal dans les chaînes subalpines méridionales au sud et à l'est de la Durance. Leur répartition géographiques*: Bull. Soc. Géol. Fr., ser. 7, t. VIII, p. 468-470, figs. 1-2 (1966).

TEMPLE, P.G. and PERRY, L.J., 1962, *Geology and Oil Occurrence, Southeast Turkey*: Bull. Amer. Assoc. Pet. Geol., vol. 46, p. 1596-1612, text figs. 1-11.

TEN DAM, A. and TOLUN, N., 1962, *Struttura e Geologia della Turchia*: Bol. Soc. Geologica Italiana, vol. LXXX, Fasc. 3, p. 45-80, 9 paleogeographic maps, unnumbered; 1 tectonic sketch map.

TERMIER, H. and G. TERMIER, 1952, *Histoire géologique de la biosphere*: p. 1-721, many illustrations including paleogeographic sketch maps, Masson & Co., Paris.

——, 1956-1957, *L'Evolution de la Lithosphere. Pt. II, Orogenese*: Fasc. 1, p. 1-498, Fasc. 2, p. 499-940, 152 figs. 49 plates, Masson & Co., Paris.

TOLLMANN, A., 1963, *Résultats nouveaux sur la position, la subdivision, et le style structurale des zones helvétiques, penniques et austroalpines des Alpes orientales*: *In* Livre à la mémoire du Professeur Paul Fallot, t. II, p. 477-490, figs. 1-2.

——, 1964, *Ubersicht über die alpinischen Gebirgsbildung-phasen in den Ostalpen und West Karpaten*: Mitt. Geol. Gesell. Bergb. Studien Wien, Bd. 14, p. 81-88, 1 table (1963).

TREVISAN, L., 1963, *La paléogéographie du Trias de l'Apennin septentrional et central et ses rapports avec la tectogénese*: *In* Livre à la Mémoire du Prof. Paul Fallot, t. II, p. 217-235, figs. 1-9.

TRÜMPY, R., 1960, *Paleotectonic Evolution of the Central and Western Alps*: Bull. Geol. Soc. Amer., vol. 71, p. 843-908, figs. 1-14, pls. 1-2.

TSOFLIAS, P., 1969, *Sur la découverte d'ammonites triasiques au front de la nappe du Pinde en Péloponnèse septentrional (Grèce)*: Comptes rendus sommaire, Soc. Géol. Fr., Fasc. 4, p. 118-119.

VIRGILI, C., 1962, *Le Trias du nord-est de l'Espagne*: *In* Livre à la mémoire du Professeur Paul Fallot, t. I, p. 300-311, figs. 1-2.

VUCKOVIC, J., FILJAK, R. and V. AKSIN, 1959, *Survey of Exploration and Production of Oil in Yugoslavia*: Proc. Fifth World Petroleum Congress, Section I, p. 1003-1021, text figs. 1-8.

WEBER, H., 1963, *Ergebnisse erdölgeologischer Aufschlussarbeiten der DEA in Nordost-Syrien*: Erdöl und Kohle, Erdgas, Petrochemie, No. 6, p. 669-682, figs. 1-11.

——, 1964, *Ergebnisse erdölgeologischer Aufschlussarbeiten der DEA in Nordost-Syrien. Teil II. Geophysikalische Untersuchungen und Tiefbohrungen in der Haute Djésirah*: Erdöl und Kohle, Erdgas, Petrochemie, vol. 17, No. 4, p. 249-261, figs. 1-8.

WETZEL, R. and D.M., MORTON, 1959, *Contribution a la géologie de la Transjordanie*: Notes et Mémoires sur le Moyen Orient, T. VII, p. 95-191, figs. 1-25.

WOLFART, R., 1967, *Geologie von Syria und dem Libanon*: Beiträge zur Regionalen Geologie der Erde, Bd. 6, p. I-XII, 1-326, figs. 1-78, tables 1-42, 1 geologic and one soil map.

YALCINLAR, I., 1954, *Les lignes structurales de la Turquie*: XIX Intern. Geol. Congress, Proc., Fasc. XIV, p. 293-299, 1 fig.

ZWART, H.J., 1967, *The Quality of Orogenic Belts*: Geologie en Mijnbouw, 46 Jaargang, No. 8, p. 283-309, figs. 1-12.

10

Reprinted from *Nature* **256**:117–119 (1975)

TRIASSIC SEAWAYS AND THE JURASSIC TETHYS OCEAN IN THE CENTRAL MEDITERRANEAN AREA

P. Scandone
Istituto di Geologia e Geofisica
Universita degli Studi di Napoli

In the central Mediterranean area phases of middle Triassic and middle Liassic rifting are recognisable. The former produced only relatively deep pelagic basins founded on thinned continental crust; the latter, however, marked the opening of the central Tethys. The Porphyrit-Hornstein and the Diabas-Hornstein formations[1] are the results of such phenomena. During Dogger and Malm times the spreading continued and a large amount of oceanic crust was formed. The Tethys Ocean opened parallel to the long axes of some Triassic deep sea basins and across the long axes of others. Consequently, the positions of Triassic seaways along the southern continental margin of the Tethys 'geosyncline' are defined by the effects of at least two main palaeotectonic events.

In the Triassic System the Germano–Andalusian facies is characterised by prevalent continental deposits, whereas the Alpine facies is characterised by marine deposits. In the circum-Mediterranean area the basements of both Germano–Andalusian and Alpine facies sediments everywhere comprise continental crust which may underlie Palaeozoic sediments and volcanics. Only in the Antalya Nappe (Turkey) are there indications that a crust of oceanic composition underlies Alpine facies sediments.

The Alpine facies, occurring mainly in the Upper Triassic, is generally represented by shallow water carbonates (dolomites and limestones), and in certain areas by evaporites. Locally, *Halobia* limestones are known, showing relatively deep sea environments. The outcrops of *Halobia* limestones form narrow belts, more or less disconnected, which stretch along the circum-Adriatic chains from Greece to Sicily.

The complete Triassic sequence containing the *Halobia* limestones is not much varied in the whole Mediterranean region. The Lower Triassic is represented by shallow water terrigenous sediments and acid volcanics, which are a prolongation in time of the Upper Permian facies. The Middle Triassic is represented by terrigenous sediments simulating flysch deposits (Monte Facito Formation in the Apennines and Sicily; Montenegro 'flysch' in Yugoslavia), by mafic volcanics (Porphyrit-Hornstein Formation in Yugoslavia and in Greece), and less frequently by nodular limestones like Ammonitico Rosso (Han Bulog Beds in Yugoslavia and in Greece). The Upper Triassic is represented by *Halobia* limestones with interbedded rare mafic lavas and hyaloclastites, and locally by nodular limestones like Ammonitico Rosso (Hallstatt Beds in the eastern Alps).

The *Halobia* limestones have been considered as markers of an oceanic Palaeo-Tethys in the central Mediterranean area[2,3]. I suggest, however, the basins to which pelagic pelecypods and ammonites migrated originated from an eastern oceanic Palaeo-Tethys and penetrated deeply into a continental Palaeo-Mediterranean gulf as 'seaways' founded on thinned continental crust. Among these Triassic seaways the most considerable is the Pindos Basin, which stretches for more than 1,200 km through Greece, Albania and Yugoslavia. On the opposite side of the Adriatic Sea, after a gap of some 100 km, the basin is recognisable in the southern Apennines–Sicily arc as unconnected outcrops stretching for about 800 km.

In Jurassic times a real ocean separated the European and the African continents. Contemporaneously, large shearing movements began between Europe and Africa as a result of the opening of the southern Atlantic[4].

Jurassic Tethyan deposits consist of mafic and ultramafic rocks, radiolarites, pelagic limestones, claystones and sandstones, frequently affected by high pressure–low temperature metamorphism.

A basic assumption in any attempt at a palinspastic restoration of Triassic and Jurassic palaeogeography[3,5] is that the volume of continental crust remained roughly the same before and after the Alpine deformation. Consequently, the kinematic system is considered a closed system. But a lot of continental crust must have disappeared into the asthenosphere[6,7] and, therefore, huge volumes (or surfaces in a two-dimensional calculation) have been taken out of the system. These lost volumes are not counterbalanced by additional small volumes of new oceanic crust. In any palinspastic restoration it is therefore necessary to smooth out all the nappes of the peri-Mediterranean Alpine systems to calculate the real shortening of the plate margins during the Alpine compression. Figures 1 and 2 attempt such a palinspastic restoration for the Upper Triassic and Upper Jurassic. The original position of the palaeogeographical units has been obtained by fixing the 'stable' European and the Saharan plates, smoothing out folds and nappes, removing the main lateral displacements and connecting those belts which are common to the African and European margins. This construction of the Western part of Fig. 2 relied upon the palinspastic restoration of Laubscher[7].

This reconstruction produces an area corresponding to the Alpine facies, forming a gulf open in the direction of the Palaeo-Tethys and surrounded by regions characterised by

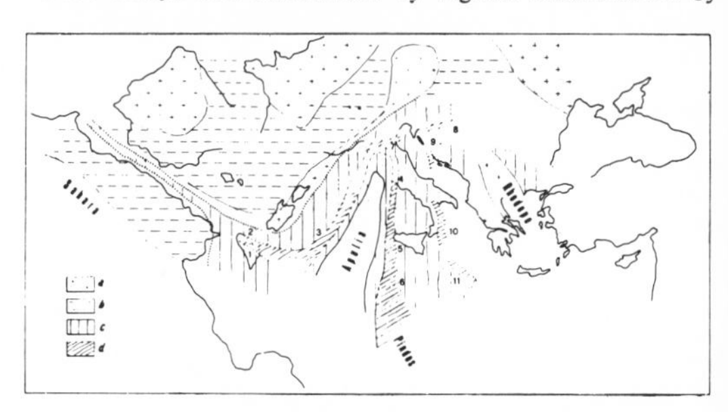

Fig. 1 Palinspastic restoration of Upper Triassic palaeogeography in the central Mediterranean area. *a*, Stable areas with slight subsidence; *b*, subsided basins of Germano–Andalusian facies; *c*, shallow-water carbonates of Alpine facies; *d*, 'seaways'; 1, Sicani zone; 2, Imerese zone; 3, Lagonegro zone; 4, Triassic pelagic sediments of Slovenia; 5, Budva-Kotor zone; 6, Pindos zone; 7, Australpine of Hallstatt; 8, Australpine of the Gemerides; 9, Australpine of the Balaton Lake; 10, Subpelagonian Triassic pelagic limestones of Serbia; 11, Subpelagonian Triassic pelagic limestones of Northern Pindos and Othrys.

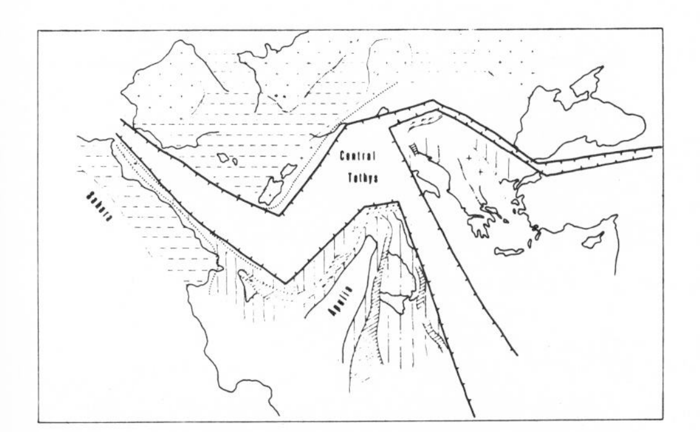

Fig. 2 The Tethys Ocean in the central Mediterranean area.

Germano–Andalusian facies. In these regions it is possible to distinguish stable zones, slightly subsident and unaffected by synsedimentary tectonics, and less stable or clearly unstable zones, characterised by high rates of subsidence and sedimentation, synsedimentary faulting and mafic volcanism.

In the area comprising Alpine facies the palaeogeography changes radically from the Lower to the Upper Triassic because of a Middle Triassic tectonic phase, which was responsible for the tectonic regime immediately before the opening of the Tethys. This tectonic phase included an attempt at rifting, which produced a complex system of tensional faults that crossed the Palaeo-Mediterranean Gulf in several directions. The faulted blocks underwent different vertical movements. The uplifted blocks often underwent severe erosion, whereas flysch-like sediments were deposited on the downshifted blocks. Contemporaneously, widespread mafic magmatic activity (comprising an alkali-basaltic series) occurred along the main faults. Deep sea basins developed along preferential directions (probably corresponding to strips of crustal thinning) within the Palaeo-Mediterranean Gulf, allowing the diffusion of the pelagic fauna from the Triassic Palaeo-Tethys through Turkey, Greece, Albania, Yugoslavia, Hungary, Czechoslovakia, Austria, Italy, and as far as western Sicily. The absolute homogeneity of the fauna proves that close intercommunications were established between all of the seaways. Figure 1 shows the region during the late Triassic.

During Liassic times, starting mainly in the Middle Liassic, important new events occurred in the central Mediterranean area: along giant tensional fractures, partly parallel to, and partly discordant with, the former seaways, the Tethys Ocean began to open. In the internal nappes of the Dinarides and of the Hellenides at least two main ophiolite assemblages deriving from the oceanic Tethys may be distinguished: the Diabas-Hornstein Formation, and the ultrabasic massifs at Zlatibor and Vourinos, with their basalts and sediments. The Diabas-Hornstein Formation comprises alternated diabases, radiolarian cherts, sandstones, claystones, graded calcareous microbreccias, allodapic and pelagic limestones, and intraformational conglomerates. This sequence may overlie Jurassic limestones founded on continental crust[8].

The generally tectonic basal contact and the occurrence of slices of serpentinites between the Jurassic limestones and the Diabas-Hornstein Formation suggest that the latter (at least in part) was possibly founded on oceanic crust. On the other hand, the occurrence of terrigenous material and of neritic calcareous turbidites suggests provenance from a very near continental platform. I suggest that the Diabas-Hornstein Formation is the product of the Liassic rifting which marked the beginning of the opening of the Tethys Ocean, so it may be considered as a marker of the continental margin or the continental rise. As the opening of the Tethys Ocean continued, the areas in which the Diabas-Hornstein Formation had been deposited were shifted laterally and, according to plate accretion models, new areas of oceanic crust were generated in their place. These new areas of oceanic ridges which are, of course, younger than the Diabas-Hornstein Formation, are today recognisable as the ultramafic massifs.

In the Alps and in the Apennines this distinction between initial rifting (Diabas-Hornstein Formation) and real ocean spreading (ultramafic massifs) of the central Tethys is not so clear as it is in the Dinarides, but it is probable that the Ligurian and Piemontese ultramafic rocks represent an ancient, oceanic ridge area, whereas parts of the Southpennine *Bündnerschiefer* represent an equivalent of the Diabas-Hornstein Formation.

During the Middle and Upper Liassic the combination between the Atlantic and the Tethyan movements produced a renewal of synsedimentary tectonic activity, and consequently the palaeogeography became modified not only around the central Tethys, but also around the former Triassic seaways. For example, in western Sicily the former Imerese seaway, which during the Upper Triassic ended at the longitude of Palermo, suddenly lengthened westwards during the Liassic, in the direction of the Maghreb along a newly generated deep furrow in which radiolarian oozes and calcareous turbidites stratigraphically overlie collapsed shallow water dolomites[9]. Contemporaneously, in areas of maximal tension (Imerese, Sicani, Pindos), flows of alkali-basalts were erupted.

During post-Jurassic times the Triassic seaways near to the margin of the Tethys Ocean, and those which were cut across by the opening of the ocean, were affected by Cretaceous compression. These orogenic phases, the earliest in the central Mediterranean area, were responsible for the reduction, and the local disappearance, of large oceanic zones. In Greece (Northern Pindos[10] and Othrys), sediments deposited very near to the Tethys Ocean were affected by Cretaceous compression, and were involved in the building of the ophiolite nappes.

On the other hand, the continuing sedimentary history of those seaways which lie more within continental areas, will cease only after the collision between Europe and Africa. In Sicily (Sicani Mountains[11]), such sedimentary deposits were affected by compression during the Upper Miocene, when Europe and Africa had already collided and a lot of continental crust from both plates had disappeared because of subduction.

1 Ciric, B., and Karamata, S., *Bull. Soc. géol. Fr.*, **2**, 276–380 (1960).
2 Wezel, F. C., *Riv. Miner. Sicily*, **21**, 187–198 (1970).
3 Dewey, J. F., Pitman, W. C., Rian, W. B. F., and Bonnin, J., *Bull. geol. Soc. Am.*, **84**, 3137–3180 (1973).
4 Pitman, W. C., and Talwani, M., *Bull. geol. Soc. Am.*, **83**, 619–646 (1972).
5 Bosellini, A., and Hsu, K. J., *Nature*, **244**, 144–146 (1974).
6 Laubscher, H. P., *Am. J. Sci.*, **271**, 193–226 (1971).
7 Laubscher, H. P., *Le Scienze*, **72**, 48–59 (1974).
8 Aubouin, J., *Bull. Soc. géol. Fr.*, **15**, 5–6, 425–462 (1974).
9 Giunta, G., and Liguori, V., *Boll. Soc. geol. ital.*, **92**, 903–924 (1973).
10 Terry, J., *C. r. Seanc. Soc. géol. Fr.*, 384–385 (1971).
11 Scandone, P., Giunta, G., and Liguori, V., C.I.E.S.M.XXVI Congr., Monaco, 1974.

11

Copyright © 1962 by the Sociéte Géologique de France
Reprinted from pages 97–101 of *Livre a la mémoire du Professor Paul Fallot consacré a l'évolution paleogeographique et structurale des domaines Méditerranéens et Alpins d'Europe,* M. Durand-Delga, ed., Vol. 1, Soc. Géol. France, 1962, 657pp.

UNE HYPOTHÈSE SUR LE RÔLE DES COURANTS MARINS DANS LE DÉPÔT DE CERTAINS CALCAIRES JURASSIQUES ET CRÉTACÉS

par Henri Termier * et Geneviève Termier **

A partir du Sinémurien et jusqu'au Néocomien compris, on peut distinguer un ensemble de faciès de type dit « alpin », en réalité méditerranéen, dont le cachet propre est dû à des sédiments et à des faunes assez spéciaux. Le dessin qu'il présente sur les cartes successives, ses caractères fauniques particuliers, la finesse du grain des sédiments qui le constituent, ont paru un faisceau de preuves permettant de voir dans cet ensemble des couches de mer profonde ou tout au moins déposées loin des côtes. Nous allons essayer de montrer qu'il peut en avoir été autrement.

Le calcaire blanc et rouge de Hierlatz (Autriche), qui représente au Sinémurien *s.l.* [1] (=Lotharingien) le type « alpin », s'étend sur l'Italie et l'Afrique du Nord jusque dans le centre du Haut-Atlas marocain. On y trouve une faune de Brachiopodes, de Gastéropodes, de Lamellibranches, d'Ammonites et de Crinoïdes qui peut être considérée comme d'un type moyen, encore relativement peu différencié. En Toscane et jusqu'à la Spezzia, un calcaire rouge à Ammonites de la même époque a reçu le nom d'« Ammonitico rosso » inférieur (il monte dans le Pliensbachien inférieur). Dans l'Apennin, commence alors le dépôt des « calcari selciosi », à nodules siliceux, qui s'est poursuivi jusqu'au Kimeridgien.

Au Pliensbachien supérieur (= Domérien), se situe le « medolo » du Monte Domaro (val Trompia, Tyrol méridional) qui comprend des calcaires à silex associés à des schistes noirs à Ammonites. Ce faciès se suit aussi bien du côté des Alpes autrichiennes (région de Salzbourg) que de l'Apennin et du S de l'Espagne, et, plus au Sud, de l'Atlas Tunisien oriental au Moyen-Atlas et au SE marocains, en passant par la chaîne numidique, la Petite Kabylie et le Djurdjura. On y connaît le Pliensbachien supérieur sous la forme de calcaires à silex. Une différenciation de faunes commence à se préciser : parmi les Ammonites dominent les Phyllocératidés, les Lytocératidés et les Harpocératidés, qui constituent la célèbre « faune italienne ». Les Brachiopodes y sont caractérisés par *Glossothyris aspasia, Koninckella, Amphiclinodonta,* genres associés à diverses Térébratules, Zeilleries et Rhynchonelles.

Au Toarcien, le faciès des calcaires à silex continue dans les régions alpines proprement dites, avec ses Brachiopodes spéciaux *(Glossothyris, Dictyothyris, Antiptychina, Aulacothyris).* Plus au Sud, on en distingue l'« Ammonitico rosso » typique qui commence d'ailleurs au Pliensbachien supérieur et monte jusque dans le Bajocien inférieur : pris au sens large il se rencontre dans l'Apennin, dans les Alpes lombardes, dans le N de l'Albanie (Scutari), en Grèce occidentale, dans l'Algérie (Ghar Rouban) et le Maroc (Prérif). Il s'agit d'un ensemble de calcaires et de marnes nodulaires rouges, roses, verdâtres ou gris, dans lequel abondent les moules d'Ammonites (surtout *Phylloceras* et *Lytoceras*), associés en Grèce à des types particuliers. Il est parfois lié aux schistes à *Steinmannia bronni* et Ammonites pyriteuses, qui forment une bande méridionale partant

* Professeur à la Sorbonne, Paris.
** Maître de Recherches du C.N.R.S., Paris.

1. Nous adoptons la classification donnée par W.-J. Arkell dans « Jurassic Geology of the World » (1956).

s Alpes lombardes et couvrant l'Apennin, le N de la Sicile, le Djurdjura. Ce faciès, largement ›arti en Europe centrale, est parfois accompagné d'oolithes ferrugineuses (par exemple en ovence et, au Maroc, dans le Moyen-Atlas) ainsi que de cherts et de radiolarites.

Au Dogger, on retrouve les calcaires à silex en Afrique du Nord (Zaghouan, Bou Taleb, .bors, Ouarsenis), dans les Cordillères bétiques (Sierra de Ricote), dans les Baléares, dans le .r et dans l'Apennin. Les Ammonites, souvent pyriteuses, sont associées à *Posidonia alpina.*

A l'Oxfordien supérieur, on rencontre de nouveau le faciès « Ammonitico rosso » et les lcaires à Radiolaires dans l'Apennin, en Tunisie, en Algérie (Beni Afeur, Petite Kabylie) et ns les Cordillères bétiques. Le marbre de Guillestre, dans le Briançonnais, suivi de calcaires radiolarites rouges, en est un exemple. Dans l'Apennin, les « calcari selciosi » de Cittiglio .i atteignent la base du Crétacé, riches en Radiolaires, fournissent des jaspes (diaspri). En tite Kabylie les calcaires ont donné de nombreux Radiolaires, surtout des *Spumellaria,* et des icules d'Eponges.

Au Kimeridgien, l'« Ammonitico rosso » se rencontre dans les Préalpes vénitiennes.

Au Tithonique, à côté des calcaires à silex qui persistent, on a distingué, en Italie, le :iès « maiolica » (= biancone) qui fait souvent suite à l'« Ammonitico rosso » : il s'agit calcaires marneux, blancs, marmoréens, « porcelanés », pauvres en Ammonites *(Phylloceras, risphinctes, Simoceras),* avec des *Pygope* et des Spatangides *(Collyrites).* Dans ce faciès, .i est bien représenté en Lombardie (du Lac Majeur au Lac de Garde) et en Toscane, on voit paraître les Calpionelles (ZIA, 1955 ; FERASIN et RIGATO, 1957) et le Bélemnitidé *Duvalia.* s calcaires à Calpionelles existent aussi dans le Briançonnais. On trouve également des calcaires Calpionelles, Radiolaires et Chlorophycées en Algérie : ils sont accompagnés par des jaspes Radiolaires. Dans les Baléares et à Sidi-Marouf en Petite Kabylie, des calcaires à lits iceux, escortés de fausses-brèches, de calcaires oolithiques et de poudingues calcaires, avec uches à « filaments » (qui à notre sens pourraient être interprétés comme des sections de ›sidonies) font partie d'un ensemble plus détritique.

Les mêmes faciès se suivent, souvent sans interruption au Néocomien (Valanginien surtout), .ns presque toutes les régions citées pour le Tithonique. Le « biancone » du Trentin, des Alpes nitiennes, du Vicentin et des Préalpes Lombardes, est l'équivalent exact de la « maiolica ». renferme des Ammonites *(Phylloceras, Lytoceras, Holcostephanus)* parmi lesquelles surgit premier des types déroulés si caractéristiques du Crétacé : *Crioceratites.* On y trouve encore *ıvalia, Pygope, Collyrites* et son voisin *Metaporhinus.* Dans l'Apennin central, il se prolonge .r des calcaires sublithographiques, à silex, renfermant de nombreux Radiolaires, ainsi que s Calpionelles et *Nannoconus.* Ce faciès se suit dans l'E de la Sicile puis dans l'Afrique du ›rd (Tell Oranais, Hodna, Aurès, Tunisie, Chaîne numidique) où il s'associe au Flysch, et dans SE de l'Espagne. Les calcaires à Calpionelles atteignent même les Alpes suisses et les .rpathes roumaines.

A la fin de l'Hauterivien, ces faciès pélagiques si particuliers disparaissent assez brutanent, avec une partie des faunes qui leur étaient attachées *(Pygope, Duvalia)* et font place des marnes.

INTERPRÉTATION

On peut s'étonner que nous ayions réuni tant de faciès lithologiques. Mais on remarquera ı'ils possèdent en commun une faune benthonique et nektonique (qui a été en se différenciant puis le Pliensbachien supérieur) ainsi qu'une richesse assez grande en organismes siliceux et plankton, alliés à des calcaires à grain fin. La teinte, blanche ou rouge, qui a frappé les emiers observateurs, est l'élément le plus variable, mais non le plus important de l'aspect ces sédiments.

Examinons successivement les points suivants : 1) la présence de la silice ; 2) la teinte uge ; 3) la finesse du grain des calcaires ; 4) la microfaune ; 5) la macrofaune.

PRÉSENCE DE LA SILICE.

La plupart des « accidents » siliceux rencontrés dans les formations calcaires ont été pportés par les auteurs au cortège d'éruptions volcaniques, le plus souvent sous-marines.

Dans les régions méditerranéennes, pendant la période considérée, on ne connaît volcanisme important qu'au Rhétien dans les Alpes occidentales, les Pyrénées et les Carpath au Bajocien en Sicile, et au Crétacé inférieur (Néocomien ou Barrémien) dans l'Apennin. (trois phases principales, qui peuvent avoir favorisé le développement des Radiolaires (memb du complexe ophiolithique), ne sont donc pas contemporaines des calcaires à silex et radiolarites associées à l'Ammonitico rosso ou aux calcaires à Calpionelles.

Si l'on observe la répartition actuelle des organismes siliceux, on doit remarquer av Riedel (1959) que : *a*) environ 40 % des sédiments pélagiques leur sont dus, qu'il s'agi de Diatomées, de Radiolaires, de Silicoflagellés, voire de Spongiaires benthoniques ; *b*) organismes sont particulièrement nombreux dans les régions pourvues de systèmes de coura équatoriaux et situées à la périphérie des masses océaniques centrales, surtout là où se produise des courants d'ascendance ; *c*) ils sont beaucoup moins nombreux là où les eaux superficiel sont plus stables, dans les aires océaniques centrales stratifiées. Un remarquable exemple celui du système équatorial de courants du Pacifique au large de l'isthme de l'Amérique centr qui coïncide presque exactement avec la distribution de vases à Radiolaires.

L'origine de la silice n'est donc pas ici dans les apports volcaniques, mais se trou uniquement dans les matériaux détritiques provenant de l'érosion continentale.

2. Teinte rouge.

Nous avons vu qu'à certains moments, les calcaires ou les radiolarites étaient colorés rouge. Cette teinte si particulière pour des calcaires marins, a suscité le vocable d'« Amm nitico rosso ». Nous connaissons des roches tout à fait comparables dès le Paléozoïque, particulier les « griottes » du Dévonien supérieur, qui présentent le même caractère nodule et une faune analogue d'Ammonoïdes associés, déjà, à des Posidonies *(P. venusta)*. Da tous ces cas, la tonalité rouge ne peut être attribuée qu'au lessivage d'un continent aya subi une évolution pédologique antérieure : ferrallitisation ou cuirassement. Le pigme rouge est incontestablement fourni par un apport continental, sous la forme de sesquioxyd de fer en solution, apport contemporain d'ailleurs de celui qui a fourni le fer des faun pyriteuses et des oolithes ferrugineuses associées aux schistes à Posidonies.

3. Finesse du grain des calcaires.

La plupart des calcaires jurassiques des faciès alpino-méditerranéens sont à grain très fi sublithographiques ou porcelanés. On ne peut parler à leur sujet d'une origine détritique sens habituel de ce terme : ce sont d'anciennes vases calcaires. D'où provient le carbonat On peut, certes, penser que les abondants dépôts de calcaire et de dolomie du Trias et l'Infralias alpins, exondés ultérieurement, ont pu constituer d'importantes réserves. D'aut part, l'évolution pédologique libère les sels alcalins et calco-alcalins en même temps que la silic De toute manière, venant de terres basses et pratiquement nivelées, au moins depuis le Tria le calcaire ne pouvait se déposer que sous forme de vase à grain très fin, quelle que s la distance des côtes.

4. Microfaune.

Nannoconus étant de position systématique incertaine, nous insisterons surtout sur l Radiolaires et sur les Calpionelles qui sont des Infusoires Tintinnoïdiens. Il s'agit essentielleme d'un zooplankton.

Contrairement au phytoplankton des types Flagellés, Cyanophycées et Diatomées, jama les Radiolaires et les Calpionelles ne semblent avoir proliféré comme des « fleurs d'eau », avoir émis de toxines provoquant des hécatombes : les sédiments auxquels nous avons affai ici sont des sédiments largement oxygénés, mais recevant en abondance des éléments solution (fer, calcium) et certainement des substances nutritives. Parmi les Radiolaires actuel les Spumellaires et les Nassellaires vivent en Méditerranée entre 0 et 50 m de profonde tant que le thermomètre ne dépasse pas + 20° C. Au-dessus de cette température, ils descende dans des eaux plus profondes (jusqu'à 400 m). C'est normalement un plankton du gran rge, mais, par exemple en baie de Villefranche, il peut s'approcher des côtes et être amené a surface par les courants d'ascendance. Les Tintinnoïdiens sont des Ciliés planktoniques mmuns dans toutes les mers ; on en connaît même dans un gyttja d'eau douce du Quaternaire andinave.

MACROFAUNE.

La macrofaune se répartit entre deux catégories écologiques :

a) un *nekton* important de Céphalopodes : Ammonites (dont les membres permanents sont s Phyllocératidés et les Lytocératidés, accompagnés de formes occasionnelles de types thysiens) et Bélemnites avec, aux seuls Portlandien et Néocomien, le genre spécial *Duvalia ;*

b) un *benthos* composé d'une part de Brachiopodes spéciaux, *Glossothyris* puis *Pygope*, ayant olué sur place, d'autre part d'Echinides irréguliers, les *Collyritidés*. Ces deux types d'organismes nt microphages et adaptés à vivre sur la vase ou en partie enfouis dedans.

Nekton et benthos groupent des organismes sténo-halins poecilosmotiques donc purement arins.

CONCLUSIONS

Les faciès pélagiques du Jurassique et de la base du Crétacé en Méditerranée nous ont permis faire les constatations suivantes : 1) leur silice est d'origine organique ; les vases à Radio- ires actuelles se déposent selon des courants équatoriaux hors des masses océaniques centrales ratifiées, surtout là où il y a des courants d'ascendance ; les Radiolaires vivent dans des ux de température inférieure à 20° C ; 2) le lessivage des continents voisins arasés leur fourni des substances en solution parmi lesquelles la silice, le calcaire et les sesquioxydes colo- nt en rouge certains niveaux ; 3) une macrofaune nektonique et benthonique s'est peu à peu fférenciée dans ces faciès particuliers dont le riche plankton lui prodiguait la nourriture.

La fréquence des couches rouges indique une certaine proximité des côtes. L'intrication de s faciès pélagiques avec le Flysch nord-africain ou avec les marnes à Posidonies incline à penser i'ils ne sont pas profonds. D'ailleurs, l'épaisseur des dépôts (le « Biancone » par exemple teint à lui seul 400 m) suggère non seulement la subsidence, mais encore le voisinage des rres émergées. On doit donc abandonner pour ces faciès l'idée qu'il s'agit de sédiments de mer ofonde ou déposés loin des côtes.

Les faciès méditerranéens envisagés constituent cependant une entité non seulement par urs caractères pélagiques mais encore par leur macrofaune qui s'est individualisée sur place ogressivement. Tout se passe donc ici comme s'il y avait eu une sorte d'isolement, d'anacho- se [2] dans un faciès cependant de mer ouverte. Il ne peut être fait appel à une séparation ır la profondeur ni par la salinité mais il demeure un autre type d'isolement : celui d'une ne d'eau immiscible au sein du reste des mers à l'entour. Les procédés employés dans s parcs zoologiques pour mettre à l'écart sans vitre ni grillage les oiseaux des îles par un ran thermique nous donnent une idée d'un tel isolement.

Dès lors nous nous trouvons devant une nouvelle hypothèse de travail : n'y aurait-il pas u de substituer à l'idée d'une fosse profonde « eugéosynclinale », celle d'un vaste système courants méditerranéens de surface, plus froid que les mers qu'il traversait. Nous avons vu ie les Radiolaires ne supportent guère des eaux de surface ayant une température périeure à 20° ; or, à certains moments, mais hors de cette zone, au N comme au il y avait des biohermes coralliens dont nous savons qu'ils ne peuvent prospérer qu'au-dessus 18°. La totale absence de récifs dans la zone pélagique envisagée, pendant tout le Jurassique diquerait donc pour cette zone une température superficielle inférieure à 18° C.

L'origine de ce courant ne semble pouvoir être recherchée que du côté atlantique dans troit passage ouvert entre le Massif hespérique et le Massif moghrabin, de part et d'autre ı massif bético-rifain. Si notre hypothèse est exacte, le début des apports d'eaux atlantiques us froides dans les eaux téthysiennes chaudes semble ainsi dater du Lotharingien. Le plankton

2. H. et G. TERMIER : Evolution et Paléogéographie (Albin Michel, éd., 1959), p. 87-89 et 121.

seul devait accompagner ces eaux, tandis que les macrofaunes évoluaient sur place à par des organismes téthysiens qui avaient pu supporter le refroidissement, compensé par un appc nutritif. Quant aux solutions fournies par les continents sur l'emplacement méditerranéen, ell furent transportées par des courants de densité au-dessous du courant de surface, et par déposées directement parmi les sédiments pélagiques, partie remontées vers la surface p des courants d'ascendance. Le benthos spécial peut donc avoir vécu hors du courant atlantiqu mais dans sa dépendance immédiate, parce qu'il en recevait une pluie nourricière.

Les caractères qui ont affecté presque constamment les sédiments pélagiques méditerranée depuis le Lotharingien, ont cessé dès l'Hauterivien. Incontestablement, l'épeirogénèse q affecta l'ensemble des terres à la limite du Jurassique et du Crétacé a eu parmi ses cons quences une érosion renouvelée des terres émergées et une sédimentation plus grossière, moi calcaire et plus argileuse. L'emplacement de notre zone pélagique a reçu alors des sédimen marneux dans lesquels se sont développés les Spatangues et d'où sont précisément exclus l organismes pélagiques.

La nouvelle interprétation que nous venons de proposer permet d'envisager le poi de vue structural de la géologie méditerranéenne au Jurassique. En Afrique du Nord, zone en question se place au S des massifs cristallins géanticlinaux de Primaire métamc phique : elle fait donc partie de l'avant-fosse du géosynclinal rifo-kabyle. Son caractère peu pr fond, voisin des côtes et subsident, s'accommode beaucoup mieux de cette disposition.

On ne peut certainement pas se servir de l'argument des faciès pélagiques pour prouv gue la distance séparant l'Afrique de l'Europe avant les plissements alpins était beaucoup pl grande qu'aujourd'hui. En effet, nous savons qu'ils se rencontrent de toutes façons près terres émergées, par exemple les massifs kabyles (Durand Delga, 1955) et nous venons voir qu'ils ont été constamment alimentés par des apports continentaux, dont la finesse qrain n'est due qu'à l'arasement des terres bordières.

(Manuscrit reçu le 10 novembre 196

12

Reprinted from *Geol. Soc. London Quart. Jour.* **126**:293–318 (1971 for 1970)

Stratigraphy in mountain belts

RUDOLF TRÜMPY

23RD WILLIAM SMITH LECTURE

CONTENTS

SUMMARY

The importance of stratigraphical research for the understanding of the structure and genesis of complex mountain chains was stressed and some of its special problems were discussed.

The main part of the lecture dealt with the sedimentary history of the Alps, from Upper Carboniferous to Lower Cretaceous time, and with its implications for the crustal development in the geosyncline. Continental Upper Paleozoic and uniformly shallow-water Triassic formations point to the existence of a normal continental crust before the onset of subsidence. Deeper basins, probably bordered by normal faults, developed gradually during the Jurassic, and in late Jurassic and early Cretaceous time the geosyncline may have been comparable to some of the present smaller ocean basins. This implies thinning of the original continental crust; nevertheless, large remnants of granitic basement are present even beneath eugeosynclinal sediments and ophiolitic volcanics. In the Alps, the stratigraphical sequence is better compatible with the hypothesis of oceanization (replacement of the lower part of the crust by denser matter) than with the mechanism of ocean-floor spreading according to the Atlantic model.

1. Introduction

On this anniversary of William Smith's birth, it may not be unfitting to talk to the glory of stratigraphy, which he furnished with the biostratigraphical method, one of its most precious tools.

One of its most precious tools, but by no means the only one. In the great revolution of geological thinking which took place around the turn of the 18th to the 19th century, the essential breakthrough was probably the recognition of the historical implications of geology, and in the achievement, James Hutton's analysis of an unconformity may rank as high as William Smith's demonstration of the geochronological use of fossils.

This is to say that stratigraphy is here understood in its broadest sense, as outlined by H. D. Hedberg (1967). Stratigraphy is the science of strata, in all their aspects, and its aim is always the unravelling of part of the Earth's history. Some harm has been done by trying to restrict stratigraphy to biostratigraphy only, and to oppose 'orthostratigraphy' based on guide fossils, to a second-rate 'parastratigraphy' employing all other means of correlation (e.g. Schindewolf 1944). This in turn has led many modern geologists to consider stratigraphers as a necessary but somewhat antiquated race, fully occupied with such innocuous pastimes as measuring type sections, establishing ammonite zones and quarrelling about terminology. But while these studies are highly useful and indispensable for the mutual understanding of geologists, they do not constitute the ultimate objective of stratigraphy, which is the elucidation of geological history. And this end is a noble and worthwhile one; it is concerned with the very core and essence of geology, with its historical aspect, which lends to our science its proper character and its glorious uncertainties.

2. Difficulties of stratigraphy in mountain belts

The first steps of stratigraphy had to be taken in lithologically well characterized, fossiliferous and flat-lying beds: in the Permian and Triassic formations of central Germany, in the Tertiary cover of the Paris basin and in the Jurassic strata of England. Horace-Benedict de Saussure's early attempts to understand the structure of the Alps were doomed to failure. Only in the second half of the 19th century, once stratigraphical methods had been well established in 'normal' regions, did it become possible to understand the main lines of Alpine stratigraphy and also of Alpine structure. The famous alpine tectonicians of the beginning of this century reaped what the patient work of three generations of little-known alpine stratigraphers had sown.

The Author must ask for indulgence for choosing his examples from his native Alps. He fully realizes that the Alps are by no means a typical mountain chain; but in this respect again they are quite typical insofar as no chain is like another in its sedimentary and structural evolution.

There are of course many pitfalls which await the stratigrapher in a complex mountain belt, and especially in its crucial, eugeosynclinal part. Some of these difficulties stem from the sedimentary formations themselves, some from the succeeding tectonic and metamorphic events.

The scarcity of fossils in the thick eugeosynclinal Schistes lustrés group as well as in certain flysch formations is not only due to subsequent metamorphism (Bolli & Nabholz 1959). In a few instances, Schistes lustrés formations have been spared metamorphism by being incorporated into décollement nappes which left the central part of the alpine edifice before metamorphism set in—which, in the central Alps, means before Early Oligocene time. Examples of such non-metamorphic (and consequently non-lustrous) Schistes lustrés are the Jurassic formations of the Breccia nappe in the Prealps (Chessex 1959), or the Jurassic and

Cretaceous shales of the Arosa nappe in northern Graubünden. These non-metamorphic formations too are almost unfossiliferous.

The tectonic instability of the geosyncline is responsible for abrupt changes in facies and thickness of formations, such as are never found in normal platform deposits except perhaps in the vicinity of reefs. These discontinuous facies changes are especially marked near sedimentary fault scarps, where also such anomalous sediments as deep-sea breccias are encountered.

There are all kinds of transitions between sedimentary mega-breccias over olistostromes to gravity-slide nappes, and often it is very difficult to decide whether abnormal contacts are due to tectonic thrusting during orogenesis or to sliding at some moment or other of geosynclinal history.

These primary difficulties—scarcity of diagnostic fossils and disorderly facies changes—are further aggravated by the complexity of the alpine structure. The origin of many cover nappes is still in dispute. The intense internal deformation of most rocks created additional hazards of course by destroying sequences, sedimentary structures and fossils. In this respect, the intense lamination which many rocks have suffered at an intermediate structural level—under an overburden of the order of 5 to 8 km—can do as much damage as fairly high metamorphism. The latter has sometimes, quite surprisingly, spared delicate fossils in high-grade rocks. As early as 1930, Raguin, in the Vanoise mountains, found *Globotruncana* preserved in albite phenocrysts; in the same area, Ellenberger (1958) has described Callovian ammonites, pierced by large glaucophane needles, but still retaining well recognizable sutures. One of the last and most spectacular finds is due to F. Bianconi (1965); echinoid spines are perfectly preserved in muscovite crystals, in a garnet-and staurolite-bearing rock (compare also, Higgins 1964).

Search for fossils in metamorphic rocks needs a great amount of time and optimism, but the rewards are worthwhile. Much of the progress made during the last two decades in our understanding of the structure of the Penninic belt is due to the discovery of fossils in the most unpromising-looking formations.

A bad fossil is more valuable than a good working hypothesis. Palaeontologists should not disdain the very unattractive remains submitted to them by the mountain geologists, but rather try their utmost to determine the most probable—even if very wide—biochronological range of this poor material.

3. Stratigraphy and structure

Some enthusiastic structural geologists will claim that they are able to unravel the structure and the structural history of a mountain chain without a single fossil. This is quite true—provided there are no fossils. In the Alps, at any rate, it is hard to imagine what the structural picture would be if we did not know the age relationship of the various formations. Some major thrust-planes would quite certainly pass undetected.

In the Glärnisch and Rautispitz mountains, above Glarus, an apparently conformable succession of shales and limestones is exposed on high cliffs. Yet there are up to seven thrust-sheets (see figures in Helbling 1938 and Oberholzer 1933).

In most cases, the thrust-planes are very inconspicous. We may also mention the section of the Serenbach waterfall, on the northern side of Lake Walenstadt (Fig. 1). Here, there is a very spectacular thrust-plane (according to Albert Heim 1921, p. 377, "one of the most beautiful of all the Alps") beneath the Upper Valanginian Betlis limestones. Generations of geologists have admired it as the boundary between the Säntis nappe above and the 'Flysch' of the Mürtschen nappe below. But this 'Flysch' turned out to be Middle Valanginian shale (Brückner 1940); the true thrust-plane lies below these shales, and it can hardly be located without the help of a microplaeontologist. What Arnold Heim (1910–17) had taken for the thrust is only the trace of a subordinate and presumably late differential movement.

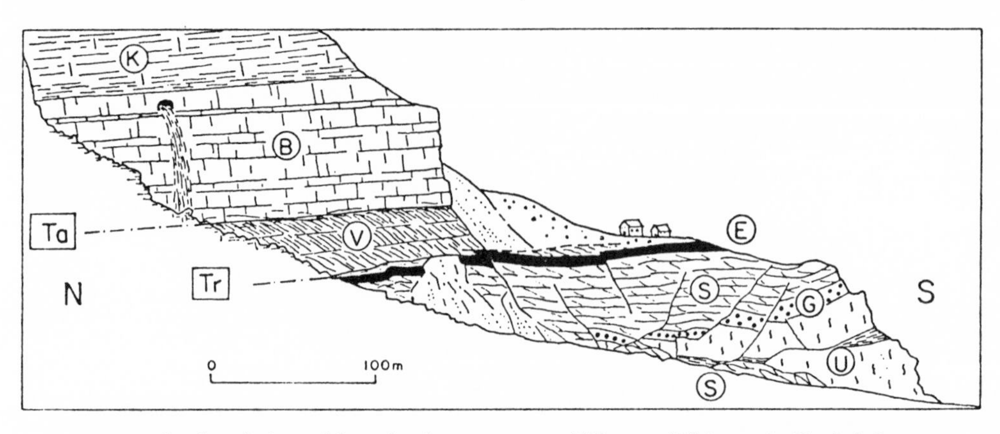

FIG. 1. Sectional view of Serenbach gorge, east of Weesen (Walensee). Copied from Heim (1921) with modifications according to Brückner (1940). V, Middle Valanginian shales; B, Betlis limestone, Upper Valanginian; K, Kieselkalk fm., Hauterivian; U, Schrattenkalk fm., Barremian; G, "Gault", Albian; S, Seewen limestones, Cenomanian-Turonian; E, Eocene Assilina greensand and Globigerina shale. Ta, apparent thrust-plane of Säntis over Mürtschen nappe; Tr, true thrust-plane.

Still more astonishing discoveries have been made in the central, metamorphic part of the chain. Here, the phenomenon of 'cover substitution' has proved to be extremely widespread. What appeared to be the normal Mesozoic cover of certain basement units has been shown to belong in fact to allochthonous décollement sheets; the true cover is extremely reduced, where it has not been stripped away before or during the arrival of the higher cover nappes. Especially in the western Alps, cover substitution takes place on a regional scale at the base of the great masses of Schistes lustrés, which the geologists of the first half of this century had regarded as original cover of the Ambin, Great St Bernard, Paradiso and Dora-Maira basement units. Painstaking stratigraphic and mapping work by Ellenberger, Gidon, Lemoine, Michard and others has thus led to a complete revolution of our views on the structure and palaeogeography of the Alps (summary in Barbier *et al.* 1963). A similar situation is found in the eastern part of the Gotthard massif, where the Jurassic formations, which seemed to form its normal cover, have been shown to be inverted (Frey 1967; Jung 1963). In most cases, these cover-nappes have moved at a (locally) early stage of structural deformation; the corresponding thrust-planes have been subsequently annealed by metamorphism and can hardly be detected by structural analysis alone.

Stratigraphical studies will not only allow us to locate otherwise unrecognizable thrust-planes, but also to replace structural units in their original position and to discuss their mutual geometrical and palaeogeographical relationships. Arnold Heim (1916) was the first geologist to show the use of detailed facies analysis for structural interpretation. This method is especially effective in the miogeosynclinal belt, where facies changes are more or less systematic and not too abrupt, and in coherent cover nappes, i.e. in nappes which are derived from a coherent sedimentary prism and where only minor gaps due to erosion or to deep burial intervene between individual sheets. Both conditions are realized in the Helvetic nappes, Arnold Heim's classical area. The Author has recently (1969) tried to work out a palinspastic development of the eastern Helvetic nappes (see Fig. 2), and Bally, Gordy & Stewart (1966) have given an excellent example of palinspastic reasoning applied to the Canadian Rockies. The merit of the method resides in that it provides partly independent sets of criteria, geometrical and palaeogeographical, for the correlation of structural units.

In the miogeosynclinal coherent cover nappes of the Canadian Rockies and of the Glarus Alps, these palinspastic reconstructions can reach a good degree of probability, and the puzzle of structural bodies can be reassembled, in section and in plan, in a quite satisfactory manner. The problem becomes much more difficult when an incoherent sequence of cover nappes, and especially units of eugeosynclinal origin with their erratic facies distribution, are studied. Such nappes are derived from widely distant parts of the geosynclinal belt, and between superposed units a breadth of several tens or even hundreds of kilometers may have lain in the original basin.

As cover nappes of this type can normally not be traced back to their roots, the discussion of their correlation bears essentially on facies analogies. But facies analogies are polymath: synsedimentary faults may cause great disparity of facies in immediately adjoining areas, and, on the other hand, similar facies associations may recur in widely distant zones of the geosyncline. In the Alps, for instance, large bodies of Jurassic and Cretaceous deep-water breccias are found in at least four zones: on the talus bordering the Valais eugeosyncline to the north (e.g. Niesen breccias, Feuerstätt nappe of Vorarlberg) and to the south (e.g. Falknis, Tarentaise), and again on the northern margin of the Piemont eugeosyncline (e.g. Breccia nappe of the Prealps) as well as on its southern one (Lower Austroalpine nappes of Engadine). Facies comparisons must of course be made on the base of the whole of the sedimentary sequence, and not on that of one formation alone. But even so, the origin of many cover nappes is debatable. Alpine literature is full of arguments between individual geologists or between schools of thought on the probable origin of some cover units; and these mandarin disputes make rather uninspiring reading for our lowland colleagues. But they have a bearing on some fundamental problems, such as the amount of crustal shortening. Even today, after many generations of geologists have left their hammermarks on the Alpine

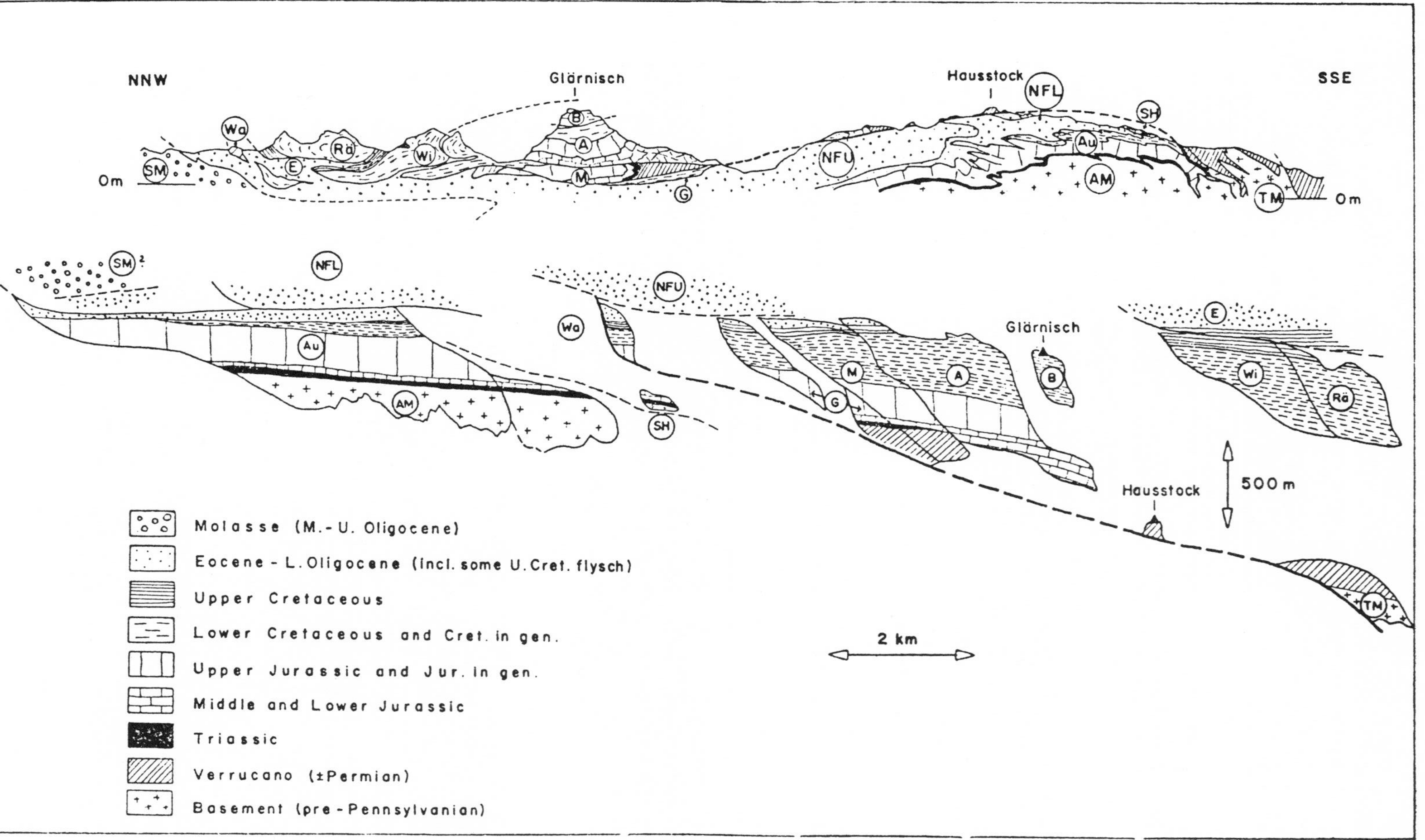

FIG. 2. Structural section and corresponding palinspastic section through the Glarus Alps. Scale 1:100 000; vertical scale not exaggerated in structural section, exaggerated $2\frac{1}{2}$ times in palinspastic section. Am, Aar massif; TM, Tavetsch massif. Au, Autochthonous cover; SH, Subhelvetic slices; W, Wageten slice; G, Glarus nappe; M, Mürtschen nappe; A, Axen nappe; B, Bächistock slice; W, Wiggis- and RäRäderten-subunit of Drusberg-

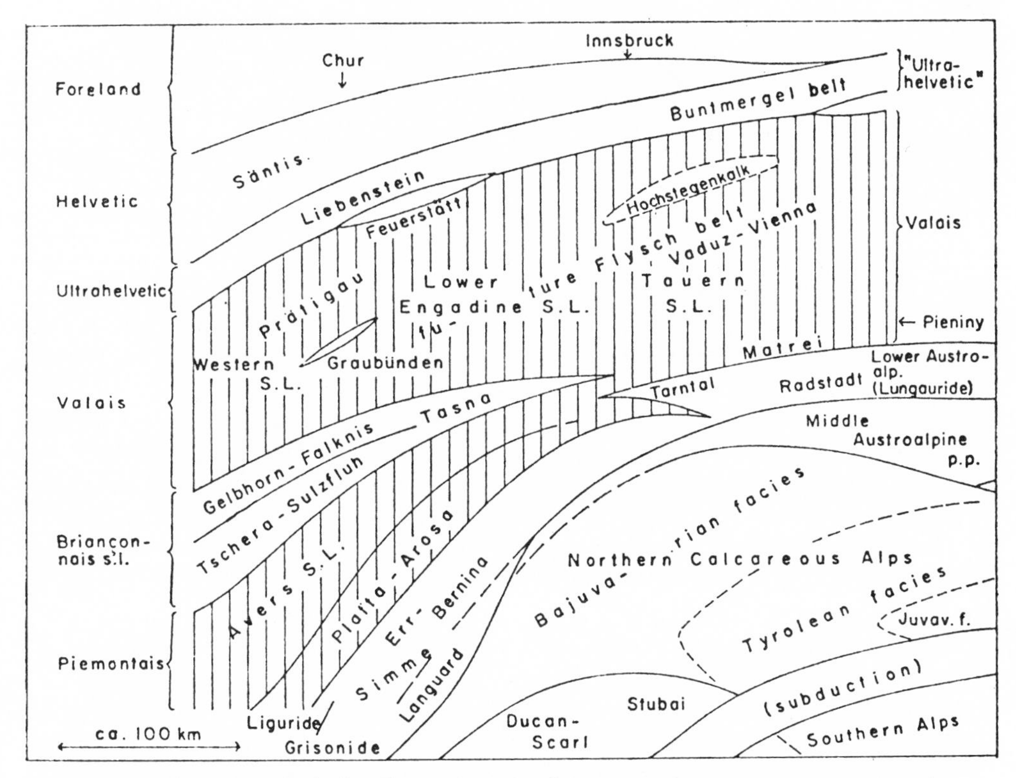

FIG. 3. Very hypothetical palinspastic map of the Alps of eastern Switzerland and western Austria. Vertical ruling: more or less eugeosynclinal belts.

rocks, the origin of some important units is being disputed. This is the case for the Simme nappe of the Prealps (see especially Elter, Elter, Sturani & Wiedmann 1966), for the strange bundle of the Schams nappes in Graubünden and for all the Northern Calcareous Alps between the Rhine and the Danube.

These doubts on the origin of certain units account for very different paleogeographic reconstructions drawn by different authors. In Fig. 3, we have tried to give a diagrammatic palinspastic map for the segment between Graubünden and the Tauern window. It will be noted that this figure differs markedly from the one drawn by Tollmann (1965) and, to a lesser degree, also from the remarkable series of reconstructions recently published by Oberhauser (1968). The main difference between Tollmann's reconstruction and our own lies in the position of the Tauern Schistes lustrés, which we attribute to the Valais rather than to the Piemont eugeosynclinal belt. In Tollmann's interpretation the Falknis nappe of central Graubünden and the Tasna nappe of the Lower Engadine are torn apart; but the facies of these two units as well as their structural position is practically identical (see Allemann 1957; Cadisch 1953; D. Trümpy 1916). These discrepancies may be mentioned merely to illustrate that some quite relevant points are at stake, and that no palaeogeographic reconstruction—not even Fig. 3—should be taken as an expression of absolute truth.

4. Early stages of geosynclinal evolution in the Alps

The second part of this lecture deals with the possible evolution of the crust underneath the Alpine geosyncline. It is merely intended to point out what contributions stratigraphy can make towards the weighing of the different working hypotheses, not to propose a solution to the problem.

This makes it necessary to give a short outline of the evolution of the Alpine geosyncline up to its oceanic (or rather 'para-oceanic') stage, in Late Jurassic and Early Cretaceous time. It may be very brief, as we have already on two occasions (Trümpy 1960, 1965) written general summaries on the subject intended for the reader not acquainted with the intricacies of Alpine geology.

As far as we know, the whole area of the future Alps was affected by the Hercynian orogeny, especially by its 'Sudetic' phase in mid-Carboniferous time (310 to 300 m.y.). Fossiliferous Ordovician to Lower Carboniferous formations are only preserved in the south-eastern part of the Alps (and in the northern Grauwacken zone, which is probably derived from a belt just to the north), where the Hercynian deformation was less pronounced; but even here (in the Carnic Alps) there are small thrusts of Carboniferous age (for a summary on the Palaeozoic of the eastern Alps see Flügel 1964). The Hercynian folding must have produced a thickened and probably unstable continental crust, in the future Alps as well as their future foreland.

Hsu (1958) has suggested that after a tectonic disturbance, the continental crust will tend towards a new equilbrium by the subsidence of local depressions. This is exactly what happened after the Sudetic folding: elongated basins formed on the site of the Hercynian mountain chain. During Late Carboniferous and Early Permian time, these basins received the detritus from the surrounding mountains. Sedimentation was accompanied by strong volcanism, of both basaltic and rhyolitic composition.

Throughout the western and central Alps, the Pennsylvanian and Permian deposits are of continental origin.[1] Marine strata of this age are again only found in the south-eastern Alps, where they are interbedded with continental sandstones and contain coral-bearing limestones as well as gypsum, thus giving clear evidence of shallow-water origin. Nothing suggests the presence of deeper seas in Upper Carboniferous or Permian times, which might be interpreted as remnants of the Palaeozoic geosynclines. This applies not only to the Alps, but to the whole of the western Mediterranean. Marine transgressions of Late Carboniferous (Late Moscovian and Gshelian) and Permian age reached westwards only to the south-eastern Alps, to Sicily and to southern Tunisia. Some reconstructions (e.g. Glangeaud 1968, fig. 6) show Permian ocean basins in this area—basins which have to be carefully placed under present seas, as on the surrounding land all Permian strata are of continental or shallow-water origin. There is no stratigraphic evidence

[1] At one time, the Author thought he had found a marine limestone bed in the Glarus Verrucano (in Brückner *et al.* 1958), containing gastropods, calcareous algae and dubious echinoderms. The presence of the latter, however, has not been confirmed, and the whole environment is more compatible with a freshwater limestone (Nio, in press).

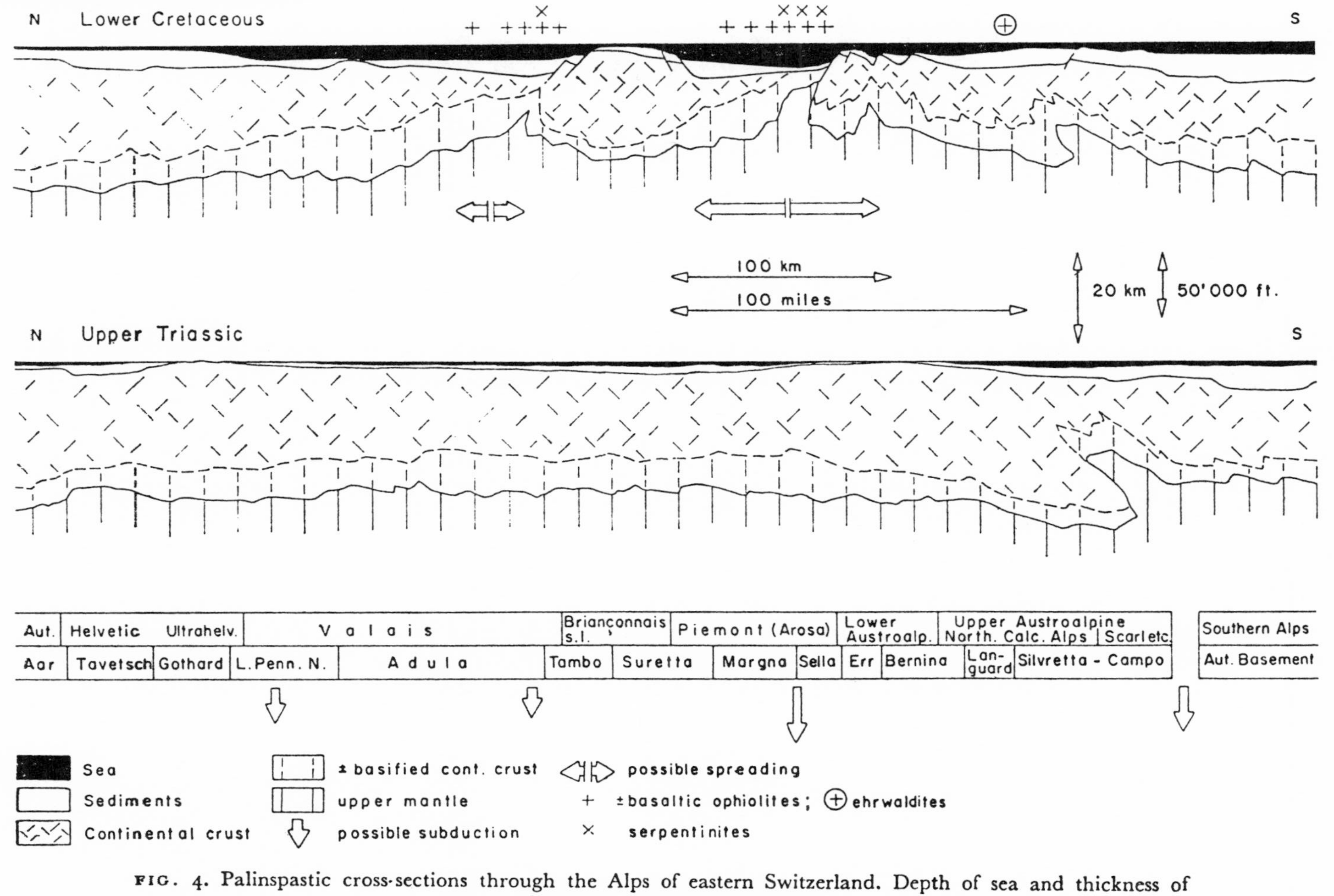

FIG. 4. Palinspastic cross-sections through the Alps of eastern Switzerland. Depth of sea and thickness of sediments somewhat exaggerated.

whatever for the existence of Permian oceans in any part of the western Mediterranean.

Another sensational idea of some geotectonicians receives little support from stratigraphical observations. This is the 'Tethys twist', put forth by van Hilten in 1964 on the ground of rather scattered palaeomagnetic evidence (deBoer 1963). This enormous Permian and Early Mesozoic movement would have been a dextral shift of the order of 3000 or 5000 km, and if we keep the northern Alps in place, the southern Alps would have lain, during the Palaeozoic, somewhere around the Persian Gulf or even in the Himalayan foreland. But in this case, the fact that marine Palaeozoic formations are only found in the eastern part of the southern Alps (Carnic Alps) on the one hand, in the original extreme south-eastern part of the Alps proper (Graz area, northern Graywacke belt) on the other, and that they show a coherent facies development in these two areas, would become an almost incredible coincidence. In the central Alps, too, the character of the Pennsylvanian and Permian continental basins is essentially the same, from the Bergamask Alps in the south to the Glarus Alps in the extreme north—and this would be very strange indeed if they had originally lain many thousands of kilometres apart. The very detailed analogies between the Triassic formations of the northern Calcareous Alps and those of the southern Alps have been recognized for more than a century, and the facies of the northern Calcareous Alps is in turn linked by transitions of that of the Penninic and Helvetic realm. Much stronger evidence for the Tethys twist must be brought forward before it can be accepted in spite of these and other stratigraphical improbabilities. This does not mean that we would exclude lateral displacements along the great disturbance of the Tonale or Insubric line, bordering the southern Alps to the north (see Gansser 1968). Only these movements would be far smaller than those postulated by Bemmelen (1966), Boer (1963), Hilten (1964), Hilten & Zijderveld (1966) and Rutten (1963), and much later (essentially Neogene).

At present, we have no stratigraphical evidence that the area of the future Alps, during Late Carboniferous and Permian time, differed in any significant way from the other parts of Hercynian Europe. Inside and outside the Alps, we find the same elongated, often fault-bordered troughs, filled with continental detritus (redbeds in the Permian), with the same kind of volcanism and affected by the same mid-Permian ('Saalic') posthumous folding.

By the beginning of the Triassic period, local subsidence, local uplift, erosion and sedimentation had apparently restored the crustal equilibrium. All over western Europe, Triassic times are characterized by extremely low relief. Sediments which must have been laid down a few metres below or a few metres above sea-level are spread out over vast areas (e.g. the Carnian Schilfsandstein from northern Germany to Provence and well into the western Alps, see Wurster 1964).

In the Alps, too, with very few exceptions, Triassic rocks have formed in extremely shallow water. Quartz sandstones and sandy shales prevail in the Lower Triassic. Evidence of a very shallow marine, frequently intertidal and supratidal environment, includes coarse cross-bedding, symmetrical ripple marks, shrinkage cracks, mudcracks, gypsum deposits, stromatolites and saurian footprints (see e.g. Fischer 1965; Somm & Schneider 1962). Sedimentary rocks formed in rather deeper water occur only quite locally, in the Ladinian Buchenstein and Wengen beds of South Tyrol, where they are linked to a region of strong submarine volcanism and concomitant faulting, and in the very highest nappes of the eastern part of the northern Calcareous Alps (Upper Triassic Hallstatt limestones). Even these formations, probably laid down under a few hundred metres of water, intertongue with shallow-water carbonates (Schlern dolomites and Dachstein reef limestones respectively).

The prevalence of Triassic shallow-water deposits in the Alps implies the presence of a stable, continental crust, presumably of normal thickness. Nowhere have the slightest indication of oceanic conditions during the Triassic been found.

The geosyncline originated during these extremely tranquil Triassic times, during this interregnum between the Hercynian and the Alpine cycles. Certain belts with greater thickness of shallow-water formations appear already in the Lower and Middle Triassic: 400m of Lower Triassic quartzites and 700m of Middle Triassic carbonates in the Briançonnais belt of the western Alps, up to 2000m of Middle Triassic in parts of the southern Alps. But the location of these zones does not conform to that of the future geosynclinal furrows, which only appear a little later, mainly at the turn from the Ladinian to the Carnian (in the Piemont belt of Western Alps, according to Michard 1966, already during the Ladinian, and probably even earlier in the south-eastern Alps). This first downwarping of future eugeosynclinal troughs must have been very gentle and was not accompanied by major tectonic movements. The Alpine geosyncline was not born with a bang, for instance by a sudden tearing apart of continental blocks; subsidence began quite slowly and only gradually became stronger.

5. The oceanic stage of the Alps (Jurassic and Lower Cretaceous)

At the beginning of the Jurassic period, the Alpine seas were still of moderate depth. Deposits of Hettangian and Sinemurin age are in many places of shallow-water character. The oldest pelagic limestones or uniform shale sequences frequently belong to the Upper Pliensbachian (Domerian). The gradual acceleration of subsidence during the Lower Jurassic is exemplified in many profiles, from the Ultrahelvetic belt (e.g. Trümpy 1951) down to the southern Alps (e.g. Bernoulli 1964).

The palaeotectonic features of Sinemurian to Bajocian or Bathonian times have been reconstructed in several different facies belts, from the Helvetic belt (Baer 1959; Schindler 1959; Trümpy 1949) over the Briançonnais zone (Debelmas 1957; Lemoine 1953) to the southern Alps (Bernoulli 1964; Wiedenmayer 1963). They are apparently governed by flexures and by normal faults, which are often associated with submarine breccias. Especially along the northern margin of the geosyncline, there are antithetic blocks, inclined to the north and limited by south-facing normal fault scarps. These early Jurassic tectonic structures are best explained by tensional stresses; at any rate they are not compatible with Argand's (1920) and Staub's (1917) notion of embryonic folding during the early stages of geosynclinal evolution.

In the Author's 1960 review, he had tentatively put forth a relatively simple scheme of sedimentary evolution in the eugeosynclines: rapid sedimentation during the Lower Jurassic and Bajocian ('older Schistes lustrés'), then a phase of reduced, mainly pelagic deposition (our leptogeosynclinal phase, Goldring's (1961) bathyal lull or Aubouin's (1958) période de vacuité), followed by the sedimentation of the 'younger Schistes lustrés' during the Lower Cretaceous. Most of the Schistes lustrés of the external, Valais eugeosynclinal very probably belong to the younger group, as they pass upwards into preflysch formations dated as Middle Cretaceous. But, at that time, we thought that most of the Piemont Schistes lustrés belonged to the older group, in spite of their great similarity to those of the Valais trough. Recent work by French geologists, especially by Lemoine (1953, 1959) and by Michard (1966), seems to indicate that the true ophiolite-bearing schistes lustrés of the Piemont trough may also be of Cretaceous age, and that they are separated by a break from the underlying, fossiliferous Liassic formations. There are some other indications for a widespread Middle Jurassic event: continental influences (bauxite, coal) occur in the Mytilus beds of the western Alps, pebbles of gneiss and rhyolite appear in the Upper Bajocian and Bathonian of the Helvetic and Ultrahelvetic belt, and flysch-like Middle Jurassic formations are known from the Schams nappes (Jäckli 1941; Schmid 1965), from the western margin of the Piemont trough (Lemoine 1953) and from the Lower Austroalpine nappes. In other sections, there is quite a gradual passage from the Liassic–Bajocian formations to the overlying pelagic beds of Bathonian or Callovian to early Cretaceous age. The significance of this mid-Jurassic event and its very existence is quite elusive; does it imply the emergence of Penninic islands, an abortive attempt at folding (Jenny 1924), or is it in some way connected with a phase of crustal spreading?

At any rate, this still badly understood mid-Jurassic disturbance is followed by quiet, pelagic sedimentation over all Alpine zones south and east of the Briançonnais platform. The time from the Callovian to the Valanginian is one of reduced sedimentation—the bathyal lull, to use the most suggestive of the three alternative terms proposed. The usual sediments are pelagic cherts and micritic limestones, which show signs of aragonite and even of calcite leaching. Autochthonous benthonic fossils are exceedingly scarce. Detritus derived from metamorphic rocks and from granites, forming breccias and graded arkose layers, occurs in certain areas, though less frequently than either before or after the period of vacuity.

The combination of several criteria—none of which are decisive in themselves—such as carbonate leaching, enrichment in manganese, absence of benthonic fossils and of any sign of high-energy environment, the presence of turbidites and above all the slow deposition, wide extent and uniform character of these formations decidedly point to deep-sea origin for the radiolarian cherts, the Aptychus limestones and the accompanying shales. It is hazardous to name a precise depth by taking, for instance, the present-day horizons of calcite and aragonite solution (see also criticism by Hudson 1967 and Sonder 1939), but the whole character of the Upper Jurassic and part of the Lower Cretaceous formations is most easily explained by assuming depths of several thousand metres. At least, there are no observations to contradict this.

The horizontal spread of the radiolarian chert–Aptychus limestone suite is quite remarkable. We find it locally in the Subbriançonnais belt, in the entire Piemont eugeosyncline, on the Austroalpine platform (broken only by the local high of the Plassen limestones, in Salzkammergut, which pass laterally, over calciturbidites, into the pelagic limestones; see Flügel & Fenninger 1966; Garrison 1967) and in almost the entire southern Alps. This makes for a N–S extension of a least 250km (and very possibly much more, if we take into account the southward extension under the Italian plain and if we assume greater shortening of the Piemont-Ligurian trough); from west to east, these rocks are found all through the chain, over 1200km. And the very same rock suite re-appears in several Mesozoic troughs of the Mediterranean area outside of the Alps.

We are forcibly led to admit the existence of an ocean-like body of water in the western Mediterranean during Late Jurassic and Early Cretaceous time (cf. Colom 1957; Steinmann 1925). In the Alps, we are confronted with the northern margin of this ocean, with an unstable, horst-like platform (Briançonnais), a marginal mud-filled trough (Valais eugeosyncline), a possible continental slope with mainly pelagic sediments laid down at depths of several hundred metres (Ultrahelvetic belt) and the bordering shelf seas (Helvetic miogeosyncline and adjoining stable shelf of the European foreland).

This Mesozoic Mediterranean para-ocean was in no way of uniform depth. Deeper troughs, with typical eugeosynclinal sediments and ophiolites lay alongside broad swells with pelagic sedimentation in moderate depths (several hundred to a few thousand metres); these broad intra-oceanic ('Blake') platforms are the 'aristogeosynclines' of Tollmann (1968), exemplified by the Austroalpine belt. Most geologists will regret the creation of this new term, but Tollmann has a certain point insofar as these unstable platforms have neither the character of true eugeosynclines nor that of miogeosynclines. Inside the Mediterranean ocean, there were also some islands and extensive platform areas with sedimentation of shallow-water carbonates. The complete reconstruction of this Jurassic Mediterranean palaeogeography requires some rather adventurous shifting around of larger and smaller continental blocks (see Fallot 1960). Interesting attempts at such reconstructions have been made (e.g. Glangeaud 1957, p. 217), but the over-all picture will still require a fair amount of discussion.

6. Reduction of the continental crust during the oceanic phase

During the oceanic stage, from Lower Jurassic to Upper Cretaceous, and especially from late Middle Jurassic to Lower Cretaceous, the continental, 'granitic' crust must have been very much reduced under large portions of the Alpine geosyncline and of the Mediterranean area in general. Several independent observations make this assumption necessary. The first is the widespread occurrence of deep-sea sediments, which has already been mentioned. The second is furnished by the ophiolites of the eugeosynclinal belts. Parts of these are submarine flows and sills of basaltic composition, part are bodies of serpentinite, the mode of emplacement of which is in most cases still in dispute. At any rate they point to the presence of peridotitic, presumably Upper Mantle, material at shallow depths below the ocean floor; in some cases, it may be suspected that pelagic sediments were laid down directly upon serpentinites.

Thirdly, it is almost impossible to understand the intensity of alpine folding if we imagine that a continental crust of normal thickness had been involved (see Closs 1964). The original width of the Alps is of the order of 600km, perhaps even considerably more—we would not balk at a figure of 800km. Taking 600km and a crust of 30km plus an average thickness of sediment of 3km, this would produce a cross-sectional area of some 20 000km^2, of which about 1500km^2 might have been removed by erosion since the Oligocene. There is no room for such an enormous root of sialic material, even if we admit considerable attrition by convection currents.

On the other hand, there is ample evidence for the preservation of continental crust under very large parts of the Tethys, and even under some of its eugeosynclinal troughs (see e.g. Nicolas 1966). This is proved by the broad platforms with pelagic or shallow-water sediments between the eugeosynclinal furrows, by the abundance of 'granitic' detritus in practically all eugeosynclinal sediments (Booy 1966, 1967) and above all, of course, by the outcrops of pre-Triassic (mainly pre-Pennsylvanian) basement. This basement complex forms, for instance, the crystalline cores of the Penninic and Austroalpine nappes. Minimum figures for the thickness of the remaining continental crust are given by the thickness of the exposed basement complex in these nappes. It can be estimated at 7km for the Bernhard nappe, 4km for the Dentblanche (and probably much more for the corresponding Sesia complex), 4km for the Suretta and 9km for the Austroalpine Silvretta–Oetztal basement nappe. Even beneath frankly eugeosynclinal Mesozoic sediments, a basement of fairly normal 'granitic' character is still present (Adula nappe: 4 to 5km; Margna nappe: 1·5km). The true figures may be much higher, as there is no indication that these nappes reach down into deeper parts of the granitic crust (except, perhaps, the Dentblanche nappe with its curious gabbro bodies, of still uncertain age, and its highly metamorphic basement rocks).

If we are searching for eugeosynclinal sediments laid down directly on oceanic crust, our only hope lies in certain décollement nappes (strip-sheets) derived from the Valais and Piemont troughs, and especially in the Ligurian eugeosyncline of the Apennines, which continues into the western Alps (Elter *et al.* 1966). Its partial equivalent in northern Graubünden is the Arosa nappe, probably identical to the Platta nappe of southern Graubünden. Even in these nappes, the evidence for oceanic crust is largely negative, as there are generally few or no basement rocks.

We can take the Arosa nappe as an example of a structural unit for which absence of continental crust has at least been suspected (Laubscher 1969). It comes from the southern part of the Piemont trough, immediately next to the Austro-alpine Err nappe. The Arosa–Platta nappe contains great masses of serpentinite (the best known of which is the Totalp serpentinite near Davos, see Peters 1963) and also basaltic pillow lavas. The latter have recently been dated by Dietrich (1967), who found Middle Cretaceous (approximately Albian) pelagic foraminifera in sediments between lava flows. Part of the serpentinites are believed to be older. The Mesozoic sediments (Dietrich 1967; Gees 1955; Grunau 1947, 1959; Richter 1957) apparently measure only a few hundred metres; they are represented by calcareous shales with calcarenite and sandstone layers (Lower Jurassic to lower Middle Jurassic), by radiolarian cherts (upper Middle Jurassic to lower Upper Jurassic), by Aptychus limestones (uppermost Jurassic to lowermost Cretaceous) and by non-calcareous Cretaceous shales with beds of the very fine-grained limestones which are called Palombino in the Apennines. There is also some Cretaceous flysch (e.g. Verspala flysch of the Rhätikon). The Jurassic formations often contain boulder beds, derived from Triassic dolomites and from acid basement rocks (gneisses and green granites); some of these boulder beds have been mistaken for gneiss or micaschist by earlier workers. Pre-Jurassic formations are quite subordinate, but nevertheless present within the Arosa-Platta nappe. They consist of ordinary gneisses, continental Lower Triassic and shallow-marine Middle and Upper Triassic (including fossiliferous Rhaetic beds). Some of these rocks may represent slumped megablocks; but in several places (e.g. east of Alp Flix in Grisons) the whole pre-Jurassic section is exposed in a continuous outcrop.

It is not impossible that part of the Jurassic deep-sea sediments of the Arosa nappe were laid down directly on oceanic crust. On the other hand, there must have been areas with a normal pre-Triassic basement complex close by (boulder beds!), and even within the basin remnants of the original continental crust and of its cover of shallow-water sediments were preserved.

In all its outcrops, the Arosa-Platta nappe shows a very intricate structure. Part of these complications may be due to large scale submarine slumping (coloured mélange type, see Gansser 1959 and Kündig 1959), but many of them are definitely tectonic. For this reason, it is rather difficult to estimate the original width of the trough; 30km may be considered as a minimum. The unit may formerly have been much more important, as it furnished large amounts of fine-grained detritus (especially grains of chrome spinel) to Upper Cretaceous flysch formations further south (see Gasser 1967; Oberhauser 1968). A first phase of folding and uplift must have taken place shortly after the outpouring of the basaltic ophiolites (Aptian to Albian), as chrome spinel from ultrabasic rocks of the Arosa zone already appears in the Cenomanian. The proponents of the ocean floor spreading theory for the Alpine geosyncline will hold out that the Arosa belt must have been much wider than it appears now. This may very well be true; but the Author would not go too far for simple space reasons and because the boulder beds show that higher blocks with indubitable, though probably diminished continental crust were rather close by; 100km might seem a reasonable maximum figure.

At the beginning of the Cretaceous, the crustal structure in the future Alps must have been a rather complex one. There were belts with a continental crust of normal thickness, belts with an abnormally thin (perhaps also abnormally dense) continental crust and also a few, probably narrow trenches which may have lacked continental crust. The over-all picture may have been comparable to that of the present-day Caribbean region (Butterlin 1956; Officer *et al.* 1959; Škvor 1969). The Triassic shallow-water formations below the eugeosynclinal Jurassic and Cretaceous of the Alps recall the salt-bearing beds underlying the present Gulf of Mexico. The comparison with the present-day Mediterranean (e.g. Glangeaud 1962; Hersey 1965) seems less satisfactory, as this sea is definitely of syn-tectonic and post-tectonic origin.

7. Possible mechanisms of crustal reduction

How was this thinning or even complete disappearance of the continental crust under parts of the Alpine geosyncline achieved?

The existence of a primary oceanic crust before the beginning of subsidence can be practically ruled out. The omnipresent Hercynian folding, the continental character of the Pennsylvanian and Permian formations and the uniform shallow-water character of the Triassic formations all point to the existence of a continental crust of more or less normal thickness at least during the Lower and Middle Triassic. This leaves us to explain how part of this continental crust was reduced in the time between the Middle Triassic and the Lower Cretaceous. Three theories have been proposed:

(1) Sub-aerial erosion prior to geosynclinal subsidence.
(2) Replacement of the lower part of the continental crust by denser material (oceanization, basification, sub-crustal erosion).
(3) Ocean floor spreading according to the Atlantic model.

We should like to discuss briefly how far these mechanisms are compatible with the stratigraphical evidence.

The first hypothesis, put forth by Hsu (1965), calls for a thermal or chemical differentiation in the upper mantle, by which its density is diminished. The crust will form a bulge and will instantly be subjected to erosion. Once normal conditions are re-established in the mantle, the thinned continental crust will start to buckle down.

If we apply this theory to the Alps, we encounter a number of difficulties. As there are Pennsylvanian and Permian sediments all through the chain, the great erosion would have to be reported to pre-Pennsylvanian time—which is precisely the moment of the main Hercynian folding. It is difficult to envisage how an abnormally thin type of continental crust would have persisted during the Triassic period with its desperately flat relief (see section 4). Furthermore, if such a considerable erosion on a broad uplift had taken place, we would not expect to find rocks of (ante-alpine) greenschist facies and even unmetamorphic Ordovician to Mississipian sediments in the Alps.

It is true that the pre-Pennsylvanian Palaeozoic rocks are not known from the western Alps and that there is one place, at least, where the Permian continental crust may have been thinner than normal. This is the region of the Ivrea zone, where only 5 to 20km of gneisses lie between the Permian sediments and the ultrabasic to basic Ivrea body, which may represent a relic of Hercynian mantle rocks (see papers in the 1968 Ivrea symposium volume, in particular McDowell & Schmid, Nicolas, Vuagnat, and also Schmid 1967). But on the whole, everything is in favour of interpreting the Carboniferous erosion as a consequence of the Hercynian folding; the uplift would be due to tectonic thickening of the continental crust rather than to a density decrease of the upper mantle.

We do not believe that Hsu's hypothesis will explain the genesis of the Alpine geosyncline, but we would be tempted to apply it to quite another sort of basin, namely syneclises ('autogeosynclines'). It is not impossible that the Paris basin started to subside after the crust had been weakened by the strong erosion it underwent during the Lower Triassic.

The second theory or group of theories invokes sub-crustal erosion or basification. This seems to be the hypothesis favoured by most of the authors who have pronounced themselves on the subject (e.g. Beloussov 1962, 1968; Bemmelen 1966, 1968; Gidon 1963; Gilluly 1955). Several mechanisms have been proposed. Most of these appear to be possible, but none can be said to be absolutely verified. The essential concept is that 'granitic' material in the deeper part of the continental crust is replaced by denser matter. Theoretically this can be achieved by phase changes, by chemical processes (metasomatism) or by mechanical processes (sub-crustal erosion by convection currents). Condensation of the upper mantle might be the initial cause (Subbotin *et al.* 1965).

It may be that geologists of the last two generations have been over-fascinated by the Mohorovičič discontinuity, and that they have been persuaded to consider it as an absolute boundary limiting two different worlds. Recent geophysical research (Afanassyev 1968; Giese 1968) rather suggests that there is a transitional zone, with interpenetration of crustal and mantle material. It may be significant in this respect that under many present smaller basins of more or less oceanic type there is an abnormally thick 'basaltic' layer in the lower part of the crust (see e.g. Menard 1967).

The concept of basification or sub-crustal erosion of the continental crust—whatever its geophysical foundations may be—is, at any rate, perfectly compatible with the stratigraphical observations. We have seen in sections 4 to 6 of this paper that everything suggests a gradual thinning of an originally normal continental crust under the Alpine geosyncline. Subsidence begins in the Middle Triassic, the first pelagic sediments appear in the upper Sinemurian or lower Pliensbachian, and at the end of the Jurassic the continental crust may have been removed altogether in one or two deep trenches. The basaltic ophiolites were emplaced during the Lower and Middle Cretaceous, and shortly afterwards, just before the Cenomanian, compression set in, closing the vents of the ophiolitic magmas. The theory of sub-crustal oceanification also accounts for the persistence of remnants of approximately normal continental crust within the geosyncline (e.g. Briançonnais platform) and for the existence of broad expanses apparently underlain by continental crust of subnormal thickness or supranormal density (Austroalpine realm), nor is it contradictory to the indications of tensional stresses during the Lower and Middle Jurassic (see section 5).

The sub-crustal erosion or basification theory is capable of explaining all the known stratigraphical facts. But it has one great weakness: it appeals to processes which have not or not yet been directly verified. It is a kind of *deus ex machina*, very satisfactory to the field geologist but less so to the theoretician.

If we look for a mechanism for the genesis of oceans and ocean-like basins, which can stand the test of physiographic uniformitarianism, we forcibly arrive at the conception of ocean floor spreading. At least for the Atlantic Ocean it has passed, within the last few years, from the stage of fruitful working hypothesis (Hess 1962) to that of a fact (see e.g. Drake & Nafe 1967; Heirtzler *et al.* 1968; Hurley 1968; A. E. Maxwell 1969; Le Pichon 1968; Vine 1968). It is not surprising that many geologists are tempted to apply the sensational results of recent geophysical and oceangraphical research to geosynclines (e.g. Yeats 1968, for the California Coast Ranges).

As early as 1916, Arnold Heim compared the Helvetic Cretaceous formations to continental shelf deposits, the Ultrahelvetic shales and pelagic limestones to those of the continental slope and the Schistes lustrés to oceanic sediments. Dietz (1961) considered the sediments off the eastern shore of the United States (cf. Emery 1965) as a model for geosynclinal sedimentation in general (see also Drake, Ewing & Sutton 1959). If we take into account only one margin of the geosyncline, there is undoubtedly a certain resemblance with some present continental margins (compare Emery 1966; Engelen 1963; Ludwig *et al.* 1966; Uchupi & Emery 1963). Recently H. P. Laubscher (1969) has gone even farther, although in cautious wording. He suggests that, before the Alpine compression, the African and European blocks might not only have lain several hundreds of kilometres but even much further apart, and that the intervening 'geosyncline'—a term he rejects on account of its non-uniformitarian context—might in reality have been a basin of truly oceanic type. The Helvetic miogeosyncline and the much broader one of the Sahara Atlas would represent the respective shelf expanses; the nappes with basement cores would be derived from the areas with tapering continental crust or from isolated remnants of granitic crust within the ocean, comparable for instance, to the Seychelles Islands in the Indian Ocean. The broadest part of the basin, underlain by oceanic crust, would not have been directly involved in the folding; only some of its 'eugeosynclinal' sediments would have been stripped off, forming, together with slivers of mantle material (serpentinites) the décollement nappes of the Arosa type. A somewhat similar model has been proposed by Dewey (1969) for the Caledonian Chains of the North Atlantic area.

The present Author confesses that he is rather reluctant to accept the 'Atlantic' model for the Jurassic Mediterranean. This ocean had no median ridge we know of and far too many Seychelles. The main question is whether the eugeosyncline, and more particularly those parts for which a granitic basement is not known, can have been thousands of kilometres wide.

In the Swiss Alps, we have seen (section 6) that the belt of the Arosa zone measures at least 30km across, and that we would accept a maximum of 100km. We would not like to go much further for several reasons. In the Jurassic of the Arosa belt there are definite indications that shoals or islands with gneissic basements stood close by, and that slivers of normal basement rocks occur even within this unit. Furthermore, the two major eugeosynclines do not continue all along the strike of the chain. The external, Valais eugeosyncline wedges out towards the southwest. Its westernmost outcrops of ophiolites are found in the region of the Lesser St Bernard col (see Antoine 1968). South of the river Isère, this facies belt disappears altogether, at least at the surface; even if the Ultradauphinois belt to the west and the Subbriançonnais belt to the east cannot be proved to have been immediately contiguous, there is nothing to suggest a broad expanse between the two. A gradual narrowing and degeneration of the Valais belt from Graubünden over Valais to the Aosta valley is quite obvious. In a similar manner, the internal, Piemont trough tapers out eastwards. It is but very poorly represented in the Lower Engadine window, and, in our opinion—which is not shared by Tollmann—it has disappeared altogether in the Tauern window (see section 3, Fig. 1). This is not surprising if the two eugeosynclines were relatively narrow trenches, a few hundreds of kilometres wide, but it can hardly be reconciled with the idea that they were oceans measuring several thousand kilometres across the trend.

There are some limitations opposing the idea of enormous relative movements of the European and African blocks (e.g. Carey 1958; Wilson 1963). They have been very clearly put forward in a remarkable paper by J. C. Maxwell (1969); see also Klemme (1958). Around the sea of Alborán, the Moroccan part of Africa and the Spanish part of Europe can never, at least not since the Permian, have lain very far apart (see summary in Durand-Delga *et al.* 1962). The very same facies belts are symmetrically represented in the Sierra Nevada to the north and in the Rif mountains to the south. At present, I believe that this firmly established geological observation cannot be disregarded, even if some palaeomagnetic data suggest very large relative displacements (see e.g. Hilten 1964, for a summary).

It will be very interesting to learn more about the earliest history of the Atlantic ocean (compare Funnel & Smith 1968), whether it started by very gentle downbuckling, like the Alpine geosyncline, or whether there was an initial rift comparable to the Red Sea. We suspect that the late Palaeozoic and early Mesozoic rift belt which runs from East Greenland over Nova Scotia to the Atlantic seaboard of the United States might be interpreted as a precocious, abortive attempt to open the Atlantic ocean. In Greenland, this rift trough was filled by Upper Permian and Mesozoic sediments, mainly of marine origin; most of the detritus was derived from a land-mass to the east, at least during the Triassic (Grasmück & Trümpy 1969). In the southern part of this rift zone, further away from the ancient Arctic Ocean, only continental Triassic redbeds were laid down.

If we apply the ocean floor spreading theory to the Alpine geosyncline, we would expect to find a narrower basin for pre-spreading times, e.g. for the Triassic, than for the times immediately preceding the compression, e.g. for the Lower

Cretaceous Schistes lustrés. This is rather difficult to verify, but we would estimate that the figures we obtain for the Lower Cretaceous formations may indeed be a little higher, say about 100km more, than those measured in the Triassic formation. This difference can also be explained by assuming subduction of certain basement units with their Triassic cover. That Triassic rocks were extremely widespread throughout the geosyncline is also attested by the abundance of dolomite and limestone debris which they furnished to the various Jurassic and Cretaceous breccia formations.

8. Conclusions

In searching for a present-day model for the Alpine geosyncline of Jurassic time, we would not so much turn towards the Atlantic ocean as towards smaller ocean basins, possibly like the Caribbean sea. Menard (1967) has shown that many of these ocean basins have a structure rather analogous to the one that has here been deduced for the Mediterranean seas of Jurassic time. The geologist is obliged to look for recent examples of his fossil objects; but the principle of uniformitarianism does not imply that such examples must be found at any price. Many authors agree that ocean floor spreading of the Atlantic type is a fairly novel feature of the Earth's evolution (e.g. Beloussov & Ruditch 1961; Beloussov 1968; Booy 1968). For this reason too, we see no harm in continuing to talk about geosynclines, provided that no rigid concept is attached to the term (e.g. Aubouin 1965); we fully agree with the criticism voiced by Debelmas *et al.* (1966) and Mac Gillavry (1961) and with the undogmatic views expressed by Glaessner & Teichert (1947), Hermes (1968) and Khain & Muratov (1968).

The stratigraphical evolution of the Alpine geosyncline between Middle Triassic and Lower Cretaceous time is compatible with gradual oceanization of continental crust and also with slight distension. Direct indications of tensional stresses are furnished by the Lower and Middle Jurassic normal faults. The amount of distension can be estimated at about 100km; a spreading comparable to the Atlantic model seems highly improbable. This statement, of course, applies only to the Alps; in other geosynclines, e.g. in the Northern Apennines, conditions may have been quite different (see Decandia & Elter 1969). Causal relationships between distension and oceanization must be expected. From stratigraphical analysis alone, it would be rash to suggest which is the triggering phenomenon or what the precise mechanism of oceanization may have been.

The branch of science which received its decisive starting impulse through the work of William Smith can and must make very relevant contributions to the understanding of mountain chains and of the evolution of our planet. This is, of course, quite obvious; but the fascination exerted by the spectacular progress of the quantitative branches of geology may lead too easily to an underrating of the importance of stratigraphical research. There is no reason at all for an inferiority complex among stratigraphers, provided that they do not close their eyes to the results obtained by other methods. A constant and frank dialogue between all branches of geology is more necessary than ever, now that we are embarking on one of the greatest adventures of our science, the discovery of ocean geology.

One of the saddest examples of this inferiority complex is the fate of the theory of continental drift; at first it was accepted with enthusiasm, quite rightly as there were, even 45 years ago, many perfectly valid stratigraphical reasons for it—along, of course, with some rather dubious ones. But in spite of this sound stratigraphical evidence, geological world opinion discarded the theory, on account of the veto issued by the geophysicists—not the highly developed geophysicists of today, but their primitive forebears of the 1920's. Paradoxically this failure of geologists to stand up for the significance of their own results was largely due to their insufficient knowledge of geophysical problems.

Stratigraphy is a noble science, and stratigraphers should state their findings confidently. But they must also indicate their margins of error, and in mountain chains, these margins are very wide indeed. Yet in every mountain chain, some facts are so well established that no geotectonic theory can afford to disregard them.

Acknowledgements. For critical reading of the manuscript and helpful suggestions, the author is indebted to A. Gansser, K. J. Hsu, H. P. Laubscher and A. G. Milnes. H. P. Laubscher also made available a preprint of an important publication.

9. References

AFANASS'YEV, G. D. 1968. On the Boundary delimiting the Earth's crust and the Upper Mantle. *Int. geol. Congr.* **23** (1), 31–41.

ALLEMANN, F. 1957. Gelogie des Fürstentums Liechtenstein (Südwestlicher Teil), unter besonderer Berücksichtigung des Flyschproblems. *Jb. Histor. Verein Fürstentum Liechtenstein,* **56** (1957), 1–244.

ANTOINE, P. 1968. Sur la position structurale de la "Zone du Versoyen". *Trav. Lab. Géol. Univ. Grenoble.* **44**, 5–26.

ARGAND, E. 1920. Plissements précurseurs et Plissements tardifs des Chaînes de Montagnes. *Verh. Schweiz. naturf. Gs.* **101**, 13–39.

AUBOIN, J. 1958. Eassai sur l'évolution paléogéographique et le développement tecto-orogénique, d'un systéme géosynclinal: Le secteur grec des Dinarides (Hellénides). *Bull. Soc. géol. Fr.* (6), **8** 731–50.

—— 1965. *Geosynclines.* Amsterdam (Elsevier).

BAER, A. 1959. L'extrémité occidentale du Massif de l'Aar (Relations du socle avec la couverture). *Bull. Soc. neuchâtel. Sci. nat.* **82**, 5–160.

BALLY, A. W., GORDY, P. L. & STEWART, G. A. 1966. Structure, seismic data, and orogenic evolution of southern Canada Rocky Mountains. *Bull. Can. Petrol. Geol.* **14**, 337–81.

BARBIER, R. ET AL. 1963. Problèmes paléogéographiques et structuraux dans les zones internes des Alpes Occidentales entre Savoie et Méditerranée. *In* Livre Mémoire Paul Fallot. Paris (Soc. Géol. France) t. 2, 331–75.

BELOUSSOV, V. V. 1962. *Basic problems in geotectonics.* London (McGraw-Hill).

—— 1968. Some general aspects of Development of the Tectonosphere. *Int. geol. Congr.* **23** (1), 9–17.

—— & RUDITCH, E. M. 1961. Island arcs in the development of the Earth's structure (especially in the region of Japan and the sea of Okhotsk). *J. Geol.* **69**, 647–58.

BEMMELEN, R. W. VAN 1966. On mega-undations: a new model for the Earth's evolution. *Tectonophysics,* **3**, 83–127.

—— 1968. On the origin and evolution of the Earth's crust and magmas. *Geol. Rdsch.* **57,** 657–705.

BERNOULLI, D. 1964. Zur Geologie des Monte Generoso (Lombardische Alpen). *Beitr. geol. Karte Schweiz.* NF **118,** 134.

BOER, J. DE 1963. *The geology of the Vicentinian Alps (NE-Itoly).* Deventer (Van de Velde).

BOOY, T. DE 1966. Ein jugendliches Alter des simatischen Untergrundes der heutigen Ozeane. *Proc. Kned. Akad. Wet.* B, **69,** 283–95.

—— 1967. Neue Daten für die Annahme einer sialischen Kruste unter den frühgeosynklinalen Sedimenten der Tethys. *Geol. Rdsch.* **56,** 94–102.

—— 1968. Mobility of the Earth's crust: a comparison between the present and the past. *Tectonophysics* **6,** 177–206.

BIANCONI, F. 1965. Resti fossili in rocce mesometamorfiche della regione del Campolongo. *Schweiz. miner. petrogr. Mitt.* **45,** 571–96.

BOLLI, H. M. & NABHOLZ, W. K. 1959. Bündnerschiefer, ähnliche fossilarme Serien und ihr Gehalt an Mikrofossilien. *Eclog. geol. Helv.* **52,** 237–70.

BRÜCKNER, W. 1940. Die geologischen Verhältnisse an der Basis der Säntis-Decke zwischen Wallenstadt und Wäggital. *Eclog. geol. Helv.* **33,** 5–25.

——, HEIM, ARNOLD., RITTER, E., STAUB, R. & TRÜMPY, R. 1958. Berich tüber die Jubiläumsexkursion der Schweizerischen Geologischen Gesellschaft durch die Glarneralpen. *Eclog. geol. Helv.* **50,** 509–28.

BUTTERLIN, J. 1956. *La constitution géologique et la Structure des Antilles.* Paris (C.N.R.S.).

CADISCH, J. 1953. *Geologie der Schweizer Alpen,* (2nd edn) Basel (Wepf).

CAREY, S. W. 1958. The tectonic approach to continental drift. In *Symposium on continental drift.* Hobart (Geol. Dept. Univ. Tasmania), pp. 177–349.

CHESSEX, R. 1959. La géologie de la haute vallée d'Abondance, Haute-Savoie (France). *Eclog. geol. Helv.* **52,** 296–400.

CLOSS, H. 1964. Die tiefere Untergrund der Alpen nach neuen seismischen Messungen. *Geol. Rdsch.* **53,** 630–49.

COLOM, G. 1957. Sur les caractères de la sédimentation des Géosynclinaux mésozoïques. *Bull. Soc. géol. Fr.* (6), **7,** 1167–87.

DEBELMAS, J. 1957. Quelques remarques sur la conception actuelle du terme "cordillère" dans les Alpes internes Françaises. *Bull. Soc. géol. Fr.* (6), **7,** 463–74.

DEBELMAS, J. et al. 1966. Quelques remarques sur le concept de géosynclinal. *Revue Géogr. phys. Géol. dyn.* (2) **8,** 133–50.

DECANDIA, F. A. & ELTER, P. 1969. Riflessioni sul problema degli ofioliti nell'Appennino Settentrionale. *Memorie Soc. tosc. Sci. nat.* A, **76,** 1–9.

DEWEY, J. F. 1969. Evolution of the Appalachian Caledonian orogen. *Nature, Lond.* **222,** 124–9.

DIETRICH, V. 1967. Geosynklinaler Vulkanismus in den oberen penninischen Decken Graubündens (Schweiz). *Geol. Rdsch.* **57,** 246–64.

DIETZ, R. S. 1961. Continent and ocean basin evolution by spreading of the sea floor. *Nature, Lond.* **190,** 854–7.

DRAKE, C. L., EWING, M. & SUTTON, G. H. 1959. Continental margins and geosynclines: the east coast of north America North of Cape Hatteras, In *Physics and Chemistry of the Earth.* vol. 3, pp. 110–94.

DRAKE, C. L. & NAFE, J. E. 1967. Geophysics of the North Atlantic region. Preprint, *symposium on continental drift, Montevideo, Uruguay,* 16–19 October 1967.

DURAND-DELGA, M. *et al.* 1962. Données actuelles sur la structure du Rif. *In* Livre Mémoire Paul Fallot Paris (Soc. Géol. France) t.50, 399–422.

ELLENBERGER, F. 1958. Etude géologique du pays de Vanoise. *Mém. Serv. Carte géol. dét. Fr.*

ELTER, G. ELTER, P., STURANI, C. & WEIDMANN M. 1966. Sur la prolongation du domaine ligure de l'Apennin dans le Monferrate et les Alpes et sur l'origine de la nappe de la Simme s.l. des Préalpes romandes et chablaisiennes. *Archs Sci., Genève.* **19.**

EMERY, K. O. 1965. Geology of the Continental Margin off the Eastern United States. *Colston Pap. Univ. Bristol.* **17,** 1–17.

—— 1966. Atlantic Continental Shelf and Slope of the United States. Geologic Background. *Prof. Pap. U.S. geol. Surv.* **529-A,** 1–23.

ENGELEN, G. B. 1963. Indications for large scale graben formation along the continental margin of the eastern United States. *Geologie Mijnb.* **42,** 65–75.

FALLOT, P. 1960. Le problème de l'espace en tectonique. *Abh. dt. Akad. Wiss. Berl.* III, Kl. 3, (1), 48–58.

FISCHER, A. G. 1965. The Lofer cyclothems of the Alpine Triassic *Bull. Kans. Geol. Surv.* **169,** 107–49.

FLÜGEL, H. 1964. Das Paläozoikum in Oesterreich. *Mitt. geol. Ges. Wien.* **56,** 401–43.

—— & FENNINGER, A. 1966. Die Lithogenese der Oberalmer Schichten und der mikritischen Plassen-Kalke (Tithonium, Nördliche Kalkalpen). *Neues. Jb. Geol. Paläont. Abh.* **123,** 249–80.

FREY, J. D. 1967. Geologie des Greinagebietes. *Beitr. geol. Kaate Schweiz.* NF **131.**

FUNNELL, B. M. & SMITH, A. G. 1968. Opening of the Atlantic Ocean. *Nature, Lond.* **219,** 1328–33.

GANSSER, A. 1959. Ausseralpine Ophiolithprobleme. *Eclog. Geol. Helv.* **52,** 659–80.

—— 1968. The Insubric Line, a major geotectonic problem. *Schweiz. miner. petrogr. Mitt.* **48,** 123–43.

GARRISON, R. E. 1967. Pelagic limestones of the Oberalm beds (Upper-Jurassic-Lower Cretaceous), Austrian Alps. *Bull. Can. Petrol. Geol.* **15,** 21–49.

GASSER, U. 1967. Erste Resultate über die Verteilung von Schwermineralen in verschiedenen Flyschkomplexen der Schweiz. *Geol. Rdsch.* **56,** 300–308.

GEES, R. 1955. *Geologie von Klosters.* Bern (Mäder).

GIDON, P. 1963. *Courants magmatiques et évolution des continents. L'hypothèse d'une érosion sous-crustale.* Paris (Masson).

GIESE, P. 1968. Die Struktur der Erdkruste im Bereich der Ivrea-Zone. Ein Vergleich verschiedener seismischer Interpretationen und der Versuch einer petrographisch-geologischen Deutung. *Schweiz. miner. petrogr. Mitt.* **48,** 261–84.

GILLULY, J. 1955. Geologic contrasts between continents and ocean basins. *Spec. Pap. geol. Soc. Am.* **62,** 7–18.

GLAESSNER, M. F. & TEICHERT, C. 1947. Geosynclines: a fundamental concept in geology. *Am. Jl Sci.* **245,** 465–82, 571–91.

GLANGEAUD, L. 1957. Essai de classification géodynamique des chaînes et des phénomènes orogéniques. *Revue Géogr. phys. Géol. dyn.* (2) **1,** 200–20.

—— 1962. Les transferts d'échelle en géologie et géophysique. Application à la Mèditerranée occidentale et aux chaînes péripacifiques. *Bull. Soc. géol. Fr.* (7), **4,** 912–61.

—— 1968. Les méthodes de la Géodynamique et leurs applications aux structures de la Méditerranée occidentale. *Revue Géogr. phys. Géol. dyn.* (2), **10,** 83–135.

GOLDRING, R. 1961. The Bathyal Lull: Upper Devonian and Lower Carboniferous sedimentation in the Variscan geosyncline. *In* Coe, K. *Some aspects of the Variscan fold belt.* (Manchester Univ. Press), pp. 75–91.

GRASMÜCK, K. & TRÜMPY, R. 1969. Triassic stratigraphy and general geology of the country around Fleming Fjord. *In* Notes on Triassic stratigraphy and paleontology of north-eastern Jameson Land (East Greenland). *Meddl. Grønland,* **168,** 2.

GRUNAU, H. 1947. *Geologie von Arosa (Graubünden), mit besonderer Berücksichtigung des Radiolarit-Problems.* Zürich (Aschmann & Scheller).

GRUNAU, H. R. 1959. *Mikrofazies und Schichtung ausgewählter, jungmesozoischer, radiolaritführender Sedimentserien der Zentral-Alpen.* Leiden (Brill).

HEDBERG, H. D. 1967. Status of stratigraphic classification and terminology. *Geol. Newsletter (I.U.G.S.)* 1967/3, 16–29.

HEIM, ALBERT 1921. *Geologie der Schweiz,* Band II/1 (Die Schweizer Alpen, 1. Hälfte). Leipzig, (Tauchnitz).

HEIM, ARNOLD 1910, 1913, 1916, 1917. Monographie der Churfirsten-Mattstockgruppe. *Beitr. geol. Karte Schweiz,* N.F., **20,** (1–4), 1–272, 1–86, 1–205, 1–88.

—— 1916. Über Abwicklung und Facieszusammenhang in den Decken der nördlichen Schweizer-Alpen. *Vjschr. Natf. Ges. Zürich.* **61,** 474–87.

HEIRTZLER, J. R. *et al.* 1968. Marine magnetic anomalies, geomagnetic field reversals, and motions of the ocean floors and continents. *J. geophys. Res.* **73,** 2119–36.

HELBLING, R. 1938. Zur Tektonik des St. Galler Oberlandes und der Glarner Alpen. *Beitr. geol. Karte Schweiz,* (N.F). **76,** 69–125.

HERMES, J. J. 1968. The Papuan geosyncline and the concept of geosynclines. *Geologie Mijnb.* **47,** 81–97.

HERSEY, J. B. 1965. Sedimentary basins of the Mediterranean Sea. *Colston Pap. Univ. Bristol.* **17,** 75–92.

HESS, H. H. 1962. History of ocean basins. *In Petrological Studies: Buddington volume.* (Geological Society of America), pp. 599–620.

HIGGINS, A. K. 1964. Fossil remains in staurolite-kyanite schists of the Bedretto-Mulde Bündnerschiefer. *Eclog. geol. Helv.* **57,** 151–6.

HILTEN, D. VAN 1964. Evaluation of some geotectonic hypotheses by paleomagnetism. *Tectonophysics* **1,** 3–71.

—— & ZIJDERVELD, J. D. A. 1966. Paleomagnetism and the Alpine tectonics of Eurasia—II. The magnetism of Permian porphyries near Lugano. *Tectonophysics* **3,** 429–46.

HSU, K. J. 1958. Isostasy and a theory for the origin of geosynclines. *Am. J. Sci.* **256,** 305–27.

—— 1965. Isostasy, crustal thinning, mantle changes, and the disappearance of ancient land masses. *Am. J. Sci.* **263,** 97–109.

HUDSON, J. D. 1967. Speculations on the depth relations of calcium carbonate solution in recent and ancient seas. *Marine Geol.* **5,** 473–80.

HURLEY, M. 1968. The confirmation of Continental Drift. *Scient. Am.* April 1968, 53–64.

JÄCKLI, H. 1941. Geologische Untersucheungen im nördlichen Westschams (Graubünden). *Eclog. geol. Helv.* **34,** 17–105.

JENNY, H. 1924. *Die alpine Faltung. Ihre Anordnung in Raum und Zeit.* Berlin (Borntraeger).

JUNG, W. 1963. Die mesozoischen Sedimente am Südostrand des Gotthardmassivs (zwischen Plaun la Greina und Versam). *Eclog. geol. Helv.* **56,** 653–754.

KHAIN, V. E. & MURATOV, M. V. 1968. Geosynclinal belts, orogenic belts, folded belts and their relation in time and space. *Int. geol. Congr.* **23,** (3), 9–13.

KLEMME, H. D. 1958. Regional geology of circum-Mediterranean region. *Bull. Am. Ass. petrol. Geol.* **42,** 477–512.

KÜNDIG, E. 1959. Eugeosynclines as potential oil habitats. *Wld. Petrol. Congr.* **5,** (1), paper 25.

LAUBSCHER, H. 1969. Mountain building. *Tectonophysics* **7,** 551–63.

LE PICHON, X. 1968. Sea-floor spreading and continental drift. *J. geophys. Res.* **73,** 3661–97.

LEMOINE, M. 1963. Remarques sur les caractères et l'évolution de la paléogéographie de la zone briançonnaise au Secondaire et au Tertiaire. *Bull. Soc. géol. Fr.* (6), **3,** 105–21.

—— 1959. Remarques à propos de quelques faits et hypothèses concernant l'âge des Schistes lustrés piémontais dans les Alpes cottiennes et briançonnaises. *Bull. Soc. géol. Fr.* (7), **1,** 90–2.

LUDWIG, W. J. *et al.* 1966. Sediments and structure of the Japan trench. *J. geophys. Res.* **71,** 2121–37.

MAXWELL, A. E. 1969. Recent deep sea drilling results from the South Atlantic. *Trans. Am. geophys. Un.* **50,** 113–5.

MAXWELL, J. C. 1969. The Mediterranean, ophiolites and continental drift. In *What's new on Earth.* Rutgers University Press.

MACGILLAVRY, H. J. 1961. Deep or not deep, fore-deep or "after-deep"? *Geol. Mijnb.* **40,** 133–48.

MCDOWELL, F. W. & SCHMID, R. 1968. Potassium-Argon dates from the Val d'Ossola Section of the Ivrea-Verbano Zone (Northern Italy). *Schweiz. miner. petrogr. Mitt.* **48,** 205–10.

MENARD, H. W. 1967. Transitional types of crust under small ocean basins. *J. geophys. Res.* **72,** 3061–73.

MICHARD, A. 1966. Etudes géologiques dans les zones des Alpes Cottiennes méridionales. Thèse, Paris (preprint).

NICOLAS, A. 1966. Interprétation des ophiolites piémontaises entre le Grand Paradis et la Dora Maïra. *Schweiz. miner. petrogr. Mitt.* **46,** 25–41.

—— 1968. Relations structurales entre le massif ultrabasique de Lanzo, ses satellites et la zone de de Sesia Lanzo. *Schweiz. miner. petrogr. Mitt.* **48,** 145–56.

OBERHAUSER, R. 1968. Beiträge zur Kenntnis der Tektonik und der Paläogeographie während der Oberkreide und dem Paläogen im Ostalpenraum. *Jb. geol. Bundesanst., Wien.* III, 115–45.

OBERHOLZER, J. 1933. Geologie der Glarneralpen. *Beitr. geol. Karte Schweiz.* **28.**

OFFICER, C. B. *et al.* 1959. Geophysical investigations in the eastern Caribbean: summary of 1955 and 1956 cruises. In *Physics and Chemistry of the Earth.* vol. 3, 17–108.

PETERS, T. 1963. Mineralogie und Petrographie des Totalpserpentins bei Davos. *Schwiz. miner. petrogr. Mitt.* **43,** 529–685.

—— 1969. Rocks of the Alpine ophiolitic suite. *Tectonophysics* **7,** 507–9.

RAGUIN, E. 1930. Haute-Tarantaise et Haute-Maurienne (Alpes de Savoie). *Mém. Serv. Carte géol. dét. Fr.*

RICHTER, D. 1957. Beiträge zur Geologie der Arosa-Zone zwichen Mittelbünden und dem Allgäu. *Neues Jb. Geol. Paläont. Abh.* **105,** 285–372.

RUTTEN, M. G. 1963. Paleomagnetism and Tethys. *Geol. Rdsch.* **53,** 9–16.

SCHINDEWOLF, O. 1944. *Grundlagen und Methoden der paläontologischen Chronologie.* Berlin (Bornträger).

SCHINDLER, C. M. 1959. Zur Geologie des Glärnischs. *Beitr. geol. Karte Schweiz.* **107.**

SCHMID, F. 1965. Zur Geologie der Umgebung von Tiefencastel (Kanton Graubünden). Diss., University of Zürich.

SCHMID, R. 1967. Zur Petrographie und Struktur der Zone Ivrea-Verbano zwischen Valle d'Ossola und Val Grande (Prov. Novara, Italien). *Schweiz. miner. petrogr. Mitt.* **47,** 935–1117.

ŠKVOR, V. 1969. The Caribbean Area: A case of destruction and regeneration of continent. *Bull. geol. Soc. Am.* **80,** 961–8.

SOMM, A. & SCHNEIDER, B. 1962. Zwei paläontologische und stratigraphische Beobachtungen in der Obertrias der südwestlichen Engadiner Dolomiten (Graubünden). *Ergebn. wiss. Unters. schweiz. Natn. Parks.* N.F. **7,** 47.

SONDER, R. A. 1939. Meerestiefe und lithologische Fazies. *Eclog. Geol. Helv.* **39,** 260–3.

STEINMANN, G. 1925. Gibt es fossile Tiefseeablagerungen von erdgeschichtlicher Bedeutung? *Geol. Rdsch.* **16,** 435–68.

STAUB, R. 1917. Ueber Faziesverteilung und Orogenese in den südöstlichen Schweizeralpen. *Beitr. geol. Karte Schweiz.* **46.**

SUBBOTIN, S. I., NAUMCHIK, G. L. & RAKHIMOVA, I. Sh. 1965. Influence of upper mantle processes on the structure of the Earth's crust. *Tectonophysics* **2,** 185–209.

TOLLMANN, A. 1965. Faziesanalyse der alpidischen Serien der Ostalpen. *Verh. geol. Bundesanst., Wien* Sonderheft G, 103–33.

—— 1968. Bemerkungen zu faziellen und tektonischen Problemen des Alpen-Karpaten-Orogens. *Mitt. Ges. Geol. Bergbaustud.* (*Wien*). **18,** 207–48.

TRÜMPY, D. 1916. Geologische Untersuchungen im westlichen Rhätikon. *Beitr. geol. Karte Schweiz* **46.**

TRÜMPY, R. 1949. Der Lias der Glarner Alpen. *Denkschr. schweiz. naturf. Ges.* **79,** 1–192.

—— 1951. Le Lias de la Nappe de Bex (Préalpes internes) dans la Basse Gryonne. *Bull. Soc. vaud. Sci. nat.* **65,** 1–22.

—— 1960. Paleotectonic evolution of the central and western Alps. *Bull. geol. Soc. Am.* **71,** 843–908.

—— 1965. Zur geosynklinalen Vorgeschichte der Schweizer Alpen. *Umschau* 65/g., **18,** 573–7.

—— 1969. Die helvetischen Decken der Ostschweiz: Versuch einer palinspastischen Korrelation und Ansätze zu einer kinematischen Analyse. *Eclog. Geol. Helv.* **62,** 105–38.

UCHUPI, E. & EMERY, K. O. 1963. The continental slope between San Francisco, California and Cedros Island, Mexico. *Deep-Sea Res.* **10,** 397–447.

—— & —— 1968. Structure of Continental Margin off Gulf Coast of United States. *Bull. Am. Ass. Petrol. Geol.* **52,** 1162–93.

VINE, F. J. 1968. Magnetic Anomalies Associated with Mid-Ocean Ridges, *In* Phinney, R. A. (Ed). *The history of the Earth's crust.* (Princeton Univ. Press), pp. 73–89.
VUAGNAT, M. 1968. Quelques réflexions sur le complexe basique-ultrabasique de la zone d'Ivrée et les ultramafites alpinotypes. *Schweiz. miner. petrogr. Mitt.* **48,** 157–64.
WIEDENMAYER, F. 1963. Obere Trias bis mittlerer Lias zwischen Saltrio und Tremona (Lombardische Alpen). *Eclog. geol. Helv.* **56,** 529–640.
WILSON, J. T. 1963. Continental drift. *Scient. Am.* **868,** 1–16.
WURSTER, P. 1964. Geologie des Schifsandsteins. *Mitt. geol. St. Inst. Hamb.* **33,** 1–140.
YEATS, R. S. 1968. Southern California structure, sea-floor spreading, and history of the Pacific Basin. *Bull. geol. Soc. Am.* **79,** 1693–702.

13

PALEOGEOGRAPHY OF THE ASIATIC LAND MASS IN THE MESOZOIC

L. B. Rukhin

This article was translated expressly for this Benchmark volume by P. Sonnenfeld, University of Windsor, from "Paleogeografiya Aziatskogo materika v mesozoe," in Doklady Soviet. Geol. Problem 12, 1960, pages 85–98. *The figures have been reprinted from the original article.*

MESOZOIC PALEOGEOGRAPHY OF ASIA

Asia is the largest continent, and a study of its geological history has a special merit. Comparing paleogeographic maps for different periods of time is one possible way of studying it. This paper attempts to analyze the geological history of the continent of Asia in the Mesozoic era.

What is a paleogeographic map? In current literature, there is no general agreement on this question. We shall here consider as such those maps which depict the physical geography prevailing at the time when the ancient sediments formed. In other words, paleographic maps depict ancient land forms.

Paleogeographic maps should not be mistaken for facies maps. Facies maps not only generalize the physical geography during sediment accumulation, but also indicate the thickness and composition of the deposits. This is not shown on paleogeographic maps.

A frequent mistake of paleogeographic maps is that they show only ancient seas. Dry land is represented by white areas left over after contouring the ancient seas. To represent land on paleogeographic maps as a white patch is not permissible. There are usually sufficient data available to generalize. The relief of the ancient land can be reconstructed from observing a pinchout of sediments of a given age, from the study of facies preserved in marine and continental deposits and, in some cases, from a buried relief. Ancient rivers can be traced from widespread alluvial deposits, from the record of former freshening in adjacent marine waters, and from terrigenous-mineralogical provinces discovered through detailed investigations. From the attitude of inclined bedding in eolian sandstones, one can reconstruct the prevailing direction of ancient winds. The study of volcanic sequences tells the location of ancient centers of volcanic eruptions. In other words, the paleogeographic characteristics of land areas need not be less detailed than those of ancient basins. In order to demonstrate the value of such reconstructions, this paper deals chiefly with continental and not with marine deposits.

The author believes that the paleogeography can be portrayed by maps or by outlines. In a way, paleogeographic maps are instant snap-

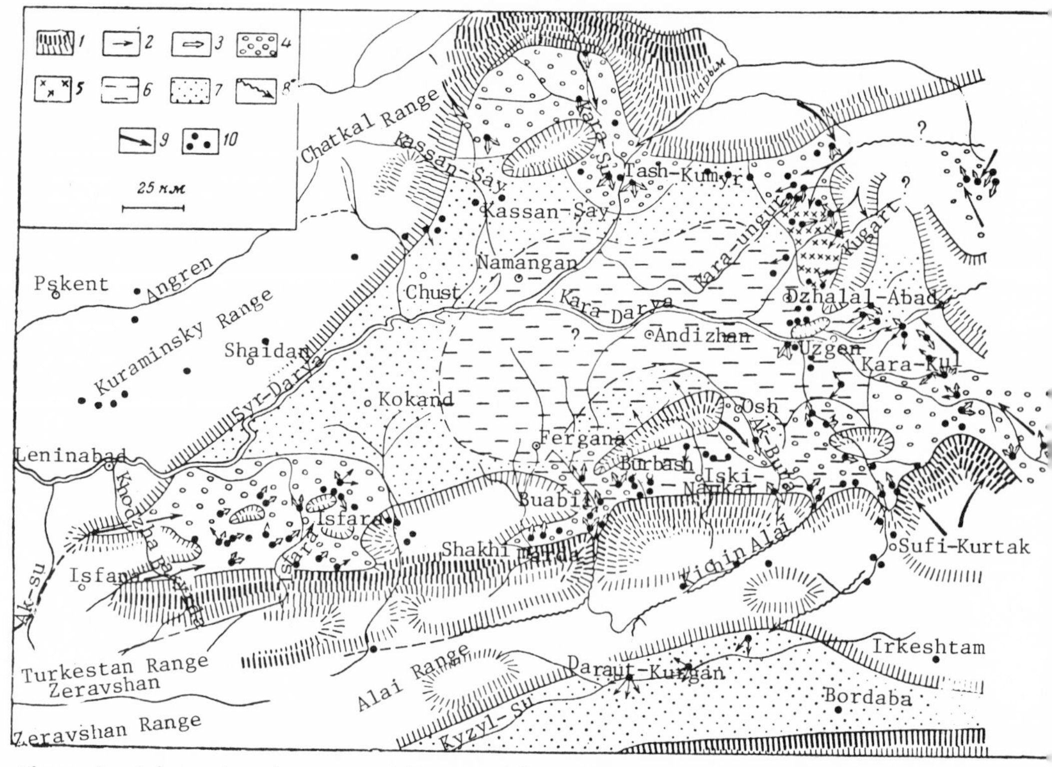

Figure 1. Schematic paleogeographic map of the Fergana Depression and the Alai Valley at the beginning of the Lower Changetian (Neocomian? Aptian).
1. Boundaries of areas of denudation and their relief (the more dissected, the thicker the lines). 2. The direction of the denudation according to prevailing dip of the inclined bedding. 3. Th[e] same according to the orientation of pebbles. 4. Fanglomerates. 5. Sandy desert. 6. Freshwater basin. 7. Sandy–clayey plains. 8. Assumed position of rivers in the areas of denudation. 9. Direction of denudation. 10. Outcrops.

shots of the past. They are compiled for very brief time intervals with sediments of identical age. A specific condition of physical geography that formerly prevailed over a given part of the earth surface can be shown.

In contrast, paleogeographic outlines are compiled for longer periods of time (periods and series, more rarely epochs). They show only the types of land forms that were persisting continuously. Paleogeographic outlines thus represent generalizations of a whole series of successive paleogeographic maps.

Paleogeographic maps are compiled after the field work is completed; only some preliminary observations can be reported before that.

Figure 1 shows a schematic paleogeographic map of the time interval during which Cretaceous continental red beds began to accumulate in the Fergana basin. Conglomerates do not occur everywhere at the base of the section; they fill individual depressions in the ancient relief. Differences in the petrographic composition and in the orientation of pebbles show that ancient boulder trains correspond to rivers. According to their location, they are precursors of present-day rivers. The map shows how the degree of dissection of the relief can be reconstructed in ancient regions of denudation from fanglomerate distribution and from fragment sizes. Finally, the distribution of alluvial plains as well as the contours of a lake basin in the central part of the Fergana depression are shown.

Figure 2 shows a schematic paleogeographic map of Neocomian (Valanginian?) continental deposits in the eastern part of Central Asia. As the preceding map, it is compiled from the author's field observations complemented by data from the literature.

The inclusion of a considerably larger area in this map brought out the assymmetric relief of uplifted massifs. (This was very typical for the Neocomian of Central Asia.) Ancient river courses and the boundaries between various climatic zones could be reconstructed. Gypsum occurs in Cretaceous sediments in several stratigraphic horizons where remains of terrestrial vertebrates are rarely found. In contrast, gypsum is absent farther north, where there are large accumulations of bones of terrestrial reptiles. Plant remains are also common; they are very rarely found in the south. It can therefore be assumed that in Neocomian time the boundaries between the hot arid zone and the more temperate climate passed through the region studied.

If we compare these paleogeographic maps with the present-day geography, we see that as early as the onset of the Cretaceous, or about 100 million years ago, elevated regions were in the same place as they are today. They can even be traced through earlier stages of the geological history right down to the Paleozoic, when the comparatively narrow Caledonian zones were formed, followed by the early Variscan massifs. Thereafter, these ancient folded structures were uplifted repeatedly and became areas of denudation as indicated by the distribution of younger sediments.

For example, in Neocomian and Cenomanian times, the early Variscan Alai-Turkestan massif had a dissected relief. Although it was thereafter peneplained, it did not turn into an area of sediment accumulation. Instead, a mountainous region formed on the same site in Neogene time. This is one of many cases where regions of denudation formed repeatedly in Mesozoic and Cenozoic times on the sites of Variscan, Caledonian, and even older massifs.

The relief is similarly inherited also in many areas of subsidence. However, not all large land forms of the Mesozoic relief in Central Asia were inherited from the Paleozoic. Some new forms also appeared. An example is the Fergana Mountain Range, and particularly its southern part, which is at the site of a Jurassic trough. The inversion occurred only at the beginning of the Cretaceous and the Fergana Range turned from the Aptian onwards into a powerful source of clastic material.

Small-scale (regional) paleogeographic maps and outlines are com-

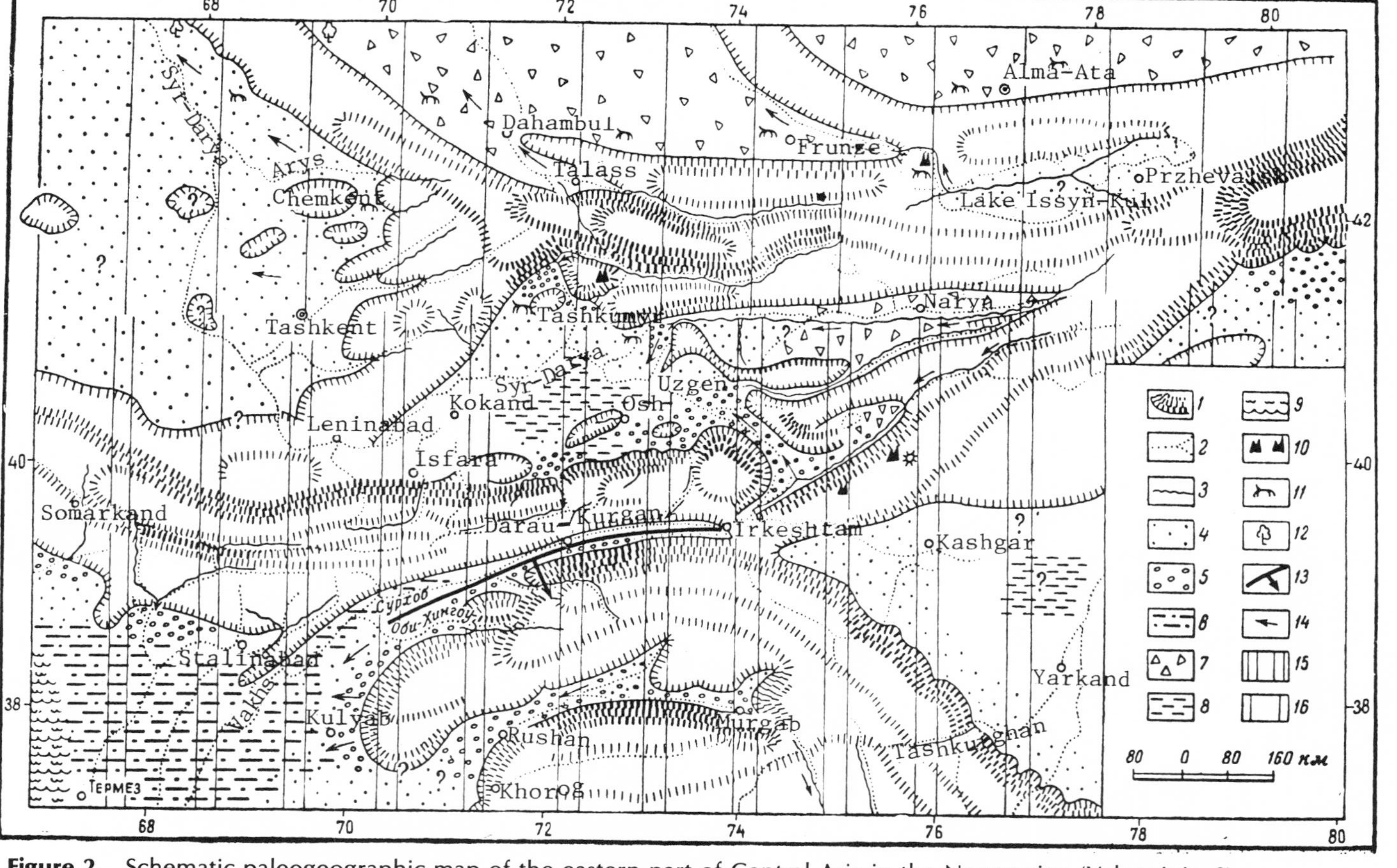

Figure 2. Schematic paleogeographic map of the eastern part of Central Asia in the Neocomian (Valenginian?). 1. Areas of denudation, their boundaries and relief (the more dissected, the thicker the lines). 2. Contours of present-day rivers and lakes. 3. Lower Cretaceous rivers. 4. Sandy–clayey alluvial plains. 5. Fanglomerates. 6. Very much freshened brackish water basins. 7. Weakly dissected hummocky topography with local accumulation of sediments. 8. Freshwater basin. 9. Basin with variable salinity. 10. Volcanoes. 11. Terrestrial vertebrates. 12. Areas of thick vegetation cover. 13. Vakhsh overthrust (the arrow indicates the distance and direction where the areas to the south of it must be shifted to, in order to reconstruct the palinspastic arrangement in the Cretaceous period).

piled by generalizing large-scale paleogeographic maps, just as this is done in producing geographic or geologic maps. Because few paleogeographic data are available, regional paleogeographic maps currently must be compiled exclusively from data in literature. Besides, the geology of large areas is still insufficiently known, and detailed stratigraphic columns are missing, particularly for continental deposits. Deep bore holes, from which one could assess the distribution of these sediments, are also scarce. Consequently, all the paleogeographic outlines presented below for the whole of Asia are schematic.

To illustrate the mode of study of ancient continental deposits, paleogeographic maps were prepared for the Lower Triassic and the Lower Jurassic. In the Cretaceous period, we selected the end of the upper part, which characterizes the final stage of the Mesozoic history of Eurasia.

Very important for the paleogeography of Eurasia are those papers which synthesize data on the paleogeography of the whole continent or of a considerable part of it. Classical paleogeographical maps of European Russia were prepared by A. L. Karpinskii (1887) and A. D. Archangel'ski (1912). Collective monographs ("Geological structure of the West-Siberian Lowlands," 1959, and "Geology and Occurrences of Petroleum in the West-Siberian Lowlands," 1958) are very important for the study of the West-Siberian Lowlands, and so are the papers by S. Bubnoff (1956), Gignoux (1952) and Wills (1951) for western Europe.

Liu-Hun-Yun (1955) gave an exhaustive description of the paleogeography of China and Arkell (1956) did a very detailed survey of Jurassic sediments in various parts of the world.

Figure 3 shows the paleogeography of Lower Triassic sediments. The configuration of Asia noticeably differed from its present one only in the Tethys zone and in the extreme northeast during Lower Triassic time. The closing of many geosynclines in Europe, Central and Eastern Asia created the core of the Asian continent at the end of the Paleozoic era. Later, during the Mesozoic era, it was only enlarged or partly modified by later tectonic uplift. To some degree, the Variscan folding revived older structures and produced many mountain ranges on the Lower Triassic land surface, most of which exist even on the present surface of the earth. However, most of them did not have a high mountain relief in the Triassic.

One of the remarkable features of these and of younger mountain systems is the arcuate outline. As classical example of this serve the Himalayas, the arcs of Iran, Turkey, the Carpathians and Alps, which were incipient already in the Triassic, as well as the island arcs fringing eastern Asia. Many older ranges also had an arcuate outline, such as the Upper Proterozoic Baikal arc, the Caledonian and Variscan arcs of Kazakhstan, the Karakorum arc, the Greater Khingan Mountains (Tahsinganling Shanmo of northeastern China) and many other systems. At first sight, straight-line ranges such as the Urals are composed of two arcs, convex towards the west (Rukhin, 1959). Regions of downwarp-

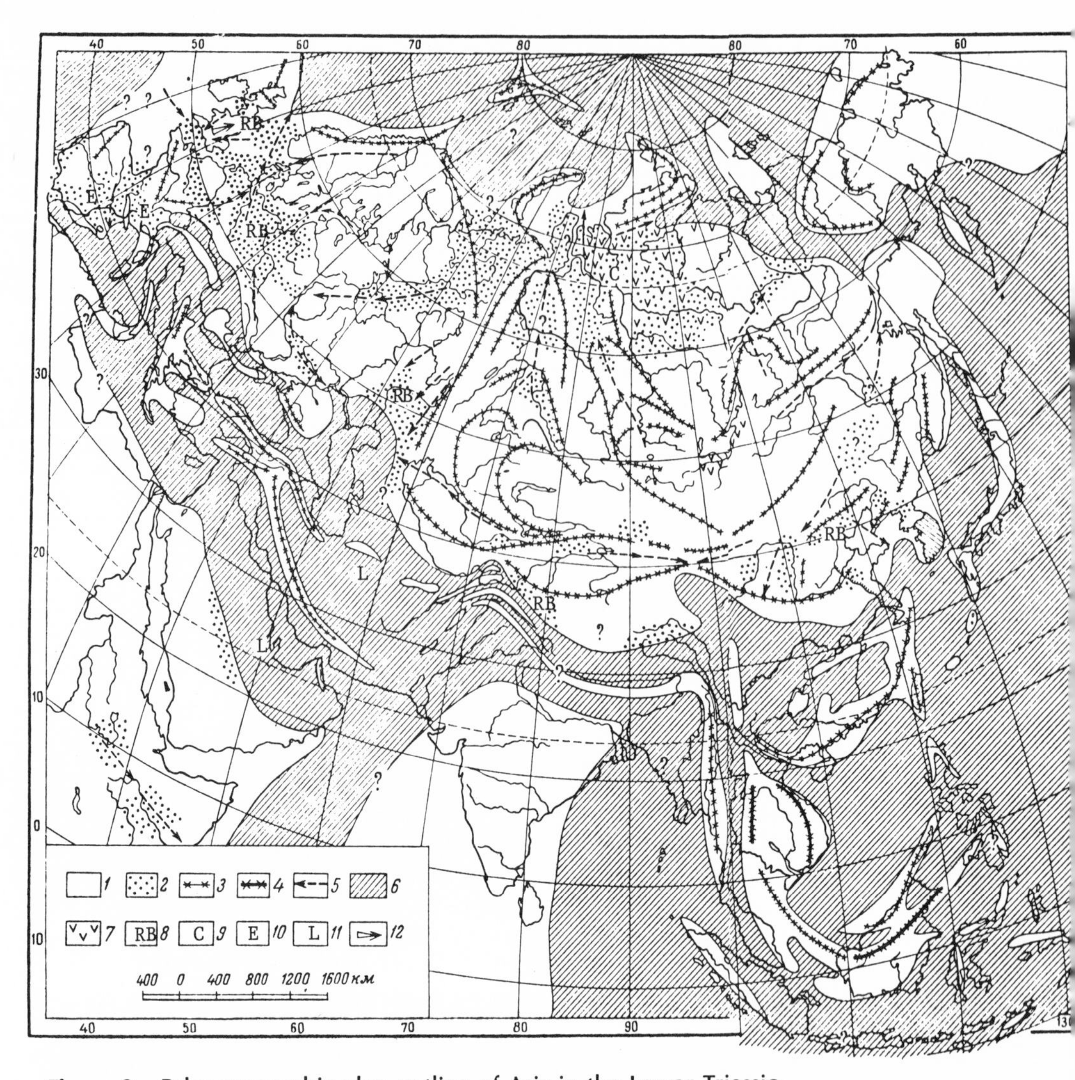

Figure 3. Paleogeographic plan outline of Asia in the Lower Triassic.
1. Dry land with rolling relief (primarily an area of denudation). 2. Alluvial- deltaic plains. 3. Low mountain ranges. 4. Mountain ranges with very dissected relief. 5. Assumed flow directions of rivers. 6. Sea. 7. Area of volcanic eruptions. 8. Red beds. 9. Coal bearing strata. 10. Evaporites. 11. Carbonates. 12. Prevailing wind direction.

ing (foredeeps) are always more active on the outer side of the arcs than on the inner side. Sediment accumulation and volcanism are always assymmetrical in arcs. The curvature of mountain ranges, which existed in Lower Triassic time in the West-Siberian Lowlands and later completely buried under a cover of younger sediments, can be conjectured from available data.

The overall rise of Asia in Lower Triassic time caused an almost complete absence of epeiric seas. This follows in part also from the large-scale erosion of clastic material. Its deposition overcompensated for the subsidence and thus prevented the entry of epeiric seas.

Lower Triassic continental sedimentation was continuous mainly in three areas on the northern border of Eurasia. One of these is western Europe where alluvial red beds accumulated from Variscan structures (convex to the north) and, probably, also from Caledonian ones (convex to the south–east).

Another area of continental Triassic sedimentation is the central part of the Russian platform. Typically, the most extensive Triassic sediment accumulation was derived from the erosion of Variscan structures of the Ural and Timan Mountains, which were convex to the west and southwest respectively. Pebbles of Uralian and Timan rocks contained in Triassic sediments of the central part of the Russian platform leave no doubt that rivers carried the bulk of the clastic material to the southwest. Only farther south, near the Caspian depression, was the transport from north to south.

Lower Triassic sediments also accumulated in a broad area of the western part of the Siberian platform. Simultaneously, there the volcanism also became intensive, and this must be considered to be a precursor of the downwarping of the West–Siberian Lowlands. Within the latter, only the uppermost horizons of Triassic sediments have so far been found, but there is good reason to believe that older horizons will also be found in the northern part.

Some changes occurred in the Lower Jurassic in the distribution of areas where continental sediment accumulated (Fig. 4). They cease to accumulate in western Europe and occur only in Poland. Evidently, they are also absent on the Russian platform, except for the Donets Basin and the Caspian Depression. East of it, there is within western Kazakhstan a new area of continental sedimentation that has not been in existence previously.

Continental sedimentation also ceased in the western part of the Siberian Platform. It continued, however, in the West–Siberian Lowlands. Continental sedimentation centered in a broad belt from the western Baikal to the Vilyuy Depression. Continental deposits were, evidently, distributed over a considerable area in Indochina. New centers of continuous sedimentation formed in China and India.

The main sources of Lower Jurassic clastic material were Caledonian and Variscan structures of Kazakhstan, the Altai-Sayan and the Baikal regions. The clastics were transported by rivers which had about the same general direction of flow every since the beginning of the Mesozoic as they have now, since the principal areas of denudation and sedimentation were similarly distributed.

In Central Asia and in Kazakhstan, there was a river flowing towards the Aral Sea even then, just as the present Syr–Dar'ya does, lengthwise between Caledonian and Variscan arcs.

Evidently, there was also another large river, which drained the Kaz-

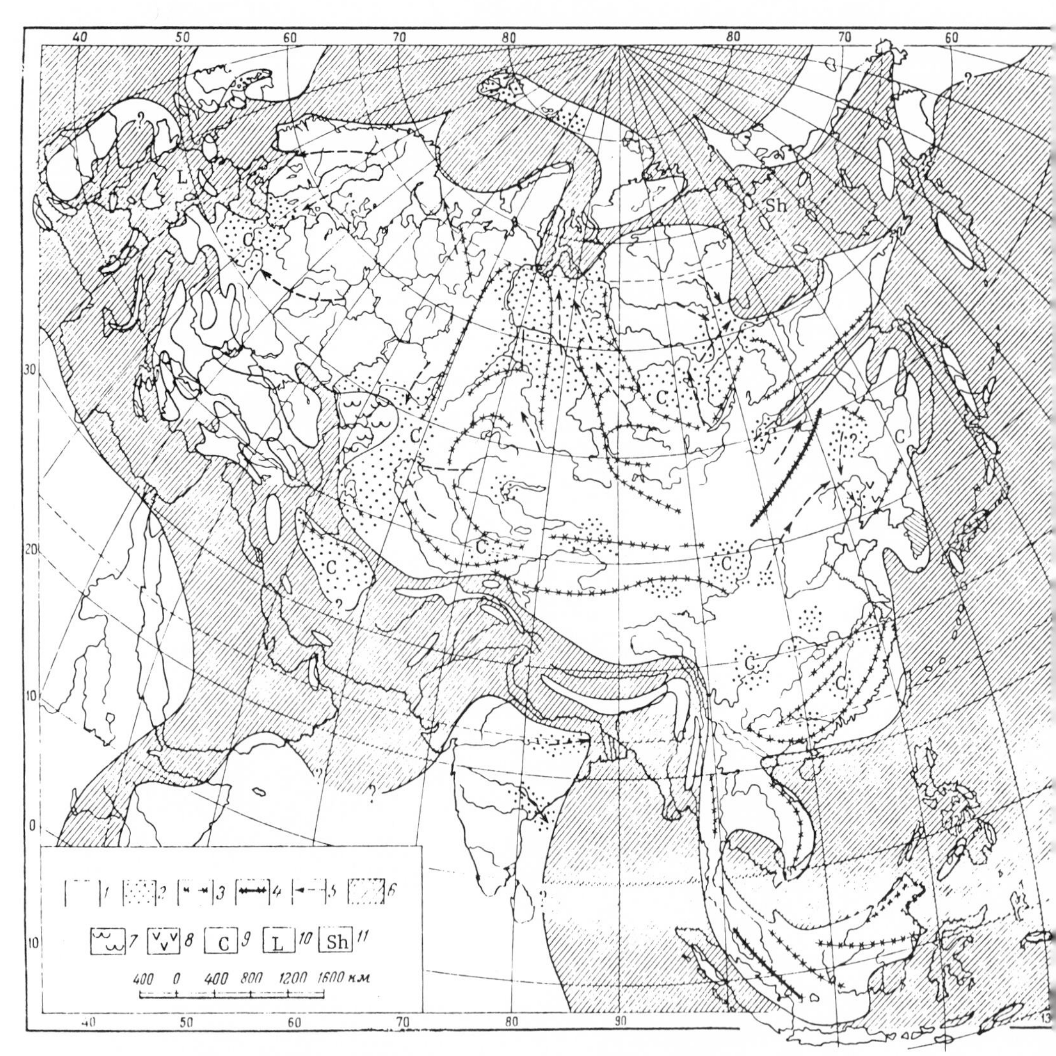

Figure 4. Paleogeographic outline of Asia in the Lower Jurassic (Middle Lias). 1. Dry land with rolling relief (primarily an area of denudation). 2. Alluvial- deltaic plains. 3. Low mountain ranges. 4. Mountain ranges with very dissected relief. 5. Assumed flow direction of rivers. 6. Sea. 7. Freshened parts of marine basins. 8. Areas of volcanic eruptions. 9. Coal bearing strata. 10. Carbonates. 11. Shales.

akh Mountains. It comprised the present-day rivers Chu, Ili, and Sarysu. Both the similar distribution of areas of degradation and of aggradation, and also the factual data derived from the study of the ancient alluvium, prove that principal present-day large rivers already existed in the lower Mesozoic (Zhilinskii, 1956).

The rivers which flow north from the Altai-Sayan Massif were formed long ago (Florensov, 1958). The upper reaches of the present-day Yenisey River existed then in part (after leaving the mountains, the river then flowed approximately along the river Keti). This is evidenced by a continuous accumulation of alluvial sediments (sometimes red in colour) in the adjacent southeastern parts of the West-Siberian Lowlands. Undoubtedly, rivers also existed in the lower Mesozoic on the site of the upper reaches of the present-day Ob and Irtysh rivers because areas of uplift and subsidence were similarly distributed, because folded structures had an identical extent and, most importantly, because alluvial Jurassic sediments were widely distributed in the West-Siberian Lowlands and because large freshwater reservoirs were feeding on river waters. By the same token, it can be assumed that the upper reaches of the Lena and possibly the Aldan rivers also existed in the lower Mesozoic.

If we compare the paleogeographic settings presented above for the beginning of the Triassic and Jurassic periods, it is not difficult to notice some displacement from west to east and southeast of areas where continental sediments accumulated continuously. This tendency becomes even more apparent in the Cretaceous period (see Fig. 5).

Not only the West-Siberian Lowlands, but even vast stretches of western Kazakhstan are covered by the sea towards the end of the Cretaceous period. Continental sedimentation continues only in the east, in the Katanga Depression, the Vilyuy Trough and the Lena Downwarp, on the outer side of the then peneplained Variscan arc running from Manchuria to southern Tyan-Shan, in the central part of that arc, and also in its inner part (Mongolia). Basins formed here in time and some mollusks spread through them from Japan to Fergana.

Cretaceous continental sediments are also known from Indochina and from central Kazakhstan. As in preceding periods, the area where continental sediments accumulated are related to the arcuate folded structures. If the latter formed concentric systems, proportional to the subsidence of the earth crust, the continental sediments are all in concentric circles closer to the center.

In the lower Jurassic of Kazakhstan, for instance, the continental coal bearing sequences accumulated on the periphery of the Variscan arc running from the Alai and Turkestan ranges to the Urals. The sea inundated this area during the Lower Cretaceous and continental deposits began to accumulate between this arc and the Caledonian arc of the Talas Range (Kara-Tau to Ulu-Tau) whose inner part continued to rise. A further transgression occurred in the Upper Cretaceous, and the sea invaded the area between the arcs. The shore line ran at that time about along the Kara-Tau-Ulu-Tau, and continental sediments began to accumulate on their inner side.

A similar, progressive "disintegration" of concentric arcuate systems occurred also in the northeast of Asia. In the Lower Triassic (see Fig. 3), the sea bordered the outer Verchoyana-Kolyma-Chukot arc. The sea then drowned the outer concentric zones in the Upper Triassic, but the inner Kolyma massif continued to remain dry. However, in the Upper Jurassic (see Fig. 4), it also is submerged, and only the highest points remain above sea level as islands.

Just as in earlier times, the character of Cretaceous continental deposits depended on the climate. It has already been mentioned (Sheinman, 1954) that the climatic belts, as delineated by features of sedimentary rocks, cut through Asia in a northwest–southeast direction. Red beds are completely missing from continental Mesozoic deposits in the northeast of Asia, but they are extensive in Europe (in the Triassic), in Central Asia and in the southern part of the West-Siberian Lowlands, and also in China and Indochina (Triassic, Jurassic, Cretaceous). The distribution of Upper Cretaceous vegetation is another indicator of the orientation of Mesozoic climatic belts.

Figure 5, after T. I. Baikovskaya (1956), shows the northern (northeastern, in present geographic coordinates) boundary of subtropical flora. It almost completely coincides at sea with the northern boundary (merely displaced a little to the north) of such markedly thermophile organisms as are the rudista. This line also marks approximately the extent of other thermophile forms, such as some ammonites (Naidin, 1956).

North of this line, coniferous–broad leaf forests of a temperate climate spread out. Based on the admixture of ferns and sago palms, T. N. Baikovskaya distinguished forests of a typical maritime climate. A similar vegetation in Sakhalin and Anadyr can, evidently, be explained by the presence of a warm current, directed in Cretaceous time from south to north along the eastern margin of Asia. Furthermore, the north pole was probably situated in the Pacific Ocean in the Mesozoic (Rukhin, 1955). This is supported by the absence of a strong drop in temperatures, such as occurs at present upon approaching the poles (Rukhin, 1958).

Another important aspect of Mesozoic continental deposits, very easy to observe in the paleogeography of Asia in the Cretaceous period, is the distribution of volcanic rocks.

It is known that they accumulate extensively, beginning with the Middle Jurassic in continental and locally also marine sediments of eastern Asia. By Upper Cretaceous time, lavas and pyroclastic rocks cover a huge area in northeastern Asia, in the Far East, in northern and southern China. That volcanic eruptions are confined to the east must, obviously, be related to the formation of the Eastern Asiatic arc.

Arcuate tectonic systems are usually twofold (Kuenen, 1950). The inner arc is older, less curved and is marked by a particularly extensive intrusive and effusive volcanic activity (Rukhin, 1959). East Asiatic mountains marginal to the continent of Asia (the Kolyma Range, Sikhote-Alin' etc.) are such inner structures in respect to the present island arcs. From the increase in volcanic activity, we can, therefore, assume that the arcs of eastern Asia were formed in the Jurassic and especially in the Cretaceous periods. The occurrence of the remains of undoubtedly shallow-water Cretaceous organisms on top of the submarine ridge (a former island arc) in the central part of the Pacific Ocean west of the Hawaiian Islands (Hamilton, 1956) confirms that the island arc reached maximum uplift in the Cretaceous.

East of Asia the Aleutian arc continues to the Kuril–Kamchatka arc,

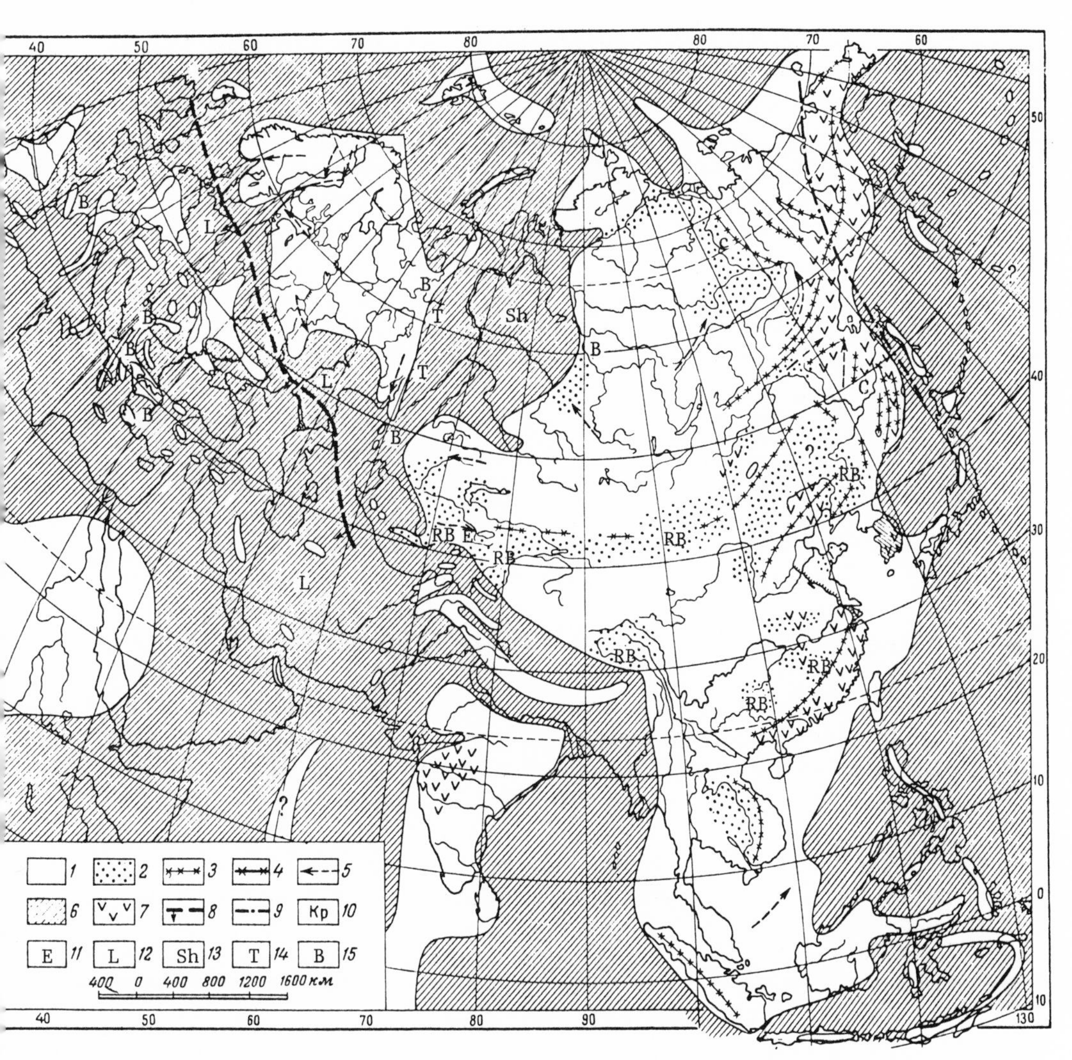

Figure 5. Paleogeographic outline of Asia in the Upper Cretaceous (Maastrichtian). 1. Dry land with rolling relief (primarily an area of denudation). 2. Alluvial- deltaic plains. 3. Low mountain ranges. 4. Mountain ranges with very dissected relief. 5. Assumed flow direction of rivers. 6. Sea. 7. Areas of volcanic eruptions. 8. Northern limit of tropical vegetation. 9. Limit of vegetation typical for a maritime temperate climate. 10. Red beds. 11. Evaporites. 12. Limestones. 13. Shales. 14. Continental beds. 15. Basins.

which in turn meets the Sakhalin–Japan arc and the latter the Ryu–Kyu arc. Accordingly, the Aleutian arc is the youngest, and the Ryu–Kyu arc is relatively older. Where the arcs meet, there is a tendency towards uplift while the central parts of the arcs sag the most by comparison. This is typical of ancient arcs where rivers flowed over or where the sea advanced primarily across the central parts.

The following conclusions can be drawn from the above:

1. The Mesozoic configuration of Asia is largely inherited. Variscan, Caledonian and also older folded zones represent in most cases also areas repeatedly uplifted in the Mesozoic era. Depressions filled by Mesozoic marine and continental sediments formed on the outer side of these ancient zones of uplift, which in most cases are distinctly arcuate. This explains the distribution of many Mesozoic areas of sedimentation. The Amur–Zeya Depression, which existed during almost the whole Mesozoic era and through which the Jurassic sea penetrated to the east of Lake Baikal is thus accounted for, as this depression extends on the northern, convex side of a Caledonian arc. It also explains the accumulation of Jurassic continental deposits north of the Sayan Caledonian arc, etc.

Not only the arcs but also some plateaus inside the arcs displayed a continuing tendency towards uplift. This is especially obvious in the case of the Patom inter–arc plateau, which is the original, ancient top of Asia. It applies to a somewhat lesser degree to the Altai–Sayan area.

The inheritance of uplifts does not mean the continuous existence of positive forms of relief. It is more often a systematic revival of the uplift on the same site, whereby the denudation could almost completely destroy these positive forms of relief in intervening periods of time.

Basic forms of relief are inherited during all the Mesozoic era by the continuous existence of large rivers. Naturally, their length changed considerably depending on transgression and regression, but their position and direction remained relatively stable outside the areas of sedimentation.

2. This astonishing stability of large forms of relief, however, persisted throughout the Mesozoic era in the face of the appearance of new forms.

One of the most striking examples of this is the downwarp of the folded structures of the West–Siberian Platform under a thick sequence of Mesozoic sediments. This has no explanation at present, especially since similar forms of relief farther south, in Kazakhstan, existed continuously as positive forms of relief after the formation of the Caledonian structures. These are evidently processes of planetary scale which, in addition to geologic structures, cause uplift or downwarp of wide regions of the earth crust.

The formation of new arcs (and of new uplifts and downwarps) is an analogous process which in future will give rise to deep faults that cut older structures.

As Suess had mentioned already, arcs generally encircle the core of Asia in concentric fashion, and they have increased the dimensions of the continent. However, subsequent folding often filled the free space inside the framework of arcs, and therefore, alternating zones of young and old folding are usually found with increasing distance from the platform. In southern China, for instance, the Mesozoic folded structures formed inside the Caledonian arcs. N. M. Sinitsyn (1957) described alternate zones of Caledonian and Variscan folding (in the south Variscan and Alpine folding) in Central Asia.

3. Broad areas of continuous accumulation of continental sediments in most cases gravitate to the platforms, but not to the geosynclinal seas (Triassic sediments of the Russian platform, Triassic and Jurassic sediments of the West–Siberian Lowlands and of the Siberian Platform, etc.). The excess compensation of the downwarp by sedimentation, preventing seaside areas from being inundated by the sea, occurred in the first place in relatively weakly subsiding areas. In addition, there must be rising massifs nearby. Continuous accumulations of continental sediments are, therefore, found around ancient arcs.

4. The most widespread landscape in areas of continuous accumulation of continental sediments are the alluvial plains, usually imperceptibly passing into a zone of lagoons and deltas, washed by shallow sea. In these alluvial–deltaic plains, there often are large lakes. The general appearance of such plains, and with it the character of the sediments, depends in a very large measure on the climate. In the zone of hot, periodically arid climate, red beds accumulated in these plains. Coal bearing sediments accumulated farther to the north. The northward replacement of red continental sediments by coal-bearing ones is very well documented on the continent of Asia.

However, the width of the zones where red and grey (coal bearing) sequences accumulated did not remain constant. The most widespread coal formation occurs in the Jurassic. Red sediments penetrate relatively far north in the Triassic and in the Cretaceous. Such a difference is explained by climatic changes that are, evidently, due to changes in the ocean currents.

5. During the Mesozoic, the areas of primary accumulation of continental sequences in Asia moved from west to east. This is due to the more extensive tectonic movements in the east and, consequently, the more frequent rejuvenation of the relief of ancient arcs. Their erosion formed the clastic material which makes up the continental accumulations.

6. The arcs of eastern Asia developed most actively towards the end of the Jurassic and especially in the Cretaceous periods. The arcs of eastern Asia which adjoin the continent become younger from north to south. This is shown by their configuration, which is caused by successive deep faults.

In his time, Archimedes promised to lift the earth if one gave him a proper lever. The paleogeographer, especially when studying Mesozoic landscapes, can in future confidently predict the distribution of land forms if he only knows the distribution of more ancient arcuate systems, some planetary principles of tectonic movements, and the climatic zonality of the past.

REFERENCES

Amantov, V. A., and G. P. Radchenko. 1959. The Continental Permo-Triassic deposits of central Mongolia. *Akad. Nauk USSR Doklady* **124**(1).

Archangel'skii, A. D. 1912. Upper Cretaceous deposits of eastern European Russia. *Mater. Geol. Russia* **25**.

Arkell, W. J. 1956. *Jurassic Geology of the World*. London.

Baikovskaya, T. N. 1956. Upper Cretaceous flora of northern Asia. *Paleobotanika* **3**.

Belousov, V. V. 1958. The principal features of the tectonics of central and southern China. *Akad. Nauk SSSR Izv. Ser. Geol.*, No. 8.

Bo-Tsin-Huan. 1952. *Principal Features of the Tectonic Structure of China*. Moscow: Foreign Language Press.

Bubnoff, S. 1956. *Einfuhrung in die Erdgeschichte*, 3rd. ed. Berlin.

Chzhman Gen, Chzhen Uin-da, and P. P. Zabarinskii. 1958. *Deposits of Petroleum and Natural Gas in the Chinese People's Republic*. Moscow: Gostoptekhizdat.

Cissarč, A. 1952. *Mineral Resources of Yugoslavia*. Moscow: Foreign Language Press.

Davydova, T. N., and U. L. Gol'dshtein. 1949. *Lithologic Studies in the Bureya Basin*. Moscow: Gosgeolizdat.

Derviz, T. L. 1959. *Jurassic and Cretaceous Deposits. Volga-Ural Petroleum Province*. Moscow: Gostoptekhizdat.

Donovan, D. T. 1957. The Jurassic and Cretaceous systems in East Greenland. *Medd. Grönland* **155**(4).

Dorn. P. 1958. Problematik des vindelizischen Landes. *Geologie*, No. 3–6.

Florensov, I. A. 1958. The paleogeography of the continental Mesozoic in the south of Eastern Siberia. *Akad. Nauk SSSR Sibirsk. Izv.*, No. 4.

Furon, R. 1941. *La Paléogéographie*. Paris.

Furon, R. 1941. Géologie du Plateau Iranien. *Mus. Natl. Histoire Nat. Mém.* **7**(2).

Furon, R. 1955. Histoire de la geologie de la France d'Outremer. *Mus. Natl. Histoire Nat. Mem.*, Ser. C, **5**.

Geological Structure of West-Siberian Lowlands (in Russian). 1959. Moscow: Geosgeoltekhizdat.

Geology and Occurrences of Petroleum in the West-Siberian Lowlands. Vses. Neft. Nauchno-Issled. Geol.-Razved. Inst. Trudy **114**, 1958.

Gignoux, M. 1952. *Stratigraphic Geology*. Moscow: Foreign Language Press.

Grachev, P. I. 1957. *Paleogeography of the Caspian Depression in Lower and Middle Jurassic Times, Geology and Geochemistry*, Vol. 1(7). Moscow: Gostoptekhizdat.

Grossgeim, V. A. 1957. The history of Mesozoic sedimentation in the northern Caucasus and Pre-Caucasus. *Moskov. Obshch. Ispytateley Prirody Byull., Otdel Geol. No. 2.*

Hamilton, E. L. 1956. Sunken islands of the Mid-Pacific Mountains. *Geol. Soc. America Mem. 64.*

Kanskii, N. E., V. P. Makridin, and B. P. Sterlin. 1956. Facies and paleogeography of Jurassic deposits on the northwestern margin of the Donets Folded Structure. *Sci. Notes, Kharkov Govt. Univ.* **73**.

Karpinskii, A. P. 1887. Outline of the physical geography of European Russia in former geologic periods. *Akad. Nauk SSSR, Leningrad, Zapiski*, vol. 55.

Khabakov, A. V. 1948. Dynamic paleogeography, its aims and possibilities. *All-Union Geol. Conf., 2nd, Proc.* **2**:115–131.

Kravchenko, K. N. 1958. New data on the stratigraphy of the Kugara Depression (Syn'ts-zyan). *Sovetskaya Geologiya*, No. 8.

Krishnan, M. S. 1954. *Geology of India and Burma*. Moscow: Foreign Language Press.

Kseszkewicz, M. and Ya. Samsonovich. 1956. *Outline of the Geology of Poland.* Moscow: Foreign Language Press.
Kossovskaya, A. G. 1958. History of Mesozoic sediment accumulation in the western Verkhoyansk area and the Vilyuy Depression. *Akad. Nauk SSSR Izv. Ser. Geol.*, No. 7.
Kuenen, Ph. H. 1950. *Marine Geology*. New York: J. Wiley & Sons, 551pp.
Li, Su-Huan, 1952. *Geology of China*. Moscow: Foreign Language Press.
Liu, Hun-Yun. 1955. *Paleogeographic maps of China* (in Chinese). Sci. Press of China.
Lyutkevich, E. M. 1955. Permian and Triassic deposits of northern and northwestern Russian Platform. *Vses. Neft. Nauchno-Issled. Geol.-Razved. Inst. Trudy*, No. 75.
Mashrykov, K. 1958. *Jurassic Coal-bearing Deposits of Northwestern Turkmenistan and Their Position in the Crimea-Caucasus-Caspian Coal-bearing Province*. Ashkhabad: Izdat. Akad. Nauk Turkmen. SSR.
Mesozoic and Tertiary Deposits of the Central Part of the Russian Platform, 1958. Moscow: Gostoptekhizdat.
Muratov, M. V. 1949. Tectonics and history of the alpine geosynclinal area of the southern part of European SSSR and contiguous areas. *Tectonics of the USSR*, vol. 2.
Naidin, D. P. 1956. [reference not given in the original].
Nikitina, Yu. P. [1948]. Paleogeographic conditions of sediment accumulation in Lower Cretaceous time in the Embensk petroleum region. *Moskov. Obshch. Ispytateley Prirody Byull., Otdel Geol.* **23**(2).
Proceedings, All-Union Conference to establish a unified scheme of Mesozoic stratigraphy of the Russian Platform, 1956. Moscow: Gostoptekhizdat.
Proceedings, Interdisciplinary Conference on Stratigraphy of Siberia. Papers on the stratigraphy of Mesozoic and Cenozoic deposits, 1957. Moscow: Gostoptekhizdat.
Rukhin, L. B. 1955. Climates of the past. *All-Union Geograph. Soc. Izvest.* **87**(2).
Rukhin, L. B. 1958. The problem of the origin of the continental glaciation. *All-Union Geograph. Soc. Izvest.* **90**(1).
Rukhin, L. B. 1959. *Problems of General Paleogeography*. Leningrad: Gostoptekhizdat, 557pp.
Sazonov, I. T. [1948]. Geological history of the Jurassic period in central areas of the Russian Platform. *Moskov. Obshch. Ispytateley Prirody Byull., Otdel Geol.* **23**(2).
Saks, V. N. and Z. Z. Ronkina. 1958. Paleogeography of the Khatanga Depression and adjacent areas during the Jurassic and Cretaceous periods. *Nauchno-Issled. Inst. Geologii Arktiki Trudy* **85**.
Sarkisyan, S. G. 1958. *Mesozoic and Tertiary Deposits on the Cis-Baikal, Trans-Baikal and Far East Regions*. Moscow: Izdat. Akad. Nauk SSSR.
Sheinman, Yu. M. 1954. Upper Paleozoic and Meso-Cenozoic climatic zones of Eastern Asia. *Moskov. Obshch. Ispytateley Prirody Byull., Otdel Geol.* **29**(6).
Sherlock, R. 1948. *The Permo-Triassic Formations*. London.
Shutov, V. D. 1958. Lithologic-stratigraphic breakdown and conditions of sediment accumulation in Permian and Lower Triassic deposits of the Verkhoyansk Ridge. *Akad. Nauk SSSR Izv. Ser. Geol.*, No. 7.
Sinitsyn, V. M. 1956. Basic elements of the structure of Gobi. *Moskov. Obshch. Ispytateley Prirody Byull., Otdel Geol.* **30**(2).
Sinitsyn, V. M. 1955. General scheme of the tectonics of High Asia, *Moskov. Obshch. Ispytateley Prirody Byull., Otdel Geol.* **30**(2).
Sinitsyn, N. M. 1957. The scheme of the tectonics of the Tyan-Shan. Leningrad Univ. Vestnik, No. 12.

Sinitsyn, N. M. and V. M. Sinitsyn. 1958. Tyan-Shan. Main elements of tectonics. *Akad. Nauk SSSR Izv. Ser. Geol.*, No. 4.
Slavin, V. I. 1956. The Mesozoic history of the Carpathians. *Sci. Notes, Moscow Govt. Univ.* **176**.
Smirnov, A. M. 1954. Basic questions of the geology of Manchuria. *Kharbin Soc. of Naturalists and Ethnographers, Proc.*, No. 13.
Sobolevskaya, V. N. 1951. Paleogeography and structure of the Russian Platform in the Upper Cretaceous epoch. *Memorial Volume of Acad. A. D. Archangel'skii.* Moscow: Izdat. Akad. Nauk SSSR.
Stchepinsky, V. 1947. Paléogéographie de la Turquie. *Rév. Sci.*, No. 12.
Tromp, R. 1958. Remarks on the pre-orogenic history of the Alps. *Geologie en Mijnbouw*, No. 10.
Tuchkov, I. I. 1957. Paleogeography of the northeastern USSR in Upper Triassic, Jurassic and Lower Cretaceous times. *Sovetskaya Geoliya*, No. 59.
Umbgrove, J. H. 1947. *The Pulse of the Earth*. The Hague.
Van Bemmelen, O. 1957. *Geology of Indonesia*. Moscow: Foreign Language Press.
Vasil'ev, V. G., V. S. Volkhonin et al. 1959. *Geologic Structure of the Mongolian People's Republic*. Moscow: Gostoptekhizdat.
Vakhrameev, V. A. 1957. Plant–geographic and climatic zonality of Eurasia in Jurassic and Cretaceous times (in Russian). *Voprosy paleobigeografii i biostratigrafii*, Vses. Paleontol. Obshchestvo, Trudy 1. Sessii.
Vakhrameev, V. A. 1952. Stratigraphy and fossil flora of Cretaceous deposits of western Kazakhstan. *Regional Stratigraphy of the USSR*, No. 1.
Yakovlev, V. N. 1957. The Cretaceous system of Sikhote-Alin." *Akad. Nauk SSSR Dal'nevostochn. Filial Trudy* **3**.
Yanshin, A. L. 1953. *Geology of the Northern Aral Region. Moskov. Obshch. Ispytateley Prirody Izdat.*
Wills, L. J. 1951. *Paleogeographical Atlas*. London.
Zhilinskiy, G. B. 1956. Ancient sediments of Central Kazakhstan (in Russian). *Razvedka i okhrana nedr* (Prospecting and Conservation of Mineral Wealth), *No. 10.*

14

Reprinted from *Geol. Mag.* **111**:369–383 (1974)

The Indus Suture Line, the Himalaya, Tibet and Gondwanaland

A. R. CRAWFORD

Summary. Misunderstanding exists about the origin of the Himalaya, Indo-Tibetan relationships and the extent and history of Gondwanaland. The Indus Suture Line is a relic of fracture to the mantle but for a period only represented by the faunas of the 'exotic blocks' of the Tibetan Himalaya, i.e. Permian–Late Jurassic. Tibet appears originally to have been part of a plate including India, but submerged while India remained continental. Associated with the fracture extending to the mantle represented now by the Indus Suture Line, shallower subparallel fractures developed within which the Gondwana sediments of the Himalaya were preserved. These, together with the salt at the base of the Tethyan marine sequence, facilitated intra-continental orogeny when, much later, India was affected by vigorous sea-floor spreading and plate movement in the NW Indian ocean after it had become attached to Asia.

The northern plate boundary was beyond Tibet, on the N side of the Tarim Basin Block, and along the Tien Shan. The Tien Shan present peculiar features, particularly very deep long intramontane depressions such as Issyk Kul in USSR and Turfan-Hami in China, and their stratigraphy shows persistent mobility. The Tarim Basin Block has moved sinistrally relative to Tibet and in Gondwanaland lay near northern Western Australia. The suggested extent of Gondwanaland explains the present lack of continental crust W of Western Australia, now in Tibet, and the intracontinental origin of the Himalaya is consistent with the absence of recent volcanism in an area of considerable seismicity. The association of Tibet with India in Gondwanaland destroys the effectiveness of arguments against continental drift based upon the extension of Indian stratigraphic sequences across the Himalaya. The hypothesis of an enlarged Gondwanaland is given support by recent Chinese discoveries of terrestrial vertebrate fossils of Gondwanic type near the Tien Shan.

1. Introduction

Much misunderstanding and confusion appears to exist about the origin of the Himalaya. This affects understanding of Indo-Tibetan relationships, the history of Gondwanaland and the development of Asia. Concepts of the significance of the Indus Suture Line (Gansser, 1959, 1964, 1966; Dewey & Bird, 1970) and of the height of the Tibetan Plateau are incompatible. If the former is the relic of the plate boundary between Gondwanaland and an Asia *minus* India, after the closing of Tethys, the latter cannot be raised up by a double thickness of crust, the lower element of Gondwanic and the upper of Asian, provenance as suggested by Carey (1958). Holmes (1944) at first suggested a different origin, with the Tibetan Plateau as a median mass forming part of a 'geosynclinal block' between the continents of Asia and India; but in the revised edition of his well-known textbook (1965, p. 1100) showed a quite different hypothetical cross-section, suggesting in the accompanying text that the 'northern prow of the advancing (Indian) shield may now extend beneath Tibet, perhaps as far as the Altyn Tag'. Any concept of a double thickness of crust built up in this way presents great geophysical difficulties, while the particular explanation of Carey (1958) is unacceptable on geological and palaeomagnetic grounds.

If closing-up of Tethys by subduction of oceanic crust in a Cordilleran-type trench on the S side of Asia was followed by continent–continent collision, as envisaged by Dewey & Bird (1970), then the relic of that trench would be expected to lie on the southern side of the Himalaya, perhaps covered by later thrust masses. Its position on the northern side is anomalous.

I suggest that the Indus Suture Line is the relic of a period of subduction, following a period of localized sea-floor spreading, itself following crustal fracture to the mantle level in one plate; that the Himalaya are not the product of continent–continent collision, but intra-continental in origin; that Tibet was an offshore part of Gondwanic India during the Palaeozoic, underlain by Precambrian continental crust; and that the main plate boundary of that era probably lay N of the Tarim Basin Block, relics of the boundary being related to peculiar features of the Tien Shan. That Block and Tibet, but with Tarim dextrally moved from its present relationship, lay with it in Gondwanaland and close to a proto-Australia.

2. The Indus Suture Line

The Indus Suture Line is a concept developed by Gansser (1959, 1964, 1966) following much field work in the Himalaya of the Indo-Tibetan border, particularly in the Kiogar region of the Kumaon Himalaya and in the area W of Lake Manasarowar (Fig. 1). Gansser (1966, p. 841) described a 'sudden root-like downbuckling', with orogenic effects on Middle Cretaceous marine sediments of Tethyan facies. Upper Cretaceous flysch contains ophiolites and huge exotic blocks (Griesbach, 1891, 1893; Diener, 1898) of sediments containing a pelagic facies unknown *in situ* in the Himalaya. These blocks of sediments are frequently attached to blocks of basic igneous rocks. Gansser (1966, p. 842) regarded the sediments as having been deposited 'in deep basins situated between the present Himalayas and the Tibetan Plateau; these original basins have completely disappeared and are now testified only by the presence of the Indus Line'. Gansser recognized that original crust had disappeared in the Upper Cretaceous and the exotic blocks squeezed out. At Amlang La he described a large sheet of peridotite which he regarded as mantle material originating from a 'deep, primary fracture zone which follows the Indus Line'. He estimated about 200 km of crustal shortening.

On his Tectonic Map of the Himalayas and Surrounding Areas Gansser (1964) shows this ophiolitic belt as extending from Lake Manasarowar north-westwards along the Upper Indus to bend W through Kargil and Dras then swinging north-westwards and finally northwards along the E side of the Nanga Parbat massif (Misch, 1949) and round its northern end to meet the Indus again near Bunji, with a branch continuing around the massif and another extending up the Gilgit river to slightly beyond that town.

Following Hess (1955), Dewey & Bird (1971) regard ophiolites as indicative of the former existence of oceanic crust. They (1970) regard the Indus Suture Line as the relic of a closed ocean, that which lay between Gondwanic India and Asia. Because India is known to have been part of Gondwanaland until the Jurassic, the ocean has been generally assumed to have been a wide one. That is questionable, particularly as the palaeopositions of major crustal units now in central Asia have been disregarded.

The restriction of the range of ages of the faunas of the exotic blocks suggests that the oceanic crust, on which the sediments containing these deep-water faunas originally lay, existed only from the Permian to the Late Jurassic, and

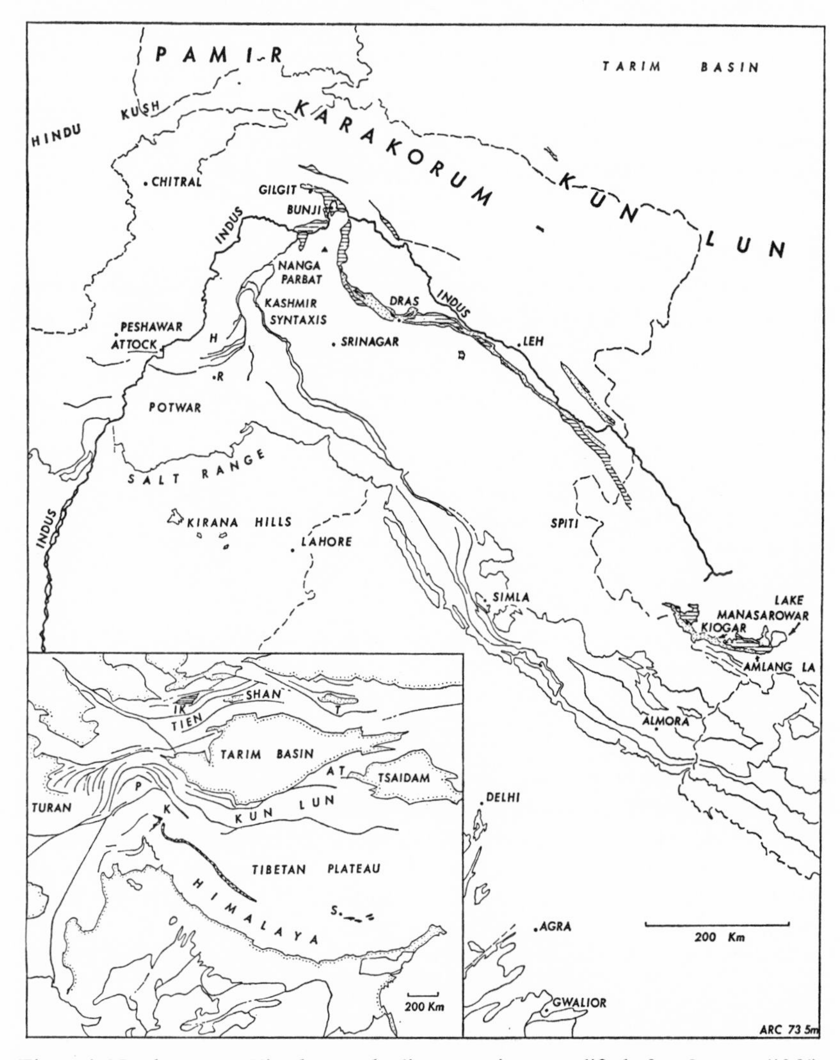

Figure 1. North-western Himalaya and adjacent regions, modified after Gansser (1964), showing position of Indus Suture Line (ophiolites, horizontal shading; flysch, dotted). Outer edge of Himalaya shown by full line, behind which full lines denote thrusts. Broken lines, international frontiers (not shown in Kashmir). H, Hazara; R, Rawalpindi. Inset shows larger regional relationships. AT, Altyn (Astin) Tagh; IK, Issyk Kul; K, Karakorum; P, Pamir; S, Shigatse; T, Turfan.

that before the Permian no break down to the mantle existed. As the Indus Suture Line has been traced eastwards only as far as Lake Manasarowar (latitude 30° 40′ N, longitude 81° 31′ E), it may be that this relatively short period of crustal opening to the mantle was limited also in regional extent. Gansser (1964, p. 254) has very tentatively suggested a possible connexion much farther E on two grounds: the occurrence of basic rocks in a tectonically disturbed zone of Jurassic sediments between Shigatse (Jih-k' o-tse, latitude 29° 18′ N, longitude 88° 50′ E) and Yandrok Tso (Yang-cho-yang-Hu, latitude 29° 00′ N, longitude 91° 30′ E) in southern Tibet, and the long trend of the Tsangpo (Upper Brahmaputra) comparable with that of the Upper Indus and almost in line with it. The second of these is of doubtful validity, as the association of the Upper Indus with the ophiolite belt, itself approximate, could be coincidental.

3. Tibet and India

Comparison of the Peninsular India, Extra-Peninsular (Himalayan) and Tibetan Precambrian and Phanerozoic sequences (Wadia, 1931; Krishnan, 1968; Ch'ang Ta, 1959) strongly suggests that the only difference between them is that Tibet and the Tethys Himalaya (i.e. the northern of the four structural zones of the Himalaya) have, like Peninsular India, a basement of complex patterns of metamorphic Precambrian rocks overlain by largely unmetamorphosed Upper Precambrian; but that while in Peninsular India the Phanerozoic is very incomplete and probably mainly undeposited, the sequence in the Tethys Himalaya and Tibet is almost complete and largely marine. It is significant that the characteristic Gondwana rocks of Upper Palaeozoic–Mesozoic age, of glacigene, sub-aerial, fluviatile or lacustrine facies, do extend across the Himalaya into western Tibet and into Sikkim and southern Tibet (Fig. 2). This alone renders suspect a major, distant separation of a Gondwanic India of approximate Peninsular size from an Asia, the area between occupied by ocean or even shallow sea at this time. The presence of these rocks in Tibet at the surface makes it unlikely that they have got there by any form of Gondwanic underthrusting, and has understandably been cited by opponents of continental drift (Meyerhoff, 1970). That the Gondwana rocks of Peninsular India are mostly found in rifts will be shown below to have particular significance; but as Mehta (1964) stated, it is not now the general view that they were confined originally to such rifts, which Fox (1931) believed were formed contemporaneously. Rather, the Gondwanas are preserved in rifts formed by later faulting, which though tensional, is locally reversed (Mehta, 1964).

Together with the strong impression from the distribution of metamorphic isograds for the Precambrian basement of the Peninsula (Pichamuthu, 1967) that the whole block is tilted northwards (Krishnan, 1968), this suggests that India and Tibet are to be regarded for most of Phanerozoic time as one huge crustal unit, the southern half emerged, the northern submerged. In this light the deep crustal fracture to mantle level now represented by the Indus Suture Line assumes its true proportions as an episode of relatively short duration and possibly merely regional. It is significant that the final closure of this fracture was about the time of opening of the major fracture between what are now Peninsular India and East Antarctica, at about the end of the Jurassic. I suspect that the opening of the earlier northern fracture in Permian time was related to major tectonic effects in the southern part of the Precambrian mobile zone along which the later fracture took place, and associated also with the development of rifts within the Indo-Tibetan block in which the Gondwana sediments are preserved (Fig. 3). These major tectonic effects included thrusts in southernmost India and Ceylon, with elevation of the area between them.

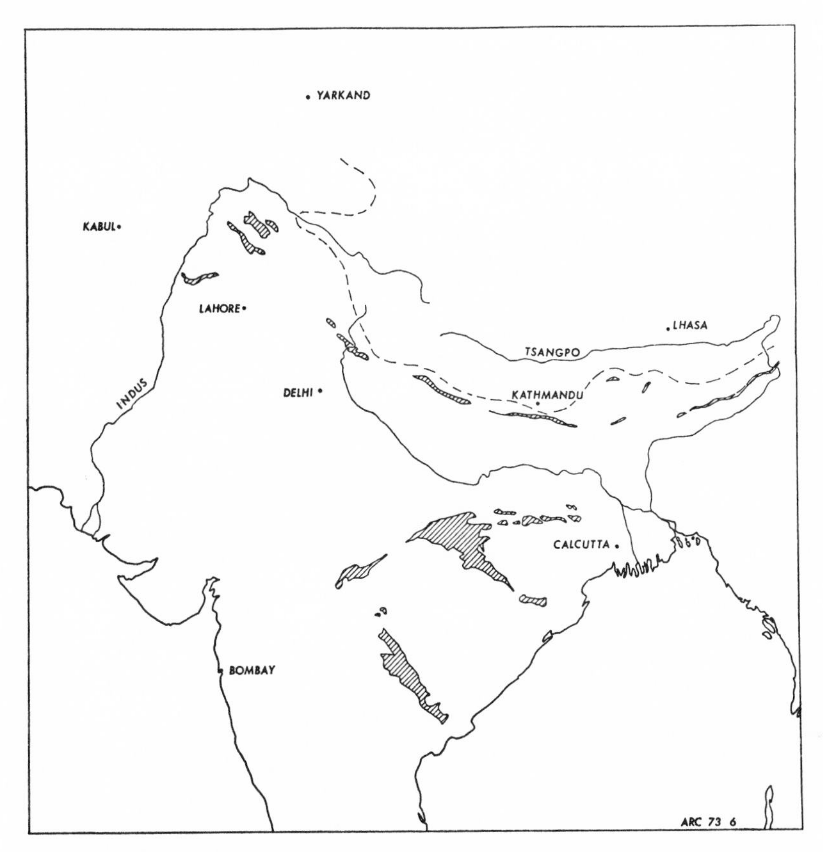

Figure 2. Distribution of Gondwana sediments with glacial influence. Outcrops, hatched. Approximate extension shown by broken line. After Gansser (1964) and Geological Survey of India maps.

4. The origin of the Himalaya

The great height of the present-day Himalaya is quite unrelated to their original formation as a geological structure. Least of all is it directly related to Tertiary collision. Popular concepts of its being related to the speed of movement of Gondwanic India, or an alleged distance that unit moved, are geologically mythical. As Sahni (1936) suggested, early human migration probably crossed a

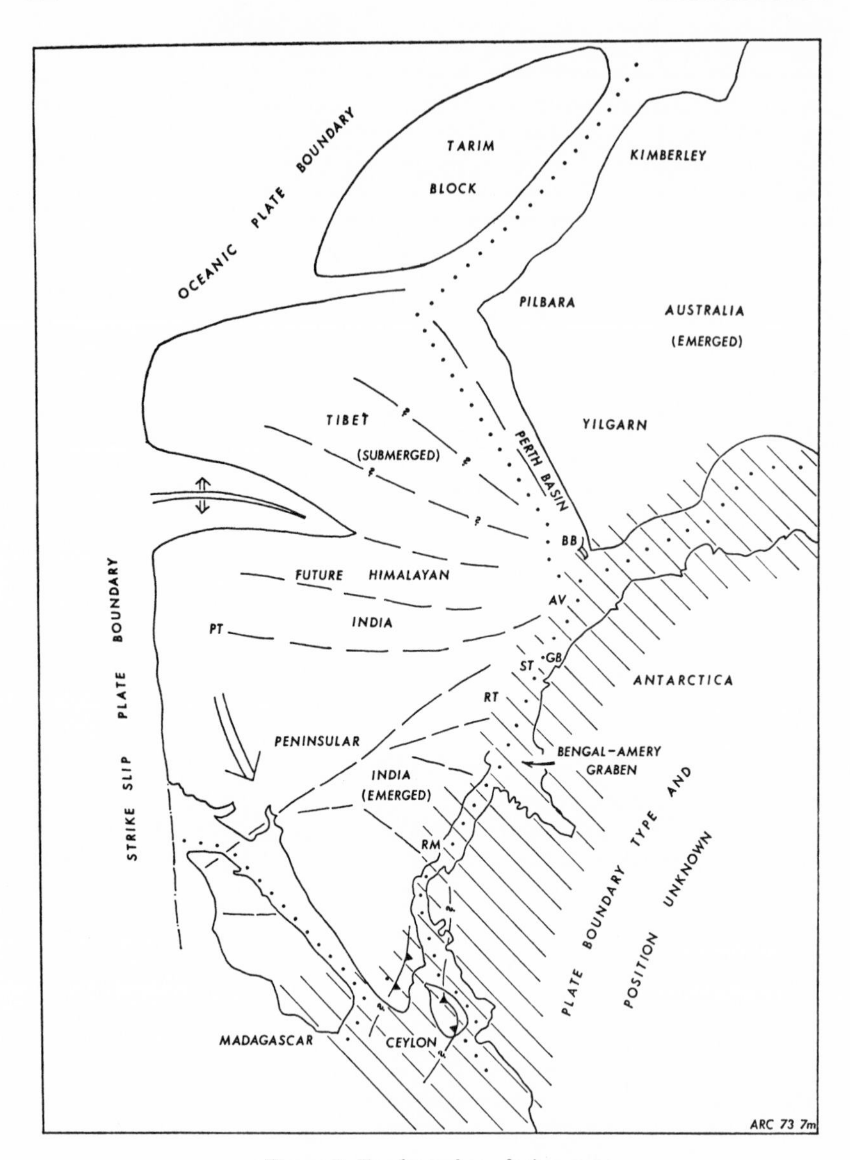

Figure 3. For legend see facing page.

much lower mountain system. Gansser (1964, p. 261) stated that the main elevation was an event 'witnessed by the earliest men'. He related the final rise to the Tibetan Plateau with the 'youngest, actually recent morphogenic rise of the Himalayas, and...its counterpart, the sinking of the Indian foreland'. This Recent uplift is clearly confirmed by stratigraphical observations, as, for example in Kashmir, where Dainelli (1922) suggested a Pleistocene uplift alone of 2000 m, and at the other end of the Himalaya the very young uplifts resulting in the great Tsangpo gorges. The Tibetan part of the Indo-Tibet plate is known, however, to have started its rise before the Main Himalayan orogeny, as the marine Eocene of Tibet passes up into continental deposits.

Geomorphologically, it can be argued, as did Wager (1937), that the great Himalayan peaks, which lie towards the northern side of the system except in the Punjab Himalaya, are merely the upturned edge of the Tibetan Plateau, following removal of enormous masses of rock by erosion in the gorges of numerous great antecedent rivers, with consequent isostatic rebound of the relatively narrow edge of the Plateau. But geologically the Himalaya are almost entirely built of what to an Indian geologist appears to be Peninsular material, and what to a Tibetan geologist – were there any – would appear Tibetan; the only exception being the material of the Indus Suture Line (Ch'ang Ta [1959], as a Chinese, compares the rocks with those of China). This is to regard the young granites as palingenetically altered old rock.

Such an interpretation of the Himalaya as made up of Peninsular material has long been accepted in India, if not always appreciated elsewhere. It was first observed by early British geologists such as Hayden (1908) and even by earlier explorers. Auden (1935) particularly emphasized that the unfossiliferous formations of the Lesser Himalaya closely resemble the Peninsular Precambrian.

Figure 3. Part of Gondwanaland during the Permian–Late Jurassic, showing diagrammatically the opening to the mantle and development of oceanic crust in the area now closed, subduced and represented by the Indus Suture Line. Present-day coastal outlines used to assist recognition, but western limit of Australia is the Darling Fault system. Opening developed by counter-clockwise rotation of the Indian part of an Indo-Tibetan-Australo-Antarctic plate, relative to the Tibetan and Antarctic parts. Associated rifts developed which in an emerged India allowed preservation of largely terrestrial Gondwana sediments. The rotation led to thrusting and some strike-slip in the Ceylon – south Indian part of the Late Precambrian – Early Palaeozoic zone of mobility (hatched) along which crustal separation between India (with Ceylon) and Australo-Antarctica took place finally in the Jurassic, along dotted lines. Periodic volcanism occurred during this whole time at localities marked: AB, Abor Volcanics of Assam (Gondwana); BB, Bunbury Basalt of Western Australia (mid-Cretaceous); GB, Gaussberg, Antarctica (age unknown); RM, Rajahmundry Traps of Godavari graben (?Jurassic, ?Cretaceous); RT, Rajmahal Traps of West Bengal-Bihar (mid-Cretaceous); PT, Panjal Traps of Kashmir (Permian-Trias); ST, Sylhet Traps of Bangladesh (pre-Upper Cretaceous). The closing-up of the oceanic zone of the Indus Suture Line started at the time crustal separation took place between the Indo-Tibetan–Tarim part of the plate and Australo-Antarctica, though the Tarim Block may have moved rather earlier. Strike-slip took place between the plate as a whole and areas to the W, and perhaps periodically between India and Madagascar before final separation of that fragment in the Cretaceous, when extensive volcanism occurred along the fracture and spread over much of India. Bengal-Amery graben probably developed during the Jurassic when widespread volcanic effusion affected Bengal.

These are separated by the Main Central Thrust from a massive crystalline sheet on which lie the Tethyan Phanerozoic sediments. The allocation to the Precambrian of some of these unfossiliferous formations in Nepal was given some early isotopic work by Krummenacher (1961), one sample of muscovite he regarded as detrital giving a K–Ar age of 728 Ma. I have since made many total-rock and mineral Rb–Sr analyses of the formations in Hazara, Kashmir, and the Simla and Darjeeling districts (to be published in detail elsewhere) which confirm the allocation beyond reasonable doubt. This in no way conflicts with the thorough and detailed structural work by Naha & Ray (1972), who demonstrate that the deformation and metamorphism at Simla and in other areas must be of Tertiary age. In the Darjeeling–Kalimpong region the undeformed Daling Series rocks give total-rock and mineral ages of up to at least 1100 Ma, while the high-grade metamorphic rocks at higher topographic levels give young Tertiary mineral ages. The unfossiliferous formations at Simla are all of Upper Precambrian age. This isotopic work is in agreement with the assessment of an Upper Precambrian age from stromatolithic evidence by Valdiya (1969).

Once the Precambrian age of the so-called unfossiliferous formations of the Lesser Himalaya is accepted, then it is no longer appropriate to contrast these as possible Phanerozoic rocks with the Tethyan Phanerozoic, as did Wadia (1957), but rather to accept that the sequences complement each other. There is, therefore, no need to assume that any 'normal' Tethyan Phanerozoic is from distant parts of the Tethys (Jain & Kanwar, 1970), but merely from its southern shore. Gansser's estimates of the amount of crustal shortening in the Himalaya, which total 300 km if that for the Indus Suture Line closure is excluded, are admitted by him to be conservative. But even if they need increasing by 100 % or 200 %, in accordance with the views of Krishnan (1953) the Himalaya are then not the product of the collision of two continents, but a pile of fractured slices of continental material in the middle of a continent, the pile lying approximately along the boundary where that continent had its lower Palaeozoic shore.

It is now desirable to consider the significance of the existence of Gondwana sediments in the Himalaya and Tibet from a different point of view, and particularly their outcrop at the foot of the Nepal and eastern Himalaya. These freshwater rocks occupy now a narrow belt from S of Kathmandu eastwards for 1200 km (Fig. 4). They are overthrust from the N by the Precambrian Daling Series. In the extreme E they are associated with the basalts of the Abor Volcanics. In Peninsular India Gondwana rocks are mostly found preserved in conspicuously linear rifts, and commonly found to be intruded by basic and ultrabasic igneous rocks. Other occurrences of Gondwana rocks are thinner and more scattered. It is highly possible then that the Gondwanas of the Nepal and eastern Himalaya, particularly as their outcrop is so linear, are also in a rift, the southern boundary of which is formed by the Peninsular rocks of high metamorphic grade which form the basement not only of the Shillong Plateau horst, but which as part of that horst occur as scattered inliers of Precambrian in the Brahmaputra alluvium between longitudes 90 and 94° E, close to the Himalayan front. That Evans (1964) postulated major dextral movement of the Shillong Plateau relative to the rest of the Peninsular shield in the Eocene does not affect the hypothesis of a rift except in so far as it necessitates acceptance of strike-slip

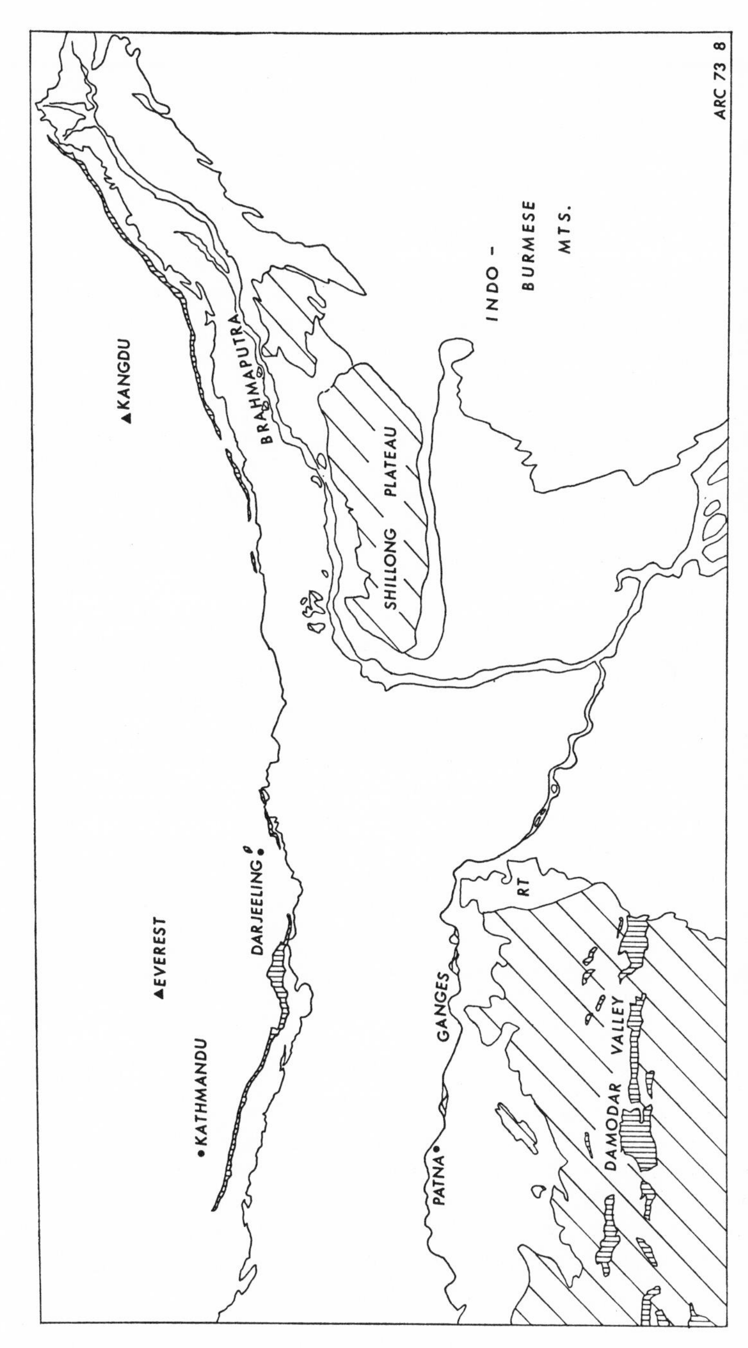

Figure 4. The Gondwanas of the Nepal and eastern Himalaya, after Gansser (1964), showing strongly linear outcrops (vertically hatched) comparable with but longer than the linear outcrops in the troughs of the Damodar Valley, typical of the Peninsula. Peninsular Precambrian, diagonally hatched.

along the southern side, a movement known to have affected tensional rifts elsewhere in the Peninsula (Bhimasankaram & Pal, 1973).

The western extension of such a rift beyond the limit of Himalayan foothill occurrences of Gondwana rocks would be covered by late thrusting, just as in the area of exposure the northern side of the rift is overthrust, with nevertheless a window into the Gondwanas in SW Sikkim (Ghosh, 1952). The existence of glacigene Gondwanas at Almora and Simla in the Kumaon Himalaya, again overthrust by Precambrian rocks, gives strong support to the suggestion of an extended rift. This concept of the structural control of the preservation of the Gondwanas has to be associated with the recognition that passing westwards to Kashmir the Gondwanas again become interbedded with volcanic rocks, the Panjal Traps, up to 2500 m thick, and that marine intercalations come in with eventually a complete vertical transition to marine conditions in the Triassic.

The existence of such a rift or rift system, formed subsequent to the deposition of the Gondwana rocks, together with the presence on its northern side of an abnormal thickness of continental crust consisting of the Precambrian basement, its unmetamorphosed cover, and the Tethyan sediments, would facilitate structural failure along it. Late in Phanerozoic time vigorous sea-floor spreading in the NW Indian Ocean led to conditions in which the whole Indo-Tibetan plate was by then close to the rest of Asia, towards which it could not approach much more. The consequent overthrusting produced the Himalaya.

The existence of such a rift can be postulated on grounds additional to the pattern of distribution of the Gondwanas. If complete crustal fracture existed to the mantle in the western half of the middle of the Indo-Tibetan plate in Permian–Late Jurassic time, the very opening of that fracture might well have been associated with parallel but shallower fractures, which at their western and eastern extremities nevertheless were deep enough to permit effusion of volcanic rocks.

Finally, it is no coincidence that the thrusting in the Salt Range, which is structurally Himalayan, is associated with a thick saline sequence at its base. This salt deposit has now been traced further E along the foot of the Punjab Himalaya and was clearly a regional feature associated with the arid conditions prevailing towards the end of the Precambrian in what is now northern Peninsular India. Gansser (1964) is almost certainly correct in suspecting the role of this salt in Himalayan tectonics as seriously underestimated.

If the Himalaya originated in this manner, one major problem of Himalayan geology is solved: the baffling absence of recent volcanic rocks of Himalayan age. No young volcanoes are known between Dacht-e-Nawar in central Afghanistan (Bordet, 1972) and central Burma (Chhibber, 1934). The suggested origin nevertheless fits the general pattern of seismicity, with a concentration of epicentres along the Himalaya, while Tibet is in comparison only mildly seismic. The suggestion of an exposed rift edge along the line of the Brahmaputra particularly fits the pattern of seismicity, for earthquakes are not confined to the Himalaya but frequently occur along the eastern part of the Indo-Gangetic Plain.

5. The Northern Plate Boundary

If Tibet was, like India, part of Gondwanaland, where was the northern boundary of the plate ? The question itself is ill-stated, as 'Gondwanaland' was an assemblage of crustal units which changed substantially during its long history, with considerable internal mobility between units within the assemblage, and the loss of outer units before inner ones, apart from possible temporary separation into smaller assemblages. But I suggest that from well back in Precambrian time Tibet and Gondwanic India were one. This unity, accepting also in Tibet such internal horizontal mobility as must surely have existed in India (for example, major long-term movement along the Narmada–Son lineament in the late Precambrian) continued throughout the Phanerozoic, with a relatively short period of complete crustal fracture either confined to an area in the middle near the western edge of the plate, or possibly extending right across it. The Tethyan sea N of emerged India, lying on Tibetan continental crust, shallowed northwards to the Tarim Basin Block, from which Tibet is now separated by the Kun Lun. This is a mountain system in which some young volcanic cones have been reported (Leuchs, 1937), which has possible Precambrian in its core (Ch'ang Ta, 1959) and which is faulted against the Tarim Basin Block. Lower Palaeozoic and Devonian sandstones and limestones overlie the possible Precambrian and pass up conformably into a thin Permo-Carboniferous marine sequence. Mesozoic rocks are few and continental, with some Cretaceous igneous rocks. The Tertiary is also continental. The main folding was post-Devonian and Permian, but the elevation Himalayan. Thus the Phanerozoic geological environment changed from a marine Tethyan at the end of the Permian and was subsequently continental.

The Tarim Basin Block is structurally quite different from Tibet and has been shown by Norin (1937) to have been stable since the Cambrian and a land area for most of the Phanerozoic, contributing little of the sedimentary material now in the Kun Lun. During the Himalayan period it appears to have remained continental and stable, though I believe that as a block it has been displaced horizontally very considerably in relation to its neighbours during much of the Phanerozoic.

Along the northern edge of the Tarim Basin a major tectonic boundary separates the block from the southernmost ranges of the Tien Shan, which are thrust over it. The Tien Shan have several remarkable features. Over 3000 km long, they include parallel mountain systems separated by long narrow depressions. In the E these even extend below sea level in the Turfan–Hami basins of Sinkiang, yet immediately N of these troughs the easternmost Tien Shan rise to over 5400 m. The depressed zones have never been satisfactorily explained and the amplitude of relief is unique in mountain systems. In the Soviet Tien Shan one depression contains the extraordinary lake known as Issyk Kul, with a deep of 702 m (Davydova *et al.* 1966) and a volume nearly twice that of the very much larger Aral Sea. Its depth is exceeded only by that of Baikal in Siberia, the world's deepest lake, yet immediately to the N and S of Issyk Kul the Tien Shan rise to 5000 m. The origin of the lake has remained obscure.

It has commonly been stated (Leuchs, 1937; Nalivkin, 1960) that the Tien

Shan are a system of Hercynian age. But Precambrian orogeny has been prominent (Gansser, 1964; Ch'ang Ta, 1959) as well as Palaeozoic orogeny. From the Late Carboniferous onwards continental deposition became dominant; this is, with the area to the N and NE, the classic region of the Angara terrestrial vertebrate faunas. Though some orogeny took place in the eastern Tien Shan in the late Jurassic, and marine Cretaceous exists, first appearing in the W, the main uplift is of Himalayan age.

The Soviet Tectonic Map of Eurasia (Yanshin, 1966) shows that the Tien Shan are regarded by its compilers as a composite mountain system, with elements of Caledonian folding faulted along the length of the system against elements of Hercynian folding. Ch'ang Ta (1959) described what appears to have been a Cambrian island arc system in the Chinese Tien Shan which continued into the Ordovician and Silurian. Volcanism went on into the Devonian, with great mobility. Though it declined in the early Carboniferous, in the Middle Carboniferous large-scale igneous intrusion occurred and the area then became stable after elevation following earlier subsidence. But in the Jurassic, 3000 m of sediment was deposited in rapidly subsiding basins and later the area became one of low relief. Finally, beginning in the late Cretaceous, block mountains developed with depressions between them.

All this history indicates great and continual mobility along a relatively narrow belt now marked by the Tien Shan mountains and intra-montane depressions, a mobility accompanied by repeated volcanism in much of the Palaeozoic. This suggests the existence of a plate boundary at least for the duration of that Era. I suggest that Issyk Kul and the Turfan–Hami depressions can be regarded as the relics of gaps between plates having slight irregularities of the edges, no doubt squeezed and distorted during the period since the plates met, and modified also in the period of block mountain formation. From the extreme seismicity of the Tien Shan it seems likely that this old plate boundary is still mobile. I believe nevertheless that at least since the formation of the Himalaya the Tien Shan have acted as a barrier. As the Turanian Block has moved dextrally relative to central Afghanistan (de Lapparent, 1972), the effects of a continued attempt by India to move NNE have been complex. I suggest that the approach of Turan towards Tarim may have produced the tight Pamir arc, with profound consequences for the tectonics of the north-western Himalaya and part of the western bordering ranges of Peninsular India, Crawford, (1974).

6. Tibet and the West Australian problem

If the northern plate boundary of Gondwanaland of Palaeozoic time was along the zone now occupied by the Tien Shan, then it becomes necessary to re-examine the problem of what lay between Gondwanic India, i.e. the Peninsula of today, and Western Australia. We know that the crust lying W of the continental shelf of Western Australia is oceanic, decreasing in age from about 140 Ma off the continental margin to about 80 Ma near the Ninety-East Ridge (Heirtzler *et al.* 1973). Thus the age of the break must be about 140 Ma. It is instructive that a reassembly with Tibet in its present position relative to India, making no allowance for Himalayan crustal shortening, shows that the space is approximately filled. The crustal shortening must be accommodated, but as the exact position of the eastern margin of the Tibetan part of the Indo-Tibetan unit is very poorly known it may be possible to accommodate this in the space without significant adjustment of any Australian crustal units.

If the northern plate boundary lay along the Tien Shan line in the early Palaeozoic, then the Tarim Basin Block has to be fitted into the reassembly. This presents no problem as there are no severe physical constraints. But it will be apparent from the Soviet Tectonic Map of Eurasia (Yanshin, 1966) that the position of the Tarim Block relative to Tibet must have changed considerably in the Phanerozoic, if only from the evidence in the Kun Lun, the regional strikes within which trend WNW–ENE while the fault boundary with the Tarim Basin cuts across them, trending WSW–ENE. The Tarim Block appears to have moved sinistrally relative to Tibet (a movement which is still going on) and its palaeoposition in Gondwanaland seems likely to have been adjacent to the north-western continental margin of Australia (Fig. 3). The movement away from Australia of the Tarim Block may have been rather earlier than the movement of Tibet with India. But it may be significant that Norin (1937) described a sequence in the Kuruk Tagh (K'u-lu-k'o Shan) of tillites and other sediments which he closely compared in type, sequence and age with those of the Adelaidean of South Australia, the only ones then known in Australia which had been adequately described. Today these Central Asian sediments might more appropriately be compared with the Adelaidean of the Kimberley Basin of northern Western Australia, to which on the reassembly they would have been close at the time of deposition. These Asian sediments lie not far S of Turfan, suggesting that perhaps the then northern plate boundary lay not far from what is at present northernmost Australia.

7. Note

Since writing this I have received a paper by Sun Ai-Lin (1972) which reports the discovery of Permo-Triassic vertebrate fossils in the Tien Shan and Turfan Basin. These are of Gondwanic type and Sun Ai-Lin infers a land connexion between Sinkiang and Antarctica, where similar fossils occur, in Permo-Triassic time.

Acknowledgements. I thank Professor A. Gansser and Drs K. S. W. Campbell, T. J. Fitch, M. W. McElhinny and P. A. Jell for comment and useful discussion.

References

Auden, J. B. 1935. Traverses in the Himalaya. *Rec. geol. Surv. India* **69**, 123–67.

Bhimasankaram, V. L. S. & Pal, P. C. 1973. Palaeomagnetism and Tectonics of the Narmada–Son Lineament. *In: Seminar on Geodynamics of the Himalayan Region*, 195–6. National Geophysical Research Institute, Hyderabad.

Bordet, P. 1972. Le volcanisme récent du Dacht-e Nawar. *Revue Géog. phys. Géol. dyn.* **14**, 427–32.

Carey, S. W. 1958. The tectonic approach to continental drift. *In: Continental Drift, A Symposium*, 177–355. University of Tasmania, Hobart.

Ch'ang Ta. 1959. *The Geology of China*. English Translation. U.S. Dept. of Commerce, Washington, 1963.
Chhibber, H. L. 1934. *The Geology of Burma*. Macmillan, London.
Crawford, A. R. (1974). *Earth. Plan. Sci. Lett.* (in press).
Dainelli, G. 1922. *Spedizione italiana de Filippi nell'Himalia, Caracorum e Turchestan cinese (1913–1914). Ser.* 2, *Risultati geologici e geografici. Vol.* 3, *Studi sul glaciale.* Zanichelli, Bologna.
Davydova, M. I., Kamenski, A. I., Nekliukova, N. P. & Tushinski, G. K. 1966. *Fizicheskaya Geografiya SSSR.* Izd, Prosveshcheniye, Moscow.
Dewey, J. F. & Bird, J. M. 1970. Mountain belts and the new global tectonics. *J. geophys. Res.* **75**, 2625–47.
Dewey, J. F. & Bird, J. M. 1971. Origin and emplacement of the ophiolite suite: Appalachian ophiolites in Newfoundland. *J. geophys. Res.* **76**, 3179–206.
Diener, C. 1898. Notes on the geological structure of the Chitichum Region. *Mem. geol. Surv. India* **28**, 1–27.
Evans, P. 1964. The tectonic framework of Assam. *J. geol. Soc. India* **5**, 80–96.
Fox, C. S. 1931. The Gondwana System and related formations. *Mem. geol. Surv. India* **58**, 1–241.
Gansser, A. 1959. Ausseralpine Ophiolithprobleme. *Ecl. geol. Helv.* **52**, 659–80.
Gansser, A. 1964. *Geology of the Himalayas.* Interscience, London.
Gansser, A. 1966. The Indian Ocean and the Himalayas – a geological interpretation. *Ecl. geol. Helv.* **59**, 831–48.
Ghosh, A. M. N. 1952. A new coalfield in the Sikkim Himalaya. *Curr. Sci.* **21**, 179–80.
Griesbach, C. L. 1891. Geology of the Central Himalayas. *Mem. geol. Surv. India* **23**.
Griesbach, C. L. 1893. Notes on the Central Himalayas. *Rec. geol. Surv. India* **26**, 19–25.
Hayden, H. H. 1908. *A Sketch of the Geography and Geology of the Himalaya Mountains and Tibet. Pt* 4. *The Geology of the Himalaya.* Govt of India Press, Calcutta.
Heirtzler, J. R., Veevers, J. V., Bolli, H. M., Carter, A. N., Cook, P. J., Krasheninnikov, V. A., McKnight, B. K., Proto-Decima, F., Renz, G. W., Robinson, P. T., Rocker, K. & Thayer, P. A. 1973. Age of the floor of the Eastern Indian Ocean. *Science* **180**, 952–4.
Hess, H. H. 1955. Serpentines, orogeny, and epeirogeny. *Spec. Pap. geol. Soc. Amer.* **62**, 391–408.
Holmes, A. 1944. *Principles of Physical Geology.* Nelson, London.
Holmes, A. 1965. *Principles of Physical Geology*, new and fully revised edition. Nelson, London.
Jain, S. P. & Kanwar, R. C. 1970. Himalayan Ridge in the light of the theory of continental drift. *Nature, Lond.* **227**, 829.
Krishnan, M. S. 1953. The structural and tectonic history of India. *Mem. geol. Surv. India* **81**.
Krishnan, M. S. 1968. *The Geology of India and Burma.* Higginbotham's Private Ltd, Madras.
Krummenacher, D. 1961. Determinations d'âge isotopique des roches de l'Himalaya du Nepal par la méthode potassium–argon. *Bull. Suisse min. petrogr.* **41**, 273–83.
Lapparent, A. F. de. 1972. L'Afghanistan et la dérive du continent indien. *Revue Géog. phys. Géol. dyn.* **14**, 449–56.
Leuchs, K. 1937. *Geologie von Asien.* Borntraeger, Berlin.
Mehta, D. R. S. 1964. *Gondwanas in India.* New Delhi, International Geological Congress, 22nd session.
Meyerhoff, A. A. 1970. Continental drift: implication of paleomagnetic studies, meteorology, physical oceanography and climatology. *J. Geol.* **78**, 1–51.
Misch, P. 1949. Metasomatic granitization of batholithic dimensions. *Am. J. Sci.* **247**, 209–45.
Naha, K. & Ray, S. K. 1972. Structural evolution of the Simla Klippe in the Lower Himalayas. *Geol. Rdsch.* **61**, 1050–86.

Nalivkin, D. V. (trans. Tomkeieff, S. I.). 1960. *The Geology of the U.S.S.R.* Pergamon Press, Oxford.
Norin, E. 1937. Geology of Western Quruq Tagh, eastern T'ien Shan. *Rep. Sino-Swedish Expedition*, 3, *Geology v.* 1. Aktiebolaget Thule, Stockholm.
Pichamuthu, C. S. 1967. *The Precambrian of India. In:* Rankama, K. (Ed.): *The Precambrian*, vol. 3. Interscience, London.
Sahni, B. 1936. The Himalayan uplift since the advent of Man: its cultural significance. *Curr. Sci.* **5**, 57.
Sun Ai-Lin. 1972. Permo-Triassic reptiles of Sinkiang. *Scientia sin.* (*Notes*) **16**, 152–6.
Valdiya, K. S. 1969. Stromatolites of the Lesser Himalayan carbonate formations and the Vindhyans. *J. geol. Soc. India* **10**, 1–25.
Wadia, D. N. 1931. The syntaxis of the North-West Himalaya: Its rocks tectonics and orogeny. *Rec. geol. Surv. India* **65**, 189–220.
Wadia, D. N. 1957. *Geology of India.* Macmillan, London.
Wager, L. R. 1937. The Arun River drainage pattern and the rise of the Himalaya. *Geog. J.* **89**, 239–50.
Yanshin, A. L. 1966. *Tektonicheskaya Karta Evrazii.* Ministry of Geology, Moscow.

Part V

THE WANING CENOZOIC TETHYS AND THE WAXING MEDITERRANEAN SEA

Editor's Comments on Papers 15 and 16

The early Tertiary Tethys has not been the subject of any recent comprehensive review. J. Dercourt presented an account of the Paleogene Tethys, the gradual deformation of the western Tethys, that is, its demise. He did so to compare the western Tethys with the framework of the North American Cordillera. Only the portion of his paper dealing with the Tethys has been reproduced here (Paper 15).

By the end of the Miocene, the next cycle in the development of the Tethys had run its course: the connection with the Atlantic Ocean had been severed due to diastrophism in both the Rif Atlas of Morocco and the Betic Cordillera in Spain. Everywhere else, the Tethys turned into mountain chains.

The late Miocene evaporite deposits in the Mediterranean and Red seas derived their waters largely from the Paratethys. Contemporaneous salt deposits in Iran and in the Persian Gulf (Stöcklin, 1968) accumulated in embayments open to the north. They also received their waters from the Paratethys, but their evaporites may well be leached and redeposited older occurrences.

The Pliocene epoch marks the beginning of a new cycle. Atlantic waters entered the Mediterranean Sea, terminating several episodes of extensive evaporite deposition. At the same time, the Mediterranean Sea began to break up into individual subsiding basins separated by swells and ridges, much as the early Mesozoic western Tethys had done. Basin subsidence and submarine basalt volcanism reach a maximum in Quaternary times.

The Paratethys was a series of evaporite pans in Tortonian time, a freshwater body in Sarmatian time, and a hypersaline brine of three times the concentration of seawater in Late Sarmatian time (Belokrys, 1968). In Pliocene times, it ceased to be the major supplier of waters to the Mediterranean Sea. The Bosporus narrowed to its present width (Cvijic, 1908), and other connections dried up. Basins in the western Paratethys, such as the Pannonian, Dacian, Transylvanian, and Volhynian basins, dried up, and the Euxinian and Caspian basins were much reduced in size (Sonnenfeld, 1976).

The last paper recounts the history of the Mediterranean Sea where salts formed later than in the Paratethys. The history starts after the demise of the Tethys Sea and after the breakup of the Paratethys due to the rise of the Carpathian chain and to differential vertical movements in the Ukraine and the Caucasus.

REFERENCES

Belokrys, L. S. 1968. Has there been any freshening of the Sarmatian Sea in southern Ukraine? *Internat. Geology Rev.* **10**:350–360.

Cvijić, J. 1908. Grundlinien zur Geographie und Geologie von Mazedonien und Altserbien. *Petermann's Geog. Mitt.* **162**:1–392.

Sonnenfeld, P. 1977. Origins of Messinian sediments in the Mediterranean region—Some constraints on their interpretation. *Annales Géol. Pays Helléniques* **27**:160–189.

Stöcklin, J. 1968. Salt deposits of the Middle East. In Mattox, R. B. et al., eds., *Saline Deposits,* Geol. Soc. America Spec. Paper 88, pp. 157–182.

15

Reprinted from pages 709–722 and 739–743 of *Canadian Jour. Earth Sci.* **9**: 709–743 (1972)

The Canadian Cordillera, the Hellenides, and the Sea-Floor Spreading Theory

JEAN DERCOURT
Université des Sciences et Techniques, Lille (France), Laboratoire Associé, CNRS No. 145

Manuscript received September 9, 1971
Revision accepted for publication February 29, 1972

Introduction

Knowledge about the overall structure of the earth, provided for example by geophysical exploration of the oceans and by analysis of data from satellites (Schwiderski 1968), is accumulating at a rate inconceivable only a few years ago. Side by side with this accumulation of data, earth scientists have been proposing theories concerning the Earth's nature, origi and evolution. The sea-floor spreading theo to be discussed in this article (see e.g. Heirtzl *et al.* 1968; Isacks *et al.* 1968; Le Pichon 196 Morgan 1968; Pitman and Hayes 1968; a Vine 1966) postulates that the Earth's cru and upper mantle are divisible into a relative small number of plates, each plate genera containing both oceanic and continental cru

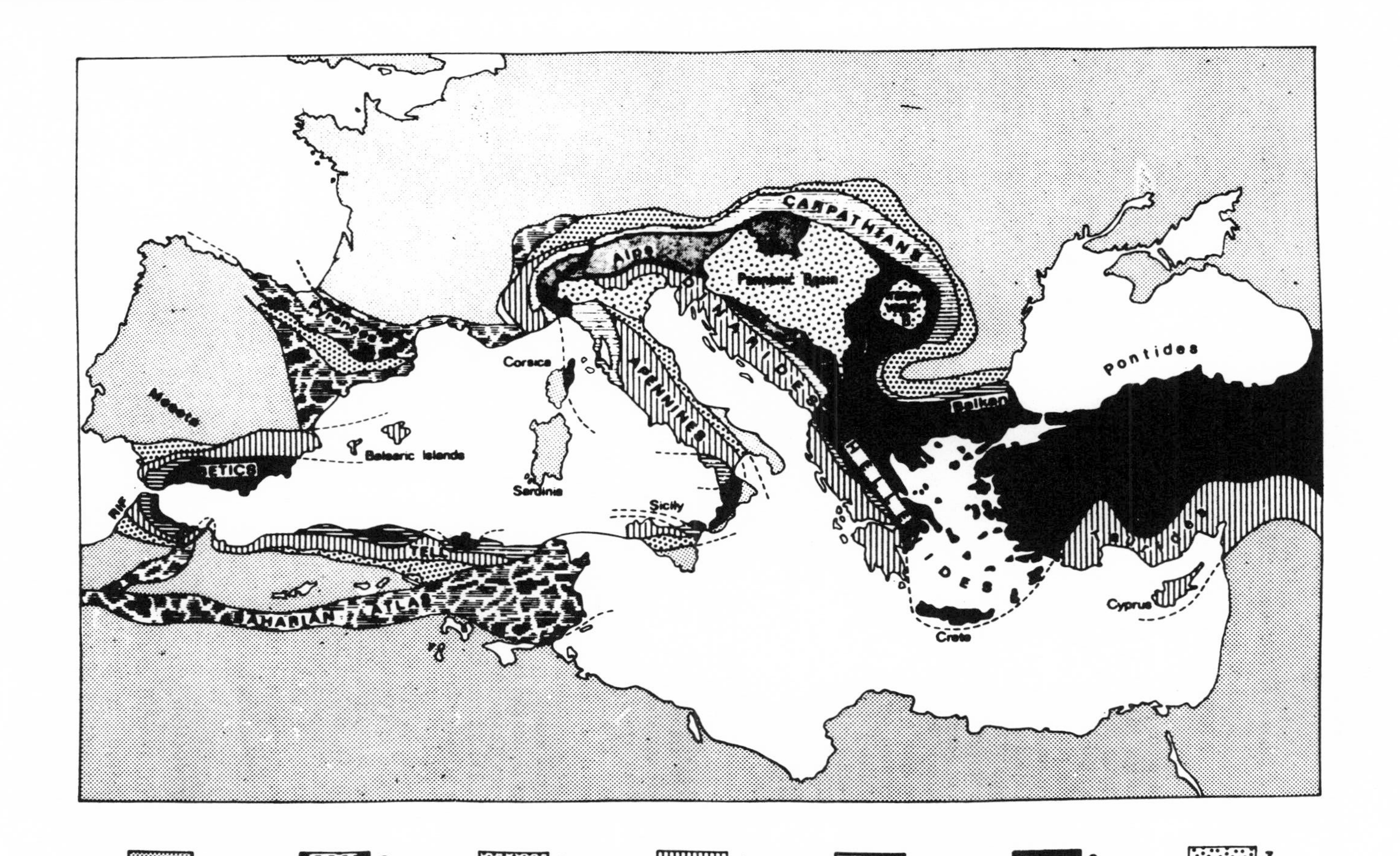

FIG. 1. The main alpine units in the Mediterranean Area (after Aubouin *et al.* 1971); 1 foreland, 2 intracratonic belt, 3 molassic fore-trough, 4 external zones, 5 Cretaceous-Paleocene flysch thrust towards the external zones, 6 internal zones, 7 molassic internal trough.

which move relative to one another. Where two plates move apart, new oceanic crust is created along the margin; where they move towards one another, oceanic crust in one plate is subducted beneath or partially obducted on top of the other plate.

The purpose of this paper is to examine the geology of the Canadian Cordillera and the Hellenides in the light of current sea-floor spreading theory. Such an examination is especially appropriate as both mountain systems contain extensive elements that can be interpreted as oceanic crust. In another way they differ, e.g. andesitic volcanism is extremely abundant in the Canadian Cordillera but is of minor importance in the Hellenides.

Part 1. Sea-Floor Spreading and the Hellenides

The Alpine chains of Europe contain a number of tectonic units (Fig. 1) that may be grouped into (1) a northern ensemble without ophiolites—the Jura, Carpathians, and Balkans, (2) a median ensemble—the internal Apennines, Alps, internal Hellenides—that is characterized by abundant ophiolites, and that tectonically overlies (3) a southern ensemble composed of the external Apennines, external Hellenides, and the Magreb Chain. This article deals only with the median and southern ensembles, and for them the following evolutionary model is proposed.

Summary of the Structure, Stratigraphy, and Metamorphism of the Hellenides

The stratigraphy and structure of the Hellenides are fairly well known. Since the publication of the most recent synthesis (Aubouin 1965), the geology of the internal and external zones[1] has been elucidated by Mercier (1966*b*) and the Inst. Fr. Pétrole (1966), respectively. In addition, Moores (1970) has proposed that the Mount Vourinos ophiolites are separated by a tectonic contact from crystalline baseme and its metamorphic cover and do not stratigraphically on them as originally suggest by Brunn (1956).

The structure of the Hellenides is dominat by a system of NNW- and ENE-trending Pli cene–Quaternary faults which cut a series older, highly eroded anticlinoria and molass filled synclinoria. From east to west, the terra of horizontally shortened strata that compri the anticlinoria is divisible into a series of i ternal and external zones, the former thru over the latter along a line marked by ophiolit (Fig. 2).

The internal zones, continuous with t Rhodopian craton, are found in their relati paleogeographic positions, and are general separated by east-dipping thrust faults. Fro east to west the following zones have be distinguished (Mercier 1966*b*): Serbo-Mac donian unit, Peonian, and Pre-Peonian thrus sheets, Païkonian anticline, Almopian thrus sheets, and the Pelagonian nappe. The extern zones, which have been thrust into their prese positions without significant alteration of pale geographic position, also face west and consi of the Parnassian nappe, the Pindus napp resting on the Gavrovian-Tripolitsian anticl norium, and the Ionian thrust-sheets; they a continuous with and gradational into t Apulian platform.

Internal zones

The internal zones, volcanic and sedimenta in origin, experienced frequent episodes plutonism and metamorphism. In the Ear Jurassic the volcanic–sedimentary Peoni trough (Figs. 3–4) formed within the platfor where previously Triassic neritic sedimentatio had been taking place. In this trough Jurass ultrabasic rocks accumulated along with son acid pyroclastics. Metamorphism and granitiz tion has since affected the Peonian rocks. Th region of the trough became emergent in th Early Cretaceous, and at the end of this epoc two new troughs developed: (1) the Sop molassic trough, which coincided approximate with the extinct Peonian trough (Fig. 4), a (2) the Almopian trough, which developed c a series of basic and ultrabasic rocks to th west. Both troughs were below sea-level duri the Late Cretaceous. At the end of the Cr taceous, terrigenous deposits were laid dow

[1]The internal side of an orogenic belt is adjacent to the hinterland (in this case the Rhodopian craton, exposed in eastern continental Greece) whereas the external side borders the foreland (Apulian platform, well developed in peninsular Italy). The internal zones in the Hellenides experienced at least one tangential tectonic phase in pre-Late Cretaceous time as well as a number of late phases. The external zones experienced their first tangential phase in Eocene time or later.

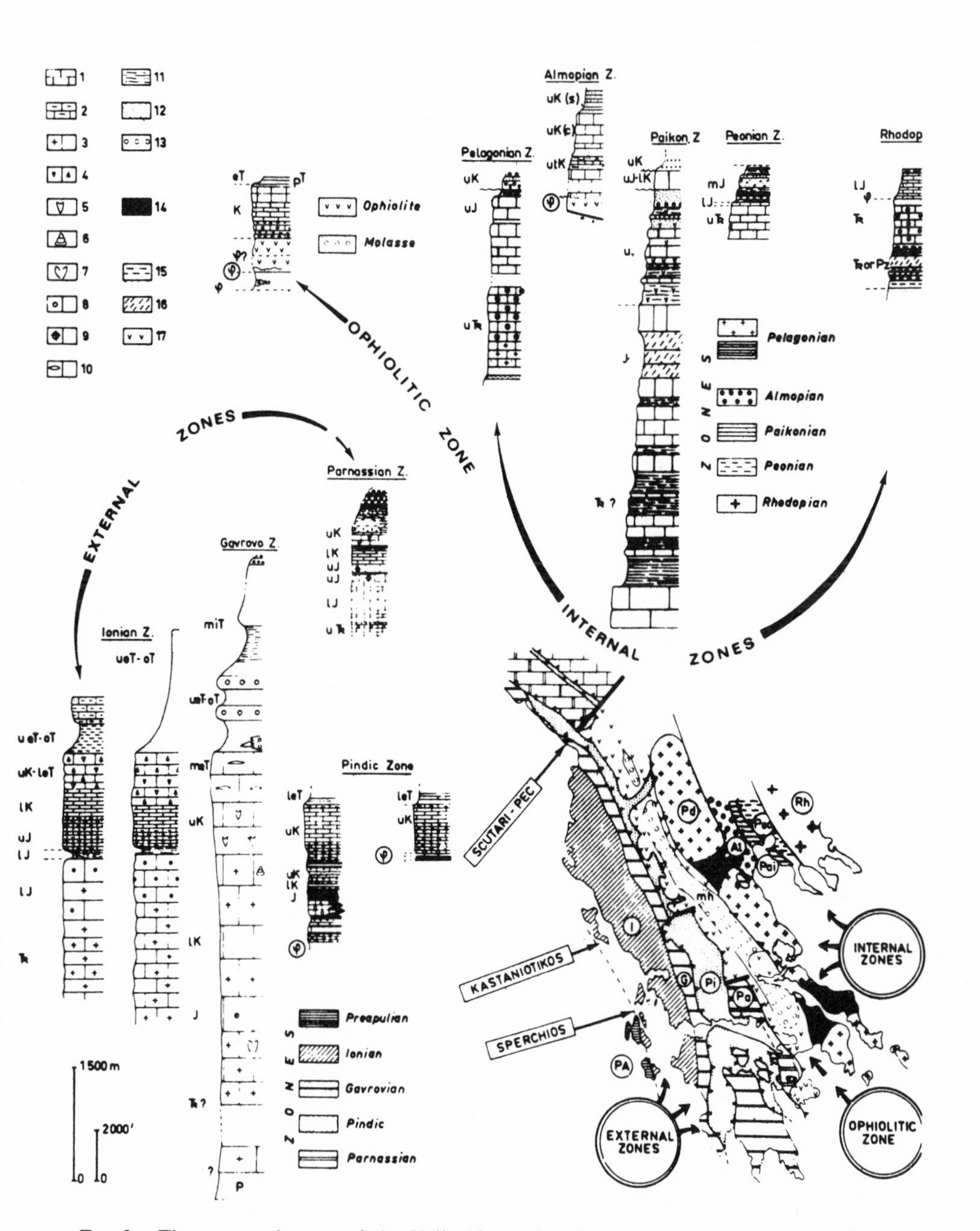

FIG. 2. The structural zones of the Hellenides and their stratigraphic successions; 1 limestones, 2 plate limestones, 3 dolomitic limestones and dolomites, 4 microbreccias, 5 rudist limestones, 6 orbitoline limestones, 7 megalodont limestones, 8 algal limestones, 9 coral limestones, 10 nummulitic limestones, 11 argillaceous rocks, 12 sandstones, 13 conglomerates, 14 radiolarian cherts, 15 tuffs, 16 porphyrites, 17 ultrabasic rocks. External zones after Aubouin (1965), Inst. Fr. Petrole (1966), Dercourt (1964), and Celet (1962). Ophiolitic zone after Aubouin (1958). Internal zones after Mercier (1966).

to the east in the Almopian trough and towards the west in the external region (Pindus trough). A second tectonic event accompanied by regional metamorphism then affected both the Peonian and Almopian zones. Erosion had destroyed the relief of the associated land areas by the end of the Middle Eocene. During the Late Eocene a third tectonic event saw the transportation of numerous thrust-sheets from these zones into the external region, as indicated by the occurrence, in the Olympus window beneath the Pelagonian nappe, of rocks belonging to a zone that remained neritic into the Paleogene (Godfriaux 1964) and which may belong to the Gavrovian or Parnassian external zones. The products of erosion accumulated in the Vardar molassic trough. Volcanic rocks, essentially basic in nature, were extruded in the Neogene.

Ophiolitic zone

A band of ophiolites along the front of the Pelagonian zone is bisected by the Eo-Oligocene, molassic, meso-Hellenic trough (Fig. 5). East of the trough the ophiolites rest with normal stratigraphic contacts on beds deposited in the Pelagonian zone (Guernet 1965; Brunn 1956) or are tectonically overlain by these rocks (Moores 1970), whereas the western ophiolites are totally allochthonous and occur on rocks belonging to one of the external zones, namely Pindus trough in northern Greece (Brunn 1956; Aubouin 1959) and in Crete (Aubouin and Dercourt 1965), and also the Parnassian shoal (Celet 1962; Dercourt 1964). It has not been possible to discover which of these two ophiolitic bands was eroded to provide the detrital sediments that were laid down at the end of Jurassic time (Cayeux 1904). The ophiolites are covered by pelagic deposits along their external margin, and by Upper Cretaceous, transgressive, neritic beds along their internal margin.

Although the ophiolites were tectonically emplaced into their present position during the Eocene, as will be discussed below, they experienced several periods of movement, the first in the late Jurassic. The meso-Hellenic trough began to form at the end of the Middle Eocene (Bizon *et al.* 1968) and remained marine until the Tortonian (Brunn 1956). According to the

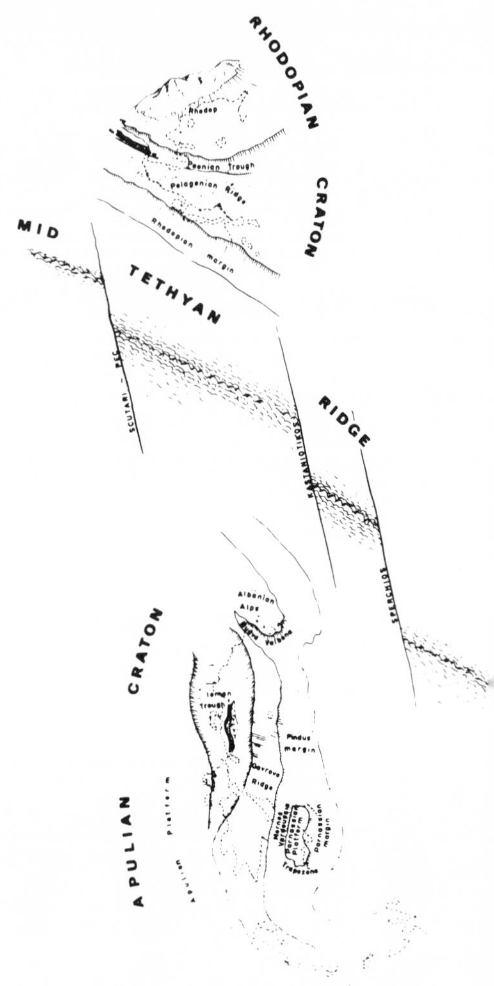

FIG. 3. Palinspastic reconstruction of the Hellenides in the Late Jurassic. On both sides of Tethyan ocean, whose floor is traversed by a mid-ocean ridge and several transform faults, the Apulian and Rhodopian cratons had margins similar to the present-day eastern margin of the North American continent. The relative positions of the two cratons cannot be established from considerations involving the Hellenides alone. This figure is an enlargement of part of Fig. 6. The reconstruction shown on the two figures takes into account the last movements of the Apulian craton which occurred along the Scutari-Pec disturbance between the Dinarides and Hellenides and which involved the eastern Alps.

hypothesis presented here, it masks a major tectonic contact between the Pelagonian zone and its ophiolitic cover on the one hand and the western ultrabasic masses on the other.

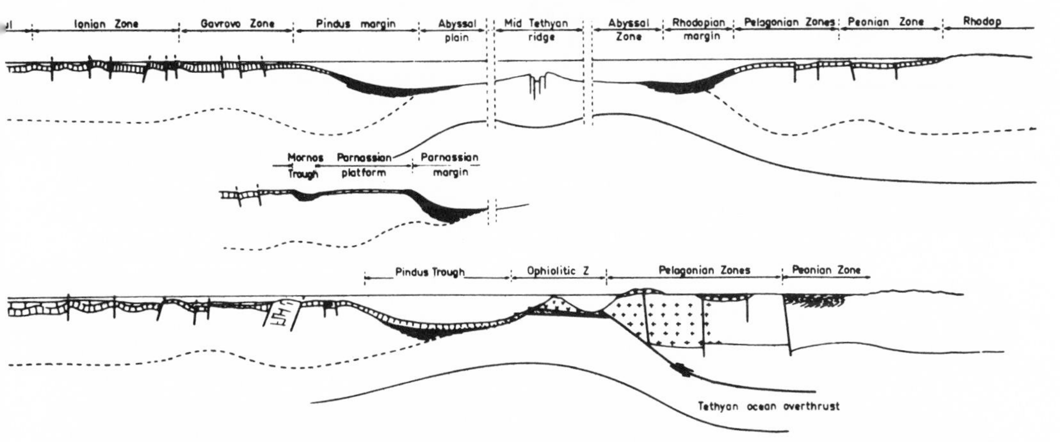

FIG. 4. Paleogeographic sections across the Hellenides. The upper two sections are for the Early Jurassic, during the existence of the mid-Tethyan ridge. The lower section is for the Late Cretaceous, during the over-riding of the Tethyan oceans—note the remnant of an oceanic trench in front of the Pelagonian zone.

lbania the ophiolites are overridden by the elagonian zone (Bourcart 1925).

External zones

The external zones are exclusively sedientary. In the Late Triassic they contained zone of pelagic sedimentation (Pindus trough), ordered by zones of neritic sedimentation. During the Early Jurassic a new pelagic zone Ionian trough) developed within the external eritic platform. Pretectonic sedimentation connued in the Pindus trough until the Middle Eocene; to the west it continued into the Aquinian. Only the Pindus zone contains sediments first flysch) that reflect the second tectonic pisode of the internal zones. From Late Cretaceous to Middle Eocene time, the Pindus rough acted as a sediment trap and prevented he spread of argillaceous and arenaceous deosits into the more external zones. The mateial in the trough then formed a nappe thrust owards the west, which was immediately subected to erosion; the resulting debris accumuated partly in the meso-Hellenic, molassic rough in the east, and partly in the more exernal zones in the west where it contributed o the external flysch.

During the Late Miocene the external zones which had so far escaped were tectonized. In my opinion the Parnassian zone is the easternnost of these zones; it is characterized by a eritic and pelagic series, Triassic to Paleocene in age, with bauxitic horizons in the Late Jurassic and Late Cretaceous. In the Pliocene, the Mediterranean was divided into horsts and grabens; molasse of very local origin accumulated in the latter.

Discussion

From this short summary four features fundamental to plate tectonics can be emphasized. (1) The ophiolites lie in two belts, one external to the Pelagonian zone and the other internal, that are bounded externally by thrusts. The external ophiolites are thrust over flysch ranging in age from Late Jurassic–Early Cretaceous (Terry and Mercier 1971) to Eocene (Brunn 1956). The internal (Almopian) ophiolites lie on all levels of the Pelagonian zone. The relationship between the Pelagonian zone and the external ophiolites is uncertain. The contact once thought to be stratigraphic (Brunn 1956) may well be tectonic. The time of tectonic emplacement of both belts must be Late Jurassic: the ophiolites are thrust over rocks of this age; in Yugoslavia they are stratigraphically overlain by Upper Jurassic detrital sediments (Blanchet 1970); and in southern Greece ophiolitic pebbles occur in clastics of Late Jurassic–Early Cretaceous age (Cayeux 1904). (2) Volcanics not associated with ultrabasic rocks, namely thin basaltic and andesitic units occur in Triassic strata in almost all zones. Internal zones contain andesites, basalts, and

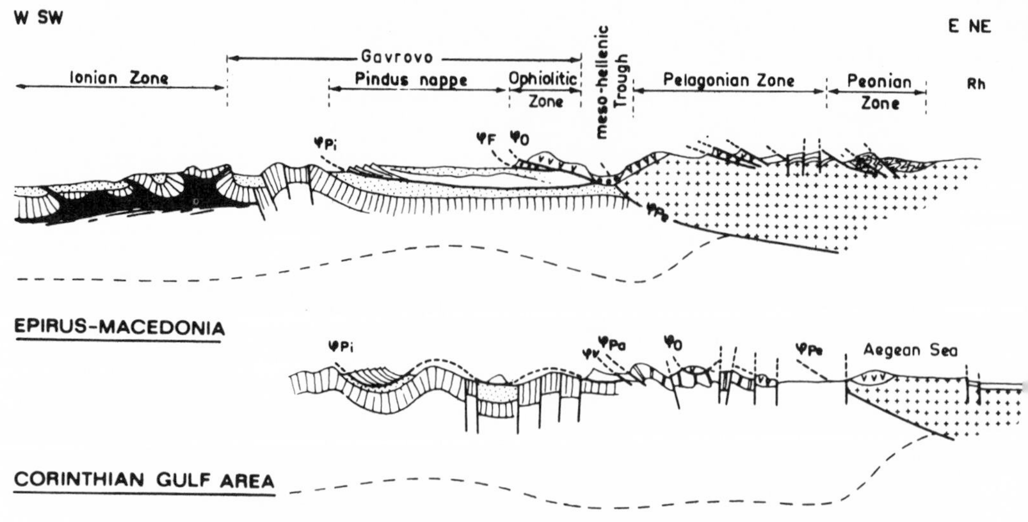

FIG. 5. Tectonic sections across the Hellenides. Ionian zone after Inst. Fr. Petrole (1966). Northern Pindus zone after Brunn (1956). Internal zone after Mercier (1966). Parnassian zone after Celet (1961). Pindus nappe after Aubouin (1959) and Dercourt (1964). φPi—Pindus thrust. φF—thrust underlying material derived from the Rhodopian craton and from oceanic crust. φO—thrust underlying ophiolites that represent fragments of crust belonging to the Tethyan ocean. φPe—thrust underlying the Rhodopian craton. φV—Vardoussia thrust within the Pindus nappe. φPa—Parnassian Plateau thrust on the Mornos depression.

dacites of late Mesozoic and Cenozoic age. None of these volcanics forms an assemblage more than 200 m thick. It should be emphasized at this point that the amount of volcanic rock in the Hellenides is very minor compared with that in the Cordillera, and that two geotectonic models appear to be necessary to interpret this difference. (3) Plutonic rocks, totally absent in the external zones, occur in the internal zones and were emplaced mainly during the Late Jurassic and Eocene, although some Miocene intrusions are known. They are mainly of the granitic-monzonitic type. (4) Metamorphism has not affected any Triassic or younger rocks of the external zones. However, in the internal zones metamorphism is widespread. At least two episodes are known: one, Late Jurassic–Early Cretaceous in age, is of greenschist grade; the other of Eocene age is very low grade but generated glaucophane (Mercier 1966).

A Geotectonic Model

During the Early to Middle Triassic a zone of sea-floor spreading was established within the Euro–African craton, and continental margins similar to the present-day eastern mar of North America were formed. This m Tethyan zone between the African craton to south and the European craton to the no continued to function until the Late Jurassic which time sea-floor spreading in the Atlan and Arctic regions was initiated. From the L Jurassic to the present, the Tethys decreased size by inward movement of the European a African cratons associated with the opening the Atlantic and Arctic Oceans. During t phase Tethyian oceanic crust was in plac overridden by the African craton and in plac by the European craton. In the Alps Apulian craton, and the Italo–Dinarid ensemb of Aubouin (1960) (southern and easte Alps, peninsular Italy and certain Ioni islands) that is considered here to be a northe extension of the African craton, played t role of "*traineau écraseur*" (Termier 1903) overthrust block, whereas in the Dinarides was the Rhodopian craton, a southern exte sion of the European craton. The oroge which accompanied this inward movement the two cratons was restricted, at first, oceanic crust, and in particular to that part

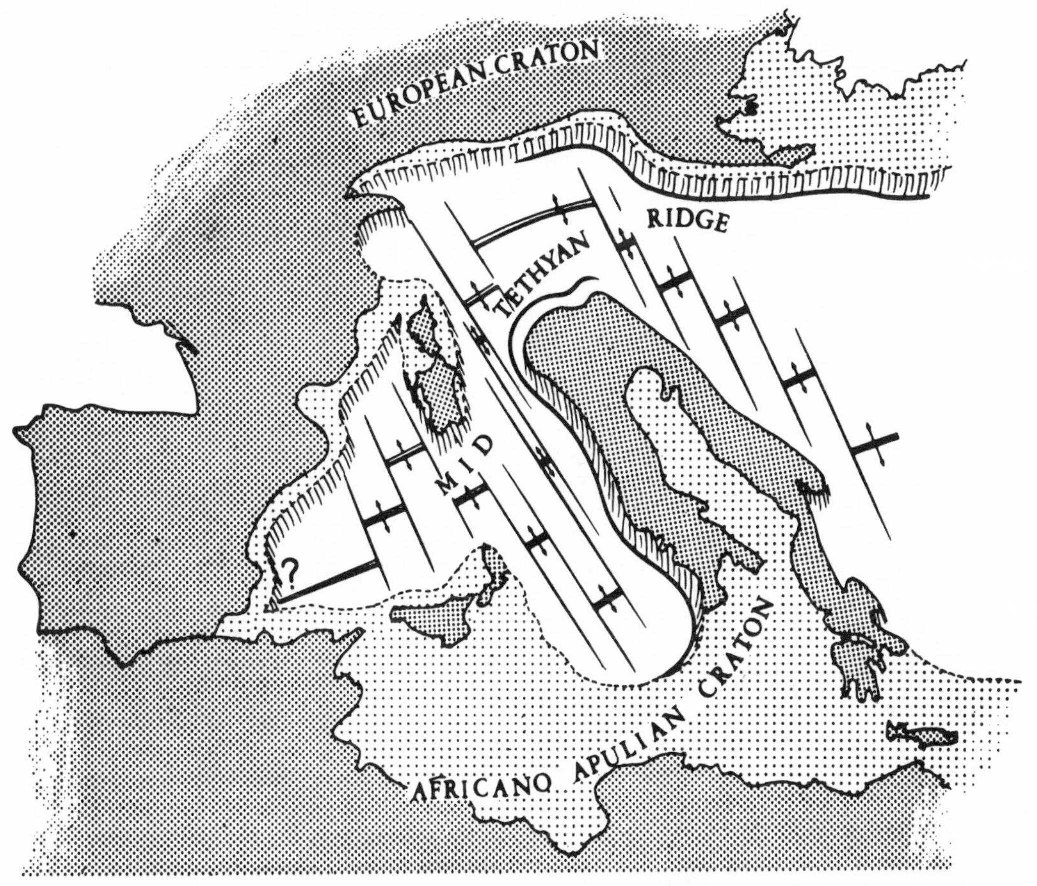

FIG. 6. The Tethyan ocean and its cratonic margins at the time the Atlantic Ocean began to open. The outline of the continents has been drawn after Bullard *et al.* (1965). The western and southern Italo-Greek ensemble has been positioned using African meridians. This reconstruction takes no account of the structure of the eastern Mediterranean.

the crust that was being overridden. After the complete disappearance of oceanic crust, orogeny migrated into the adjacent craton.

Late Triassic–Middle Jurassic: The Mid-Tethyan Ridge

By Permian and Triassic time the Hercynian orogenic belt had experienced considerable erosion, and some closed basins within it began to accumulate evaporites (Bornovas 1960; Dercourt 1964; Inst. Fr. Petrole 1966) and sandstones (Terry 1969). During the Late Triassic the Euro–African craton was traversed by a zone of sea-floor spreading which saw the development of two neritic platforms separated by a pelagic sedimentary trough. The transitional zones between the neritic and pelagic environments received red, nodular sediments of the ammonitico rosso type (Renz 1906; Dercourt 1964; Tsoflias 1969) that were associated with rhyodacitic tuffs.

During the Late Triassic the margins of the two cratons acquired a character similar to that of the present-day eastern margin of North America (see e.g. Heezen *et al.* 1959). Thus within the Apulian craton the shallow-water Gavrovian ridge and Apulian platform were separated by the deeper water Ionian trough (Fig. 4). Normal faults cut this craton, and movement along them was responsible for its horizontal extension and vertical thinning; these faults which cut both basement and cover were also responsible for the mobile paleogeography of the Ionian trough. A similar profile consisting of the Pelagonian ridge, Peonian trough, and Rhodopian platform can be recognized within the Rhodopian craton. The Rhodopian craton also experienced faulting related to the extrusion and intrusion of basic igneous rocks with tholeiitic affinities (Mercier 1966*b*) during the Rhaetic and Early Jurassic and also during the Early and Middle Jurassic. The bilateral symmetry of the Hellenides expressed by this arrangement of miogeosynclinal troughs and miogeanticlinal ridges is far from perfect: e.g. the thickness of the sedimentary cover on the Apulian craton is by far the greater. Basaltic rocks were extruded in the vicinity of the crest of the mid-Tethyan ridge. J. Terry (pers. comm. 1971) has recently discovered Triassic fossils (*Halobia* and conodonts) in sediments interbedded with basalts associated with the ophiolites that crop out west of the meso-Hellenic molassic trough. Sedimentation on the adjacent abyssal plains was very slow, resulting in 75–150 m of Triassic, siliceous limestones and Jurassic radiolarites. Calcareous breccias accumulated on the continental rises adjacent to the cratons.

One of the transform faults affecting the mid-Tethyan ridge in the north separated the zone of sea-floor spreading into two segments, the more northerly branch being in what are now the Dinarides and the more southerly in the Hellenides. This fault coincides with the Scutari-Pec disturbance, which separates the Albanian Alps (part of the Apulian craton) from the Pindus zone but which does not displace the Pindus and Gavrovo zones (Aubouin and Ndojaj 1964; Dercourt 1968). A characteristic feature of transform faults, that they affect only those rocks which have been produced along the crests of mid-oceanic ridges, distinguishes them from other varieties of wrench fault. A paleogeographic reconstruction for the Jurassic, characteristic of the extensional phase in the development of the Hellenide geosyncline, is shown in Figs. 4–7.

Late Jurassic–Recent: The Mid-Atlantic Ridge

The mid-Atlantic ridge which extends northwards into the Arctic ridge became active in the Late Jurassic. As this happened, activity ceased on the mid-Tethyan ridge, and throughout the opening of the Atlantic and Arctic Oceans, the Apulian and Rhodopian plates reacted passively. The resulting movement of these plates, presumably directly controlled b the periodicity of activity along the Atlantic Arctic zone of sea-floor spreading, was SE an NNW for the Rhodopian and Apulian plat respectively (Smith 1971). These passiv movements can be resolved into two com ponents: one of contraction normal to the zon of Tethyan sea-floor spreading, and the othe of shear parallel to this zone. The contraction movements resulted first of all in a break withi the Rhodopian plate between oceanic and con tinental crust (Fig. 4). In the Alps, the initi break occurred in the middle of the Apulia plate with the result that the continental pa of this plate overthrust both oceanic crust an the European craton.

1. Late Jurassic–Early Cretaceous: Disappearance of Eastern Abyssal Plain under Rhodopian Craton

The disappearance of the eastern abyss plain (Fig. 4) occurred during the Late Jura sic, and as a result, in Kimmeridgian time, th Rhodopian craton emerged periodically abov sea-level as did elements of oceanic crust whic had piled up into thrust sheets in front of th craton. During this period faults develope within the Rhodopian craton and along ther acid and basic lavas were extruded. Plutonis also occurred at this time (Borsi *et al.* 1966) Throughout its movement towards the SE, th Rhodopian plate did not encounter much re sistance and as a result experienced few struc tural modifications. Low-grade regional meta morphism accompanied by small-scale, syn metamorphic, isoclinal folding was restricted t rocks that had accumulated in the Peonia trough.

Sedimentation on the Apulian craton and i the Pindus trough was controlled by wate depth, distance from the source of the homo geneous breccias (Aubouin 1959), and vertica movements of the sea floor associated wit normal faulting. The sediments that accumu lated in the eastern abyssal plain were probabl largely restricted to the region of the conti nental rise. They were probably thin, simila to those of the Pindus zone, and are equivalen to horizon B in the crust of the present-da western Atlantic Ocean (Ewing *et al.* 1966 whose eastern limit is far away from the axi of the sea-floor spreading. The present-da whereabouts of these sediments are not know

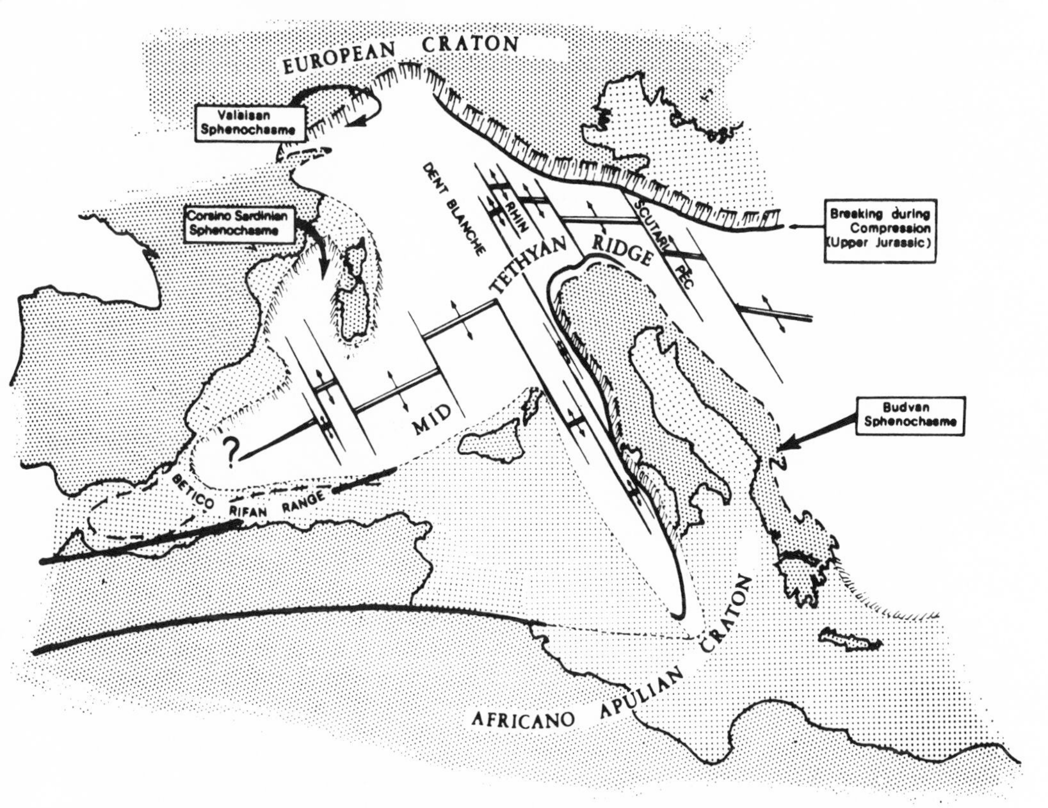

FIG. 7. The Tethyan ocean and its cratonic margins during the Late Jurassic, at the time of opening of the Atlantic Ocean. The European and African continents have been positioned according to the reconstruction of Drake and Nafe (1968). Africa is assumed to have moved along the south Atlantic fault at the time of opening of the South Atlantic Ocean. The position of the axis of rotation of the South Atlantic is such that rotation produces (1) overriding of the eastern and middle Tethys, (2) disruption of the European craton leading to the formation of the Gulf of Gascony, (3) underthrusting of the Iberian Meseta and African shield by the Betico–Rifan Range. The resumption of spreading in the North Atlantic about a different axis of rotation, during Maestrichtian and early Cenozoic time, began after the European and African cratons had been reunited in the eastern Alps, and produced (a) underthrusting of western Europe by the Austro–Alpine block, (b) the major Pyrenean phase, (c) emplacement of the Betic and Magreb units, (d) faulting in the western Mediterranean and in the Apulian craton leading to the creation of the Apennines. In this reconstruction and in that of Fig. 6, most of the crust under the western Mediterranean is oceanic and derived from the Triassic–Jurassic Tethys, whereas that of the eastern Mediterranean is almost entirely cratonic. In Figs. 6 and 7 the illustrated termination of the mid-Tethyan ridge inside the Betico–Rifan Range is unsatisfactory and will have to be modified by specialists of the Beltics and North African Ranges.

with any degree of certainty. They could have been overthrust by the Rhodopian craton in the way that sediments along the Pacific border have been overthrust by the North American craton (Gilluly 1969; Hamilton 1969). Alternatively they could have been thrust along with the thrust sheets belonging to the ophiolitic zone.

2. *End of Early Cretaceous: Accentuation of Stress*

The craton continued to transmit stress generated by the zones of Arctic and Atlantic sea-floor spreading. The Rhodopian craton was uplifted and the products of the ensuing erosion accumulated as flysch—the first flysch of Aubouin (1959)—in the Pindus trough and Almopian zone, and as molasse in the Sopur trough whose deposits are discordant on the folded sediments of the Peonian trough.

3. Late Cretaceous: Quiescence

The ophiolitic and Pelagonian ridges, previously emergent, were progressively and discordantly covered by neritic deposits, mainly reefal limestones with rudists. Sedimentation in the Pindus trough and on the Apulian craton continued as before. The western margin of the Gavrovo ridge sank and received very thick reefal sediments (Aubouin 1959). The margins of the Pindus and Ionian troughs were affected by turbidity currents, and homogeneous calcareous breccias invaded the troughs as far as their axes. By late Campanian time the entire Hellenides were covered by water.

4. Maestrichtian: Renewal of Activity

During the Maestrichtian the Rhodopian craton was once more exposed to horizontal compression which was manifested first by uplift along its older faults. Subsequent erosion led to deposition of flysch in the Pindus and Almopian troughs. During the Maestrichtian and Paleocene these clastic deposits gradually encroached onto the Apulian craton (Aubouin and Dercourt 1965; Fleury 1970). Up to the end of the Middle Eocene the Pindus trough continued to play the role of a "*barrière en creux*" (Aubouin 1959). When this trough was full the detrital sediments were transported onto the Gavrovian ridge and into the Ionian trough beyond. On the Gavrovian ridge the thickness of the flysch is not very great except in grabens such as that of Langadia (Peloponnese) where it may reach several hundred meters.

The compressive stresses were responsible not only for the uplift but also for tectogenesis within the Rhodopian craton. Some recumbent folds in the cover and thrust sheets in the basement were produced in the Peonian zone. Isoclinal syn-metamorphic folds occur in the western craton (Marinos 1957; Mercier 1966*a*; Papastamatiou 1963; Brunn 1956; Aubouin and Guernet 1963). Zwart (1967) has emphasized that a certain amount of overburd is required before regional metamorphism c take place. The rocks on the Rhodopian crat that experienced metamorphism in Late C taceous and Early Paleocene time do not, first sight, appear to have been overlain by appreciable thickness of overburden. Howev the 1000–1500 m of flysch that accumulat in the 200–300 km wide Pindus trough duri the Late Cretaceous and Paleocene must, ev if some longitudinal transportation occurre have been derived from a comparable thickne of rock on the Rhodopian craton. Thus t metamorphism is explicable when one co siders this thickness of rock, presumably t result of the superposition of thrust sheets.

In the same way, the base of the Gavrovi series within the Apulian craton in the Pelopo nese rests on a basement which has been met morphosed (Thiebault 1968). This occurr along the axis of the Langadia trough whe thick flysch accumulated. Thus there appea to be a relationship between the thickness sediments in the Gavrovian trough and t regional metamorphism that has occurred ther

During the Late Eocene the stresses whi began in the Maestrichtian reached their max mum and produced in the Rhodopian craton new thrust sheet involving basement. Th thrusting continued the tectonics of Ear Eocene time in the Peonian zone and initiate tectogenesis in the Almopian zone. At abo this time the ophiolitic mass was separate from its substratum along a serpentinite hor zon. The ophiolitic mass pushed by the Rh dopian craton resulted in the ejection sediments belonging to the Pindus trough on the Apulian craton. The Pindus nappe sep rated from its substratum along the base of th Triassic siliceous limestones in the east, an moved partly as the result of tectonic stre and partly as the result of gravity. This di placement was facilitated by the existence an extremely incompetent layer of Jurassic an Lower Cretaceous radiolarites. The front of th Pindus nappe was imbricated where thes radiolarites are thicker stratigraphically an where they have been thickened tectonically.

The ophiolitic mass and the Rhodopia craton overthrust the Apulian craton. Th over-thrusting of one craton by another ha been established in the Olympus window (Godfriaux 1964), situated in the middle of the Pelagonian zone, where metamorphosed nummulitic limestones probably belonging to the Parnassian or Gavrovian zone occur. Since the Late Eocene the two cratons have been in contact and the longitudinal stresses could no longer be relieved by movement within oceanic crust. Some longitudinal wrench faults slightly oblique to the isopic zones developed in the cratons, and zones of breccia formed along them (Mercier 1966*b*). Where they cut the Almopian zone they are post-Late Cretaceous and pre-Late Eocene, i.e. approximately synchronous with the thrusting of one craton on the other. Although more wrench faults exist in the hinterland (Brunn 1960; Pavoni 1961), they are not responsible for the major movement of the Apulian craton towards the NNE which occurred before the welding of the two cratons. For example, the major movements in the Alps are pre-Late Eocene: namely Barremian, pre-Cenomanian, or pre-Gosau in the Austrian Alps, and Eocene in the Franco-Italian Alps.

5. Oligocene: Period of Quiescence

Throughout the Oligocene no significant tectonic activity occurred. On the eastern craton the Vardar molassic trough was inactive although deposition in the meso-Hellenic trough continued with the accumulation of debris from the Pindus and Ophiolitic nappes and from the more internal zones. On the western craton the Pindus nappe continued to advance and underwent erosion thereby feeding the Ionian flysch. In this way it overthrust its own debris. The clastic deposits did not reach the Apulian platform because of the existence of the Ionian trough which at this time was not full and thereby played the role of a "*barrière en creux*".

6. Miocene: Renewal of Tectonic Activity

This renewal was marked by three types of movement.

(a) The Rhodopian craton transmitted stress to the Apulian craton. The cover of the Ionian zone, at this stage the only one not to have experienced tectogenesis, was deformed and separated from its substratum along a zone of Permo-Triassic gypsum (Inst. Fr. Petrole 1966). The craton itself was folded into a series of very large gentle anticlines and synclines, their existence assisted the gravitational sliding of the Pindus nappe whose movement terminated by overthrusting Oligocene conglomerates (Mansy 1969). The debris eroded from the new chain accumulated in the Molise molassic trough and in long narrow intermontane basins within the Ionian zone.

(b) Although this folding absorbed some of the stress, when all the rocks had been folded and the stresses continued to be applied, general uplift of the Hellenides occurred in Late Miocene (Pontian) time.

(c) A system of faults cut the entire region, by now entirely cratonized, and divided it into horsts and grabens. These faults strike NNW and NNE, and are related only to the causative stresses and not to the pre-existing zones of weakness inherited from the Tethyan period. Then during the Pliocene the Mediterranean Sea formed contemporaneously with the sinking of the Aegean and Adriatic Seas, the formation of the Gulf of Corinth, etc. The Albanian and continental Greek sectors have not been elevated so much and are thus less eroded than the Dodecanese and the Peloponnese where the Apulian craton crops out to a very large extent (Aubouin and Dercourt 1965). A period of volcanism, generally basaltic, accompanied this episode of sinking. This was not the result of the relaxation of stress, but represents a new form of reaction on the part of the completely cratonized region.

Summary

All the foregoing reconstructions, from the Jurassic to the present, are linked with the Atlantic and Arctic zones of sea-floor spreading; the chronology derived from field evidence agrees reasonably well with that from geophysical data by Le Pichon (1968) for the Atlantic Ocean and may be summarized as follows.

1. *Late Jurassic*: opening of Atlantic and Arctic Oceans–overthrusting of oceanic part of Rhodopian plate; subsequent isostatic readjustment (uplift and erosion of Rhodopian craton); plutonism.
2. *Aptian-Albian*: active expansion in Atlantic and Arctic Oceans—compression of Rhodopian craton, origin of tensional structures, metamorphism; uplift and formation of flysch and molasse.

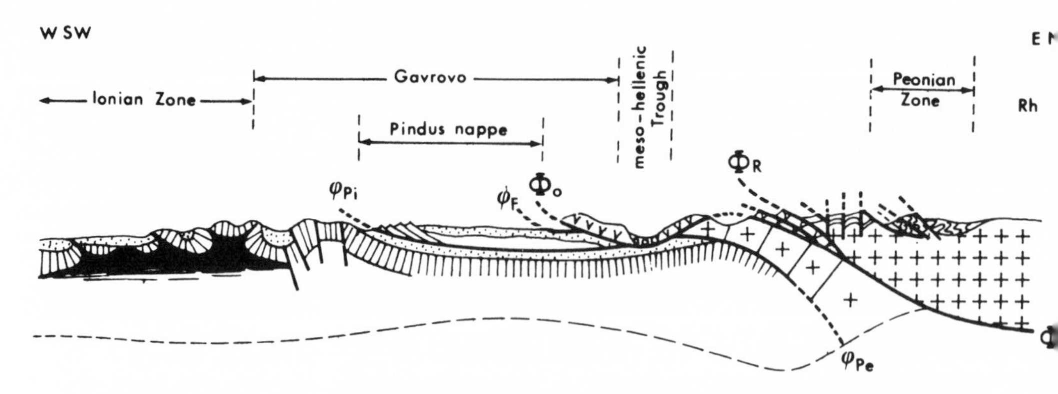

FIG. 8. Alternative tectonic section across the Hellenides (cf. Fig. 5). The ophiolites are shown here as forming a single thrust sheet that originated between the Pelagonian and Païkon–Peonian zones. The thrust was initiated in the Late Jurassic and stabilized in the Early Tertiary.

3. *Late Cretaceous*: cessation of expansion in the Atlantic and Arctic Oceans–inundation of previous relief.
4. *Paleocene to Late Eocene*: renewal of expansion in Atlantic and Arctic Oceans–tectogenesis of internal zones and uplift from Maestrichtian to Middle Eocene; thrusting of Rhodopian craton preceded by emplacement of Pindus and Ophiolitic nappes.
5. *Oligocene period of stability*: sedimentation of flysch in zones not as yet tectonized and molasse elsewhere.
6. *Miocene*: renewal of expansion in Atlantic and Arctic Oceans–tectonization of zones not as yet effected; general uplift with emergence during Late Miocene (Pontian); brittle failure of crust caused by application of stresses.

Discussion: A Single Ophiolitic Thrust Sheet

In the present paper the basal contact of the ophiolitic series has been assumed to be tectonic in character, as suggested by Moores (1970), Zimmermann (1970), and Dercourt in Aubouin (1970), rather than stratigraphic as suggested by earlier workers, i.e. all ophiolitic masses are considered here to be thrust sheets.

To conserve the currently accepted model of the Hellenides (see Aubouin 1965) I have proposed that there are two ophiolitic thrust sheets, both coming from internal zones and both thrust onto relatively more external zones: (1) the Almopian ophiolites thrust onto the Pelagonian zone and (2) the Subpelagonian ophiolites thrust onto the Pindus zone.

The oldest tectonic emplacement is fair well documented as Late Jurassic. A secor phase which affected the ophiolites and the substrata displaced all masses externally. Du ing these phases the Almopian and Pelagonia zones were thrust over the Gavrovian–Parna sian neritic ridge, as demonstrated by th Olympus window (Godfriaux 1964), and th Gavrovian ridge was overthrust by the Pind nappe (Cayeux 1903).

A drastic change in the above geologic model can be proposed (Fig. 8); namely th existence of one unique ophiolitic thrust shee which originated as a series of slices of oceani crust located between a Pelagonian–Apulia craton on the one hand and a Païkonian Rhodopian craton on the other. This oceani nappe covered both the Pelagonian and Pindu zones in Late Jurassic time. Subsequently, Cre taceous erosion separated the nappe into tw masses but klippes of the nappe remain on th Pelagonian zone. The chronology of events an the later part of the history is not changed i this proposal which is similar to one made b Stille and subsequently retracted.

As has been shown by Dercourt (1970 and Dewey and Bird (1970), the geosyncline concept is entirely compatible with that c collision in plate tectonics. For this reason i may be useful to consider the way the drasti change in the geological model for the Hel lenides proposed above leads to a change i geosynclinal terminology (recognition of mio and eugeosynclinal zones) for the Hellenides The following zones may be considered as mio geosynclinal: Apulian platform, Ionian trough, Gavrovian ridge, Parnassian shoal, Pindus trough, Pelagonian ridge, and the following as eugeosynclinal: the ophiolites and their sedimentary cover, Païkonian ridge, Peonian trough, Rhodopian platform.

In such a model the following comparisons may be made between paleogeographic zones in the Hellenides and the corresponding zones in the Franco–Italian Alps (Debelmas and Lemoine 1970).

"Dauphinois" zone—Ionian, Gavrovian, Parnassian, and Pindus zones
"Brianconnais" zone—Pelagonian zone
"Piemontais" zone—ophiolites and their sedimentary cover
Austro–Alpine zones—Païkonian, Peonian, and Rhodopian zones

These comparisons take into consideration structural features, tectonic phases, metamorphic characters, granitic plutonism, etc.

As far as plate tectonics are concerned this geological alternative leads to a few changes: e.g. relocation of the Tethyan ocean between the Pelagonian and Païkonian ridges instead of between the Gavrovian and Pelagonian ridges. The location of the paleo-transform faults does not change because their positions were deduced from field evidence. The choice between the more traditional geosynclinal model in the Hellenides and the one proposed here will have to be based on further analysis of the field data.

[*Editor's Note:* Part 2 on seafloor spreading and the Canadian Cordillera has been omitted.]

Acknowledgments

The work leading to this publication was carried out during two visits to Canada in 1968 and 1969 when I spent three months in the Canadian Cordillera and four months at the Geology Department, University of Alberta, Edmonton. During my visits to the Cordillera I was given invaluable assistance by S. Blusson, R. B. Campbell, G. Eisbacher, W. W. Hutchison, J. W. H. Monger, J. E. Muller, J. G. Souther, J. A. Roddick, H. W. Tipper, and J. O. Wheeler of the Geological Survey of Canada, and P. S. Simony and H. A. K. Charlesworth of the Universities of Calgary and Alberta. The manuscript was submitted at various stages to H. A. K. Charlesworth, J. Aubouin, J. W. H. Monger, and J. O. Wheeler. I should like to thank them for their constructive criticism. I gratefully acknowledge the assistance of the Canada Council who awarded me a fellowship and the French Ministry of Foreign Affairs who supported me while in the field. The French manuscript was modified and translated into English by H. A. K. Charlesworth who, moreover, during my visit to Edmonton let me profit from his experience of the geology of the Canadian Cordillera.

J. L. Mansy at Lille has helped me prepare the text-figures which were drawn by M. Prouvot.

AITKEN, J. D. 1968. Pre-Devonian history of the southern Rocky mountains. Alb. Soc. Petrol. Geol. 16th ann. field conf., pp. 15–28.

——— 1969. Documentation of the sub-Cambrian unconformity, Rocky mountains Main Ranges, Alberta. Can. J. Earth Sci., **6**, pp. 193–200.

ATWATER, T. 1970. Implications of plate tectonics for the Cenozoic Tectonic evolution of western North America. Geol. Soc. Am., Bull. 87, pp. 3513–3536.

AUBOUIN, J. 1959. Contribution à l'étude de la Grèce septentrionale: les confins de l'Epire et de la Thessalie. Ann. Géol. Pays Hell., **X**, 525 p.

——— 1959. Granuloclassement vertical (graded bedding) et figures de courant (current marks) dans les calcaires purs: les brêches de flanc des sillons géosynclinaux, Bull. Soc. Géol. de France, **1**, (7), pp. 578–582.

——— 1965. Geosynclines. Elsevier, Amsterdam, 335 p.

AUBOUIN, J. et DERCOURT, J. 1965. Sur la géologie de l'Egée: regard sur la Crète (Grèce). Bull. Soc. Géol. de France, (7), **VII**, pp. 787–821.

AUBOUIN, J. et DURAND-DELGA, M. 1971. Mediterranéenne (Aire). Encycl. Universalis, **X**, pp. 743–745.

AUBOUIN, J. *et al.* 1960–63. Esquisse de la géologie de la Grèce. Livre à la memoire de P. Fallot. Mém. h. sér. Soc. Géol. France, **II**, pp. 583–610.

AUBOUIN, J. et GUERNET, C. 1963. Sur la stratigraphie et la tectonique de l'Eubee moyenne (Grece). Bull. Soc. Géol. de France, (7), **V**, pp. 821–827.

AUBOUIN, J. et NDOJAJ, I. 1964. Regard sur la géologie de l'Albanie et sa place dans la géologie des Dinarides. Bull. Soc. Géol. de France, **VI**, (7), pp. 593–625.

AUBOUIN, J. *et al.* 1970. Données nouvelles sur la géologie de zone subpélagonienne, du Pinde et du Gavrovo. Ann. Soc. Géol. Nord, **XC**, pp. 277–306.

BAER, A. J. 1967. Bella Coola and Laredo Sound map-areas, British Columbia. Geol. Surv. Can., Pap. 66-25, 13 p.

——— 1968. Model of evolution of the Bella Coola Ocean Falls Region, Coast Mountains, British Columbia. Can. J. Earth Sci., **5**, pp. 1429–1441.

BALLY, A. W., GORDY, P. L., and STEWART, G. A. 1966. Structure, seismic data and orogenic evolution of Southern Canadian Rocky Mountains. Bull. Can. Petrol. Geol., **14**, pp. 337–381.

BIZON, G., BIZON, J. J., LALECHOS, N., et SAVOYAT, E. 1968. Présence d'Eocène transgressif en Thessalie. Incidences sur la paléogéographie régionale. Bull. Soc. Géol. de France, (7), **X**, pp. 36–38.

BLANCHET, R. 1970. Sur un profil des Dinarides, de l'Adriatique (Split-Omis, Dalmatie) au Bassin pannonique (Banja Luka-Dobbj, Bosnie). Bull. Soc. Géol. de France, (7), **XII**, (sous presse).

BLANCHET, R. *et al.* 1969. Sur l'existence d'un important domaine de flysch tithonique-crétacé i ferieur en Yougoslavie: l'unite du flysch b niaque. Bull. Soc. Géol. de France, (7), X pp. 871–880.

BORNOVAS, J. 1960. Observations nouvelles sur géologie des zones préapulienne et ionien (Grèce occidentale). Bull. Soc. Géol. de Fran (7), **II**, pp. 410–414.

BORSI, S., FERRARA, G., MERCIER, J., et TONGLORGI, 1966. Age stratigraphique et radiométrique jur sique superieur d'un granite des zones intern des Hellénides (granite de Fanos) Macédoir Grèce. Rev. Géogr. phys. et Géol. dyn., (2 **VIII**, pp. 279–287.

BOURCART, J. 1925. Observations nouvelles sur tectonique de l'Albanie moyenne. Bull. Sc Géol. de France, (4), **XXV**, pp. 391–428.

BREW, D. A., LONEY, R. A., and MUFFLER, L. J. 1966. Tectonic history of Southeastern Alask Can. Inst. Min. and Metal., Spec. Vol., **8**, p 149–170.

BROWN, S. 1969. Geology of the Queen Charlo Islands (British Columbia). B.C. Dep. of Mi and Petrol. Res., **54**, 226 p.

BRUNN, J. H. 1956. Contribution a l'étude géologiq du Pinde septentrional et d'une partie de Macédoine occidentale. Ann. Géol. Pays He **VII**, 358 p.

——— 1960. Les zones helléniques internes et le extension. Reflexions sur l'orogenèse alpine. Bu Soc. Géol. de France, (7), **II**, pp. 470–486.

BUDDINGTON, A. F. and CHAPIN, T. 1929. Geology ar mineral deposits of southeastern Alaska. U. Geol. Surv., Bull. 800, 396 p.

BULLARD, E., EVERETT, J. E., and SMITH, A. G. 196 The fit of the continents around the Atlantic. *I* A symposium on continental drift. Roy. So Phil. Trans., **1088**, pp. 41–51.

CAMPBELL, R. B. 1968. Canoe river, British Columbi Geol. Surv. Can., Map 15.

——— 1970. Structural and metamorphic transitio from infrastructure to suprastructure, Caribo mountains, British Columbia. Geol. Ass. Ca Spec. Pap., **6**, pp. 67–72.

CAMPBELL, R. B. and MANSY, J. L. 1970. Stratigrap and structure of the Black Stuart synclinoriu Quesnel Lake map area—British Columbia. *I* Report of Activities, Geol. Surv. Can., Pap 70-1.

CAREY, S. W. 1954. The rheid concept in geote tonics, J. Geol. Soc. Aust., **1**, pp. 67–117.

——— 1962. Folding. J. Alberta Soc. Petrol. Geo **10**, pp. 95–144.

CAYEUX, L. 1903. Sur les phénomènes de charria en Méditerranée orientale C.R.Ac.Sc. Paris (A **136**, pp. 474–476.

——— 1904. Géologie des environs de Naupl Existence du Jurassique superieur et de l'inf crétacé en Argolide (Grèce). Bull. Soc. Gé de France, (4), **IV**, pp. 87–105.

CELET, P. 1962. Contribution à l'étude géologique Parnasse-Kiona et d'une partie des régions mé dionales de la Grèce continentale. Ann. Gé Pays Hell., **XIII**, 446 p.

CHARLESWORTH, H. A. K. *et al.* 1967. Precambri

geology of the Jasper region, Alberta. Res. Counc. of Alberta, Bull. 23, pp. 1–74.

CHRISMAS, L. *et al.* 1969. Rb/Sr, S, and O isotopic analyses indicating source and date of contact metasomatic copper deposits, Craigmont British Columbia, Canada. Econ. Geol., **64**, (5), pp. 479–488.

CHURKIN, M. JR. 1969. Paleozoic Tectonic History of the Arctic Basin, North of Alaska. Sci., **165**, pp. 549–555.

CLAPP, C. H. 1913. Geology of the Victoria and Saanich Map-Areas, Vancouver Island, British Columbia. Geol. Surv. Can., Mem. 36, 143 p.

COCKFIELD, W. E. 1961. Geology and mineral deposits of Nicola Map-area, British Columbia. Geol. Surv. Can., Mem. 249, 164 p.

COLOM, G. 1955. Jurassic–Cretaceous pelagic sediments of the western Mediterranean zone and the Atlantic area. Micropaleontol., **1** (2), pp. 109–124.

DALY, R. A. 1951. A Geological Reconnaissance between Golden and Kamloops, British Columbia, along the Canadian Pacific Railway. Geol. Surv. Can., Mem. 68, 620 p.

DEERE, R. E. and BAYLISS, P. 1969. Mineralogy of the lower jurassic in West Central Alberta. Bull. Can. Petrol. Geol., **17**, pp. 133–153.

DEBELMAS, J. and LEMOINE, M. 1970. The Western Alps. Paleogeography and structure. Earth Sci. Rev., **6**, pp. 221–256.

DERCOURT, J. 1964. Contribution à l'étude géologique d'un secteur du Péloponnèse septentrional. Ann. Géol. Pays Hell., **XV**, 417 p.

——— 1968. Sur l'accident de Scutari-Pec, la signification paleo-geographique de quelques series condensées en Albanie septentrionale. Ann. Soc. Geol. Nord, **LXXXVIII**, pp. 109–117.

——— 1970. L'expansion océanique actuelle et fossile; ses implications géotectoniques. Bull. Soc. Géol. de France, (7) **XII**, pp. 261–317.

DEWEY, J. F. 1969. Continental margins: a model for conversion of Atlantic type to Andean type. Earth and Planet. Sci. Lett., **6**, pp. 189–197.

DEWEY, J. F. and BIRD, J. M. 1970. Plate tectonics and geosynclines. Tectonophys., **10**, pp. 625–638.

DIETZ, R. S. 1963. Alpine serpentine as oceanic rind fragments. Geol. Soc. Am., Bull. 74, pp. 947–952.

DOUGLASS, R. C. 1967. Permian tethyan fusulinids from California. U.S. Geol. Surv., Prof. Paper 593-A.

DRAKE, C. L. and NAFE, J. E. 1968. Geophysics of the North Atlantic region. Symposium on continental drift emphasizing the history of South Atlantic Area.—Montevideo—Uruguay, Oct. 1967. *In*: Drake, C. L. *et al.*, 1968—Can. J. Earth Sci., **5**, pp. 993–1010.

DUFFEL, S. and MCTAGGART, K. E. 1951. Ashcroft map area British Columbia. Geol. Surv. Can., Mem. 262, 122 p.

EWING, J., EWING, M., and LEYDEN, R. 1966. Seismic profiles survey of Blake Plateau. Bull. Am. Assoc. Petrol. Geol., **50**, pp. 1948–1971.

EWING, JOHN, WORZEL, J. LAMAR, EWING, MAURICE, and WINDISCH, CHARLES. 1966. Age of horizon A and the oldest Atlantic sediments. Sci., **154**, no. 3753, pp. 1125–1132.

FAIRBAIRN, H. W., HURLEY, P. M., and DINSON, W. H. 1964. Initial $^{87}S/^{86}S$ and possible sources of granitic rocks in Southern British Columbia. J. Geophys. Res., **64**, pp. 4889–4893.

FINDLAY, D. C. 1969. Origin of the Tulameen ultramafic-gabbro complex southern British Columbia. Can. J. Earth Sci., **6**, pp. 399–425.

FLEURY, J. J. 1970. Le sillon du Pinde à la limite Crétacé-Eocène. Analyse des "couches de passage au flysch" et essai d'interpretation paléogéographique. Bull. Soc. Géol. de France, (7), **XII**, (sous presse).

GABRIELSE, H. and REESOR, J. E. 1964. Geochronology of plutonic rocks in two areas of the Canadian Cordillera. Roy. Soc. Can., Spec. Publ., **8**, pp. 96–138.

GABRIELSE, H. and WHEELER, J. O. 1961. Tectonic Framework of Southern Yukon and Northern British Columbia. Geol. Surv. Can., Pap. 60-24.

GODFRIAUX, Y. 1964. Etude géologique de la région Olympe, Grèce. Ann. Géol. Pays. Hell., **XIX**, 282 p.

GILLULY, J. 1969. Oceanic sediment volumes and continental drift. Sci., **166**, pp. 992–994.

GUERNET, C. 1965. Contribution à l'étude des "roches vertes". La base du complexe ophiolitique en Eubée. C.R. Somm. Soc. Géol. de France, pp. 334–335.

GUNNING, H. C. (*Ed.*). 1966. A symposium on the tectonic history and mineral deposits of the Western Cordillera. Can. Inst. Min. and Metal., Spec. Vol., **8**, 353 p.

HAMILTON, W. 1969. Mesozoic California and the underflow of Pacific mantle. Geol. Soc. Am., Bull. 80, pp. 2,490–2,429.

HAMILTON, W. and MYERS, W. B. 1966. Cenozoic tectonics of the Western United States. Rev. of Geophys., **4**, pp. 509–549.

HEEZEN, B. C., THARP, M., and EWING, M. 1959. The floors of the ocean: the North Atlantic. Geol. Soc. Am., p. 65.

HEIRTZLER, J. R. *et al.* 1968. Marine magnetic anomalies, geomagnetic field reversals, and motions of the ocean floor and continents. J. Geophys. Res., **73**, pp. 2,119–2,136.

HENDERSON, G. G. L. 1959. A summary of the regional structure and stratigraphy of the Rocky mountains trench. Trans. Can. Inst. Min. and Metal., **62**, pp. 156–161.

HUTCHISON, W. W. 1970. Metamorphic framework and plutonic styles in the Prince Rupert region of the Central Coast Mountains, British Columbia. Can. J. Earth Sci., **7**, pp. 376–405.

HYNDMAN, D. W. 1968. Mid-Mesozoic multiphase folding along the border of the Shuswap metamorphic complex. Geol. Soc. Am., Bull. 79, pp. 575–587.

INST. FRANÇAIS PETROLE AND GEOL. INST. OF SUBS. RES. (GREECE). 1966. Etude geologique de l'Epire (Grece Nord occidentale). Technip, Paris, 306 p., 101 Fig., 9 plates, 2 cartes au 1/100 000.

ISACKS, B., OLIVER, J., and SYKES, L. R. 1968. Seismology and the New Global Tectonics. J. Geophys. Res., **73**, pp. 5,855–5,899.

JELETZKY, J. A. 1950. Stratigraphy of the West coast of Vancouver Island between Kyuquot Sound and Esperanza Inlet, British Columbia. Geol. Surv. Can., Pap. 50-37.

KANASEWICH, E. R. 1966. Deep crustal structure under the plains and Rocky Mountains. Can. J. Earth Sci., **3**, pp. 937–945.

LE PICHON, X. 1968. Sea-floor spreading and continental drift. J. Geophys. Res., **73**, pp. 3661–3697.

LEECH, G. B. 1966. The Rocky Mountains trench. Geol. Surv. Can., Pap. 66-14, pp. 307–329.

MANSY, J. L. 1969. Etude géologique des Monts de Kiparissia (Messénie, Grèce). D.E.A., Université Sciences et Techniques, Lille, 1969, 75 p.

MARINOS, G. 1957. Zue Gliederung Ostgriechenlands in Tektonishe Zonen. Geol. Rundschau, **46**, pp. 421–426.

MCGUGAN, A. *et al.* 1968. Permian and Pennsylvanian biostratigraphy and permian depositional environments, petrography, and diagenesis, southern Canadian Rocky Mountains. Alberta Soc. Petrol. Geol., 16th ann. field conf., pp. 48–66.

MCKENZIE, D. P. et MORGAN, W. J. 1969. Evolution of triple junctions. Nat., **224**, pp. 125–133.

MCKENZIE, D. P. and PARKER, R. L. 1967. The North Pacific: and example of tectonics on a sphere. Nat., **216**, pp. 1276–1280.

MERCIER, J. 1966*a*. Sur l'existence et l'âge des deux phases régionales de métamorphisme alpin dans les zones internes de Hellénides en Macédoine centrale (Grèce). Bull. Soc. Géol. de France, (7), **VIII**, pp. 1014–1017.

——— 1966*b*. Paléogeographie, orogenèse, métamorphisme, et magmatisme des zones internes des Hellénides en Macédonie (Grèce): °vue d'ensemble. Bull. Soc. Géol. de France, (7), **VIII**, pp. 1,020–1,049.

MONGER, J. W. 1969. Stratigraphy and structure of Upper-Paleozoic rocks, Northeast Dease Lake Map-Area, British Columbia. Geol. Surv. Can., Pap. 68-48, 41 p.

MONGER, J. W. H. and ROSS, C. A. 1971. Distribution of fusulinaceans in the western Canadian Cordillera. Can. J. Earth Sci., **8**, pp. 259–278.

MOORES, E. M. 1970. Petrology and structure of the Vourinos ophiolitic complex. Northern Greece, Thesis Univ. California, pp. 1–71.

MORGAN, W. J. 1968. Rises, trenches, great faults, and crustal blocks. J. Geophys. Res., **73**, pp. 1959–1982.

MUFFLER, L. J. P. 1967. Stratigraphy of the Keku Islets and neighboring parts of Kuiu and Kupreanof Islands Southeastern Alaska. U.S. Geol. Surv., Bull. 1 241-C, pp. C1–C52.

MULLER, J. E. and CARSON, D. J. J. 1969. Geology and minerals deposits of Albernie map area. Vancouver Island and Gulf Islands, British Columbia. Can. Geol. Surv., Pap. 68-50.

PAPASTAMATIOU, J. 1963. Les bauxites de l'ile de Skopelos (Sporades du Nord). Bull. Geol. Soc. Greece, **5**, pp. 52–74.

PAVONI, N. 1961. Faltung durch Horizontalversch bung. Ecl. Geol. Helv., **54**, p. 515–534.

PITMAN III, W. C. and HAYES, D. E. 1968. Sea-fl spreading in the Gulf of Alaska. J. Geoph Res., **73**, pp. 6,571–6,580.

PRICE, R. A. et MOUNTJOY, E. W. 1970. Geolo structure of the Canadian Rocky Mountains tween Bow and Athabasca rivers. A progr report. Geol. Assoc. Can., Spec. Pap., **6**, 7–25.

RENZ, C. 1906. Uber neue Trias-Vorkommen Argolis. Centr. fur Min. Geol. und Pal., **9**, 270–271.

RODDICK, J. A. 1965. Vancouver North, Coquitl and Pitt Lake map-areas. British Columb Geol. Surv. Can., Mem. 335, 276 p.

——— 1966. Coast crystalline Belt of British C lumbia. Can. Inst. Min. and Metal., Spec. V **8**, pp. 73–82.

RODDICK, J. A., WHEELER, J. O., GABRIELSE, H., a SOUTHER, J. G. 1967. Age and nature of Canadian part of the circum-Pacific oroge belt. Tectonophys., **4**, pp. 319–337.

ROSS, J. V. 1968. Structural relations at the easte margin of the Shuswap Complex, near Rev stoke, southeastern British Columbia. Can. Earth Sci., **5**, pp. 831–849.

SCHWIDERSKI, E. W. 1968. Mantle convection a crustal tectonics inferred from a satellite's biti: a different view of sea-floor spreadi J. Geophys. Res., **73**, pp. 2,828–2,833.

SIMONY, P. and WIND, G. 1970. Structure of the D tooth Range and adjacent positions of the Roc Mountains Trench. Geol. Ass. of Can., Sp Pap., **6**, pp. 53–65.

SMITH, A. G. 1971. Alpine Deformation and Oceanic Areas of the Tethys, Mediterranean a Atlantic. Geol. Soc. Am. Bull., **82**, pp. 203 2070.

SOUTHER, J. G. and ARMSTRONG, J. E. 1966. No Central Belt of the Cordillera of British Colu bia. Can. Inst. Min. and Metal., Spec. Vol., pp. 171–184.

STACEY, R. A. and STEPHENS, L. E. 1969. An interp tation of gravity measurements on the w coast of Canada. Can. J. Earth Sci., **6**, pp. 46 474.

TERMIER, P. 1903. Les nappes des Alpes orientales la synthèse des Alpes. Bull. Soc. Géol. de Fran (4), **III**, pp. 711–765.

TERRY, J. 1969. Etude géologique d'un secteur de Messénie septentrionale (Grèce). D.E.A., U versité Sciences et Techniques, Lille, 1969, 122

TERRY, J. and MERCIER, M. 1971. Sur l'existen d'une série détritique berriasienne intercalée en la nappe des ophiolites et le flysch eocène de nappe du Pinde (Pinde Septentrional, Grèce C.R. Somm. Soc. Geol. de France, pp. 71–72.

THAYER, T. P. 1969. Peridotite–gabbro complexes keys to petrology of mid-oceanic ridges. Bu Geol. Soc. Am., **80**, pp. 1515–1522.

THIEBAULT, F. 1968. Etude préliminaire des sér épimetamorphiques du Taygete septentrio (Péloponnèse méridional-Grèce). Ann. Soc. Gé Nord, **LXXXVIII**, pp. 209–214.

TIPPER, R. H. W. 1959. Revision of the Hazelton and Takla groups of central British Columbia. Geol. Surv. Can., Bull. 47.

——— 1963. Taseko Lakes. Geol. Surv. Can., Map 29—1963.

TOBIN, D. G. and SYKES, L. R. 1968. Seismicity and tectonics of the Northeast Pacific Ocean. J. Geophys. Res., **73**, pp. 3,821–3,845.

TSOFLIAS, P. 1969. Sur la découverte d'Ammonites triasiques au front de la nappe du Pinde en Péloponnèse septentrional (Grèce). C.R. Somm. Soc. Géol. de France, pp. 118–119.

VINE, F. J. 1966. Spreading of the Ocean Floor: New Evidence. Sci., **154**, pp. 1,405–1,415.

WHEELER, J. O. 1961. Whitehorse map area—Yukon territory. Geol. Surv. Can., Mem. 312, 156 p.

——— 1965. Big-Bend map area, British Columbia. Geol. Surv. Can., Pap. 64-32.

——— 1967. Tectonics. Canadian Upper mantle report. Geol. Surv. Can., Pap. 67-41, pp. 3–23.

WHEELER, J. O. (*Ed.*). 1970. Structure of the Southern Canadian Cordillera. Geol. Assoc. Can., Spec. Pap., **6**, 166 p.

WHITE, W. H. *et al.* 1967. Isotopic dating of the Quichon batholith British Columbia. Can. J. Earth Sci., **6**, pp. 677–690.

WILSON, J. T. 1965. Transform Faults, Oceanic Ridges, and Magnetic Anomalies Southwest of Vancouver Island. Sci., **150**, pp. 482–485.

WISE, D. U. 1963. An outrageous hypothesis for the tectonics of the Northeast Pacific Ocean. J. lera. Geol. Soc. Am., Bull. **74**, pp. 357–362.

YOLE, R. W. 1969. Upper paleozoic stratigraphy of Vancouver Island, British Columbia. Geol. Assoc. Can., **20**, pp. 30–40.

ZIMMERMAN, J. 1968. Structure and petrology of rocks underlying the Vourinos ophiolitic complex, northern Greece. Princeton Univ., Ph.D., Unpubl.

ZWART, H. J. 1967. The duality of orogenic belts: Geol. en Mijnbouw, 46 ste, pp. 283–309.

16

Reprinted from *Geol. Rundschau* **63**:1133–1172 (1974)

The Upper Miocene evaporite basins in the Mediterranean Region — a study in paleo- oceanography

By Peter Sonnenfeld, Windsor, Ont. *)

With 8 figures

Zusammenfassung

Das Mittelmeer ist ein Verdunstungsbecken, welches seinen Wasserverlust durch Einströmung in der Straße von Gibraltar und im Bosporus wettmacht. Salzüberschüsse werden durch einen Bodengegenstrom abgeführt. In kühleren plio-pleistozänen Zeiten hat ein Wasserüberschuß einen hinausfließenden Oberflächenstrom verursacht und eine nach innen gerichtete Bodenströmung, die im kanarischen Strome aufwallende Gewässer hereinsog.

Der Wasserkreislauf in modernen Verdunstungsbecken kann als geeignetes Modell dienen, um im Mittelmeergebiet vorkommende ältere Evaporite zu erklären. Die letzte solche Versalzung führte zu obermiozänen Evaporiten, die sich von Südost-Spanien bis zum Kaspischen Meer, vom Karpatenvorland bis Jemen ausbreiten. Sie entstanden in einer Reihe von untereinander verbundenen Becken, welche die zirkulierenden Bodenströmungen vorkonzentrierten oder örtlich versüßten. Zu einer Zeit, als die Straße von Gibraltar geschlossen war, wurde die Wasserzufuhr von brackischen ponto-kaspischen Gewässern gestellt und nicht von normalem ozeanischem Salzwasser.

Abstract

The Mediterranan Sea is an evaporite basin that compensates its water deficit by inflow through the Straits of Gibraltar and the Bosporus. Excess salinities are discharged through a bottom counter current. In Plio-Pleistocene cooler periods a water surplus produced a surface outflow and a bottom inflow bringing in waters upwelling in the ancestral Canary current.

Water circulation in modern evaporite basins can serve as an adequate model to explain ancient evaporites in the Mediterranean region. The last such high-salinity event comprises Upper Miocene evaporites stretching from southeastern Spain to the Caspian Sea, from the Carpathian Foreland to Yemen. They formed in a series of interconnected basins which pre-concentrated or locally diluted circulating bottom currents. Not normal oceanic saltwater but brackish Ponto-Caspian waters were the source of supply during a period when the Straits of Gibraltar were closed.

Résumé

La Meditérrané est un bassin évaporitique, qui comble son déficit d'eau par un influx à travers le détroit de Gibraltar et le Bospore. Un surplus d'eaux salées est déchargé par un contre-courant de fond. Dans les périodes plus froides du Plio-Pleistocène, un excès d'eau produisait un courant de sortie à la surface et un influx au fond, qui faisait entrer des eaux profondes remontées dans le courant de Canarie.

La circulation d'eau dans les bassins modernes évaporitiques peut servir comme un modèle approprié pour expliquer les anciennes évaporites de la région méditeranéenne. La dernière phase de haute salinité est indiquée par les évaporites du Miocène supérieur, qui s'étendent de l'Espagne sud-orientale jursqu'à la mer Caspienne, des

*) Author's address: Dr. P. Sonnenfeld, Professor of Geology, University of Windsor, Windsor N9B 3P4, Ontario, Canada.

avant-pays des Carpates jusqu'à Yémen. Elles se formaient das une série de bassins liés entr'eux, dans lesquels se produisait une préconcyentration ou une dilution locale des eux profondes. Lors d'une période de fermeture du détroit de Gibraltar la source d'eau n'était pas l'eau océanique normale, mais l'eau saumâtre de la lac-mer ponto-caspienne.

Краткое содержание

Средиземное море — бассейн испарения; эта потеря компенсируется в нем притоками воды через проливы Босфор и Гибралтар. Избыток солей выносился наддонными течениями. В более холодный плиоценово-плейстоценовый период избыток воды вызывал вытекание поверхностных слоев воды и направленное к середине бассейна придонное течение.

Круговорот воды в современных бассейнах испарения может послужить моделью для объяснения появления древних эвапоритов. Последнее такое осолонение образовало верхнемиоценовые эвапориты, которые распространяются от юго-восточной Испании до Каспийского моря и от Предкарпатья до Иемена. Они образовались из ряда связанных между собой бассейнов, в которых циркулирующие придонные течения или насыщались, или же разводились пресной водой. Когда Гибралтарский пролив был закрыт, приток воды осуществлялся солоноватыми понто-каспийскими водами, а не обычными водами океана.

Introduction

The Mediterranean Sea is a remnant of the ancient Tethys Sea shaped by the gradual approach of the Eurasian to the African Plate since mid-Jurassic time (Glangeaud et al., 1970; Galanopoulos, 1970; Burke & Wilson, 1973). It has been separated from another body of water, the Paratethys, for most of Cenozoic time. The Paratethys constituted a large inland sea which ultimately broke up into residual waters in the Aral and Caspian Lowlands in the east, the Black Sea and the Hungarian Plains with remnant Lake Balaton in the west. The Paratethys varied in areal extent and in salinity, fluctuating between a brackish and a fresh body of water. A narrow land bridge comprising parts of Iran, Turkey and the Balkan Peninsula separated it from the Tethys Sea. As the latter gradually altered its configuration towards a semblance with the present-day Mediterranean, connections to the world oceans changed in time. The results of the Glomar Challenger voyage through the Mediterranean Sea prompt a re-assessment of the late Miocene history of southern Europe. Sufficient work has been published in recent years on various local areas to permit a synthesis of available data. Much of it appeared in regional publications, often without summary in a foreign language; to provide a comprehensive bibliography would be beyond the scope of this paper.

This account will attempt to correlate Upper Miocene evaporites wherever they may occur in the Mediterranean region. In order to do so on an actualistic or uniformitarian basis, we shall first review briefly Holocene and Plio-Pleistocene conditions in the area.

Holocene conditions

The Mediterranean Sea is today an evaporitic basin in a summer-dry, winter-wet climate, in which only a fraction of the water loss is made up by rainfall of winter cyclones and by runoff. According to Tixeront (1970), the daily evaporative water loss of 8.2 km³ is compensated by 3.8 km³ of precipitation and runoff,

0.5 km³ net gain from the Black Sea and 3.9 km³ net gain from the Atlantic Ocean. Other estimates of water loss and of water exchange with the Atlantic Ocean and the Black Sea vary over wide ranges because they are reflections of recent fluctuations in the local climate.

Today the Canary Current moves along the Atlantic coast of Morocco and is partially fed by slowly upwelling deep waters. Seasonally the area of upwelling migrates along the West African coast, Agadir is plagued by summer fog and

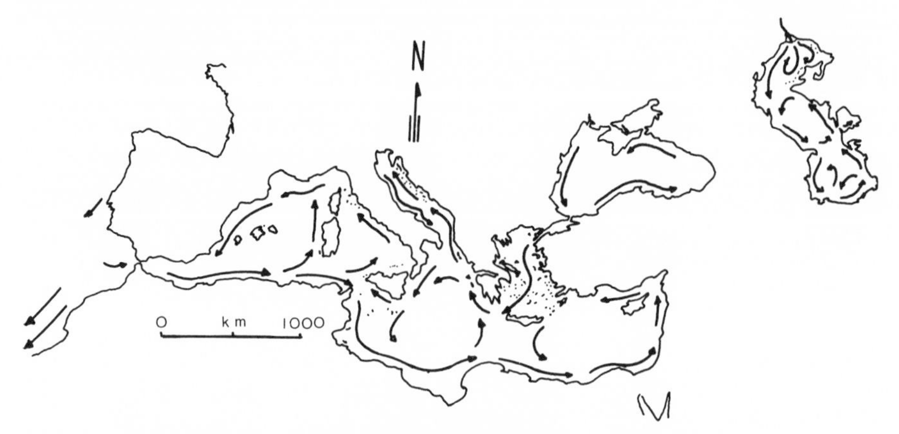

Fig. 1. Major surface current distribution (generalized picture).

this fog extends to 40° N. Lat. along the Portuguese west coast (KRÜMMEL, 1911). Such upwelling is also known from the California coast of similar latitude and is typical of the eastern margins of tropical oceans on either side of the equator, where longshore currents move with the Trade Winds towards the equator. The rising evaporation rates take their toll and the water deficit is made up from below.

Both the Atlantic Ocean and the Black Sea deliver a low-density surface current to the Mediterranean Sea, of about 18.5—20 ‰ salinity through the Bosporus and of about 34—37 ‰ salinity through the Straits of Gibraltar. Velocity fluctuations of the surface current through the Dardanelles match diurnal and annual Trade Wind velocity (KRÜMMEL, 1911). A barometric Low over the Mediterranean Sea has a suction effect and increases the intake (OREN, 1971). The surface current then follows the North African coast, spinning off geostrophic branches into the eastern Balearic, the Tyrrhenian and the Adriatic Seas (see fig. 1). Because of evaporation losses the current progressively increases in density. Lower salinity values remain, however, always on the right side of the observer looking in the direction of the current (LACOMBE & TCHERNIA, 1960).

Since a water loss through evaporation leads to a gradual increase in salt content of the residual fluid, the Mediterranean Sea has to dispose of some of the accumulating salts by outflow. A higher density undercurrent (38 ‰ salinity) has been recorded in the Straits of Gibraltar below a depth of about 200 m and it forms a tongue into deeper Atlantic waters of equal density (NEUMANN, 1968) to a depth of 1,200 m (LACOMBE & TCHERNIA, 1973). It is this tongue flowing out of the Straits of Gibraltar which prevents the waters upwelling in the Canary Current from entering the Straits (cf. fig. 2).

The interface between inflow and outflow is tilted in the south in consequence of the rotation of the earth, leaving a greater cross sectional area for outflow along the Spanish than along the Moroccan coast (LACOMBE, 1961). If the high-salinity undercurrent reaches to higher levels on the north shore, it also enters indentations of the coast there more easily, is more readily flushed out of

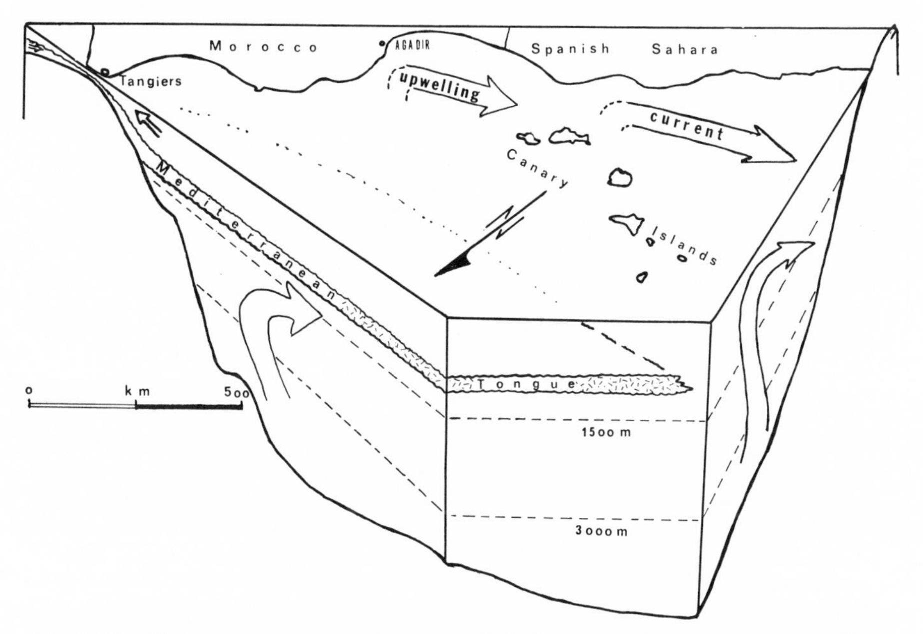

Fig. 2. The Mediterranean tongue. A tongue of Mediterranean brine prevents upwelling Canary Current waters from entering the Straits.

indentations along the south shore. This affects salinity concentrations in any bays or lagoons.

An undercurrent of about 38 ‰ salinity is also found in the Bosporus below a depth of about 10—30 m (KRÜMMEL, 1911), at the shallower levels along the European coast. It forms a tongue into the Black Sea (BOGDANOVA, 1969) analogous to the one extending from the Straits of Gibraltar. The outflow through the Dardanelles has been well known to Procopius of Caesarea in the sixth century A. D., but is now detected only in years of lower Black Sea levels. ULLYOT & ILGAZ (1946) found no Mediterranean waters entering the Black Sea. PECTAS (1954), however, found these entering seasonally. BOGDANOVA (1969) proved the continued existence of the undercurrent and ROJDESTVENSKY (1971) again showed a distinct seasonal preference to be in existence similar to SERPOIANU (1967) who linked reduction of penetration of undercurrent to highest Black Sea levels (March-June). At the turn of this century it has been estimated that a further rise in Black Sea water levels by 29 cm would increase the force of the surface current to the point where the wedge of Black Sea waters would press against the bottom of the Straits and no Mediterranean waters would reflux (KRÜMMEL, 1911). A substantial reduction in the salinity of the Black Sea would then occur.

The contribution of the Suez Canal to the water budget is negligible. Because of its near-saturation with salts dissolved from evaporites exposed in the Bitter Lakes it represents a density barrier and forms a small tongue at depth both north- and southward.

The high water loss in the Mediterranean Sea is caused by the Trade Wind regime which rules the whole Mediterranean area in the summer, all but its northern parts in the winter. As the northeasterly winds warm up, their water carrying capacity increases at rapidly accelerating rates and evaporation rates rise correspondingly.

This has not always been a constant picture. Ptolemy noted in the first century A. D. more cyclonic thunderstorms in Alexandria than are recorded there today. The Numidian, Getulian and Garamantian cattle kingdoms south of Punic coastal settlements bear witness that North Africa was wetter in Roman times and thus the water loss smaller. Hannon describes in the sixth century B. C. pastoral tribes (the Lixites) from what is now northern Mauretania. After some post-glacial oscillations the Mediterranean climate is getting drier for about the last 2,700 years.

Plio-Pleistocene conditions

During Pleistocene ice advances the Trade Winds belt was displaced farther south, rates of evaporation over the Mediterranean Sea had declined dramatically, the Sahara experienced a Pluvial Age. Water influx from the Black Sea and from circum-Mediterranean rivers fed not only by meltwaters in the north, but by increased runoff in Pluvial regions was substantially larger than it is today. Doubling the rates of precipitation would not only double the runoff and increase cloudiness, but would also reduce rates of water loss.

HUANG et al. (1972) found sedimentological evidence in the Alboran Sea that some 9—11,000 years B. P., at the end of Würm glaciation, a current reversal took place. Prior to this time, a low-density surface current flowed out of the Mediterranean Sea into the Atlantic Ocean. This is a picture similar to the flow today out of the Black Sea, the Baltic Sea and other extra-tropical seas with low or moderate evaporation losses. Surface waters in such seas are invariably of less than normal oceanic salinity.

In Pliocene time, when the Mediterranean Basin was suddenly inundated by ocean waters transgressing from the Atlantic Ocean, fossil evidence indicates that a good portion of these waters was of deeper Atlantic origin (CITA & RYAN, 1973; BENSON & SYLVESTER-BRADLEY, 1971). Since the Pliocene was an age of gradual pre-glacial cooling, the Mediterranean Sea at some time achieved a positive water balance with an outflowing surface current. The inflowing bottom counter current then swept into the Mediterranean Sea the waters upwelling in the ancestral Canary Current and with it an Atlantic deepwater fauna from depths significantly below 1,200 m. It follows that the fauna faithfully depicts current distribution, but that it says little about actual depths of the Mediterranean Basin at the time of transgression (SONNENELD, 1974). In a period of cool climate even a coldwater fauna from deeper Atlantic waters will briefly survice if sucked in by the bottom current flowing into a shallow basin; the fauna cannot be found in the Ionian or Levantine Seas.

The same phenomenon of a current-dependent faunal distribution can also be observed in the Bosporus. Whenever the water level of the Black Sea is lowered by increased aridity, more penetration of the undercurrent results in more fauna swept from the Mediterranean into the Black Sea (POUSANOV, 1965). A migration of Mediterranean fauna to Caucasian shores is reported from earlier interglacials (FEDOROV, 1967). In Pleistocene regression stages, when a positive water budget lowered salinities, the Crimean fauna in turn migrated to the Dardanelles, to Corinth and the Gulf of Megara west of Athens, and to the lakes of central Greece (GILLET, 1957 a).

Evaporite formation

The mildly evaporitic conditions in the Mediterranean Sea today are not a unique phenomenon. The different evaporite horizons bear witness that much more stringent conditions have prevailed in the Mediterranean area on several occasions during Mesozoic or Cenozoic history. The behaviour of the present Mediterranean Sea and of some of its lagoons in Tunisia or Cyprus, of the Gulf of Karabugaz off the Caspian Sea, or of the Mexican Ojo de Liebre lagoon is the key to the past.

Currents at basin entrance

In general, concentration or dilution of a brine will take place when the influx of waters into an evaporitic basin carries more or less salts into the basin per unit of time than can be carried out by bottom currents of higher or surface currents of lower density. In most oceans and bays the two currents are in equilibrium and thus no concentration or dilution occurs other than through seasonal fluctuations. The equilibrium can be disturbed if the entrance to a basin is partially blocked by a barrier. This barrier can be a narrow and shallow strait, a reef chain, a mud flat, a sand bank, or a variety of tectonically or volcanologically controlled features.

As long as a common sea level is maintained, a supply of runoff and rainfall in excess of evaporation losses per unit of time will result in a low-density brackish surface current flowing out, while a head of ocean waters will spread into the basin below the interface with surface waters. Friction effects at the interface increase with increasing rates of outflow until an overall freshening of the restricted basin ensues. Such happened in the Black Sea in Würm times.

If the supply of water in the recharge area is smaller than the loss produced by evaporation per unit of time, an inflowing surface current must equalize the sea level surface otherwise quickly depressed by evaporation losses. Such a current establishes an interface with denser (e.g. Mediterranean) waters underneath and these produce a bottom counter movement of a dense fluid head displacing laterally less dense waters. An increasing rate of inflow establishes a disequilibrium in favour of inflowing waters which results in a gradual concentration of the brine (see fig. 3) even in cases where initial salinity is as low as in the Caspian Sea, the source of waters for the Gulf of Karabugaz.

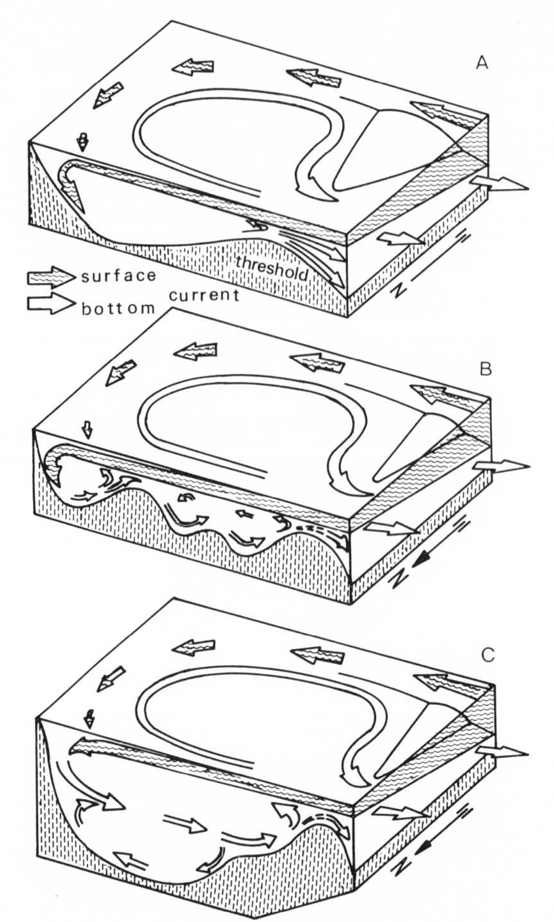

Fig. 3. Water circulation in an evaporitic basin. An anticlockwise surface current generated in response to evaporative water loss drags the clockwise undercurrent back into the restricted basin. Salinity in the surface current increases to the left. Cross sectional area of bottom current is greatest at north end of threshold. Dashes indicate where a strong surface current will interrupt the undercurrent over the swells or thresholds. A. single shallow basin; B. multiple basin and swell topography pre-concentrating bottom brines; C. deep basin (e. g. present Mediterranean Sea).

The degree of concentration of the brine is a function of the ratio between inflowing and outflowing amounts of solute maintained over a period of time. The shallower the initial basin, the shorter the time required to reach a specific concentration; deeper basins require conditions to remain favourable for longer periods. Highly soluble potash salts are commonly restricted to multiple basins composed of several depressions which act as pre-concentrators (see fig. 3 B).

The climatic factors thus control the water balance in the basin, namely the rate of evaporation over the area of the basin and the rate of precipitation in recharge areas of the drainage region. Two geologic factors control the development and the size of an evaporite deposit: either the slow rate of subsidence of a threshold barrier (allowing e. g. a reef to keep up its growth), or the slow rate of uplift (allowing the current in the strait to keep up erosion), controls the water intake from adjacent bodies of water which have to maintain a common sea level at least at high tide. If the rate of evaporite precipitation becomes greater than the rate of subsidence, or if evaporation losses per unit of time exceed the influx, the basin dries out in places farthest from water supply: The water level in the

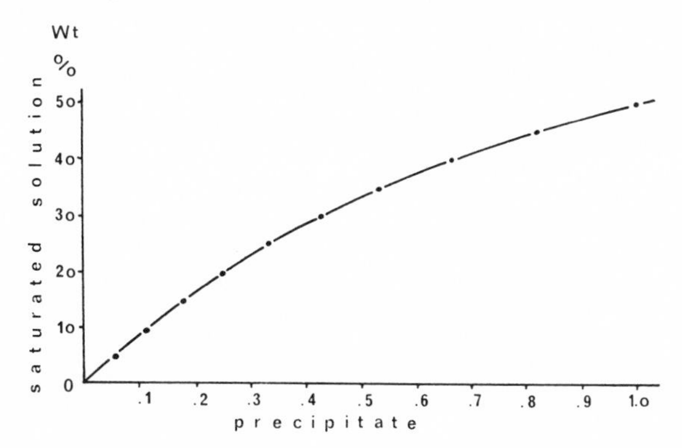

Fig. 4. Computed amount of precipitation from an evaporating saturated solution (grams of precipitate per gram of water lost to one kilogram of saturated solution). The higher the required concentration for saturation, the more precipitate is produced per unit of evaporation loss per unit of time.

basin then drops below the water level outside the threshold, the basin becomes a dying lake.

In recent decades the Gulf of Karabugaz off the Caspian Sea has become such a salt lake: In 1902 the outflow had a salinity of 164 ‰ (SVERDRUP et al., 1946, p. 149), but half a century later, outflow had ceased and the water level had dropped. Salinities in the Gulf now approached double that figure (STRAKHOV, 1962, III, p. 158). The area of the Gulf had significantly shrunk in the same period of time, the cross sectional area of the threshold had been reduced by over 40 per cent (SEDELNIKOW, 1958).

The evaporite precipitation per unit of time is proportional to the degree of saturation required and to the water loss per unit of time. In other words, at a given water loss, once the respective saturation is reached, the most soluble salts will produce greater quantities of precipitate per unit of time than the less soluble ones (see fig. 4). This explains why the less soluble gypsum produces much thinner annual layers, an observation discussed by STRAKHOV (1962, III, p. 446).

Currents inside evaporite basins

Two major systems of currents are bound to develop within the restricted basin. A geostrophic current of counterclockwise direction will distribute inflowing waters along the shores. As evaporation progressively increases the density of the circulating surface waters, they tend to sink and produce a vertical circulation system. RICHTER-BERNBURG (1953) noted that the near-shore shallow waters are warmer and subject to greater evaporation losses: calcium carbonate and gypsum are easily stripped from the circulating waters. Vertical water exchange is fostered in the winter season. High-salinity waters, chilled by cold northeasterly winds, sink to 2,000 m along the French coast, 1,000 m in the Adriatic Sea, but only 200—300 m in the Rhodes-Cyprus region, the area of greatest salinity concentrations. It is here that descending waters are found (LACOMBE & TCHERNIA, 1960). Eventually the geostrophic current can no longer complete the

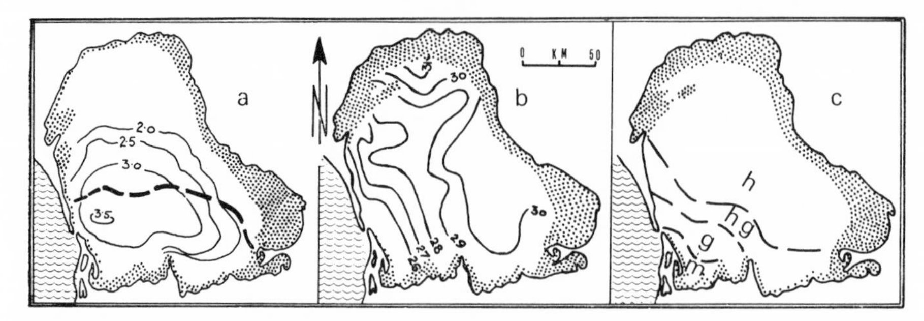

Fig. 5. The Gulf of Karabugaz. Stippled areas were desiccated land in the dry period of the mid-fifties. a) Zone of mixing along the south shore before the inflow fans out into the central zone of concentration; depth contours in meters (after SEDELNIKOW, 1958). b) Salinity isopleths in the surface layer of the brine in 1956 (after STRAKHOV, 1962). c) Sediment facies in 1956: h) halite without and with epsomite; hg: halite with glauberite; g: glauberite; m: glauberite with mirabilite (after STRAKHOV, 1962).

journey around the basin because the inflow fans out over the denser fluids within the basin. This is now happening in the Gulf of Karabugaz (SEDELNIKOW, 1958), where the inflowing current moves along the south shore in a zone of mixing and then fans out into a central zone (see fig. 5 a).

A clockwise undercurrent attempts to remove from the basin excessive quantities of solute. It is slowed underneath by the friction along the basin floor and above by friction with the inflowing current. The effects of this friction can be seen in flow velocities at various depths or flow diagrams, published for the Dardanelles (KRÜMMEL, 1911) or for the Straits of Gibraltar (LACOMBE, 1961). Surface waters entering the eastern Mediterranean Basin still are Atlantic waters mixed with waters of the counterflowing intermediate layer (LACOMBE & TCHERNIA, 1960) which are being dragged back at the Sicilian threshold into the Ionian and the Levantine Seas. The counter bottom current in the Gulf of Karabugaz, although it is trapped, can still be measured at one third of the surface current velocity before it mixes near the entrance with new supplies of surface waters (SEDELNIKOW, 1958).

In deep basins, anticlockwise flow and clockwise counter flow, upwelling and sinking, mainly involve a surface and an intermediate layer of water. Each time a current is deflected, its velocity is sharply curtailed. Bottom waters between Malta and Sicily serve as concentrators of the northeast flowing brine (MOREL,

1971), but are then drained. Bottom waters in deeper depressions of the Mediterranean Sea seem to be trapped (MEDOC GROUP, 1970). They represent the cell of highest possible concentration at present influx and outflow conditions.

In the Gulf of Karabugaz the inflowing surface current moves along the south shore and reaches highest salinity values along the north end of the Gulf (see fig. 5 b); this geostrophic distribution is related to earth rotation. However, bottom sediments show the opposite sequence (see fig. 5 c), going from halite precipitation in the north to glauberite and mirabilite deposition in the southwest. This demonstrates that increasing concentration in the bottom counter current determines deposition, n o t the salinity of the surface current. Mirabilite deposition in the zone of mixing may be influenced also by the initial temperature difference between undercurrent and influx, particularly in the winter time. In contrast, the Gulf of Karabugaz deposited only gypsum at the turn of the century when bottom reflux was still unhindered (SVERDRUP et al., 1946).

In the El Melah lagoon near Zarzis in southeastern Tunisia, an elongate recent evaporite basin, one can find no gypsum at the deeper centre, bu maximum gypsum precipitation along the rims. In the innermost reaches of the bay this concentrated brine produces carnallite and polyhalyte (JAUZEIN & PENTHUIZOT, 1972). Similarly, highest concentrations occur in the innermost depressions of the Ojo de Liebre lagoon in Baja California (PHLEGER, 1969). Both are shallow enough lagoons to have no counter current of any consequence.

If we accept the present-day Mediterranean Sea as a suitable model for a basin with a water deficit, we find:

1. Salinities in the surface layer are lowest along the south coast, highest along the north coast, specifically along the south coast of Asia Minor, the Aegean Sea, the Gulf of Lyon.

2. Salinities are highest in north shore embayments farthest from the water intake at Gibraltar. Only local, small shallow basins (e. g. El Melah) form an exception.

3. The undercurrent is found at shallower depths along the north shore than along the south shore, resembling a tilted vessel.

4. The undercurrent is found at shallowest depths in those embayments along the north shore that are farthest from the water intake.

5. Density differences are smaller in winter than in summer. The cold northeasterly wind (called Vardarats in Macedonia) churns and cools the surface layer of the sea and extracts moisture as it itself warms up.

7. Bay mouth thresholds may be more effective in trapping deep waters in the southern half of the sea, but deep waters are more likely to form in the northern half where the intermediate layer occurs at a higher level.

These points are also borne out by the Gulf of Karabugaz, from which case the following may be added:

8. The undercurrent controls type and quantity of evaporite deposition; highest bottom concentrations occur near the threshold barrier, where s u r f a c e salinities are the lowest.

9. Bottom counter currents develop even in waters only 1—3 m deep. Internal friction traps the bottom current, once inflow velocities exert a sufficient suction.

10. Discontinuation of a common sea level with the parent water body progressively leads to desiccation along the shallow fringes, mainly in the regions farthest from the intake of the basin, under concurrent development of salt flats.

Evaporites in the Mediterranean region

The Mediterranean region has experienced repeated arid periods. Permian, Triassic, Jurassic and Upper Cretaceous evaporites have been reported from various localities ranging from Spain in the west to Rumania and Yougoslavia or Turkey in the east. Even if we discount stratigraphic miscorrelations (examples cited by NEBERT, 1956; PAUCA, 1968, 1971; ORBOCEA, 1972), evaporite conditions were probably at least locally present in Eocene, Oligocene and Lower to Middle Miocene beds from the Iberian Peninsula to the Aegean Sea and Turkey.

The Upper Miocene strata contain evaporites spread over virtually the whole Mediterranean Basin and many adjacent depressions. They developed over an area stretching from the Alboran Sea to Yemen, from the Adriatic Sea to the Ukraine and Turkmenistan and beyond. Upper Miocene evaporites todate represent the last evaporitic event in the Mediterranean region, because Pliocene or Pleistocene interglacial climatic and geomorphological factors never again reached Upper Miocene conditions.

The precise configuration of each basin cannot be mapped yet. However, we can construct a preliminary picture of evaporite distribution. It shows that there obviously was not one single large evaporite basin, but a series of basins, each of a different rate of subsidence, each composed of a series of subsidiary basins. Each basin was separated from the next one by a barrier. This barrier could range from being merely a swell with carbonate deposition (as the Balearic from the Ionian Basin) to a rapidly rising mountain chain or at least a land mass with resulting dumping of fine clastics (as the Volhyno-Podolian evaporites along the north rim of the Carpathian Mountains from those in Rumanian Transylvania or from those in the Vienna Basin).

Since many of the faunal assemblages are endemic, correlation problems from basin to basin have prevented a complete picture to be developed. A tremendous amount of work carried out in the last fifteen years has reduced the number of breaks and steps in correlation lines that commonly occur at the jurisdictional boundary faults of state geological surveys. Evaporites occurring in pockets along the rim of the Pannonian [1]) Basin, in Transylvania and in an embayment between the Carpathian Mountains and the Podolian Platform had been assigned to a central European substage Badenian-M_4 c (SENNEŠ et al., 1971), an equivalent of the evaporitic phase of the Messinian stage of Italy. Other evaporites appear to be slightly younger.

In most localities there is only one evaporite formation recorded. However, where a sufficiently detailed study has been carried out, often more than one set of gypsiferous beds is distinguished, such as five in the Suez area (KERDANY,

[1]) In order to avoid misunderstandings, the terms "Pannonian", "Dacian", "Pontian", "Levantine", Carpathian", "Volhynian", "Tyrrhenian" are used in the remaining text strictly as geographical terms with no time-stratigraphic connotations, unless specifically qualified.

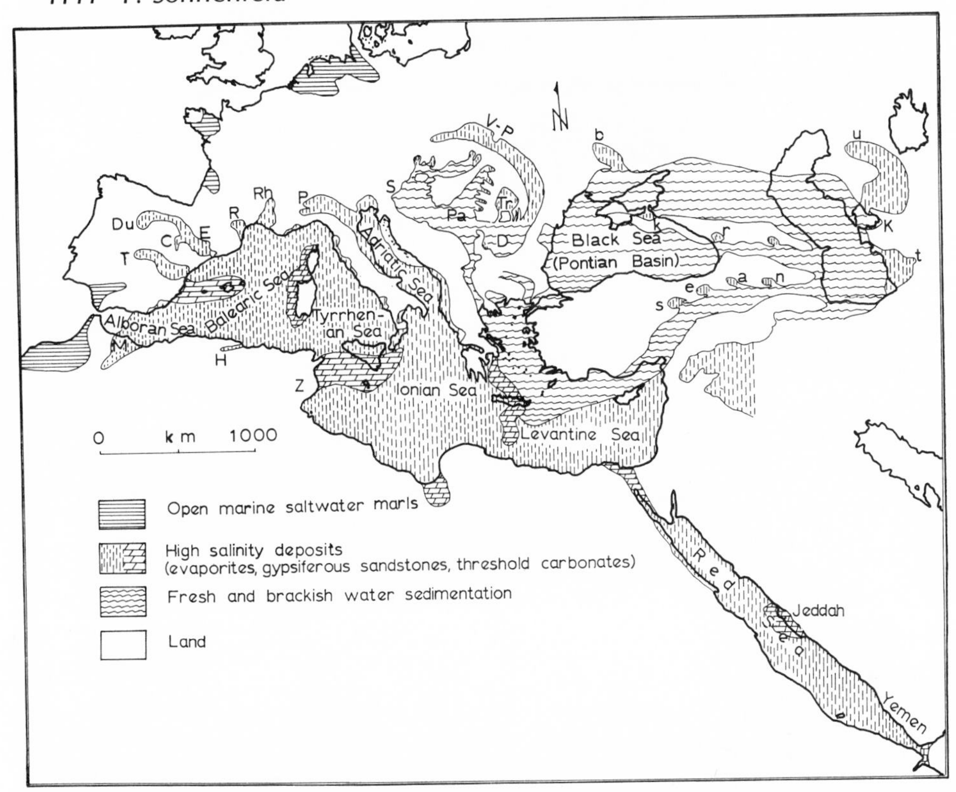

Fig. 6. Upper Miocene evaporites in the Mediterranean region (generalized distribution). C: Calatayud-Teruel; D: Dacia; Du: Duero; E: Ebro and Catalania; H: Hodna; M: Melilla; P: Piedmont-Po; Pa: Pannonia; R: Roussillon; Rh: Rhône; S: Styria; T: Tajo; Tr: Transylvania; V: Vera; V-P: Volhyno-Podolia; Z: El Melah at Zarzis; a: Armenia; b: Borisfen; e: Erzerum and Tekman-Karayazi; K: Karabugaz; k: Krasnodar; n: Nakhichevan; r: Rioni; s: Sivaş; t: Turkmen basin.

1967) and five in Poland (Poborski, 1952), two in Slovakia (Slávik, 1971) and two in southern Italy (Selli, 1973), and one evaporite series unconformably on another in Sicily (Decima & Wezel, 1973). Not all of these sets contain halite or other salts. Individual evaporitic events were estimated to have lasted $0.5 - 2 \times 10^5$ years (Pauca, 1968; Trashliev, 1969) or $5 - 9 \times 10^5$ years for the total evaporitic periods (Ogniben, 1957). The accumulation of any one gypsiferous bed may not have been completely synchronous to an evaporite bed in another basin.

Nonetheless we can bracket an interval of time or a sequence of time-rock units which encompass all Upper Miocene evaporites. They are distinguished as a group from non-evaporitic sequences underneath and above. The distribution of these evaporites is shown in fig. 6. Since it would have been onerous to plot each source of information, references used in the compilation of this map were incorporated into the bibliography and marked with an asterisk. All specific localities discussed have been indicated on figs. 6 or 7.

It should be mentioned that the Upper Miocene evaporites coincide with one of the Neogene periods of increased volcanic activity. Vast quantities of tuffs and ash accumulated in several regions from the Carpathian to the Atlas Mountains. If any crude ratio exists between the amount of eruptive production and the associated solfataria activity, then substantial amounts of jevenile or recirculated sulphur and other elements were released into the drainage system at that time. It cannot entirely be ruled out that tectonic processes exposed to leaching also some of the older evaporites.

The main question is the source of the water supply and its route to the individual basins, not the source of sulphur or chlorine. Throughout Tertiary times various channels opened and in turn closed: it is as yet difficult to assess the precise life span of each of these channels. Since potash salts require the greatest pre-concentration of the brine, we can work back from their occurrence to the source of the waters. The current distribution in the individual basins is not too difficult to understand if we press the analogies to the present state.

Distribution of Upper Miocene evaporites

Red Sea Basins

The potash deposits in the southern Red Sea area (Ross et al., 1973) would suggest that the feeder current probably came over a barrier today betrayed by exposed reefal carbonates on the Red Sea coast near Jeddah in central Saudi Arabia (HEYBROEK, 1965). This barrier trapped the brine which deposited halite in Yemen in the southern Red Sea.

Gypsum deposits are also known from the Egyptian Red Sea coast (STRAKHOV, 1962), probably as flank deposits to evaporites now buried in the central portion of the northern Red Sea. Another barrier existed near Suez, where HASSAN & EL-DASHLOUTY (1970) found reefal carbonates against a gypsiferous, but mostly halite-free sequence. Halite does occur, though, farther south at the west coast of the Sinai Peninsula. Evidently, salinity values decreased northward, in part because the basin shallowed towards the barrier and in part because of a freshwater influx near the west side of the Gulf of Suez (SAID, 1962).

The Mediterranean Sea

A potash belt similar to the one above occurs in the western part of the Mediterranean region with such beds exposed on Sicily northwest of the remnants of a swell separating the Tyrrhenian from the Ionian Sea (DECIMA & WEZEL, 1973).

Although it has been argued that Atlantic waters entered the region and supplied the salts (HSÜ et al., 1973), it is more reasonable to assume that the Straits of Gibraltar were closed. This has first been suggested by GLANGEAUD some 35 years ago (cf. GLANGEAUD et al., 1967), then again by RUGGIERI (1962); SELLI (1954) considered the western Mediterranean a closed or semi-closed basin.

No direct evidence can be produced either way since the environs of Tangiers are the site of a Pleistocene nappe thrust southward (CHOUBERT et al, 1961, 1964). Miocene sediments are only exposed to the south of this overthrust. However, the following indirect evidence supports the notion of a closed western margin:

1. The evaporite series encountered at the Deep Sea Drilling Project Sites is underlain by brackish water sediments (NESTEROFF, 1973). This indicates a predominance of the Pontian freshwater supply in the layers in immediate contact with the base of the evaporites; only the Paratethys Sea could have supplied adequate quantities of low-salinity waters. This observation speaks against a gradual desiccation of a marine body of water, initially of normal salinity.

2. Gypsiferous sequences can be traced laterally over short distances into freshwater deposits, into freshwater limestones that are often lignite-bearing. Such lateral correlations are possible in Crete, in Dacia, Pannonia, Albania, western Greece and eastern Anatolia.

3. In Tuscany the gypsiferous sequence is intercalated with freshwater deposits carrying not a local river fauna, but a fauna known from the Odessa region (GILLET, 1957 a). Not only does this indicate an extant connection to Pontian waters, but periodic total freshening of the brine also occurred, rather than a return to normal marine conditions.

4. The evaporitic sequence is then overlain again by freshwater deposits represented by the Sarmatian stage in the northeast, elsewhere this time interval is missing and the covering beds are those of the Pliocene marine transgression. The hiatus covers thus the peak of regression, the maximized continental conditions or the reduction in the rate of subsidence below the rate of deposition. There was no hiatus in the northeast where fresh waters merely returned (BUDAY et al., 1961). Post-evaporite Sarmatian beds reach a thickness of 800 m in Transylvania (VANCEA, 1960). Post-evaporite freshwater beds also occur in northern Tunisia (BUROLLET & BYRAMJEE, 1973).

5. Whereas Miocene-Pliocene marine deposition is continuous in western (Atlantic) Morocco (LECOINTRE, 1952; CHOUBERT et al., 1961) and in Spain west of Gibraltar (PERCONIG, 1964; MONTENAT & MARTINEZ, 1970), the two stages are separated by an angular unconformity near Melilla on the Mediterranean coast of Morocco (CHOUBERT et al., 1964) or at drilling sites within the Mediterranean Sea (NESTEROFF, 1973). Somewhere between Melilla on the Alboran Sea and the Atlantic coast of Morocco must have been the shore line of dry lands exposed in the diastem.

6. The dolomite interfingering with the evaporitic sections on the floor of the Mediterranean Sea shows a scatter of oxygen and carbon isotope values consistent only with a derivation from dominantly meteoric waters (FONTES et al., 1973).

7. The penetration of a Black Sea fauna into the Pelopponesus and into eastern Sicily (GILLET, 1960; RUGGIERI, 1967) suggests an at least periodic water supply from the east. The presence of Volhynian mollusks at Syracuse as well as in Greece and in Tuscany indicates the same (GILLET, 1957 a, b, 1963). Messinian migration paths from Rumania and Transcarpathian basins led to Zagreb, Yugoslavia, and to Hungary, included Thessaloniki and Athens in Greece, the Adriatic and Tyrrhenian coasts of Italy, the Rhône valley and Barcelona in Catalania in the west, Russian basins in the east (GILLET, 1959).

8. Conversely, the complete Mediterranization of brackish faunas along the southern portion of the Bulgarian Black Sea coast and thence as far as Turkmenia and Georgia in latest Messinian times (SHASHIMIROV, 1971) indicate a later

migration in the opposite direction. Even at this stage no Atlantic normal-salinity benthonic faunas intermingled.

9. A large part of the emergent Alboran Sea was utilized by terrestrial vertebrates as an intercontinental land bridge (GLANGEAUD et al., 1967; DE BRUIJN, 1973).

The presence of some planktonic foraminifera of oceanic provenance in the Mediterranean Basin is not tied to a continuing connection between Atlantic and Mediterranean waters. Zones corresponding in time to the hiatus of the angular unconformity at Melilla, Morocco, would suggest an alternate route of communication, since many of the species have been recognized in the Paratethys domain. We need to have a clearer picture of precisely when the connection of the Paratethys to the Persian Gulf was severed.

Internal water circulation in the western Mediterranean Sea could not have been too different in Upper Miocene times from what it is today. A counterclockwise surface current sweeping along the shores of the Adriatic Sea moved through Calabria, depositing gypsum along the west coast of Italy, halite in the Balearic and Ebro depressions. The threshold to the latter may have been narrower than shown in fig. 6 (KLEMME, 1958). The outflow led to potash layers in Sicily (DECIMA & WEZEL, 1973), in front of the shallow water carbonate treshold back into the Ionian Sea. Had these potash laminae been followed by desiccation and a transgression of new oceanic waters, they would have been quickly dissolved. Potash beds can only be preserved if water salinities merely drop from potash saturation to a point above precipitation floor of the less soluble halite and gypsum, which then form a protective caprock layer. A counterclockwise longshore current is also postulated for an Upper Miocene deltaic flood plain northwest of Barcelona (CASANOVAS CLADELLAS, 1972).

If an entrant through Istria into the Pannonian Basin had still existed at this time (SENEŠ, 1961), it probably would have channelled currents flowing north along the east coast of the Adriatic Sea. Since gypsum deposits are known from Montenegro (MILANOVIĆ, 1969), Albania (PAPA, 1970; PASHKO et al., 1969) and from wester Greece (HAGN et al., 1967; LEONTARIS, 1970; MARAGKOUDAKIS, 1967), such a current would have been fairly high in salinity, a feature not in keeping with the brackish nature of southern Pannonian deposits in the Drava-Sava depression near Zagreb and in the Styrian Basin (SENEŠ, 1961). Moreover, Adriatic species would be swept into Hungary; instead, the fauna in the Mátra Foreland is completely endemic (HAJÓS, 1968) suggesting no direct connection to the Adriatic or Mediterranean Seas. SENEŠ et al. (1971) correctely assume the Istrian passage to be closed in Upper Miocene times. The absence of gypsum beds along the Dalmatian coast is partly due to local drainage from the karst terrain, partly to the more recent drowning of an ancestral coast line; evaporites occur in the Italian part of the peri-Adriatic trough (PAREA & RICCI-LUCCHI, 1972).

Source of Mediterranean waters

If the connection to the Atlantic Ocean via the Straits of Gibraltar or the earlier operative North Betic and South Rif Straits were closed, where did water come from?

In the eastern Mediterranean Sea we find two freshwater sources in direct juxtaposition to Miocene evaporites: one entrant through the Aegean Sea and another one in the Levantine Sea (see fig. 7). Precise paleogeographic interpretations of freshwater occurrences are still difficult. Very few areas around the eastern Mediterranean Sea have been studied in sufficient lithostratigraphic detail to separate limestones growing in small lakes on a continent or on an island, from freshwater limestones growing in the low-salinity Paratethys Sea or in its various freshwater arms, bays and straits. Both types of occurrences are referred to as lacustrine (limnic) limestones, since their fauna is a freshwater assemblage. Nevertheless, if all the freshwater limestones hitherto reported were indeed lacustrine, the Upper Miocene lake carbonates would be several orders of magnitude more frequent than lake limestones are today in similar climates. Lakes would have been as common as those in the boreal climate of the Finnish or Canadian Shields.

A freshwater flow through the Dardanelles deposited limestones along the Thrakian Black Sea coast (KESKIN, 1971), along both shores of the Dardanelles (KOPP, 1966, 1969), in the Balya region south of the Marmara Sea (AYGEN, 1956), on the island of Skyros (GREKOFF et al., 1967) and on the Attic Peninsula around Athens (GUERNET & SAUVAGE, 1970; v. FREYBERG, 1951).

Southward the Aegean waters remained fresh as far as a line between the coral reefs south of Athens (FUCHS, 1877) and the corals of eastern Crete (FREUDENTHAL, 1969; DERMITSAKIS, 1970). Westward the limit seems to have been the Gulf of Sérrai, the Peninsula of Cassandra (GILLET, 1957 a, b) and the west coast of the Gulf of Thessaloniki (CVIJIĆ, 1908 b).

The magnitude of a freshwater flow necessary to sustain a common water level in the Mediterranean Sea can be appreciated, if we look at the bottom sediments along the path of this flow, which are sandstones, greenish grey conglomerates of coral debris and of pre-Neogene material, and occasionally greenish grey shales. The current apparently touched Khios (BESENECKER et al., 1968), moved south through Naxos, eastern Paros and the Koufenisia Islands in the central Aegean Sea where sandstones abut against lithographic limestones (ROESLER, 1972). It then curved, only to be deflected southwestward by a peninsula composed of present-day Carpathos and Rhodes Islands, to pass over the Straits of Kasos swell to the immediate east of Crete. A strong current probably emanated from Kermes Bay and possibly another bay farther north near Izmir, since these were the mouths of major rivers (ERENTÖZ & ÖZTEMUR, 1961; ERENTÖZ, 1966).

Bedload clastics are exposed on Rhodes (CVIJIĆ, 1908), can be found in easternmost Crete (GRADSTEIN & VAN GELDER, 1971) where the current prograded into the quiet waters of the evaporite basin in a mainly southwesterly flow direction, possibly aided by a geostrophic longshore current from the northern Levantine Sea. In passing Crete the flow obviously had no bottom counter current in its final stage. Sedimentary structures betray a southwesterly flow direction; the current scoured the floor, judging by the poorly sorted and poorly rounded material in the upper parts of the sequence which even override coral reefs. The conglomerates correspond to highest current velocities in response to a growing water demand of a desiccating area.

It is of more than historical interest that CVIJIĆ (1908 b) postulated such a pre-Sarmatian drainage channel from the Dardanelles past Rhodes Island primarily on geomorphological grounds; he also observed current-produced Neogene gravel terraces along the western rim of the Gulf of Thessaloniki, nowadays cut at right angles by local drainage. Much of the sedimentological evidence from Aegean islands was then not yet available.

The freshwater flow into the Levantine Sea came through a second channel located in the Eastern Anatolian part of Turkey (ERENTÖZ, 1966) and suffered some pre-concentration by leaving behind gypsiferous beds in an embayment north of Iskenderum (ERENTÖZ & TOLUN, 1955) and in several other embayments that formed in intramontane basins connected to the channel, such as the Sivaş, the Erzerum and the Tekman-Karayazi basins (KURTMAN & AKKUŞ, 1971) and the Nakhichevan Basin of Soviet Armenia (STRAKHOV, 1962). Evaporites are thus confined to the northern shore of the channel. Gypsum predominates in the western embayments in Turkey (KURTMAN & AKKUŞ, 1971; ERENTÖZ, 1966) and halite and gypsum in the eastern ones in Armenia and Azerbaidzhan (STRAKHOV, 1962), an arrangement that is similar to the one along the north rim of the Pannonian-Dacian basin (see below) and is probably related to the southward tilt of any surface flow, to the gradual eastward concentration of any undercurrent.

In southeastern Anatolia we find 1,000—1,400 m of silty to conglomeratic clastics with boulders up to 1 m in diameter; locally they abut against coral and stromatolitic patch reefs (ROLOFF, 1972). The size of these clastics provides an indication of episodic current velocities in the Anatolian channel; again ultimately the flow was strong enough in a westward direction to cut out any bottom counter current in the terminal stage.

Within the Mediterranean Sea this water seems to have flowed westward along the Turkish south coast where several localities of freshwater limestones have been recorded (ERENTÖZ, 1966). The mouth of the Eastern Anatolian channel was laced by a reef chain (ROLOFF, 1972), the remnants of which are found today near Iskenderum (ERENTÖZ & TOLUN, 1955) and farther northwest in the Bolkardağ section of the South Anatolian Taurus Mountains (BLUMENTHAL, 1955). This reef chain must mark the boundary between brackish and hypersaline waters. In the modern Mediterranean Sea highest surface salinities occur along the Turkish south coast, thus an ancestral evaporite basin also would have circulated highest salinities into this area, unless deflected by a fast flowing current.

Water supplied through eastern Anatolia mixed Aegean waters, circulated the Levantine Sea; one branch swept into the Adriatic Sea, the other turned towards the Red Sea. This circulation is responsible for sea floor evaporites off southern Levantine coasts (GVIRTZMAN, 1969) or those in western Greece and Albania mentioned above. Gypsum in a small bay near Lapatso on Cyprus (HENSON et al., 1949) and gypsiferous reefal limestones fringing the island (ZOMENIS, 1972) belong into this group.

The thick gypsum sequence exposed on the banks of the river Euphrates and elsewhere in northern Syria (DUBERTRET et al., 1933) was derived from another larger body of water that covered portions of northern Iraq and had no direct communication with the Mediterranean region. It is probably related to the gypsum deposits known from cis-Caspian Iran (STRAKHOV, 1962), separated by a swell from the brackish waters extending south of the Caspian Sea.

Lower Miocene (Burdigalian) evaporites in souther Iraq were followed by

Upper Miocene continental clastics which were being dumped in copious quantities from the Zagros Mountains in the east. These thick sands and shales provided a barrier between Syria and the Persian Gulf. Later, Pliocene evaporite deposition again resumed in southern Iran (cf. ILLING, 1953). The evaporite occurrences in the Middle East as well gypsum deposits in the ancestral Borisfen Bay near the Dniepr River (SENEŠ, 1961) and on the Krasnodar Peninsula of the Azov Sea, on the southeastern shore of the Black Sea along the Rioni River, in the Ustyurt and Turkmen Plateaus on the eastern rim of the Caspian Sea, although indicated on fig. 6, are no longer part of the Mediterranean drainage region and thus beyond the scope of this paper.

The Pannonian — Dacian Basins

A separate example of freshwater supplying Upper Miocene evaporite basins is furnished by the embayments of the Pannonian-Dacian basins. Central Hungary was an Upper Miocene land mass surrounded by fresh or brackish waters, the Pannonian Sea. A connection existed to the fresh or brackish waters covering Rumania south of the rising Carpathian Mountains and Bulgaria north of the Balkans, the Dacian Basin. Some benthonic faunas otherwise endemic to the Mediterranean region are found in these areas in large numbers (GILLET, 1959) and suggest that this set of basins formed an integral, albeit enclosed part of the Mediterranean water circulation system.

A very substantial river flow must be postulated for an ancestral Danube River draining the steep slopes of the Alpine Foreland with lignite swamps in both Bavaria and Austria on the one side and on the Bohemian Massif on the other. Even this flow was not sufficient to keep the Vienna Basin a freshwater basin. Discontinuous gypsiferous beds occur at its northern rim (CICHA, 1961). Some islands southeast of Vienna (METWALLI, 1971; PANA & RADO, 1972) slowed the currents to clay flocculation and connected at least seasonally through vast mud flats to the Central Hungarian land mass with very common lignite swamps. Drainage down the eastern slopes of the Alps seems to have kept the Styrian Basin and opposite shores of the Central Hungarian land mass free of evaporite precipitation. The Styrian Basin had neither a sufficiently narrow nor a sufficiently shallow threshold to initiate evaporite precipitation. Low-salinity waters continued along the south coast of the Pannonian Basin where Bosnian shores were also covered with lignite forests. A study of reefal limestones led EREMIJA (1970) to conclude that the sea floor was tilted to the south.

On the shores of the Central Hungarian land mass we find coral reefs in small coastal bays (HÁMOR, 1970). These bays were probably too open to have a restricted circulation, e.g. Hasznos Bay in the Mátra Foreland (HAJÓS, 1968). Elsewhere the offshore areas were covered with sands and clays (JÁMBOR, 1967; SZÉPESHÁZI, 1968; METWALLI, 1971).

The volume of water generated on the western shores is further evidenced by the ability of the Danube gathering system to keep abreast of the uplift of the southern Carpathians occurring since the end of Miocene times, by cutting the gorge of the Iron Gate, todate remaining a dangerous set of rapids. CVIJIĆ (1908 a) described abandoned Miocene, Pliocene and Pleistocene terraces and speculated on the analogy between the ancestral Miocene straits in the Iron

Gate and the Bosporus with its bottom counter current. The present Danube River course gradually slides northward onto older coarse clastics deposited by ancestral currents and deltas.

The northern rim of the Pannonian and Dacian basins contains a number of individual evaporitic basins which appear to have progressively increased in salinity from west to east. The sub-Carpathian basins between the southern spurs of the West Carpathian Mountains contain gypsiferous beds in the west (SENEŠ, 1961), gypsum and halite in the east (SLÁVIK, 1971), suggesting that the bottom current concentrated in the latter direction. Volcanic debris pouring into the sea prevented the development of a continuous series of embayments along this coastline.

A potential connection between Vienna Basin and the Carpathian Foreland (SENEŠ, 1961, 1971) has not been drawn (along with BUDAY et al., 1965), because the fauna in the Sandomierz depression in Poland contains only Volhynian elements in the Tortonian stage, suggesting waters derived from the southeast (BRZEZIŃSKA, 1961). Thus the gypsiferous beds near Opava in Moravia (CICHA & SENEŠ, 1970), northeast of Cracow (KRACH, 1947) and along the margin of the Podolian Platform (NEY, 1969) could be the rim deposits of the halite occurrences mined east of Cracow at the towns of Wieliczka and Bochnia (POBORSKI, 1952) and in Upper Silesia west of Cracow (KWIATKOWSKI, 1972). Halite accumulated here only in the deepest depression centres with gypsum along the margins, in analogy to the situation in the El Melah lagoon in Tunisia. Occurrences of gypsum around the Transylvanian salt (BALTREŞ, 1968) with a reef rim (IONESU, 1973) are also analogous.

The evaporite basins of Zarand, Beius, Vad, Sylvania and Baia Mare in western Rumania may be a group of embayments extending into the Carpathian Ukraine, separated from halites in eastern Slovakia (as shown in fig. 7, after PAUCA, 1954). Alternatively, the evaporites in the Carpathian Ukraine may have had a westward rather than southward connection (after SENEŠ et al., 1971).

The Transylvanian Basin had no direct connection to the aforementioned bays (PAUCA, 1968) or had been connected around a chain of islands between Mureş and Sameş Rivers (DRAGOŞ, 1969). Whether the Transylvanian Basin thus received its bottom current through an island chain on its northwestern margin of from a gypsum bearing basin in southwestern Rumania leads to the same result: in either case pre-concentration of the bottom current had progressed to the point where halite but not gypsum were deposited; in either case the surface current moved through the Transylvanian Basin in a counterclockwise direction, as postulated by PAUCA (1968).

The southeastern entrant of the Transylvanian Basin is marked by very finely grained, laminated dolomites (PAUCA, 1968). Such dolomite laminae are known today only from extremely shallow environments. Their presence at the eastern entrant into the Transylvanian Basin suggests that the barrier for outflow of bottom currents was very effective, the threshold depths shallow enough to reach at least seasonally into the inflowing current, and the Mg/Ca ratio favourable through earlier depletion of calcium ions. The latter argument favours an out-flowing bottom current at this point; one could argue the same on the basis of absent coral reefs otherwise so ubiquitous in the system.

Gypsum was deposited even in narrow, shallow valleys leading into south rim

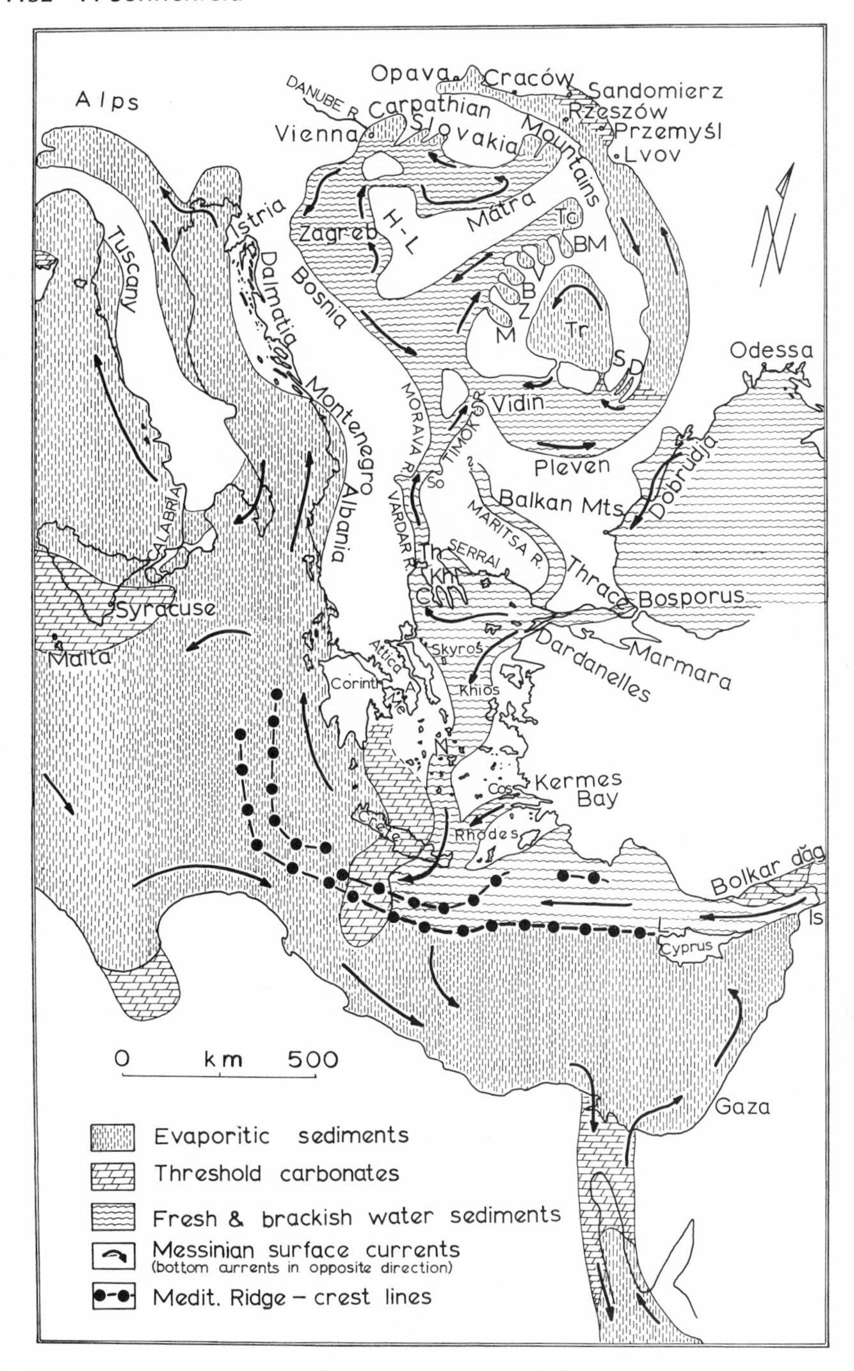

Fig. 7. (Legend see p. 1153)

of the Eastern Carpathian Mountains, such as the Slanic and Drajna valleys (PHILIPESCU, 1968). This is in sharp contrast to the valley on the opposite shores, on the north slopes of the Balkan Mountains or the Dinarides where only freshwater sediments can be found. Likewise, bottom waters are more concentrated along the north shore of the Dacian Basin than along the south shore of the Central Hungarian land mass farther west.

The easternmost Volhyno-Podolian Basin seems to have obtained the pre-concentrated effluent from the Transylvanian halite basin. This explains why potash and halite occur exclusively along the western margin of the basin. A branch of the surface current flowing eastward along the southern flank of the Dacian depression evidently entered this basin and kept the eastern margin less saline than the western one. A continuous chain of reefal limestones occur as far inside the basin as the town of Frampol north of Przemyśl (HARASIMIUK et al., 1969) and require a continuous influx of nutritious plankton. The water was saline enough to deposit gypsum all along the eastern shores (PISHVANOVA & TKACHENKO, 1971) and also on the west shore (BIGU, 1965). Only local areas of reefs are known on the west side, possibly related to drainage from the north slopes of the Carpathian Mountains, e. g. near the town of Rzeszów (AREŃ, 1957). A nearby freshwater inlet from the sea is not indicated by these reefs, merely the influx of some slightly less concentrated brines. Occasional traces of halite along the eastern margin south of the cities of Lvóv and Przemyśl (PISHVANOVA & TKACHENKO, 1971; GARLICKI, 1973) prove that the salinity was quite high. Again a progression is evident from gypsiferous basins in western Rumania to halites and gypsum in Transylvania and to gypsum and halites with potash beds in Volhynia in the east.

Source of Pannonian waters

So far, most studies assume a direct eastern source of Pannonian waters. However, the waters depositing halite and potash salts require considerable pre-concentration in transit basins, considerable exposure to evaporation. For such a sequence one is not very likely to find an entrant of slightly brackish or fresh water supplies through a gap north of Odessa (GILLET, 1963) or even farther south along the Rumanian Black Sea coast (DRAGOŞ, 1969) opposite the Transylvanian salt basin (PAUCA, 1968). An evaporite basin directly facing such a freshwater entrant would have exerted a considerable suction effect with resulting dilution of basin waters. The Tortonian beds in the coastal Dobrudja (CHIRIAC, 1970) appear to be grading from neritic deposits at the Black Sea coast to Ostrea-bearing littoral material towards their northern depositional limit, but also

---

Fig. 7. Upper Miocene flow channels and sediment distribution in the eastern Mediterranean region (generalized picture). Arrows indicate surface flow; bottom currents as a rule move in the opposite direction. A: Athens; B: Beiuş; BM: Baia Mare (Someş); C: Cassandra Peninsula; H-L: Central Hungarian land mass; Is: Iskenderum; L: Lapatso; M: Mureş River; Me: Gulf of Megara; N: Naxos Island; P: Paros Island; S: Sylvania basin; S-D: Slanic and Drajna basins; So: Sokobanja basin; Tc: Transcarpathian basin; Th: Thessaloniki; Tr: Transylvanian basin; V: Vad basin; Z: Zarand basin.

westward towards the present-day Danube River knee, indicating a westward shoreline and thus a separation from the Volhynian and Dacian Basins.

All evaporite basins in the Pannonian-Dacian depression are facing south, the very small ones to the west. We must conclude that the geostrophic counter current was depressed on the south side of these basins and was thus easily flushed out of any embayments with low enough threshold. On the north shores this current was reaching to higher levels and could enter local bays. The eastern or southeastern promontory of a bay would then deflect part of the passing current into a bayward eddy of low velocity and thus longer exposure to water loss. The small bays in western Rumania, if drawn according to PAUCA (1954), opened northwestward and had their southeastern promontory extending farther into the current than the northeastern ones. The presence of a highly concentrated bottom current along the northern margin of the Dacian Basin or of the Pannonian Basin northeast of Budapest, under concurrent movement of low-salinity waters along the south shores of the same basins can probably be explained by the poor mixibility of currents of different salinity (RICHTER-BERNBURG, 1970).

The whole of the southern rim of the Pannonian and Dacian Basins was within the domain of low-salinity waters. Consequently, it is here that any source of supply must be located: Limestone occurrences in central Macedonia (MERCIER & SAUVAGE, 1963), along the western margin of Khalkidiki Peninsula (GILLET, 1957) and thence through the Vardar depression (BRUNN, 1956) would suggest a strait here leading north. ČIČULIĆ (1964) and STEVANOVIĆ (1964) suggested just that on the basis of early Pliocene marshland distribution. The Timok Basin and the Sokobanja Basin were fed by northward flowing waters from the Morava River area that were supplying nutrients for local limestone growth (NOVKOVIĆ et al., 1970; DŽODŽO-TOMIĆ, 1963). The Morava River is the direct northward extension of the Vardar depression.

Such an original northward flow direction from the Aegean Sea through the Vardar and Morava valleys is supported by the observation that today most tributaries of the Vardar River make a sharp northward turn before reaching their mouts, rather unusual for a southward moving main river.

The crucial position of a small island south of the present Iron Gate (ČIČULIĆ, 1964), formed by present-day Timok, Morava and Danube River valley, served as deflector of freshwater currents into the Dacian and into the south-Pannonian basins. The Coriolis force would have merely turned the entering current to the right. At each swell the main current then split off a subsidiary northward branch, leaving lower salinity values always to the right side of an observer looking in the direction of the current. This is in strict analogy to present-day conditions.

Benthonic fauna in adjacent northwestern Bulgaria migrated south and west (KOYUMDZHIEVA, 1969), evidently along with the bottom counter current. This bottom current explains the presence of Volhynian forms at Sérrai in Greece and thence in Tuscany and Sicily (GILLET, 1961). It also explains why ostracods on the island of Skyros in the northern Aegean Sea belong to an assemblage known from the Vienna Basin as well as from the Greek islands of Kos and Crete farther south (GREKOF et al., 1967). Similarly, the individual interconnected tectonic depressions west of Athens carry an endemic fauna with Dalmatian and Bosnian affinities (KÜHN, 1951) known from the eastern Adriatic and south-Pannonian shores. A case could also be presented for a northward strait along the Maritsa River in Bulgaria where BONČEV (1896) and NIKOLOV (1971) found freshwater limestones exposed. Documentation of this channel or bay is, however, rather scanty.

KOYUMDZHIEVA (1969) estimated an average salinity of 16—18 ‰ for the Vienno-Pannonian and pre-Carpathian Dacian basins, compared to 14—15 ‰ for the Crimeo-Caucasian basins. Higher Tortonian beds in eastern Slovakia were laid down in waters ranging from 18—20 ‰ to 25 ‰ (BUDAY et al., 1965), no more than half the values of normal sea water, but much higher than present salinity values supplied by the Caspian Sea to the Gulf of Karabugaz. In the gypsiferous bays this could rise to 117 ‰ (KWIATKOWSKI, 1972). Climatic oscillations produced variations in the salinity and the marine series at the Yugoslav-Bulgarian border was periodically sweetened (POPOVIĆ & GAGIĆ, 1969). At the village of Gaïtanzi near Vidin, northwestern Bulgaria, TRASHLIEV (1971) found gypsum that seemed to be derived from the interface between a brine and a freshwater influx; he associated (TRASHLIEV, 1969) carbonate deposition with the rainy winter and spring, gypsum precipitation with the dry summer and autumn. STURANI (1973) postulated a similar seasonality for the laminated gypsum beds in the Piedmont of Italy. If the gypsum depositing basin at Gaïtanzi was enclosed on the north, east and south sides, it must have received its bottom current from the Timok channel. The current circulation was fresher when stronger in wet periods and later acquired a marine aspect as the water supply was progressively reduced. ATANASOV et al. (1971) find farther east in the Pleven area a progression form lacustrine beds to transitional calcareous sandstones to marine limestones, a gradual upward increase in salinity. STANCU et al. (1971) finds coeval sandstones with fresh and saltwater faunas in the Dacian Basin north of Pleven.

In Upper Messinian times the Pannoian-Dacian basins appear to have been as much an appendage of Mediterranean water circulation, as were the Balearic, Adriatic and Red Seas.

Aspects of Upper Miocene climate

The ubiquitos occurrence of lignite beds indicates higher rainfall figures than are common in the area today. STRAKHOV (1962) has shown that Mesozoic and Cenozoic coal deposits are predominantly derived from forests of temperate climates. Their juxtaposition to evaporite deposits appears at first sight to be a problem similar to the one of freshwater deposits intercalating evaporites. CVIJIĆ (1908 b) even noted gypsum crystals in fractures of Miocene lignites at Sérrai.

The lower Upper Miocene is the most important coal-forming phase of western Anatolia (BENDA, 1971). Coeval lignites are known from coastal areas of western Turkey, from the gypsiferous Sivaş Basin, from Crete, mainland Greece and Macedonia, from the southern slopes of the Transylvanian Alps and the Central Hungarian land mass, from the Alpine Foreland of Austria and Bavaria, from central and northern Poland, a span of at least 20° in latitude. From Spain (DE BRUIJN, 1973) to Turkey (BRENDA, 1971) they span 30° in longitude. To map their distribution would be a separate undertaking: it would elucidate important paleoclimatic trends.

Much of the mothern portion of the evaporite region must have had a climate humid enough to support the deciduous forests that are the precursors of the lignite beds. The shorelines could not have been in barren desert or semi-desert regions. Instead, the climate was dry enough to sustain large-scale evaporation over water bodies and seasonally wet enough to replenish the groundwater in order to sustain the extensive forests and local grasslands in the dry period; in other words, it was a Mediterranean-type climate with a very sharp contrast between a long dry summer and a short, very wet winter, not unlike the present climate of northern California or of present-day forest regions in the interior of Macedonia. It is well known that the absence of the more accessible Illyrian (Montenegro-Albania) and Macedonian forests is not a natural phenomenon, but a consequence of Roman naval engineering needs.

A closure of the Straits of Gibraltar would presently lead to salinity concentration even if rainfall and runoff rates were doubled. Upper Miocene rainfall rates could, thus, quite easily have been higher. Higher rates of precipitation mean higher cloud cover and lower radiation, lower rates of evaporation in winter on land. It is thus not surprising that upon close examination we find the lignites predominantly along those bays and their respective hinterlands where the shores were washed by brackish or by freshwaters. The absence of evaporites along the eastern margin of the Alps has been mentioned; the lignite deposits flank only that part of the Pannonian Basin which contains the freshwater, or in those parts of the Balkan Peninsula and of Asia Minor, where runoff ended in the freshwater section of the present Aegean Sea, the Black Sea or the northern Levantine Sea. The significant surplus of water to be drained from these swamps reduced the salinities in the open waters. Yet it could not counterbalance evaporation losses sufficiently to prevent evaporite deposition.

Evaporation losses at sea are higher in winter: Cool air masses moving south are warmed up and increase their water carrying capacity; warm summer air coming from an overheated continent onto cooler sea does not increase its water carrying capacity correspondingly. Upper Miocene mean annual air and water surface temperatures near the head of the Volhyno-Podolian embayment have been estimated at 19° C, bottom temperatures under 15° C, to account for glauconitic occurrences (Kwiatkowski, 1972). Only a very slight northward shift of present isotherms would be necessary to achieve such figures.

Final desiccation

The amount of water flowing through the Bosporus and through the entrant at Iskenderum would not have been enough to sustain a common water level throughout the Mediterranean region enlarged by the Pannonian-Dacian and Red Sea basins as well as by numerous smaller (Po, Rhône, Ebro, Calatayud-Teruel etc.) depressions. Reefs in the Vera Basin of southeastern Spain are regressive eastward; the westernmost ones are covered by continental debris (Völk, 1964). This would suggest that waters were retreating there into the interior of the Balearic Sea and at least some of the postevaporite hiatus is a consequence of the inadequate water supply. It follows that the hiatus covers the largest time span in the west; the upper parts of multiple evaporite events recorded in Poland, Slovakia, the Suez region, Syria and elsewhere in the east, actually correspond in time to this hiatus. SLÁVIK (1971) assigned only early evaporites to the Badenian-M_4c stage, the equivalent of Messinian beds; the upper evaporite sequence he considered lower Sarmatian in age. In studying Macedonian outcrops southwest of Thessaloniki, GILLET & FAUGERES (1970) observed that the receding Lower Sarmatian sea produced a temporary separation of Pannonian and Black Sea waters. Final desiccation occured significantly later at the Greek than at the Spanish coasts.

Towards the end of the Miocene a new direct connection of the Pannonian Basin with the Black Sea in Bessarabia west of Odessa brought a Sarmatian transgression of freshwater into the Volyno-Podolian and Dacian-Pannonian basins. This water surplus produced an evident current reversal in the Timok-Vardar depression; the Pliocene current is here directed southward. CVIJIĆ (1908 a) suspected such a current reversal also through the Iron Gate when studying generations of Miocene and Pliocene terraces. PAUCA (1972) assumed a final Upper Pliocene draining of the Transylvanian Basin to the west.

Whereas latest Miocene and lowermost Pliocene beds are absent in the western Mediterranean (NESTEROFF, 1973), lower Pliocene (Pontian stage) faunas collected near Athens and Tessaloniki are identical to those known from the Odessa area. The same is also true of lower Pliocene faunas from Corsica, the Rhône valley, the Plains of Roussillon and Catalania (GILLET, 1957 a). Pannonian and Aegean faunas moved east: EBERZIN (1945, quoted by GILLET) considered the Russian early Pliocene (Pontian stage) fauna to have an Aegean origin; possibly all the eastern fauna had their first cradle in the Pannonian Basin (GILLET, 1957 a). A Pliocene transgression of Atlantic waters through a reopened Straits of Gibraltar would, indeed, cause a migration of brackish faunas from the shores of the Balearic Sea to the Aegean and Black Seas.

Tectonics and paleogeography

The Mediterranean region has been mobile throughout the Neogene. Both vertical and horizontal movements of great magnitude have taken place. BURKE, KIDD & WILSON (1973) assumed a southeasterly movement of western Europe under concurrent slight anticlockwise rotation: Whether Africa is unterthrusting Europe (GALANOPOULOS, 1970) or Africa a stationary plate (BURKE & WILSON, 1972) and Europe overriding it, may be a moot point in the present context. Large parts of the circum-Mediterranean terrain have, no doubt, been subjected to compressional stresses between the two continents (see fig. 8).

Orogenic stresses on the African Plate concentrated mainly in the northwest, spilling over into the Betic portion of Iberia. BUROLLET (1971) found that a succession of Pliocene and younger vertical movements in Tunisia reached 3,000 m. In northern Morocco the lateral compression resulted in southward overthrusting (CHOUBERT et al., 1961, 1964, 1966). Vertical differences in levels of Miocene beds in the Sierra Nevada of Spain are of the order of 4,000 m (GLANGEAUD et al., 1967).

The Mediterranean Ridge, a rugged submarine elevated area rising over 2,500 m in its highest points between Hellenic Trough and Nile Abyssal Plain is now dated as mid-Pleistocene or younger (HSÜ et al., 1973). If all of the ridge is that young, it represents a vertical uplift of about 0.27 cm/year. The relative

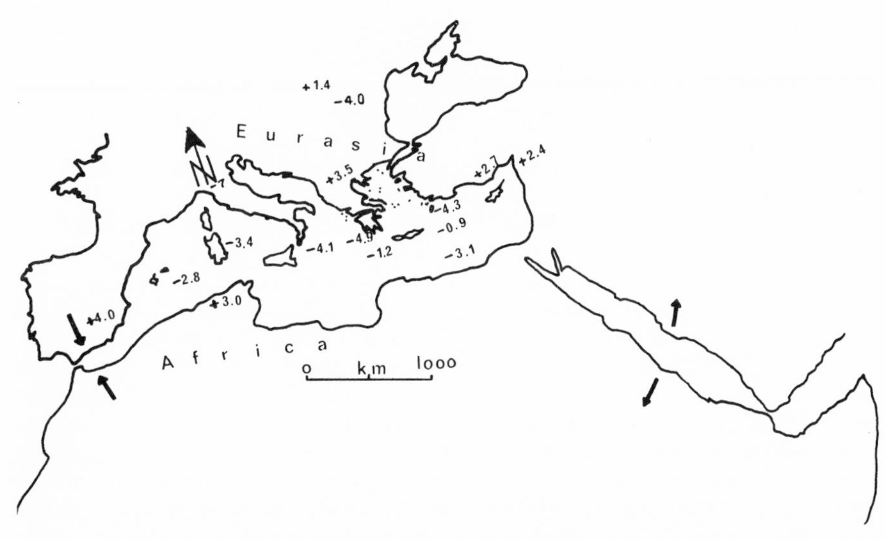

Fig. 8. Anticlockwise rotation of Eurasia leads to tension in the Red Sea and to compression in the Alboran Sea near Gibraltar. Additional compression occurs in the eastern Mediterranean Sea. Some significant post-Miocene vertical displacements are given in km.

elevation of the Ridge southwest of Crete reaches 3,760 m which would represent 0.40 cm/year. It follows that some of the downbuckling of the abyssal plains on either side of the Ridge is of the same vintage.

In the Amanus Mountains (Gâvur Naglari) north of Syria, a post-Miocene uplift (ROLOFF, 1972) on strike with the Mediterranean Ridge, the Neogene beds plunge steeply to the south (70—75°), but dip only gently (15—20°) to the north (ERENTÖZ & TOLUN, 1955). Similarly, Upper Miocene marine beds in the Taurus Mountains of southern Turkey have since been uplift to 2,300—2,700 m above sea level (BLUMENTHAL, 1955). In the Vardar area, Neogene vertical movements of up to 3,500 m were completed before the Pleistocene (ZAGORCHEV, 1970) and the connection between Balkan Mountains and the Carpathians is post-Miocene in age. The Aegean Sea has achieved its present configuration only in the last interglacial (GILLET, 1957 a; FEDOROV, 1967).

LEPICHON (1968) has suggested that plate movement may be episodic rater than continuous. Ross et al. (1973) finds that sea floor spreading resumed in the Red Sea only about two million years ago, following a period of quiescence extending into the Miocene. The episodic compression should correspond to episodic orogenic impulses on land.

The importance of the tectonic history for the evaporite formation is obvious. To bring in initially enough water to cover the whole area of interconnected basins requires a broad entrant. Support for such an idea commes from the apparent width and magnitude of the current flowing past Crete and from abandoned terraces in the Dardanelles. CVIJIĆ (1908 b) believed the Bosporus to have been three times its present width.

Even if the current only had to freshen a pre-existing Mediterranean Sea, it had to be strong enough to convert the pre-evaporite seas to brackish waters as far as the Balearic Sea. Despite a dry climate it had to freshen evaporitic bays periodically in Tuscany and it had to maintain a common water level throughout the Mediterranean and adjacent seas in the face of rising evaporation losses. In contrast, present-day influx currents are not strong enough to decrease salinities or to maintain normal oceanic salinities everywhere in the Mediterranean Sea alone.

An entrant of required width is not evident in the present configuration of either the East Anatolian channel or the Dardanelles. Either the Bosporus and the Dardanelles have to be palinspastically widened prior to Walachian (mid-Pleistocene), Rhodanian (end-Miocene), Attic (late Miocene) or Latest Styrian/Moldavian (upper Miocene) orogenic phases, or for similar reasons the entrant in eastern Turkey was wider than present knowledge permits us to conclude. At least in the northern Aegean region the Attic phase is associated with a retreat of the Mediterranean Sea (GUERNET & SAUVAGE, 1970).

A southward motio of Crete averaging 2.6 cm/year has been suggested by LEPICHON (1968). Depending on how continuous or episodic the movement has been, this can represent 50—130 km since the end of Miocene. This southward movement affected the position of the Mediterranean Ridge, the width of the Bosporus, the configuration of the northeastern Mediterranean region.

The southeastward motion of Spain is calculated to be 1.2 cm/year according to BURKE, KIDD & WILSON (1973) whereas LEPICHON (1968) suggested 1.9 cm/year, somewhat less than applies to Crete. The land bridge between Africa and Iberia was much wider in Upper Miocene times than the Straits of Gibraltar are today, possibly at least six to ten times as wide. As this assumed convergence continues, the Straits must be getting narrower: they are today only 12.5 km wide at their narrowest point between Punta Oliveros and Punta Cios. If the movement continues unabated, they could be closed in 0.65×10^6 years. The sea floor in the Straits, now 320 m below water at its shallowest point, would respond to this compression in the intervening period by further buckling and would thus commence to interfere with in-and outflowing currents much before that time. This interference may already have commenced: In the Gulf of Lyon coarse grained coldwater deposits are overlain by fine grained recent sediments. Immigrant Atlantic species disappear in higher beds and at the same time the salinity of interstitial waters is increasing (GADEL & MONGIN, 1973). This indicates a recent increase in aridity; the Messinian stage is also marked initially by a replacement of oceanic benthonic faunas by endemic ones.

A ten per cent reduction in the cross sectional area of the Straits achieves a corresponding increase in the velocity of the inflowing current, a much more substantial drag on the outflow. A change in radiation leading to a ten per cent change in rates of evaporation and concomitant changes in precipitation and runoff lead to an almost thirty per cent change in the water throughput in the Straits per unit of time. A consequence of accepting a convergence of Iberia and Africa by plate motion is thus the prediction of a repetition of Upper Miocene evaporitic event in the geologically near future. However, oscillations in climatic factors could more than easily counteract or reinforce this trend.

The interglacial periods of climates warmer than recorded at the present time have not left a record of salinity maxima in the sedimentary sequence that would be readily evident. The water inflow of the Straits of Gibraltar was adequate to flush out any high-density cells that would lead to increased salinity of interstitial waters in bottom sediments. It appears to follow that the restriction of the intake was never as severe as it is today, if we consider the history since the great Pliocene transgression.

A compression of the Mediterranean region by the convergence of the Eurasian and African Plates does not only produce a positive response of the continental mass, but also a negative response or downbuckling on the margins. Deep canyons of early Pliocene age on the Mediterranean coasts of France and Spain led DENIZOT (1951) to assume a Mediterranean sea level then lowered by at least 1,000 m, but DULEMBA (1970) rejected the idea of submerged excavated valleys for canyons on the west coast of Corsica.

A river canyon near Gaza in Palestine entered the Levantine Sea some 40 km east of the present shore line and was partially filled with earliest Miocene sediments (GVIRTZMAN, 1969). The canyon thus predates the Messinian evaporitic event, but remained a progressively shallower topographic feature into early Pleistocene times. HSÜ et al. (1973) note a deep Nile canyon upriver from the present Nile mouth filled with Pliocene sediments. Supply of fill material has increased in late Neogene times.

DICESARE et al. (1963) found Upper Miocene or Pliocene southeastward drainage in north-central Libya while BARR & WALKER (1973) noted it reactivated in the opposite direction. This is in line with the buckle and swell topography suggested by BURKE & WILSON (1972) as a consequence of plate motion. The cataracts of the Nile and the ruptured courses of Zambesi, Orange, Congo and Niger Rivers demonstrate similar Plio-Pleistocene buckling of Africa.

A canyon under the mouth of the Loire River (DENIZOT, 1952) and marine Pliocene beds cored in the subsurface of the Rhineland some 170 km from the present coast (OPPENHEIM, 1917) suggest downbuckling also along the Atlantic coasts. Numerous submarine canyons traceable from the Turkish Black Sea shelf into the central deep sea flat (ERINÇ, 1973) are early Pliocene to late Pleistocene in age.

Mediterranean canyons are thus no proof of an ancient desiccated deep basin (HSÜ et al., 1973) coexistent with Upper Miocene evaporite basins on the Eurasian continent, but part of a world-wide Plio-Pleistocene canyon development stretching from the Hudson River of New York to the Java Sea (KUENEN, 1950). They are a response to renewed plate motion.

Conclusion

The Mediterranean Sea is today an evaporitic basin that makes up its evaporation losses by compensatory influx through the Straits of Gibraltar. The higher density bottom outflow flushes excess salinities and keeps out waters upwelling in the Canary Current along the Atlantic coast.

In cooler periods, such as during Pleistocene glacial advances and during the Pliocene, the Mediterranean Sea had a water surplus like most temperate confined seas. A surface outflow was matched by a bottom influx and this bottom current sucked in the deep Atlantic waters rising in the slow upwelling of the Canary Current.

If we bracket the individual Upper Miocene evaporitic phases, we find that they encompass an interconnected system of evaporitic basins stretching from the western Mediterranean Sea to the eastern shores of the Caspian Sea, from the Carpathian Foreland to Yemen.

Several reasons speak against a connection to the Atlantic Ocean. Everywhere the evaporites are in juxtaposition laterally, underneath and above, to freshwater deposits. A connection to an ocean would have supplied normal marine salinities to some sediments. Bottom dwelling faunas would have been oceanic and not endemic or brackish Paratethys types. Threshold sediments separating freshwater and evaporite deposits are only exposed in few places, are either reefal or fine grained laminated carbonates, occasionally clastics. The source of the low-salinity water lies in the Black and Caspian Seas from where it was channelled through the Dardanelles and through an East Anatolian connection. The supplied water quantities were not sufficient to prevent an eventual drying up of large areas in the west and south. Current distribution was not dissimilar to present conditions and the present Mediterranean Sea can serve as a model.

An evaporitic basin is produced by the interaction of geologic and climatic factors. Both have been working in late Quaternary times towards a revival of evaporite conditions.

Acknowledgements

Special thanks are extended to Dr. C. G. van der Meer Mohr for many helpful suggestions and to Geological Institute, University of Leiden, Netherlands, for the use of their facilities.

Selected bibliography

Papers used in compiling figures 6 and 7 are marked with an asterisk

* Alexandrowicz, S. W.: Regional stratigraphy of the Miocene in the Polish part of the Fore-Carpathian Trough. — Acta Geol. (Acad. Sci. Hungar.), **15**, 49—62, Budapest 1965.

* Anić, D.: Das Alter der braunkohlenführenden Ablagerungen von Bosnien, der Herzegovina und Dalmatien. — Geol. Vjesnik, 2—4, Zagreb 1951.

* Areń, B.: Facies of Lower Sarmatian in the Carpathian Foredeep. — In: Znosko, J.: Geological Atlas of Poland, Geol. Inst., Warsaw, 1968.

* —: Upper Miocene facies in Poland. — In: Geological Atlas of Poland, Stratigraphic and Facies Problems, Fasc. 11, Geol. Inst., Warsaw 1957.

* Atanasov, G., et al: Caractère litho-stratigraphique et faciès du Sarmatien près de Pleven — God. Univ. Sofia, Fak. Geol. Geogr., **63**, (1), 191—216, Sofia 1971.

* Aubouin, J., & Darcourt, J.: Sur la géologie de l'Egée: regard sur le Dodècanèse méridional (Kasos, Karpathos, Rhodos). — Bull. Soc. Géol. France, Sér. 7, **12**, (3), 455—472, Paris 1970.

* Auzende, J. M., et al.: Upper Miocene salt layer in the western Mediterranean basin. — Nature, **230**, 82, London 1971.

* Aygen, T.: Étude geologique de la région de Balya. — Publ. Maden Tetkik Arama Enst., Ser. D, No. 11, 95 pp., Ankara 1956.

* Bajanik, S., & Salaj, J.: Données nouvelles sur la zone de cicatrice de la région d'Oued Zarga (Tunisie septentrionale). — Tunis. Serv. Géol. Notes, **32**, 3—23, Tunis 1971.

Baltreş, A.: Sur certains gypses diagénétiques du Sarmatien de la partie NE du bassin de la Transylvanie. — Dări de Seamă ale Şedinţelor, **55** (1), 230 (Abstr.), Inst. Geol. Romania, Bucarest 1968.

Barr, F. T., & Walker, B. R.: Late Tertiary channel system in northern Libya and its implications on Mediterranean sea level changes. — In: Ryan, W. B. F., Hsü, K. J., et al., Initial Reports of the Deep Sea Drilling Project, Govt. Print. Off., **13**, Pt. 2, 1244—1250, Washington 1973.

Benda, L.: Grundzüge einer pollenanalytischen Gliederung des türkischen Jungtertiärs (Känozoicum und Braunkohlen der Türkei, 4). — Beih. Geol. Jb., **113**, 45 pp., Bundesanst. f. Bodenforsch., Hannover 1971.

Benson, R. H., & Sylvester-Bradley, P. C.: Deep-sea ostracodes and the transformation of ocean to sea in the Tethys. — Bull. Centre Rech. Pau. — S.N.P.A. suppl., **5**, 53—91, Pau 1971.

* Besenecker, H., et al.: Geologie von Chios (Ägäis). — Geol. et Paleont., **2**, 121—150, Marburg 1968.

* Besenecker, H., & Otte, O.: Zur postalpidischen Sedimentation und Tektonik in der Ost-Ägäis. — Z. Dtsch. Geol. Ges., **123**, 527—539, Hannover 1972.

* Bigu, G.: Variaţiile litofaciale ale formaţiunilor tortoniene din nordul Moldovei in comparaţie cu acelea din partea de vest a URSS şi R.P. Polonia. — Soc. Ştiinţe Natur. Geogr. Rom., Comun. Geol., **3**, 219—228, Bucarest 1965.

Bizon, G., & Bizon, J. J.: Microfaunes planctoniques du Paléogène supérieur et du Néogène marins en Grèce occidentale. — Giorn. Geol., Ser. 2, **35** (2), 313—330, (1967), Bologna 1968.

Blumenthal, M. M.: Geologie des Hohen Bolkardag, seiner nördlichen Randgebiete und westlichen Ausläufer (Südanatolischer Taurus). — Publ. Maden Tektik Arama Enst., Ser. D, No. 7, 169 pp., Ankara 1955.

Bogdanova, A. K.: Particularités de l'échange d'eau par la Bosphore. — Rapports et Proces-verbaux des Réunions, Comm. Int. Explor. Scient. Mer Méditerr., **19** (4), 693—696, Monaco 1969.

* Bončev, S.: Das Tertiärbecken von Haskovo (Bulgarien). — Jb. k. u. k. Geol. Reichsanst., **46**, 309—384, Vienna 1896.

Brunn, J. H.: Contribution à l'étude géologique du Pinde Septentrional et d'une partie de la Macédonie occidentale. — Ann. Géol. Pays Hellèn., **7**, 358 pp., Athens 1956.

Brzezińska, M.: Miocen z pogranicza Roztocza zachodniego i Kotliny Sandomierskiej. — Biul. Inst. Geol. Polska, **158**, 5—111, Warsaw 1961.

Buday, T., et al.: Miozän der Westkarpaten. — Geol. Inst. D. Štúra, 293 pp., Bratislava 1965.

—: Die Stellung der Westkarpaten in der Paratethys. — Cursillos y Conf., **9**, 109—116 (1961), Inst. „Lucas Mallada", C.S.I.C. (España), Madrid 1964.

Burke, K., Kidd, W. S. F., & Wilson, J. T.: Plumes and concentric plume traces of the Eurasian Plate. — Nature Phys. Sci., **241** (111), 128—129, London 1973.

Burke, K., & Wilson, J. T.: Is the African Plate stationary? — Nature, **239** (5372), 387—390, London 1972.

Burollet, P. F.: Remarques géodynamiques sur le Nord-Esa de la Tunisie. — C. R. Somm. Soc. Géol. France, 8 (Nov. 22, 1971), Paris 1971.

—: Remarques sur le Néogène de Tunisie et de Libye et ses relations avec les bassins voisins. — Cursillos y Conf., **9**, 199—202 (1961), Inst. „Lucas Mallada", C.S.I.C. (España), Madrid 1964.

Burollet, P. F., & Byramjee, R. S.: Upper Miocene evaporites of the Mediterranean region — the land record. — Europ. Geophys. Soc., First Mtg., Sept. 25—29, 1973, Abstr., p. 18, Zürich 1973.

CASANOVAS CLADELLAS, M. L.: Contribución al estudio del Mioceno del Penèdes (sector Gelida). — Acta Geol. Hispan., **7** (5), 143—148, Madrid 1972.

CHINTA, R.: Date noi asupra faunei tortonian-sarmaţiene din vestul Depresiunii Transilvaniei. — Studii şi Cercetari Geol., Geofiz., Geogr., Ser. Geol., **18** (1), 283—288, Bucarest 1973.

CHIRIAC, M.: Raspîndirea şi faciesurile tortonianului în Dobrogea de Sud. — Dări de Seamă ale Şedinţelor, **56** (4), 89—112, Bucarest 1970.

CHOUBERT, G., et al.: Le prétendu « pliocéne de Charf el Akab près de Tanger est d'âge tortonien ». — Notes et Mém. Serv. Carte Géol. Maroc, **198** (27), 29—33, Rabat 1966.

—: Le Néogène du Bassin de Melilla (Maroc Septentrional) et sa signification pour définir la limite mio-pliocène au Maroc. — Com. Medit. Neogene Stratigr., Proc. 3rd Sess., Berne (1964), 238—249, E. J. Brill, Leiden 1966.

—: Stratigraphie et micropaléontologie du Néogène au Maroc septentrional. — Cursillos y Conf., **9,** 229—257 (1961), Inst. „Lucas Mallada", C.S.I.C. (España), Madrid 1964.

CICHA, I.: Stratigraphical problems of the Miocene in Europe. — Rozpravy, **35,** 134 pp., Ústř. Ústav Geol., Praha 1970.

—: Zur Oligozän-Miozän-Grenze und zur Stratigraphie des Miozäns der Westkarpaten und seiner nomenklatorischen Vereinheitlichung. — Geol. Práce, **60,** 113 —135, Bratislava 1961.

CICHA, I., & SENEŠ, J.: Stratigrafické členění mladšího terciéru Západních Karpat a jeho problematika. — Sbornik Geol. Věd, **18,** 101—123, Praha 1970.

* ČIČULIĆ, M.: Über das mögliche Vorhandensein einer zweiten transkarpathischen Verbindung zwischen dem pannonischen und gethischen Becken der Paratethys während des oberen Miozäns. — In: Comm. Medit. Neogene Stratigr., Proc. 3rd Sess. Berne (1964), p. 200—205, Leiden (E. J. Brill) 1966.

CITA, M. B., & RYAN, W. F. B.: Time scale and general synthesis. — In: RYAN, W. F. B., HSÜ, K. J., et al., Initial Reports of the Deep Sea Drilling Project, **13,** Pt. 2, 1405—1415, Govt. Print. Off., Washington 1973.

CVIJIĆ, J. (a): Entwicklungsgeschichte des Eisernen Tores. — Petermann's Mitteil., Erg.-H., **160,** 218 pp., Gotha 1908.

— (b): Grundlinien der Geographie und Geologie von Mazedonien und Altserbien. — Petermann's Mitteil., Erg.-H., **162,** 392 pp., Gotha 1908.

DE BRUIJN, H.: Analysis of the data bearing upon the correlation of the Messinian with the succession of land mammals. — In: DROOGER, C. W. (ed.), Messinian events in the Mediterranean, Kon. Nederl. Akad. Wetensch., Geodynamics Sci. Rep. No. 7, North-Holland, Amsterdam 1973.

* DECIMA, A., & WEZEL, F. C.: Late Miocene evaporites of the Central Sicilian Basin, Italy. — In: RYAN, W. F. B., HSÜ, K. J., et al., Initial Reports of the Deep Sea Drilling Project, **13,** Pt. 2, 1234—1240, Washington 1973.

* DE FRANCESCO, A., & TEDESCHI, D.: Note stratigrafiche sul Miocene superiore della pianura padana. — Giorn. Geol., Ser. 2, **35** (3), 341—356 (1967), Bologna 1968.

* DEMARCQ, G.: Un example de Tortonien terminal marin: celui de la région de Murcia, dans le Levant espagnol — Comm. Medit. Neogene Stratigr., Proc. 3rd Sess., Berne (1964), 254—266, Leiden (E. J. Brill) 1966.

* DERMITAŞH, E., & PISONI, C.: The geology of Ahlat-Adilcevaz area (north of Lake Van). — Bull. Maden Tetkik Arama Enst., **64,** 24—39, Ankara 1965.

DENIZOT, G.: Le Pliocène dans la vallée du Rhône. — Rév. Géogr. Lyon., **27** (4), 327 —357, Lyon 1952.

—: Les anciens rivages de la Méditerranée française. — Bull. Inst. Océan., No. 992, 56 pp., Monaco 1951.

* DERMITSAKIS, M. D.: Geologikai epeunai epi ton Neogenous tis eparkhias Ierapetras nisou Kritis. — Ann. Géol. Pays Hellèn., **21,** 342—484, Athens 1970.

DICEA, O., & DUTESCU, P.: Contribution à l'étude du prépannonien du bassin de Zalău (Dépression Pannonique). — Acta Geol. Acad. Sci. Hungar., **15,** 91—100, Budapest 1971.

DI CESARE, F., et al.: The Pliocene-Quaternary of Giarabub Erg region. — Rev. Inst. Fr. Pétrole, **18** (10—11), 30—37.

* DORA, O. Ö: Geologisch-lagerstättenkundliche Untersuchungen im Yamaular-Gebirge nördlich von Karşiyaka (Westanatolien). — Publ. Maden Tetkik Arama Enst., **116,** 63 pp., Ankara 1964.

* DRAGOŞ, V.: Contribuţii la cunoaşterea genezei evaporitelor din Basinul Transilvaniei. — Studii şi Cercetari Geol., Geofiz, Geogr., Sér. Geol., **14** (1), 163—180, Bucarest 1969.

* DUBERTRET, L., et al.: Le Miocène en Syrie et Lebanon. — Notes et Mém. Haut-Commiss. Rép. Fr. en Syrie et au Liban, Serv. Trav. Publ., Sec. d'études géol., **1,** 61—208, Paris 1933.

DULEMBA, J. L.: Quelques remarques sur l'origine des canyons sous-marins situés au large des côtes ouest de la Corse. — Geol. Rdsch., **59,** 601—604, Stuttgart 1970.

* DŽODŽO- TOMIĆ, R.: The microfauna of the Buglovian horizon of Timočka Krajina with special reference to its stratigraphic position. — Bull. Geol., Ser. A, **21** (1963), 121—148, Inst. Geol. Geophys. Res., Belgrade 1965.

EBERZIN, A. G.: O faune ponticheskykh otlozhenii raiona Keschan (Dardanelles). — Dokl. Akad. Nauk SSSR, **50,** 475—476, Moscow 1945.

EDELSTEIN, O., et al.: Le Miocène du versant méridionale du sommet Varatac (monts Gutii). — Dari Seama Şedinţelor, **57** (4), 43—53, Inst. Geol. Romania, Bucarest 1971.

EREMIJA, M.: Miotsenski mekushtsi Prnjavorskog basena (Bosna). — Ann. Géol. Pénin. Balkan., **36,** 51—86, Belgrade 1971.

—: Neogen izmedzy Motajice i L'ubića — biostratigrafsko-paleogeografski prikaz. — Ann. Géol. Pénin. Balkan., **35,** 25—107, Belgrade 1970.

* ERENTÖZ, C.: Contribution à la stratigraphie de la Turquie. — Bull. Maden Tetkik Arama Enst., **66,** 1—22, Ankara 1966.

* ERENTÖZ, C., & TOLUN, N.: Géologie détaillée de la structure pétrolifère de Kişlaköy (Iskenderum — Hatay). — Bull. Maden Tetkik Arama Enst., **46—47,** 1—16, Ankara 1955.

* ERENTÖZ, L.: Stratigraphie des bassins néogènes de Turquie, plus spécialement d'Anatolie Méridionale et comparaisons avec le Domaine Méditerranéen dans sou ensemble. — Bull. Maden Tetkik Arama Enst., Ser. C, No. 3, 53 pp., Ankara 1956.

* ERENTÖZ, L., & ÖZTEMÜR, C.: Aperçu général sur la stratigraphie du Néogène de la Turquie et observations sur les limites inférieure et supérieure. — Cursillos y Conf., **9,** 259—266 (1961), Inst. „Lucas Mallada", C.S.I.C. (España), Madrid 1964.

ERINÇ, S.: Some geomorphic consequences of neo-tectonics in Turkey. — Europ. Geophys. Soc., First Mtg. Sept. 25—29, 1973, Abstr., p. 34, Zürich 1973.

FEDOROV, P. V.: New data on relationships between the Paleoeuxinian and the Uzunlarian terraces on Caucasion shores of the Black Sea. — Dokl. Akad. Nauk SSSR, **174** (4), 924—926, Moscow, 1967. Transl. A.G.I. (1967), 69—71.

* FILIPESCU, M. G.: Géologie des Carpates Orientales. — Rev. Roum. Géol., Géophys., Géogr., Sér. Géol., **12** (2), 183—190, Bucarest 1968.

FONTES, J. C., et al.: Oxygen, carbon, sulfur and hydrogen stable isotopes in carbonates and sulfate mineral phases of Neogene evaporites, sediments and in interstitial waters. — In: RYAN, W. F. B., HSÜ, K. J., et al., Initial Reports of the Deep Sea Drilling Project, **13,** Pt. 2, 788—795, Govt. Print. Off., Washington 1973.

* FREUDENTHAL, T.: Stratigraphy of Neogene deposits in the Khania Province, Crete, with special reference to Foraminifera of the family Planorbilinidae and the genus Heterostegina. — Utrecht Micropaleont. Bull., No. 1, 208 pp., Utrecht 1969.

v. FREYBERG, B.: Das Neogengebiet nordwestlich Athen. — Eidikai Meletai epi tis Geologias tis Ellados (The Geology of Greece), No. 1, 1—86, Athens 1951.

* FUCHS, T.: Studien über die jungtertiären Ablagerungen Griechenlands. — Denkschr. k. u. k. Akad. Wiss., Math.-Nat. Cl., **37** (2), Vienna 1877.

GADEL, F., & MONGIN, D.: Analyses sédimentologiques, géochimiques et malacologiques d'une carotte prélevée dans le Golfe du Lion (Méditerranée française). — Paleogeogr. Paleoclimatol. Palaeoecol., **13** (1), 49—64, Amsterdam 1973.

GAGIĆ, N.: Tortonska i donjosarmatska microfauna šire okoline Koceljeva (zapadna Srbija). — Vesnik Geol., Ser. A, **27,** 229—241, Inst. Géol., Géophys. Res. Belgrade 1969.

GALANOPOULOS, A. G.: The earthquake activity in the physiographic provinces of the eastern Mediterranean Sea. — Ann. Géol. Pays Hellèn., **21,** 178—209, Athens 1970.

* GARAGUNIS, C.: Geologie und Tektonik im Bereich des Kanals und der Umgebung von Korinth. — Ann. Géol. Pays Hellèn., **18,** 147—154, Athens 1967.

GARLICKI, A.: Wyniki badán miocenu solonošnego na poludnie od Przemyśla. — Kwartalnik Geol., **17** (1), 92—105, Inst. Geol. Polska, Warsaw 1973.

* GILLET, S.: Histoire de la mer Noire. — Ann. Géol. Pays Hellèn., **14,** 337—347, Athens 1963.

* —: Affinités orientales des mollusques messiniens. — Soc. Toscana Sci. Nat. Atti, **66** (2), 415—417, Pisa 1959.

* —: (a) Relations entre bassins euxin et méditerranéen au Néogène et au Quaternaire. — C. R. Acad. Sci. France, **244,** 1803—1805, Paris 1957.

* —: (b) Contribution à l'histoire du bassin euxinique et méditerranéen en Néogène et au Quaternaire. — Bull. Serv. Carte Géol. Alsace et Lorr., **10** (2), 49—59, Strasbourg 1957.

GILLET, S., & FAUGÈRES, L.: Contribution à l'étude du Pontien de Macédoine: analyse géologique et sédimentologique des dépôts du Trilophos (SW de Salonique). — Rév. Géogr. Phys. et Géol. Dynam. (2), **12** (1), 9—24, Paris 1970.

* GLAÇON, G., & GUIRAUD, R.: La série Mio-Pliocène du Hodna et des régions voisines (Algérie du Nord). — C. R. Acad. Sci. France, Sér. D, **271,** 945—948, Paris 1970.

GLANGEAUD, L., et al.: Les structures mégamétriques de la Méditerranée: la mer d'Alboran et l'« arc » de Gibraltar. — C. R. Acad. Sci. France, Ser. D, **271,** 473—478, Paris 1970.

—: Évolution néotectonique de la mer d'Alboran et ses conséquences paléogéographiques. — C. R. Acad. Sci. France, Ser. D, **265,** 1672—1675, Paris 1967.

* GORETSKIY, V. A.: Nektoryye voprosy stratigrafii srednogo miotsena Volhyno-Podolii v svyazi s izucheniyem razvitiya kompleksov faun v otlozheniakh. — Paleont. Sb. (3), Pt. 2, 128—131, Lvóv, 1966.

GORETSKIY, V. A., & VENGLINSKIY, I. V.: O sloyakh zalegayushchikh v osnovanii miotsenovogo razreza okrestnostney L'vova. — Paleont. Sb. (3), Pt. 2, 128—131, Lvov, 1966.

GORETSKIY, V. A., et al.: O faune iz litotamniyevykh izvestnyakov miotsena Zakarpat'ya. — Paleont. Sb (2), Pt. 1, 96—98, Lvov 1965.

* GRADSTEIN, F. M., & VAN GELDER, A.: Prograding clastic fans and transition from a fluviatile to a marine environment in Neogene deposits of eastern Crete. — Geologie en Mijnbouw, **50** (3), 382—392, The Hague 1971.

* GREKOFF, N., et al.: Existence du miocène marin au centre de la mer Égée dans

l'île de Skyros (Grèce). — C. R. Acad. Sci France, Ser. D, **265** (18), 1276—1277, Paris 1967.

* GUERNET, C., & SAUVAGE, J.: Observations nouvelles sur le Néogène de la région de Pikermi et Raphina (Attique, Grèce). — Bull. Soc. Géol. France, Ser. 7, **12** (2), 241—245, Paris 1970.

GVIRTZMAN, G.: The Saqiye Group (late Eocene to early Pleistocene) in the coastal plain and Hashephela regions: Israel. — Bull. Geol. Surv. Israel, **51,** 2 vols. (1969), Jerusalem 1970.

* HAGN, H., et al.: Zur Neogen-Stratigraphie von Kephallinia und Ithaka. — Giorn. Geol., Ser. 2, **35** (2), 179—188 (1967), Bologna 1968.

HAJÓS, M.: Mátraalja miocén üledékeinek diatomai. — Geol. Hungar., Ser. Paleont., **37,** 401 pp., Budapest 1968.

* HÁMOR, G.: A kelet-Mecseki Miocén. — Ann. Hungar. Geol. Inst., **53** (1), 483 pp., Budapest 1970.

* HARASIMIUK, M., et al.: Rozwój zjawisk krasowych okolic Frampola w pliocenie i czwartorzędzie. — Ann. Univ. Curie-Skłodowska, **24** (4), 152—193, Lublin 1969.

* HARDY, L. A., & EUGSTER, H. P.: The depositional environment of marine evaporites: a case for shallow clastic accumulation. — Sedimentology, **16** (3—4), 187—220, Amsterdam 1971.

HASSAN, F., & EL-DASHLOUTY, S.: Miocene evaporites of Gulf of Suez region and their significance. — Bull. Amer. Assoc. Petrol. Geol., **54** (9), 1686—1696, Tulsa 1970.

HEGEDUS, G., & JANKOVICH, I.: Récif corallien du Badénien à Markhaza. — Ann. Hungar. Geol. Inst. (1970), 39—53, Budapest 1972.

—: Tortonai korallok Herendrol. — Földtani Közlöny Bull. Hungar. Geol. Inst., **100** (2), 185—190, Budapest 1970.

* HENSEN, F. R. S., et al.: A synopsis of the stratigraphy and geological history of Cyprus. — Quart. J. Geol. Soc. London, **105,** 1—41, London 1949.

* HEYBROEK, F.: The Red Sea Miocene evaporite basin. — In: Salt basins around Africa, The Institute of Petroleum, pp. 17—40, London 1965.

* HSÜ, K. J., et al.: The origin of the Mediterranean evaporites. — In: RYAN, W. F. B., HSÜ, K. J., et al., Initial Reports of the Deep Sea Drilling Project, **13,** Pt. 2, 1203—1232, Govt. Print. Off., Washington 1973.

* HUANG, T., et al.: Sedimentological evidence for current reversal at the Straits of Gibraltar. — J. mar. techn. Soc., **6** (4), 25—33, 1972.

IONESU, D.: O nouă ivire de depozite badeniene in NV Transilvaniei. — Studii şi Cercetari Geol., Geofiz. Geogr., Ser. Geol., **18** (1), 273—276, Bucarest 1973.

ILLING, V. C. (ed.): The world's oilfields. — Oxford U.P., London 1953.

* IRRLITZ, W.: Lithostratigraphie und tektonische Entwicklung des Neogens in Nordostanatolien. — Beih. Geol. Jb., **120,** 111 p., Bundesanst. f. Bodenforsch., Hannover 1972.

JÁMBOR, Á.: A Budapest környéki neogén képzödmények ösföldrajzi vizsgálata — Paläogeographische Untersuchung der Neogenablagerungen in der Umgebung von Budapest. — Magyar Állami Földtani Intézet Évi Jelentése, 135—142, Budapest 1967.

JAUZEIN, A., & PERTHUIZOT, J. P.: Quelques réflexions sur la sédimentation marine à propos de la Sebkha el Melah (Zarzis, Tunisie). — C. R. Somm. Soc. Géol. France, Sér. 2, 86—87, Paris 1972.

* JIŘÍČEK, R.: Problém hranice sarmat/panon ve Vídeňské, podunajské a východoslovenské pánvi. — Miner. Slovaca, **4** (14), 39—81, Bratislava 1972.

* KALAFATÇIOGLU, A.: Geology around Ezine and Bozcaada; the age of the limestones and serpentines. — Bull. Maden Tetkik Aarama Enst., **60,** 61—70, Ankara 1963.

KERIANY, M. T.: Note on the planktonic zonation of the Miocene in the Gulf of Suez region, U.A.R. — Giorn. Geol., Ser. 2, **35** (3), 157—166, Bologna 1967.

KESKIN, C.: Pinarhisar alaninim jeolojisi. — Bul. Turk. Jeol. Kurumu, **14** (1), 31—75, Ankara 1971.

KLEMME, H. D.: Regional geology of circum-Mediterranean region. — Amer. Assoc. Petrol. Geol. Bull., **42** (3), 477—512, Tulsa, Okla., 1958.

* KOPP, K. O., et al.: Geologie Thrakiens IV — Das Ergene Becken. — Beih. Jb., **76**, 136 pp., Bundesanst. f. Bodenforsch., Hannover 1969.

* —: Geologie Thrakiens III — Das Tertiär zwischen Rhodope und Evros. — Ann. Géol. Pays Hellèn., 16, 315—362, Athens 1966.

* KORECZ-LAKY, I.: A keleti Mecsek miocén foraminiferái. — Ann. Hungar. Geol. Inst., **52** (1), 200 pp., Budapest 1968.

* KOROLYUK, I. K.: Podolskiye toltry i usloviya ikh obrazovaniya. — Tr. Inst. Geol. Akad. Nauk SSSR, **110** (56), 121 pp., Moscow 1952.

KOYUMZHIEVA, E.: Formes hémisténohalines marines du Sarmatien inférieur en Bulgarie du Nord-Ouest. — Bull. Geol. Inst. Bulgar. Acad. Sci., **18**, 5—12, Sofia 1969.

* KRACH, W.: Miocen okolic Miechowa. — Biul. Geol. Inst. Polska, **43**, 95 pp., Warsaw 1947.

* KRASHENINNIKOV, V. A.: Correlation of the Miocene deposits of the eastern Mediterranean to stratotypical sections of the Miocene stages. — Giorn. Geol., Ser. 2, **35** (3), 167—178 (1967), Bologna 1968.

* KRSTIĆ, S. W.: On the Sarmatian of the broader vicinity of Lapova and Rača Kragujevačka. — Bull. Inst. Geol. Geophys. Res., Ser. A, 21, 89—100 (1963), Belgrade 1965.

KRÜMMEL, O.: Handbuch der Ozeanographie. — 2 vols., Stuttgart (J. Engelhorn Nachf.) 1911.

KUENEN, Ph. H.: Marine Geology. — New York (J. Wiley & Sons) 1950.

KÜHN, O.: Süßwassermiozän von bosnischem Typus in Griechenland. — Ann. Géol. Pays Hellèn., **3**, 185—192, Athens 1951.

* KURTMAN, F., & AKKUŞ, M. F.: Intermontane basins in Eastern Anatolia and their oil possibilities. — Bull. Maden Tetkik Arama Enst., **77**, 1—9, Ankara 1971.

* KWIATKOWSKI, S.: Sedimentacja gipsów mioceńskich południowej Polski. — Prace Muz. Zemi Polska, **19**, 3—94, Warsaw 1972.

LACOMBE, H.: Contribution à l'étude du régime du Détroit de Gibraltar. — Cah. Océanogr., **13** (2), 73—107, Paris 1961.

LACOMBE, H., & TCHERNIA, P.: Caractères hydrologiques et circulation des eaux en Méditerranée. — In: STANLEY, D. J. (ed.), The Mediterranean Sea: A Natural Sedimentation Laboratory, 25—36, Stroudsburg, Pa. (Dowden, Hutchenson & Ross Inc.) 1973.

—: Quelques traits généraux de l'hydrologie méditerranéenne d'après diverses campagnes hydrologiques recentes en Méditerranée, dans la proche Atlantique et dans le détroit de Gibraltar. — Cah. Océanogr., **12** (8), 527—548, Paris 1960.

* LECOINTRE, G.: Recherches sur le Néogène et le Quaternaire marins de la côte Atlantique du Maroc. — Notes et Mém. Serv. Carte Geol. Maroc, **28**, 80 pp., Rabat 1952.

* LEONTARIS, S. N.: Die Tektonik des Beckens von Agrinion und das Alter seiner Genese. — Ann. Géol. Pays Hellèn., **21**, 652—660, Athens 1970.

LE PICHON, X.: Sea-Floor Spreading and Continental Drift. — J. Geophys. Res., **73** (12), 3661—3697, Richmond, Va., 1968.

* LISZKOWSKI, J., & MUCHAWSKI, J.: Morfologia, budowa wewnętrzna oraz geneza masywów wapieni biogenicznych dolnego sarmatu strefy progów zewnętrznych południowej krawedzi wyżyny lubielskej. — Biul. Geol. Inst. Polska, **11**, 5—36, Warsaw 1969.

* MAGNÉ, J., & VIGUER, C.: Stratigraphie du Néogène de la bordure méridionale de la

Sierra Morena entre Huelva et Carmona (Espagna du Sud-Ouest). — Bull. Soc. Géol. France, **12** (2), 200—209, Paris 1970.

* Magnier, P.: Le Néogène du bassin de Syrie et du Sud de la Cyrénaïque (Libye). — Cursillos y Conf., **9** (1961), 193—198, Inst. „Lucas Mallada", C.S.I.C. (España), Madrid 1964.

* Maragkoudakis, N. T.: Geologia kai mikropalaiontologia tis Notiou Kerkyras. — Bull. Inst. Geol. Subsurf. Res., **12** (1), 132 pp., Athens 1967.

* Mascle, G., & Mascle, J.: Aspects of some evaporite structures in western Mediterranean Sea. — Bull. Amer. Assoc. Petrol. Geol., **56** (11), 2260—2267, Tulsa Okla., 1972.

Medoc Group: Observations of formation of deep water in the Mediterranean Sea 1969. — Nature, **227** (5262), 1037—1040, London 1970.

* Melentis, J. K.: Studien über fossile Vertebraten Griechenlands: 18. *Stenofiber jaegeri* aus Ligniten von Serrae und die Datierung der Fundschichten. — Ann. Géol. Pays Hellèn., **17**, 289—297, Athens 1967.

* Mercier, J., & Sauvage, J.: Remarques sur la géologie de la Macedoine centrale: Les calcaires à pollens et spores de la basse vallée de l'Axios. — Ann. Géol. Pays Hellèn., **14**, 330—338, Athens 1963.

Metwalli, M. H.: Stratigraphic setting of the subsurface Miocene sediments in Transdanubia (Hungary). — Acta Geol. Acad. Sci. Hung., **15**, 155—172, Budapest 1971.

* Milaković, B.: Paleogeografski značaj dinamike razvoja brakičnih sredina paratetisa. — Vesnik Geol., Ser. A, **27**, 121—127, Inst. Geol. Geophys. Res., Belgrade 1969.

Miller, A. R., & Charnock, H.: Oceanographic problems of the Mediterranean. — Rapports et Procès-verbaux des Réunions, Comm. Int. Explor. Sci. Mer Mediterr., **20** (4), 617—618, Monaco 1972.

* Milovanović, L.: Stratigrafski položaj evaporitske formacije u dubokim istražnim buštinama u Crnoj Gori (priobalna zona). — Vesnik Geol., Ser. A, **27**, 159—165, Inst. Geol. Geophys. Res., Belgrade 1969.

* Montenat, C., & Martinez, C.: Stratigraphie et micropaléontologie du Néogène au Sud de Murcia (Chaînes bétiques, Espagne). — C. R. Acad. Sci., France, Ser. D, **270** (4), 592—595, Paris 1970.

Morel, A.: Caractères hydrologiques des eaux échangées entre le bassin oriental et le bassin occidental de la Méditerranée. — Cah. Océanogr., **23** (4), 329—342, Paris 1971.

* Muldini-Mamužić, S., & Tomić-Džodžo, R.: Microfaunistic investigations of the Tortonian and Sarmatian deposits of the Pannonian basin in the regions of Croatia and Serbia (Yugoslavia). — Giorn. Geol., Ser. 2, **35** (3), 419—425 (1967), Bologna 1968.

* Müller-Miny, H.: Beiträge zur Morphologie und Geologie der mittleren ionischen Inseln. — Ann. Géol. Pays Hellèn., **9**, 73—89, Athens 1958.

* Nebert, K.: Zur stratigraphischen Stellung der Gipsserie im Raume Zara-Imranli (Vilâyat Sivaş). — Bull. Maden Tetkik Arama Enst., **48**, 79—87, Ankara 1956.

* Nesteroff, W. D.: Mineralogy, petrography, distribution and origin of the Messinian Mediterranean evaporites. — In: Ryan, W. F. B., Hsü, K. J., et al., Initial Reports of the Deep Sea Drilling Project, **13**, Pt. 2, 673—694, Govt. Print. Off., Washington 1973.

Neumann, G.: Ocean currents. — New York (Elsevier) 1968.

* Nevesskaya, L. A., & Ilyina, L. B.: On the scope and position of the Maeotic stage and on the Miocene/Pliocene boundary in the Ponto-Caspian basin. — Giorn. Geol., Ser. 2, **35** (4), 27—39 (1967), Bologna 1968.

* Ney, R.: Miocen południowego Roztocza między Horyńcem a Łowczą i przyległego

obszaru zapadliska przedkarpackiego. — Prace Geol., Akad. Nauk Polska, **60,** 94 pp., Warsaw 1969.

* Nikolov, I.: Neue Vertreter der Gattung Hipparion in Bulgarien. — Izvest. Geol. Inst., **20,** 107—122, Sofia 1971.

* Novković, M., et al.: Geološka studija Sokobanjskog tercijarnog basena I. — Vesnik Geologija, Ser. A, **23,** 249—299, Inst. Geol. Geophys. Res., Belgrade 1970.

* Ogniben, L.: Petrografia della Serie Solfifera siciliana e condizioni geologische relative. — Mem. descritt. Carta Geol. d'Italia, **23,** 275 pp., Rome 1957.

Oppenheim, P.: Über das marine Pliozän der Bohrung von Nütterden bei Cleve. — Jb. kgl. Preuss. Geol. Landesanst., **36** (2), 421—434, Berlin 1917.

Orbocea, M.: Contribuţii la stratigrafia depozitelor miocenului mediu dintre valea Cernei (jud. Vlîcea). — Studii şi Cercetări Geol. Geofiz. Geogr., Ser. Geol., **17** (2), 453—460, Bucarest 1972.

Oren, O. H.: The Atlantic water in the Levant Basin and on the shores of Israel. — Cah. Océanogr., **23** (3), 291—297, Paris 1971.

Pană, I., & Rado, G.: Die Biostratigraphie des Neogens im Beiuş-Becken. — Rév. Roum. Géol., Géophys. Géogr., Sér. Géol., **16** (1), 59—76, Bucarest 1972.

* Pannekoek, A. J.: Uplift and subsidence in and around the western Mediterranean since the Oligocene: a review. — Verh. Kon. Ned. Geol. Mijnbouwk. Gen., **26,** 53—77, The Hague 1969.

* Papa, A.: Conceptions nouvelles sur la structure des Albanides. — Bull. Soc. Geol. France, Ser. 7, **12** (6), 1096—1109, Paris 1970.

Parea, G. C., & Ricci-Lucchi, F.: Resedimented evaporites in the periadriatic trough (Upper Miocene, Italy). — Israel Jour. Earth Sci., **21** (3—4), 125—141, Jerusalem 1972.

* Pashko, P., et al.: Të dhëna paraprake mbi stratigrafinë dhe faciet o mesinianit në zonën jonike. — Përmbledhje studimesh, **12,** 23—34, Tirana 1969.

Pauca, M.: Etapele genetice ale Depresiunii Transilvaniei. — Studii şi Cercetari Geol. Geofiz. Geogr., Ser. Geol., **17** (2), 235—244, Bucarest 1972.

—: Criterii noi in stratigrafia miocenului din Bazinul Transilvaniei. — Studii şi Cercetari Geol., Geophys., Geogr., Ser. Geol., **16** (2), 397—404, Bucarest 1971.

* —: Beiträge zur Kenntnis der miozänen Salzlagerstätten Rumäniens. — Geol. Rdsch., **57,** 514—531, Stuttgart 1968.

—: Neogenul din Basinele externe ale Munţilor Apuseni. — An. Com. Geol., **27,** 259—336, Bucarest 1954.

Pectas, H.: Bogazicin de Satih. Alti Akintical ve Su Karisimlari. — Hidrobiol., Ser. A, **2,** 1—10, Istanbul 1954.

* Perconig, E.: Sull'esistenzia del Miocene superiore in facies marina nella Spagna meridionale. — Comm. Medit. Neogene Stratigr., Proc. 3rd Sess., Berne 1964, 288—302, Leiden (E. J. Brill) 1966.

* Petrović, M. V.: Tortonski foraminiferi okoline Štubika (istočna Srbija) i njihov biostratigrafski značaj. — Vesnik Geol., Ser. A, 27, 232—253, Inst. Geol. Geophys. Res., Belgrade 1969.

—: Stratigrafiya srednyet Miotsena Yardarskog basena. — Ann. Géeol. Pénin. Balkan, **34,** 45—86, Belgrade 1969.

* Pezzani, F.: Studio micropaleontologico di un campione della serie messiniana di Tabiano Bagni (Parma). — Riv. Ital. Paleont., **69** (4), 559—662, Milano 1963.

Phleger, F. B.: A modern evaporite deposit in Mexico. — Bull. Amer. Assoc. Petrol. Geol., **53** (4), 824—829, Tulsa, Okla., 1969.

* Pishvanova, L. S., & Tkachenko, O. F.: Paleogeograficheskie karty miocena zapadnykh oblastei Ukr.S.S.R. — Földtani Közlöny, Bull. Hung. Geol. Soc., **101,** 265, —276, Budapest 1971.

* Pishvanova, L. S., et al.: Stratigrafija i korelacija miocena zapadne oblasti Ukr.S.S.R.,

Bulgarske i Jugoslavije na osnovu foraminifera. — Vesnik Geol., **27,** 59—67, Inst. Geol. Geophys. Res., Belgrade 1969.

* —: On the zonation of the Miocene by means of planktonic foraminifera. — Giorn. Geol., Ser. 2, **35** (3), 233—254 (1967), Bologna 1968.

* Poborski, J.: Złoże solne Bochni na tle geologicznym okolicy. — Biul., Inst. Geol. Polska, **78,** 160 pp., Warsaw 1952.

* Popović, R., & Novković, M.: Donjokongerijske naslage slatkovodnih basena Zapadne Morave i Gruže sa osvrtom na starost ugljenih slojeva. — Vesnik Geol., **27,** 83 —104, Inst. Geol. Geophys. Res., Belgrade 1967.

Pousanov, I. I.: Successive stages of the Mediterranization of the Black Sea fauna. — New data. Gidrobiol. Zh., **2,** 1—10, Moscow 1965.

* Quirantes Puertas, J.: Estudio sedimentologico de las calizas del Terciario continental. — Brev. Geol. Astur., **11** (1—4), 107—111, Oviedo, 1967.

* Quirantes Puertas, J., & Martinez, G. E.: Un tipo de roca poco conocido: Las arenitas yesiferas de los Monegros. — Brev. Geol. Astur., **11** (1—4), 113—123, Oviedo 1967.

* Rangheard, Y.: Principales données stratigraphiques et tectoniques des îles d'Ibiza et de Formentera (Baléares); situation paléogéographique et structurale de ces îles dans les Cordillères bétiques. — C. R. Acad. Sci. France, Ser. D, **270** (9), 1227—1230, Paris 1970.

* Reuss, A. E.: Die fossile Fauna der Steinsalzablagerung von Wieliczka in Galizien. — Sitz. d. k. u. k. Akad. d. Wiss., Math.-Natur. Cl., 55 (1), 17—182, Vienna 1867.

Richter-Bernburg, G.: As quoted in Raup, O. B.: Brine mixing, an additional mechanism for formation of basin evaporites. — Bull. Amer. Assoc. Petrol. Geol., **54** (12) 2246—2259, Tulsa, Okla., 1970.

—: Über salinare Sedimentation. — Z. Dtsch. Geol. Ges., **105,** 593—645, Hannover 1953.

* Rios, J. M.: Saline deposits of Spain. — Geol. Soc. America, Spec. Pap. 88, 59—74, New York 1968.

* Roesler, G.: Das Neogen von Naxos und den benachbarten Inseln. — Z. Dtsch. Geol. Ges., **123,** 523—525, Hannover 1972.

Rojdestvensky, A. V.: Le diversement dans la mer Noire des eaux de la mer de Marmara. — Cah. Océanogr., **23** (3), 283—289, Paris 1971.

Roloff, A.: Die jungkretazisch-tertiäre Entwicklung am W-Rande des Amanosgebirges (Südtürkei). — Geotekt. Forsch., **42,** 97—129, Stuttgart 1972.

* Ross, D. A., et al.: Red Sea Drillings. — Science, **179,** 377—380, Washington 1973.

Ruggieri, G.: The Miocene and later evolution of the Mediterranean Sea. — In: Adams, C. G., & Ager, D. V. (eds.), Aspects of Tethyan Biography. Syst. Assoc. Publ., No. 7, 283—290, London 1967.

—: La serie marina pliocenica e quaternaria della Romagna. — Publ. Camera di Commercio di Forli, 77 p., Forli 1962.

* Said, R.: The Geology of Egypt. — New York-Amsterdam (Elsevier) 1962.

Scholz, G.: A visegradi Fekete-hegy tortonai korall faunaja. — Földtani Közlöny, Bull. Hung. Geol. Soc., **100** (2), 192—201, Budapest 1970.

Sedelnikow, G. S.: Über die hydrochemischen Verhältnisse bei der Salzbildung in der Karabugaz-Bucht. — Freiberger Forschungsh., **A-123,** 166—174, Berlin 1958.

* Selli, R.: Il bacino del Metauro. — Giorn. Geol., Ser. 2, **24,** 268 pp., Bologna 1954.

—: An outline of the Italian Messinian. — In: Drooger, C. W. (ed.), Messinian events in the Mediterranean. Kon. Nederl. Akad. Wetensch., Geodynamics Sci. Rep. No. 7, p. 150—171, North-Holland, Amsterdam 1973.

* Seneš, J., et al.: Korrelation des Miozäns der zentralen Paratethys (Stand 1970). — Geol. Zborník, **22** (1), 3—9, Bratislava 1971.

* —: Les problèmes interrégionaux de paléogéographie de la Paratéthys. — Giorn. Geol., Ser. 2, **35** (4), 333—339, Bologna 1967.

* —: Paläogeographie des westkarpathischen Raumes in Beziehung zur übrigen Paratethys im Miozän. — Geol. Práce, **60**, 159—195, Bratislava 1961.

SERPOIANU, G. H.: Considérations sur la pénétration des eaux méditerranéennes dans le bassin de la Mer Noire. — Hydrobiologia, **8**, 239—251, Bucarest 1967.

* SINDOWSKI, K. H.: Der geologische Bau von Attika. — Ann. Géol. Pays Hellèn., **2**, 163—218, Athens 1949.

* SLÁVIK, J.: Kamenná sol'. — Miner. Slovaca, **3** (12—13), 421—437, Bratislava 1971.

SONNENFELD, P.: A Mediterranean catastrophe? — Geol. Mag., **101** (1), 79—80, Cambridge 1974.

SPAJIĆ-MILETIĆ, O.: Biostratigraphie des oberen Miocäns Serbiens. — Ann. Géol. Pénin. Balkan., 34, 11—20, Belgrade 1969.

SPAJIĆ, O., & DŽODŽO-TOMIĆ, R.: Rezultati stratigrafskikh istrazhivanya Miotsena Srbiye (Das Miozän von der Drina bis zu den Karpaten). — Ann. Géol. Pénin. Balkan., **36**, 28—37, Belgrade 1971.

STANCU, J., et al.: Studii stratigrafice asupra Miocenului din versantul nordic al Dunarii intre Dubova şi Pojejena (Carpaţii Meridionali). — Dări de Seamă ale Şedinţelor, 57 (4), 119—133, Bucarest 1971.

* STEVANOVIĆ, P. M.: Position stratigraphique du calcaire à cérithes près de Belgrade et ses propriétés sédimentologiques. — Zbornik Red. Rudarsko-Geol. Fak. Univ. Belgrade, **13**, 101—112, Belgrade 1970.

* —: Facies und Horizonte des älteren Pliozäns in Jugoslawien und die Möglichkeiten einer Korrelation mit dem Mittelmeerbecken. — Comm. Medit. Neogene Stratigr., Proc. 3rd Sess. Berne, 1964, 188—190, Leiden (E. J. Brill) 1966.

* —: Contribution à la connaissance de l'étage Pontien de Grèce avec une remarque speciale sur le Pontien de la Mer Noire. — Cursillos y Conf., **9** (1961), 93—100, Inst. „Lucas Mallada", C.S.I.C. (España), Madrid 1964.

* STRAKHOV, N. M.: Principles of Lithogenesis (Osnovy teorii litogeneza). — Izdat. Akad. Nauk SSSR, Moscow, 1962. Transl. Consultants Bureau, New York, 1967—1970, 3 vols.

STRASHIMIROV, B.: O nekhodkakh stenogalinnoi fauny v otlozheniyakh srednemiotsenovogo otdela yuzhnee goroda Burgas (Bolgaria). — Acta Geol. Acad. Sci. Hungar., **15**, 250—280, Budapest 1971.

STURANI, C.: A fossil eel (*Anguilla* sp.) from the Messinian of Alba (Tertiary Piedmontese Basin): Paleoenvironmental and paleogeographical implications. — In: DROOGER, C. W. (ed.): Messinian events in the Mediterranean. Kon. Nederl. Akad. Wetensch. Geodynamics Rep. No. 7, 243—255, North-Holland, Amsterdam 1973.

SVERDRUP, H. U., et al.: The Oceans. — 1087 pp., New York (Prentice Hall) 1946.

* SYMEONIDIS, N., & KONSTANTINIDIS, D.: Beobachtungen zu den Neogenablagerungen des Zentralgebietes der Insel Kreta. — Ann. Géol. Pays Hellèn., **19**, 657—688, Athens 1970.

SZÉPESHÁZI, K.: Tiszántúl középsö részének miocén képzödményei a szénhidrogénkutató mélyfúrások adatai alapján. — Magyar állami földtani intézet évi jelentése (1968), 297—326, Budapest 1971.

* TERMIER, H., & TERMIER, G.: Atlas de Paléogéographie. — Carte 23, Paris (Mason et Cie) 1960.

TIXERONT, J.: Le bilan hydrologique de la mer Noire et de la mer Méditerranée. — Cah. Océanogr., 22 (3), 227—238, Paris 1970.

* TORRAS FOULON, A., & RIBA ARDERIU, O.: Contribution al estudio de los limos yesiferos del centro de la depresion del Ebro. — Brev. Geol. Astur., 11 (1—4), 125—137, Oviedo 1967.

* TRASHLIEV, S.: Basanite et sulfate anhydre de calcium dans l'horizon de gypse en Bulgarie du Nord-Ouest. — Bull. Geol. Inst., Bulgarian Acad. Sci., Ser. Metallic & Non-Metallic Min. Dep., **21—22,** 147—158, Sofia 1971.

* —: Usloviya i produlzhitelnost na obrazuvane na evaporitnata formatsiya v tortona ot severozapadna Bulgariya. — Bull. Geol. Inst. Bulgarian Acad. Sc., **18,** 155 —156 (Abstr.), Sofia 1969.

* TRIKKALINOS, J. K.: Über den tektonischen Bau und die Entstehung der Erdöllagerstätten des Polylophos-Dragopsa-Gebietes von Epirus. — Ann. Géol. Pays Hellèn., **3,** 154—184, Athens 1951.

ULLYOTT, P., & ILGAZ, O.: The hydrography of the Bosporus: an introduction. — Geogr. Rev., **36** (1), 44—66, New York 1946.

* URBANIK, J.: Wstepne wyniki badán geologicznych miocenu przedkarpatskiego nad Dunajcem. — Kwartalnik Geol., 16 (2), 347—360, Warsaw 1972.

VANCEA, A.: Neogenul din basinul Transilvaniei. — Edit. Acad. Sci. Romania, 262 pp., Bucarest 1960.

* VENZO, S.: Les nouvelles connaissances sur le Tortonien-Messinien de l'Émilie occidentale et ses comparaisons avec les sédiments de la region circumpadane. — Comm. Medit. Neogene Stratigr., Proc. 3rd Sess. Berne, 1964, 223—224, Leiden (E. J. Brill) 1966.

* VÖLK, H. R.: Zur Geologie und Stratigraphie des Neogenbeckens von Vera, Südostspanien. — Diss. Univ. Amsterdam, 160 pp., Amsterdam 1967.

* —: Riffbildung im Jungtertiär der betischen Kordilleren (Südostspanien). — Comm. Medit. Neogene Stratigr., Proc. 3rd Sess. Berne 1964, 250—253, Leiden (E. J. Brill) 1966.

* VÖLK, H. R., & RONDEEL, H. E.: Zur Gliederung des Jungtertiärs im Becken von Vera, Südostspanien. — Geol. Mijnbouw, **11,** 13—14, The Hague 1964.

* ZAGORCHEV, I.: On the neotectonic movements in a part of SW Bulgaria. — Bull. Geol. Inst. Bulgarian Acad. Sci., Ser. Geotectonics, **19,** 152 (Abstr.), Sofia 1970.

* ZAKHIDOV, A. U.: Platform phase of evolution and certain aspects of oil-gas potential of the Karabugaz area. — Internat. Geol. Rev., **15** (1), 1—11, Washington 1973.

* ZBYSZEWSKI, G.: Carta geológica de Portugal na escala de 1/50,000. Notícia explic. fohla 22 D, Marinha Grande. — Port. Serv. Geol., 45 pp., Lisbon 1965.

* —: Les rapports entre les milieux miocènes marins et continentaux au Portugal. — Cursillos y Conf., 9 (1961), 103—108, Inst. „Lucas Mallada“, C.S.I.C. (España), Madrid 1964.

ZOMENIS, S. L.: Stratigraphy and hydrogeology of the Neogene rocks in the northern foothills of the Troodos Massif, Cyprus. — Bull. Cyprus Geol. Surv. Dept., **5,** 22—90, Nicosia 1972.

AUTHOR CITATION INDEX

SUBJECT INDEX

About the Editor

PETER SONNENFELD was graduated from the universities of Bratislava and Prague, Czechoslovakia. Early field work encompassed what was then British India and is now India and Pakistan. Later, he worked in Newfoundland and Ontario as a field geologist for mining interests and in Saskatchewan, Alberta, and British Columbia as a petroleum geologist for several major integrated oil companies. In 1963, he commenced his teaching career at Texas A & I University in Kingsville, Texas; three years later he started the Department of Geology and Geological Engineering at the University of Windsor. A sabbatical leave in 1972/1973 permitted some examination of localities along the whole south rim of Asia. Dr. Sonnenfeld is a member of the Committee on Marine Geology and Geophysics of the International Commission for the Scientific Exploration of the Mediterranean Sea, Monaco.